AF342089

THERMOELECTRICS FOR POWER GENERATION - A LOOK AT TRENDS IN THE TECHNOLOGY

Edited by **Sergey Skipidarov** and **Mikhail Nikitin**

Thermoelectrics for Power Generation - A Look at Trends in the Technology

http://dx.doi.org/10.5772/62753
Edited by Sergey Skipidarov and Mikhail Nikitin

Contributors

John Stockholm, Yaniv Gelbstein, Ran Ang, Mohammad Arjmand, Soheil Sadeghi, Fuqiang Cheng, Miguel Angel Olivares-Robles, Henni Ouerdane, Alexander Vargas Almeida, Bertrand Lenoir, Christophe Candolfi, Anne Dauscher, Yohan Bouyrie, Salma Sassi, Nianduan Lu, Ling Li, Ming Liu, Aljoscha Roch, Lukas Stepien, Roman Tkachov, Tomasz Gedrange, Tomoko Inayoshi, Patrick Taylor, Theodorian Borca-Tasciuc, Adam Wilson, Jay Maddux, Diana Borca-Tasciuc, Samuel Moran, Eduardo Castillo, Patricia Aranguren, David Astrain, Xin Song, Terje G. Finstad, Yaron Amouyal, Alexander Burkov, Jose Antonio Alonso, Federico Serrano-Sánchez, Lidia Lukyanova, Yuri Boikov, Oleg Usov, Mikhail Volkov, Viacheslav Danilov, Zhiting Tian, Eurydice Kanimba, Andrei Shevelkov, Aamir Shahzad, Grigory N. Isachenko, Vladimir Zaitsev

CBS, Edition **2018**

Published by InTech

Janeza Trdine 9, 51000 Rijeka, Croatia

Notice

Statements and opinions expressed in the chapters are these of the individual contributors and not necessarily those of the editors or publisher. No responsibility is accepted for the accuracy of information contained in the published chapters. The publisher assumes no responsibility for any damage or injury to persons or property arising out of the use of any materials, instructions, methods or ideas contained in the book.

Publishing Process Manager Iva Lipović

Technical Editor SPi Global

Cover Designer InTech Design Team

Additional hard copies can be obtained from orders@intechopen.com

Thermoelectrics for Power Generation - A Look at Trends in the Technology, Edited by Sergey Skipidarov and Mikhail Nikitin
p. cm.
Print ISBN 978-953-51-2845-8
Online ISBN 978-953-51-2846-5

Contents

Foreword

Dear reader,

Greetings from the president of the Ferrotec Group.

Despite the fact that the first successful demonstration of thermoelectric generators was more than 70 years ago, this type of electricity generation is not widely used even today due to low conversion efficiency, small electrical power generated by unit thermoelectric generation module, reliability issues, and cost issues. There also exist many issues in thermoelectric materials science and the manufacturing technology of thermoelectric generating modules.

A new era in thermoelectric power generation is becoming a reality as a result of the synergistic effect of recent years' success both in thermoelectricity and in related areas of science and technology, such as materials science, microelectronics, electronic devices design, and manufacturing. Impressive advances in microelectronics and electronic device technology have led to a radical reduction in energy consumption of individual appliances and their cost, making such devices affordable for all segments of the population in the world.

As a result, a great potential market niche for personal generators (hundreds of millions of pieces) for charging USB devices and LED lighting for people residing in off-grid territories is now opening.

This market will give opportunity to greatly improve access to education for children and to information and digital services for off-grid residents by connecting to communication nets (cell telephony, Internet, and broadcasting) through personal electronic devices (smartphones, tabs (pads), notebooks) that will result in improving living standards of up to 1.2 billion people.

Now, after decades of hibernation, thermoelectric power generation has a chance to make a quick jump in development and widespread application. Moving ahead requires coordinating the efforts of the entire thermoelectric community, including a wide exchange of research and development results.

The Ferrotec Group, being one of the largest thermoelectric companies in the world, is aware of its responsibility to support the efforts that may lead to the widespread industrial application of thermoelectricity. We hope that this series of chapters, published on our initiative, can provide information to support researchers and engineers in the area of thermoelectric power generation.

A. Yamamura,
President, Ferrotec Group

Preface

At the beginning of the 21st century, problem of providing sustainable electricity for all inhabitants of the World is still very sharp. More than a billion people resides in off-grid territories. Meanwhile, mankind has faced another scourge - global warming, which is the result of the indomitable consumption of fossil fuels and has resulted in gigajoules of low-grade waste heat and huge amount of greenhouse gases. Waste heat energy is estimated to be from 60 to 70 % of primary energy produced from the burning of fossil fuels. Waste heat is conditionally free-of-charge energy, but it is difficult to use, and is thrown away - a pity.

So, mankind should solve both problems of the deficit of electricity and the recovery of waste heat simultaneously. It is possible in principle. And, thermoelectric power generation can help here.

In accordance with thermodynamic and physical laws, the generation of electromotive force (EMF) by direct conversion of an alternative form of energy into electrical energy is possible in nonequilibrium systems only. The more alternative energy and the greater the deviation from the equilibrium is in the system, the higher is the possible conversion efficiency. There are two main ways of direct conversion: photovoltaic (PV) effect and thermocouple effect (widely used type is Seebeck thermoelectric effect). Due to physical reasons, it is impossible to create PV converters, effectively generating EMF through absorption of thermal radiation photons. But, PV converters, operating with solar radiation and thermoelectric generator modules (TEGs), operating with thermal energy can work in couple.

The efficient recovery of low-grade waste heat is an important and nontrivial task. Sources of low-grade heat are plentiful everywhere. As a rule, low-grade heat is simply dissipated without any benefit to people. This is caused by the fact, that low-grade waste heat is strongly localized near heat sources. Therefore, waste heat is difficult to use in a cost-effective manner for an intended purpose. TEGs are small-sized items that can be placed as close as possible to heat (thermal) energy sources, which can have temperatures of hundreds of degrees Celsius. TEGs can operate anywhere (including indoors) and at any time of the day These factors are decisive for applying TEGs to recover low-grade heat. Therefore, in practical applications, the TEG's structure will be exposed to systematic long-term heavy temperature gradients, mechanical stresses (thermo-mechanical stresses) and high temperatures on one (hot) side. TEGs and, hence, the thermoelectric materials forming legs of thermopiles must withstand the abovementioned shock. To become attractive and affordable to customers, TEGs should have a service life of at least 5000 hrs, with thousands cycles ON-OFF and, of course, be cheap as well.

Competition in the field of TEGs in the future may be as high as it currently is for thermoelectric coolers (TECs). However, nowadays, we see a lack of innovative and affordable products on the market.

Despite the long history of thermoelectric power generation, there are many pressing issues in the thermoelectric materials science and the manufacturing technology of TEGs. Unlike TECs, where the maximum temperature in module is typically less than 60 °C, in TEGs, due to high temperatures (hundreds of degrees Celsius) on the hot side and heavy thermo-mechanical stresses in module, many processes become active, leading to a quick or gradual degradation in the performance of the thermoelectric materials and the TEG itself. These degradation processes are namely, interdiffusion, recrystallization, alloying, dissolution, phase transitions, phase separation, phase segregation, sublimation, oxidation, mechanical damage of legs, commutation and interconnections, and other phenomena.

This book is an attempt to arrange the interchange of research and development results concerned with hot topics in TEGs research, development and production, including:

1. Prospective thermoelectric materials for TEGs. The important theme here is obtaining effective p-type materials for low-, mid-, and high temperature ranges of TEGs operation.
2. Theoretical study and calculations of key parameters of inorganic and organic thermoelectric materials.
3. Research results in innovative construction nanomaterials.
4. Novel methods and apparatus for measuring performance of thermoelectric materials and TEGs.
5. Thermoelectric power generators simulation, modeling, design and practice.

We think, that the information presented in this book will be helpful to scientists and engineers involved in research and industry projects related to innovative thermoelectric power generation devices.

We are grateful to all the authors from the many countries of the World for their contributions. We hope, that coordinating these efforts in thermoelectric power generation will result in enhancements to the human living standards on our planet.

Sergey Skipidarov
PhD, Member of International Academy of Refrigeration
CEO, Ferrotec Nord Corporation
Moscow, Russia

Mikhail Nikitin
PhD, Science and Technology Adviser
Ferrotec Nord Corporation
Moscow, Russia

Advanced Thermoelectric Materials

Chapter 1

Layered Cobaltites and Natural Chalcogenides for Thermoelectrics

Ran Ang

Additional information is available at the end of the chapter

http://dx.doi.org/10.5772/65676

Abstract

We have systematically investigated thermoelectric properties by a series of doping in layered cobaltites $Bi_2Sr_2Co_2O_y$, verifying the contribution of narrow band. In particular, Sommerfeld coefficient is dependent on charge carriers' density and as function of density of states (DOS) at Fermi level, which is responsible for the persistent enhancement of large thermoelectric power. Especially for $Bi_2Sr_{1.9}Ca_{0.1}Co_2O_y$, it may provide an excellent platform to be a promising candidate of thermoelectric materials. On the other hand, high-performance thermoelectric materials require elaborate doping and synthesis procedures, particularly the essential thermoelectric mechanism still remains extremely challenging to resolve. In this chapter, we show evidence that thermoelectricity can be directly generated by a natural chalcopyrite mineral $Cu_{1+x}Fe_{1-x}S_2$ from a deep-sea hydrothermal vent, wherein the resistivity displays an excellent semiconducting character, while the large thermoelectric power and high power factor emerge in the low x region where the electron-magnon scattering and large effective mass manifest, indicative of the strong coupling between doped carriers and localized antiferromagnetic spins, adding a new dimension to realizing the charge dynamics. The present findings advance our understanding of basic behaviors of exotic states and demonstrate that low-cost thermoelectric energy generation and electron/hole carrier modulation in naturally abundant materials is feasible.

Keywords: layered cobalt oxides, narrow band contribution, natural chalcopyrite mineral, thermoelectricity generation, electron-magnon scattering

1. Introduction

Layered cobaltites with CdI_2-type CoO_2 block provide an excellent platform for investigating thermoelectric properties. A key to unveil mysterious thermoelectric properties lies in the two-dimensional (2D) conducting CoO_2 layer. For layered Bi-A-Co-O (A = Ca, Sr, and Ba), it also contains analogous conducting CoO_2 layer [1]. In particular, layered $Bi_2Sr_2Co_2O_y$ (BSC)

exhibits a rather large thermoelectric power S (~100 μV/K) at room temperature, which makes $Bi_2Sr_2Co_2O_y$ one of promising thermoelectric materials from the viewpoint of potential applications, analogous to other misfit-layered cobaltites, such as $NaCo_2O_4$ and $Ca_3Co_4O_9$ [2–5]. However, most studies of $Bi_2Sr_2Co_2O_y$ system are mainly focused on the thermoelectric improvement [2, 3, 6]. The transport mechanism based on resistivity ρ and thermoelectric power S has not been clarified. Moreover, large S is totally different from conventional value (<10 μV/K) based on a broad band model [7]. In this chapter, we will show evidence on a narrow band contribution in doped $Bi_2Sr_2Co_2O_y$ [8]. And what's more, exotic enhancement of large S is related to local density of states (DOS) near Fermi level (E_F) [9]. It could be effectively modulated thermoelectric performance by utilizing different doping. It is plausible to distinguish, which thermoelectric materials in doped $Bi_2Sr_2Co_2O_y$ could be regarded as potential candidates.

On the other hand, ternary chalcogenides serve as an ideal platform for investigating intricate physical and chemical characteristics controlling the efficiency of thermoelectric materials, and also are promising materials for potential applications in photovoltaics, luminescence, as well as thermoelectric and spintronic devices [10–13]. Ternary chalcopyrite-structured chalcogenides, such as $CuFeS_2$, have attracted particular attention owing to their unique optical, electrical, magnetic, and thermal properties [14–28]. Studies on chalcopyrite ($CuFeS_2$) have primarily focused on its electronic states [14, 15, 29–31]. However, the microscopic mechanism of electronic structure and thermoelectric character in $CuFeS_2$, which presumably arises from some scenarios such as delocalization of the Fe $3d$ electrons, charge-transfer-driven hybridization between Fe $3d$ and S $3p$ orbitals, or density of the conduction band electron states, still remains highly controversial [17, 30, 32]. The intrinsic mechanism of good thermoelectric properties is still a vital question which needs to be clarified. Another important issue is that the fabrication of artificial chalcopyrite itself requires expensively complex synthesis procedures and relatively high cost of constituent precursors, thereby potentially limiting the large-scale applications in the thermoelectric field.

In this chapter, we confirm that an unexpected thermoelectricity can directly be generated in a natural chalcopyrite mineral $Cu_{1+x}Fe_{1-x}S_2$ from a deep-sea hydrothermal vent, and demonstrate that doped carriers have strong coupling with localized antiferromagnetic (AFM) spins, which greatly enhance the thermoelectric power S and power factor, revealing the significance of electron-magnon scattering and large effective mass [33]. This will open up another useful avenue in manipulating low-cost thermoelectricity or even electron/hole carriers via the natural energy materials abundantly deposited in the earth.

2. Thermoelectric properties and narrow band contribution of $Bi_2Sr_{1.9}M_{0.1}Co_2O_y$ and $Bi_2Sr_2Co_{1.9}X_{0.1}O_y$

2.1. Crystal structure and valence states of Co ions

The crystal structure of $Bi_2Sr_2Co_2O_y$ is shown in the inset in **Figure 1**, where conducting CoO_2 layer with triangular lattice and insulating rocksalt $Bi_2Sr_2O_4$ block layer are alternatively stacked along c-axis, similar to the case of high-temperature superconductors like $Bi_2Sr_2CaCu_2O_y$.

Scanning electron microscopy (SEM) characterization of $Bi_2Sr_2Co_2O_y$ indicates surface morphology of plate-like grains. **Figure 1** shows X-ray diffraction (XRD) patterns of selected samples $Bi_2Sr_2Co_2O_y$, $Bi_2Sr_{1.9}Ca_{0.1}Co_2O_y$, and $Bi_2Sr_2Co_{1.9}Mo_{0.1}O_y$ with single phase, in agreement with XRD result of $Bi_{1.4}Pb_{0.6}Sr_2Co_2O_y$ [34]. The average Co valence was determined based on energy dispersive spectroscopy (EDS) measurement for all samples. For $Bi_2Sr_2Co_2O_y$, average Co valence is +3.330. For $Bi_2Sr_{1.9}M_{0.1}Co_2O_y$ (M = Ag, Ca, and Y), average Co valence is +3.380, +3.330, and +3.280, respectively. For $Bi_2Sr_2Co_{1.9}X_{0.1}O_y$ (X = Zr, Al, and Mo), average Co valence is +3.295, +3.347, and +3.189, respectively. X-ray photoemission spectroscopy (XPS) spectra (see **Figure 4a**) also show the valence states of Co $2p_{3/2}$ and $2p_{1/2}$ for selected $Bi_2Sr_{1.9}Ca_{0.1}Co_2O_y$ sample. Photon energy of Co $2p_{3/2}$ and $2p_{1/2}$ is 779.4 and 794.2 eV, respectively, demonstrating mixed Co valence between +3 and +4.

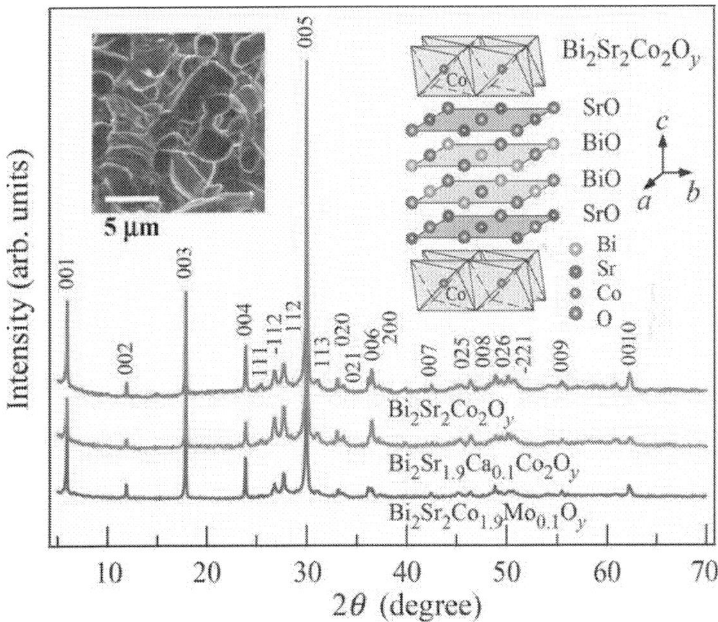

Figure 1. Powder XRD patterns for $Bi_2Sr_2Co_2O_y$, $Bi_2Sr_{1.9}Ca_{0.1}Co_2O_y$, and $Bi_2Sr_2Co_{1.9}Mo_{0.1}O_y$ samples at room temperature. Inset: crystal structure and SEM image of $Bi_2Sr_2Co_2O_y$.

2.2. Resistivity and transport mechanism

Figure 2a and **d** shows temperature dependence of resistivity $\rho(T)$ of all samples. For parent $Bi_2Sr_2Co_2O_y$ sample, an upturning point at T_p (~75 K) is observed. Metallic behavior above T_p appears, demonstrating existence of itinerant charge carriers. Compared with $Bi_2Sr_2Co_2O_y$, $\rho(T)$ of all doped samples (except $Bi_2Sr_{1.9}Ca_{0.1}Co_2O_y$) display total increase in view of the disorder effect. Furthermore, enhanced random Coulomb potential because of the doping induces the obvious shift of T_p toward higher temperature. On the other hand, $\rho(T)$ of $Bi_2Sr_{1.9}Ca_{0.1}Co_2O_y$ presents an overall decrease due to introduction of hole–type charge carriers into conducting CoO_2 layers.

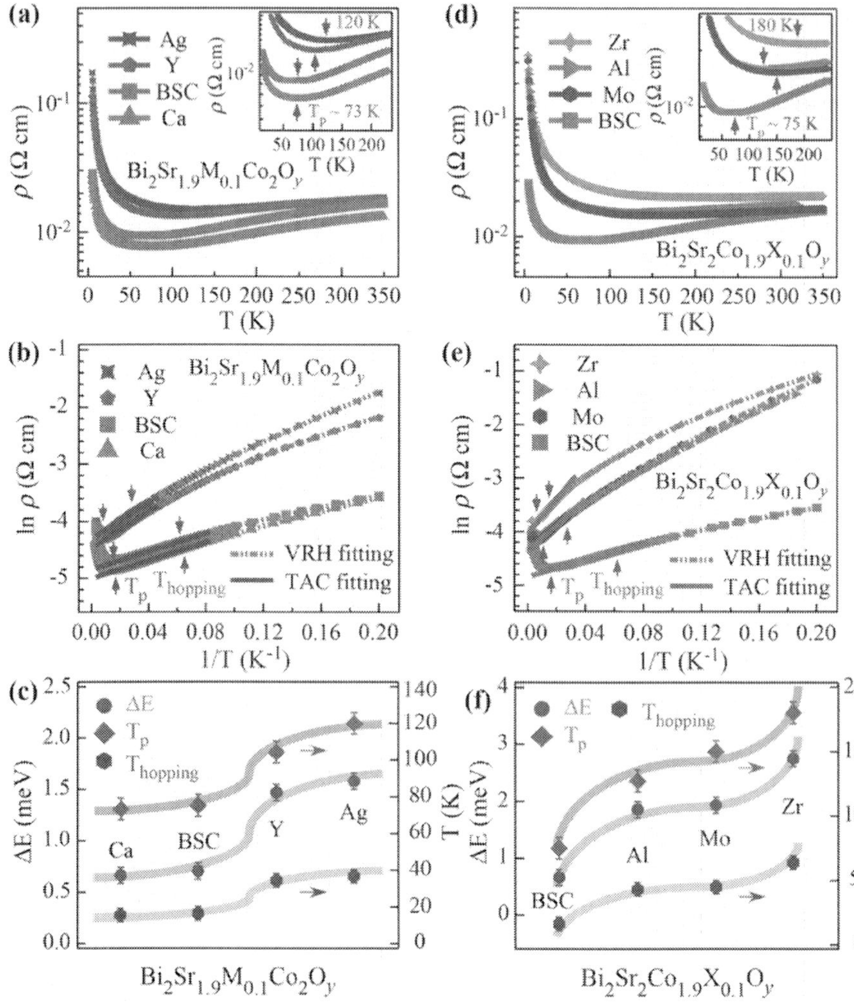

Figure 2. (a) Temperature dependence of resistivity $\rho(T)$ and inset: magnification plot of $\rho(T)$ for $Bi_2Sr_2Co_2O_y$ (BSC) and $Bi_2Sr_{1.9}M_{0.1}Co_2O_y$ (M = Ag, Ca, and Y) samples. (b) Plot of ln ρ against T^{-1} for $Bi_2Sr_2Co_2O_y$ and $Bi_2Sr_{1.9}M_{0.1}Co_2O_y$ samples. Solid lines stand for TAC fitting. Dashed curves express VRH fitting. (c) $Bi_2Sr_2Co_2O_y$ and $Bi_2Sr_{1.9}M_{0.1}Co_2O_y$ dependence of activation energy ΔE, onset temperature T_p of TAC, and onset temperature $T_{hopping}$ of VRH. The shadow in bold is guide to the eyes. (d)–(f) are similar to (a)–(c) but for $Bi_2Sr_2Co_{1.9}X_{0.1}O_y$ (X = Zr, Al, and Mo) samples.

To get insight into the conduction mechanism below T_p, dependences of ln ρ on T^{-1} are plotted in **Figure 2b** and **e**. At the beginning, it is found that thermally activated conduction (TAC) law matches $\rho(T)$ data well below T_p, namely [35], $\rho(T) = \rho_0 \exp(\Delta E/k_B T)$, where ΔE is activation energy. Interestingly, $\rho(T)$ apparently deviates from the TAC behavior with decreasing temperature further, and it follows Mott's variable-range-hopping (VRH) model described by equation

[35]: $\rho(T) = \rho_0\exp[(T_0/T)^n]$. As seen from **Figure 2c** and **f**, obtained values of ΔE and onset temperature $T_{hopping}$ of $Bi_2Sr_{1.9}Ca_{0.1}Co_2O_y$ (0.66 meV and 15.3 K) are the respective minimum, even smaller, than those of parent $Bi_2Sr_2Co_2O_y$ (0.70 meV and 16.2 K), while ΔE and $T_{hopping}$ of $Bi_2Sr_2Co_{1.9}Zr_{0.1}O_y$ (2.75 meV and 63.6 K) are both maximum among all samples.

2.3. Thermoelectric power and narrow band model

Figure 3a and **b** shows temperature dependence of thermoelectric power $S(T)$ for all samples. Positive values of S reflect electrical transport feature dominated by holes. Values of S at room temperature for all doped samples produce a substantial increase, especially for $Bi_2Sr_2Co_{1.9}Mo_{0.1}O_y$ (~117 μV/K), compared with pristine $Bi_2Sr_2Co_2O_y$ (~92 μV/K). Particularly, with decreasing the temperature until below $T_{hopping}$, $S(T)$ behavior follows with VRH model [36]: $S_{VRH}(T) \sim aT^{1/2}$, where a is factor determined by density of localized states at Fermi level $N(E_F)$. The inset in **Figure 3b** reveals Anderson localization of $Bi_2Sr_2Co_{1.9}Mo_{0.1}O_y$, in correspondence with low-temperature resistivity.

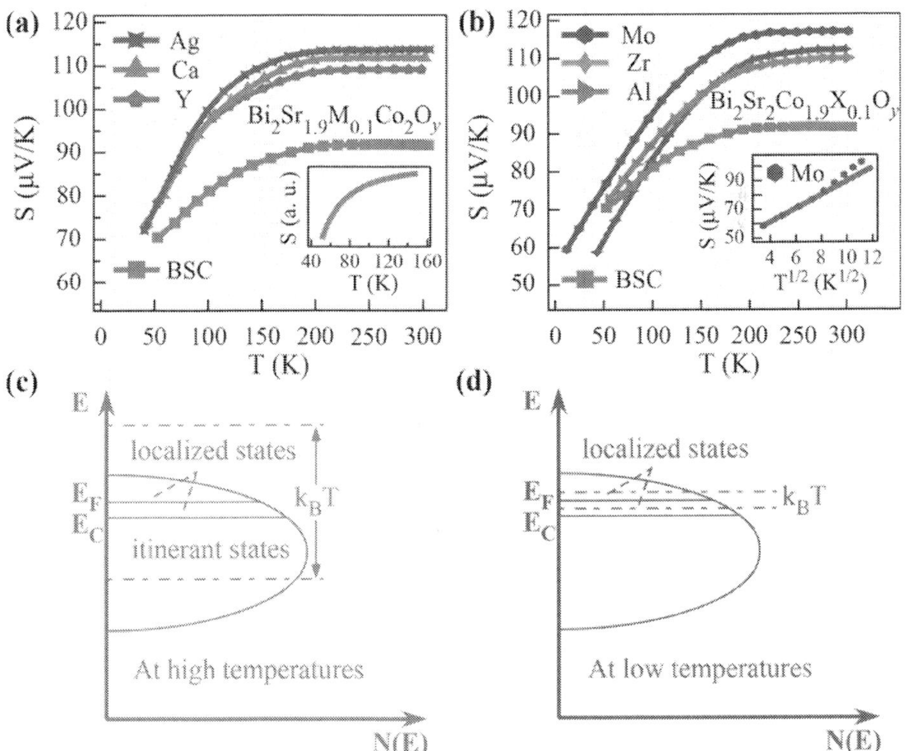

Figure 3. Temperature dependence of thermoelectric power $S(T)$ for $Bi_2Sr_2Co_2O_y$, (a) $Bi_2Sr_{1.9}M_{0.1}Co_2O_y$ (M = Ag, Ca, and Y), and (b) $Bi_2Sr_2Co_{1.9}X_{0.1}O_y$ (X = Zr, Al, and Mo) samples. Inset: calculated and fitted results of (a) Boltzmann formula and (b) VRH model for $Bi_2Sr_2Co_{1.9}Mo_{0.1}O_y$ sample, respectively. Schematic diagram of density of states in a narrow band with Anderson localization at (c) high temperatures (metallic or TAC region) and (d) low temperatures (VRH region).

In general, S is extremely small (<10 μV/K) and presents a metallic behavior in a broad band [7]. Taking into account the huge difference, large S at high temperatures (above 200 K) in a narrow band matches Heikes model [37]: $S = k_B/e\{\ln[d/(1-d)]\}$, where d is concentration of Co^{4+}. The enhanced S at high temperatures is attributed to the competition between d and spin entropy. It is noted that $S(T)$ is also described by narrow band model at intermediate temperatures. $S(T)$ follows with Boltzmann formula [38]: $S(T) = 1/eT\{\int(E-E_F)E^2 dE/[e^{(E-E_F)/2k_B T} + e^{-(E-E_F)/2k_B T}]^2\}/\{\int E^2 dE/[e^{(E-E_F)/2k_B T} + e^{-(E-E_F)/2k_B T}]^2\}$. Calculated $S(T)$ indicates monotonous increase with increasing T, as well as experimental result as plotted in the inset in **Figure 3a**, revealing the validity of narrow band model.

Actually, activation energy ΔE is equal to $E_F - E_C$, where E_C is the upper mobility edge. As $k_B T/2 > \Delta E$, conduction mainly determined by contribution of excited holes in itinerant states as specified in **Figure 3c**. At high temperatures, the majority of acceptor-like states are fully ionized, that is, occurs complete excitation of holes, that resulting in metallic behavior of $\rho(T)$ and diffused $S(T)$ (Heikes formula). As $k_B T/2$ is near to ΔE, TAC conduction forms (Boltzmann dispersion). As $k_B T/2 < \Delta E$, VRH conduction dominates the transport mechanism as shown in **Figure 3d**.

2.4. X-ray photoemission spectroscopy and thermal conductivity

In order to further verify the narrow band model, we carried out XPS spectra for $Bi_2Sr_{1.9}Ca_{0.1}Co_2O_y$. As shown in **Figure 4b**, XPS spectra present an intense peak at ∼ 0.95 eV, in line with large S and metallic behavior. Between E_F and ∼2.0 eV, Co 3d and O 2p orbitals play an important role, similar to pristine $Bi_2Sr_2Co_2O_y$ [39]. Moreover, strong hybridization between Co 3d and O 2p forms [39, 40]. Namely, antibonding t_{2g} narrow bands contribute to intense peak at ∼0.95 eV, while bonding e_g broad bands are responsible to peak within 3–8 eV. In addition, calculated $S(T)$ is also consistent with experimental value based on magnitude and temperature dependence [39]. Therefore, the narrow band model is very suitable for explaining all experimental and theoretical results.

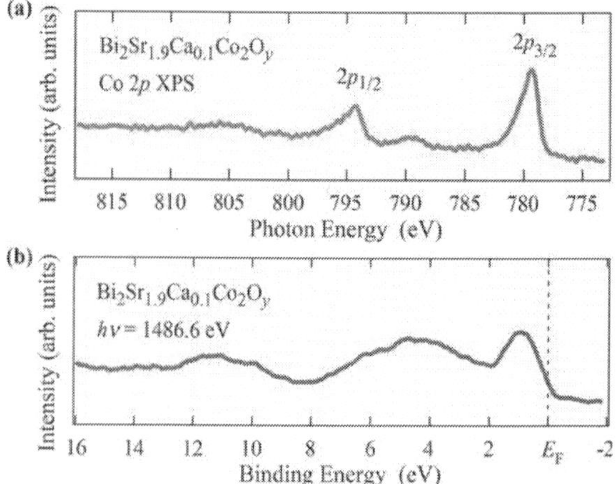

Figure 4. (a) Co 2p XPS spectra and (b) XPS spectra in wide binding-energy range for selected $Bi_2Sr_{1.9}Ca_{0.1}Co_2O_y$ sample at room temperature.

Temperature dependence of total thermal conductivity $\kappa(T)$ for all samples are shown in **Figure 5a** and **d**. $\kappa(T)$ can be expressed by the sum of phononic component $\kappa_{ph}(T)$ and mobile charge carriers' component $\kappa_e(T)$ as $\kappa(T) = \kappa_{ph}(T) + \kappa_e(T)$. Value of $\kappa_e(T)$ can be estimated from the Wiedemann-Franz law, $\kappa_e(T) = L_0 T/\rho$, where $L_0 \sim 2.44 \times 10^{-8}$ V^2/K^2 stands for Lorenz number. In **Figure 5b** and **e**, $\kappa_{ph}(T)$ dominates the thermal conductivity because CoO_2 layer and Bi-Sr-O block layer induces the interface scattering. Dimension less figure of merit $ZT = S^2 T/\rho\kappa$ reflects total thermoelectric performance (see **Figure 5c** and **f**). For pristine $Bi_2Sr_2Co_2O_y$, ZT value reaches ~ 0.007 at 300 K, while ZT value reaches 0.19 at 973 K, indicative of promising thermoelectric material for $Bi_2Sr_2Co_2O_y$ at high temperatures [2]. Especially for $Bi_2Sr_{1.9}Ca_{0.1}Co_2O_y$, ZT value reaches maximum ~ 0.012 at 137 K. Therefore, it is reasonable to predict that $Bi_2Sr_{1.9}Ca_{0.1}Co_2O_y$ could be considered as one of potential ultra-high temperature thermoelectric materials, as well as pristine $Bi_2Sr_2Co_2O_y$.

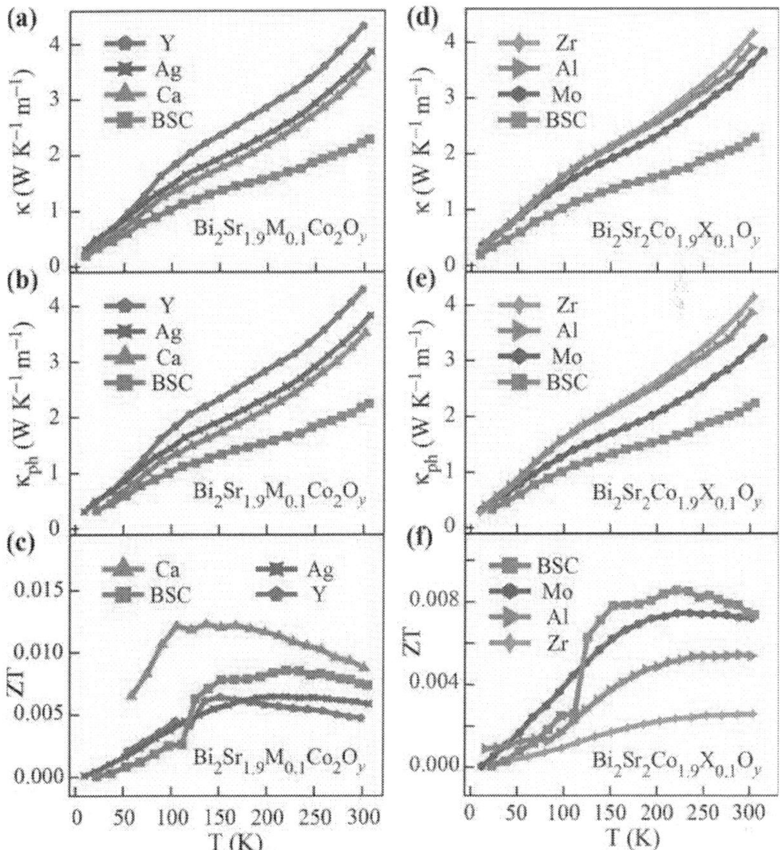

Figure 5. Temperature dependence of (a) total thermal conductivity $\kappa(T)$, (b) phononic component $\kappa_{ph}(T)$, and (c) dimensionless figure of merit ZT for BSC and $Bi_2Sr_{1.9}M_{0.1}Co_2O_y$ (M = Ag, Ca, and Y) samples. (d)–(f) are similar to (a)–(c), but for $Bi_2Sr_2Co_{1.9}X_{0.1}O_y$ (X = Zr, Al, and Mo) samples.

3. Exotic reinforcement of thermoelectric power in layered $Bi_2Sr_{2-x}Ca_xCo_2O_y$

3.1. XRD patterns and electrical transport properties

The crystal structure of $Bi_2Sr_2Co_2O_y$ is shown in **Figure 6a**. **Figure 6b** shows XRD patterns of all Ca-doping samples with single phase in $Bi_2Sr_{2-x}Ca_xCo_2O_y$ ($0.0 \leq x \leq 2.0$). With increasing Ca content, diffraction peak along [003] direction distinctly shifts to higher angle as shown in the inset in **Figure 6b**, confirming the smaller ionic radius of Ca^{2+}, than that of Sr^{2+}. SEM characterization indicates surface morphology of plate-like grains and regular grain sizes for selected samples with $x = 0.0$ and 1.0, respectively.

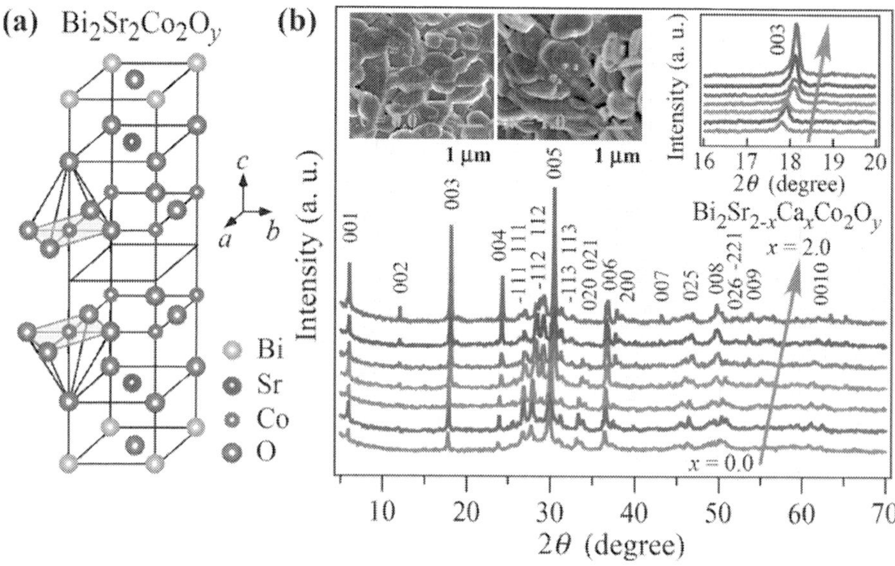

Figure 6. (a) Crystal structure of $Bi_2Sr_2Co_2O_y$. (b) Powder XRD patterns for $Bi_2Sr_{2-x}Ca_xCo_2O_y$ ($0.0 \leq x \leq 2.0$) samples at room temperature. Insets: magnified powder's XRD patterns along [003] direction for all samples and SEM images for selected samples with $x = 0.0$ and 1.0, respectively.

Figure 7a and **b** shows resistivity $\rho(T)$ of all samples in $Bi_2Sr_{2-x}Ca_xCo_2O_y$. For the present $x = 0.0$ polycrystalline sample, upturning point at T_p ($\sim$150 K) appears. Metallic behavior above T_p is observed, demonstrating the existence of itinerant charge carriers. In comparison, for $x = 0.0$ single crystal [41], in-plane resistivity ρ_{ab} also shows metallic behavior around room temperature, while it arises minimum near 80 K and diverges with further decreasing the temperature. Resistivity ρ_{ab} value of single crystal for $x = 0.0$ at room temperature is $\sim$4 mOhm×cm and is smaller than that of our polycrystalline sample ($\sim$15 mOhm×cm). On the other hand, compared with $x = 0.0$, $\rho(T)$ of all Ca-doped samples produce total increase due to disorder effect. For the samples with $x \leq 0.5$, enhanced random Coulomb potential because of Ca doping induces the shift of T_p toward higher temperature. Interestingly, for the samples with $x \geq 1.0$, the signature of transition at T_p completely vanishes and $\rho(T)$ only presents an insulating-like behavior.

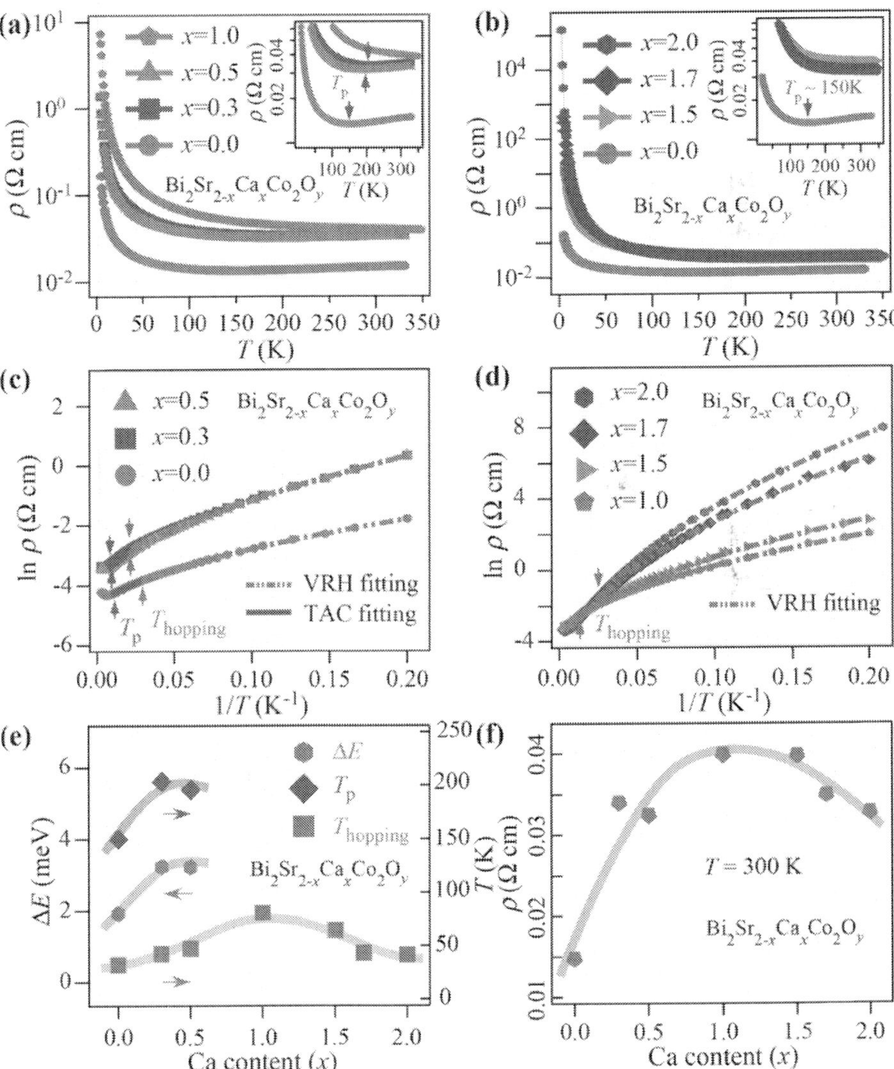

Figure 7. (a) and (b) Temperature dependence of resistivity $\rho(T)$. Insets: magnification plot of $\rho(T)$ for $Bi_2Sr_{2-x}Ca_xCo_2O_y$ samples. (c) and (d) Plot of $\ln\rho$ against $1/T$. Solid lines present TAC fitting. Dashed curves stand for VRH fitting. (e) Ca concentration x dependence of activation energy ΔE, onset temperature T_p of TAC, and onset temperature $T_{hopping}$ of VRH. (f) Ca concentration x dependence of resistivity $\rho_{300\,K}$ at room temperature.

To discern conduction mechanism below T_p, relationship of $\ln\rho$ against $1/T$ is plotted in **Figure 7c** and **d**. As for $x \leq 0.5$, at the beginning, it is found that TAC law matches $\rho(T)$ data well below T_p, namely [35], $\rho(T) = \rho_0\exp(\Delta E/k_B T)$, where ΔE is activation energy. But $\rho(T)$ apparently deviates from TAC behavior with decreasing the temperature further, and it follows Mott's VRH model described by equation [32]: $\rho(T) = \rho_0\exp[(T_0/T)^n]$. However, as for $x \geq 1.0$, $\rho(T)$ meets VRH model only, in agreement with the insulating feature of $x = 2.0$ single crystal [1, 42, 43]. Obtained values of ΔE and onset temperature $T_{hopping}$ are plotted in **Figure 7e**. Basically, ΔE increases with Ca content, as well as T_p for $x \leq 0.5$. In comparison, the present value of ΔE based on sintering temperature 800°C is larger than the previous one of $x = 0.0$ at 900°C [8], revealing the difference of grain size effect. It is worth noting that values of $T_{hopping}$ and ρ_{300K} at room temperature first increase and then decrease in whole Ca-doped range (see **Figure 7e** and **f**).

3.2. Enhancement of thermoelectric power driven by Ca doping

Figure 8a shows thermoelectric power $S(T)$ for all samples. Positive values of S demonstrate that majority of charge carriers are hole type. In addition, S exhibits a nearly T-independent behavior above 200 K, while S strongly depends on T peculiarly below 150 K. Ca doping obviously boosts S_{300K} at room temperature especially for heavy Ca contents (see **Figure 8b**). Large S_{300K} value monotonously increases from 105 μV/K$(x = 0.0)$ to 157 μV/K $(x = 2.0)$. In general, the change of S should be related to variation of n. For $x = 0$ single crystal [38], Hall coefficient (R_H) is positive and strongly dependent on the temperature in the range from 300 to 0 K. Increase of R_H toward the lowest temperature is not simple due to the decrease of n, but rather due to anomalous Hall effect. It is noted that variation of R_H with Pb doping is also similar to that of ρ_{ab}. Pb doping slightly reduces the magnitude of R_H, but the increase in number of charge carriers is much smaller than expected from chemical composition [41, 44].

As we know, S is rather low (<10 μV/K) with a metallic behavior in a broad band [7]. Taking into account the tremendous discrepancy, large S of $Bi_2Sr_{2-x}Ca_xCo_2O_y$ with a nearly T-independence at high temperatures in a narrow band should follow the so-called Heikes formula [37]: $S = k_B/e\{\ln[(g_3/g_4)d/(1-d)]\}$, where d is concentration of Co^{4+}, and g_3 and g_4 are spin orbital degeneracies for Co^{3+} and Co^{4+} ions, respectively. Concentration d at room temperature can be deduced from charge carriers' density n. As visible in **Figure 8c**, as for $x< 1.5$, d decreases, while S_{Heikes} (deriving from Heikes formula) increases, which is consistent with the trend of S_{300K}. But for $x \geq 1.5$, reduced S_{Heikes} is reverse to persistent enhancement of S_{300K}. Thus, we have to consider other possible reason of enhanced S for heavily doped samples.

3.3. Specific heat and Sommerfeld coefficient

Next we will check whether the enhanced S originates from the increased effective masses through electronic correlation. To test this point, we performed measurement of specific heat $C(T)$, which is plotted as C/T versus T^2 (see the inset in **Figure 8d**) for selected samples with $x = 0.0$, 0.5, 1.5, and 2.0. $C(T)$ at low temperatures can be described as $C(T) = \gamma T + \beta T^3$ [45], where γT and βT^3 denote electronic and lattice contribution to $C(T)$, respectively. We can get electronic coefficient γ by the linear fitting according to $C/T = \gamma + \beta T^2$ [45]. Here, we need to explicitly interpret Sommerfeld coefficient γ. For the present system, unit formula should involve two cobalt atoms. For our polycrystalline sample with $x = 0.0$, a conventional way to get γ by extrapolating high-temperature linear part of C/T versus $T = 0$ gives very large value of ~ 135 mJ mol^{-1} K^{-2} (see **Figure 8d**), comparable with that of $x = 0.0$ single crystal (~ 140 mJ mol^{-1} K^{-2}) [41]. However, it is observed that γ rapidly decreases with increasing Ca doping. For our sample with $x = 2.0$, value of γ is ~ 85 mJ mol^{-1} K^{-2}. Differently, it is noted that value of γ is only 50 mJ mol^{-1} K^{-2} for Bi-Ca-Co-O system, while such a unit formula merely includes one cobalt atom [45].

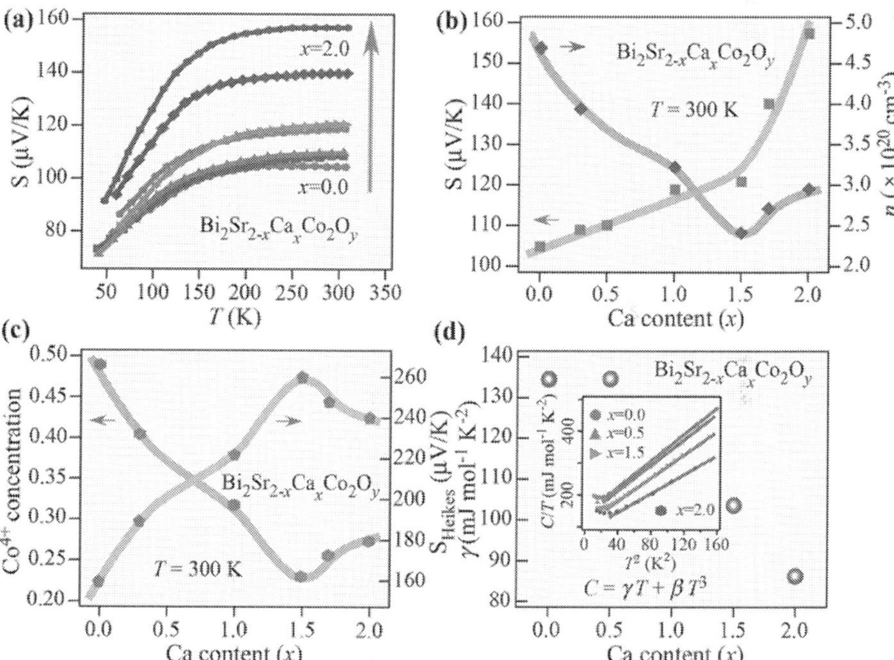

Figure 8. (a) Temperature dependence of thermoelectric power $S(T)$ for Bi$_2$Sr$_{2-x}$Ca$_x$Co$_2$O$_y$ samples. (b) Ca concentration x dependence of S and charge carriers' density n at room temperature, respectively. (c) Ca concentration x dependence of Co^{4+} ion (deduced from charge carriers' density n) and corresponding S_{Heikes} (originating from Heikes formula) at room temperature, respectively. (d) Ca concentration x dependence of electronic coefficient γ deriving from specific heat $C(T)$. Inset: temperature dependence of $C(T)$ plotted as C/T versus T^2 based on fitting lines for $x = 0.0$, 0.5, 1.5, and 2.0, respectively.

Now we discuss the underlying implications of enhanced S with Ca doping. As mentioned above, as for $x < 1.5$, decreased d based on Heikes formula should be responsible for the enhanced S. But for $x \geq 1.5$, local modification of DOS and band structure near E_F could play crucial role. $S(T)$ can be defined by Mott formula [39]: $S(T) = (\pi^2 k_B T)/(3e)[\text{dln}\sigma(E)/\text{d}E]_{E=E_F}$, where $\sigma(E)$ is electrical conductivity with $\sigma(E) = n(E)ev(E)$, $v(E)$ is mobility, $n(E)$ is charge carriers' density with $n(E) = D(E)f(E)$, $D(E)$ is DOS, and $f(E)$ is Fermi function. Apparently, in terms of Mott formula, the enhancement of S for $x \geq 1.5$ should be attributed to the increase of local DOS near E_F. In details, with decreasing A-site ionic radius (i.e., with increasing Ca content), tolerance factor decreases (not shown here), which leads to changes of lattice distortion in CoO_2 layer and local band structure near E_F, reminiscent of layered perovskite cobaltite $SrLnCoO_4$ (Ln stands for different rare earth elements) [46]. Ultimately, value of S for $x \geq 1.5$ would be enhanced. Based on all of above results, one should emphasize that Sommerfeld coefficient γ is dependent on n, and also as function of DOS at E_F, which leads to continuous enhancement of large S.

4. Thermoelectricity generation and electron-magnon scattering in a natural chalcopyrite mineral

4.1. Crystal structure and SEM characterization

A series of natural chalcopyrite minerals, $Cu_{1+x}Fe_{1-x}S_2$ (x = 0.17, 0.08, and 0.02), were obtained from a hydrothermal vent site named Snow Chimney in the Mariner field of Lau Basin [47]. Basically, mineral composition obtained from intact natural sulfide chimneys has no variation. Subsamples with x = 0.02 and 0.08 were obtained from the most interior chimney part, whereas subsample with x = 0.17 was obtained from the middle chimney wall region. The highly fluctuated and variable physicochemical conditions lead to obvious differences in mineral composition [48]. **Figure 9** shows sketches of its crystal structure and atomic planes, in which chalcopyrite crystallizes in a tetragonal lattice with space group of I-42d and produces honeycomb structure characteristic [49]. Each Fe and Cu atom is encircled by tetrahedron of S atom. The highlighted planes indicateatomic zig-zag pattern, which is likely responsible to phonon scattering. XRD Rietveld refinement of power pattern indicates that three natural samples are single phase with standard chalcopyrite structure. For x = 0.08, refined lattice parameters a and c are 5.278 and 10.402 Å, respectively (see **Figure 10**).

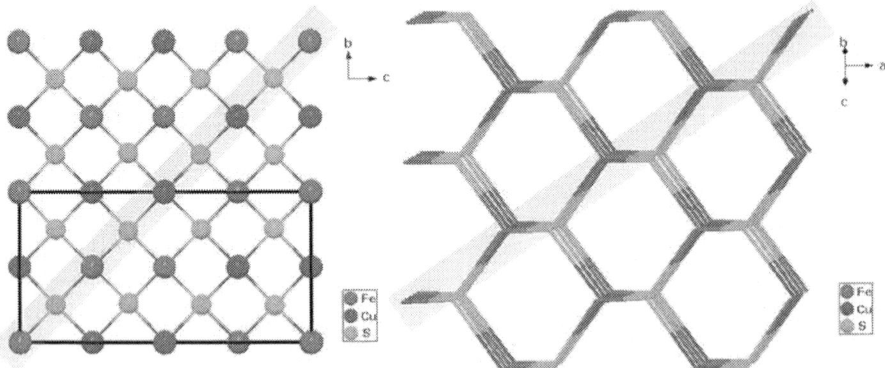

Figure 9. Crystal structure of $Cu_{1+x}Fe_{1-x}S_2$. Ball-and-stick model of the crystal structure (left) viewed along a-axis with black lines indicating unit cell. Stick model (right) showing characteristic honeycomb structure of chalcopyrite. Identical atomic arrangement is highlighted in gray in both structures, but projection is along different axes.

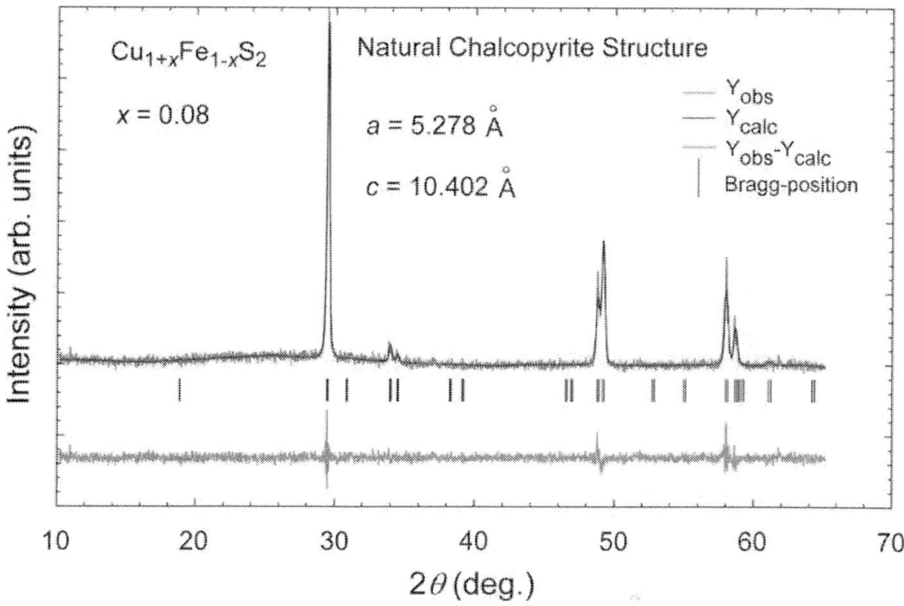

Figure 10. Powder XRD patterns with Rietveld refinement for natural sample of $Cu_{1+x}Fe_{1-x}S_2$ ($x = 0.08$). Red line indicates experimentally observed data, and black line overlapping them refers to calculated data. Vertical tick is related to the Bragg angles positions in space group I-$42d$. The lowest profile shows the difference between observed and calculated patterns. Rietveld refinement indicates that it is standard chalcopyrite structure.

To probe the microstructures of natural $Cu_{1+x}Fe_{1-x}S_2$, we performed SEM characterization (**Figure 11**). SEM analysis revealed that natural chalcopyrite with $x = 0.08$ had layered structure. Three examined natural samples were found to contain morphological diversity, which is characteristic of chalcopyrite minerals, and suggest different physical and chemical behaviors of various microstructures. The SEM observation may provide important insights of the relevance between physical and chemical functions and behaviors of chalcopyrite minerals.

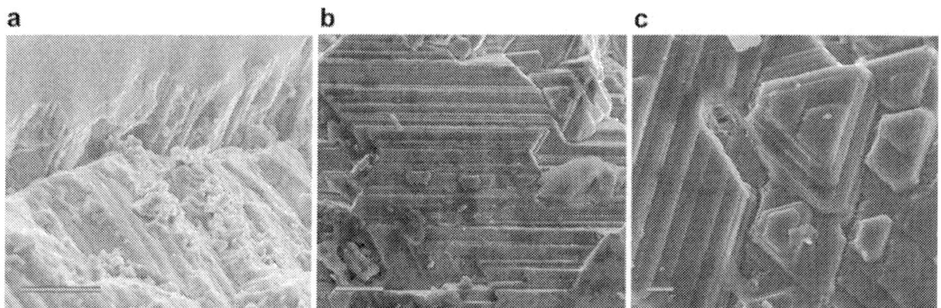

Figure 11. Surface morphology of natural sample of $Cu_{1+x}Fe_{1-x}S_2$ ($x = 0.08$) showing characteristic layered structure. (a) Areas showing cracked layered structure in natural sample $Cu_{1+x}Fe_{1-x}S_2$ ($x = 0.08$), scale bar: 10 μm. (b) Densely layered structure, scale bar: 5 μm. (c) Triangular pattern surrounded by layered structure, scale bar: 1 μm.

4.2. Thermoelectricity generation and electronic states

To examine the functional properties of natural $Cu_{1+x}Fe_{1-x}S_2$ samples, we first measured resistivity (ρ) as function of temperature (T). Three examined natural samples exhibited excellent conductive behavior with semiconductive characteristics (**Figure 12a**). With the reduction of x, the overall resistivity decreased due to the emergence of doped charge carriers. Value of ρ_{300K} for $x = 0.17$, 0.08, and 0.02 was 4.97, 0.11, and 1.01 Ohm×cm, respectively. Compared with $x = 0.08$, the increase of resistivity for $x = 0.02$ stems from the enhanced random Coulomb potential owing to the natural doping.

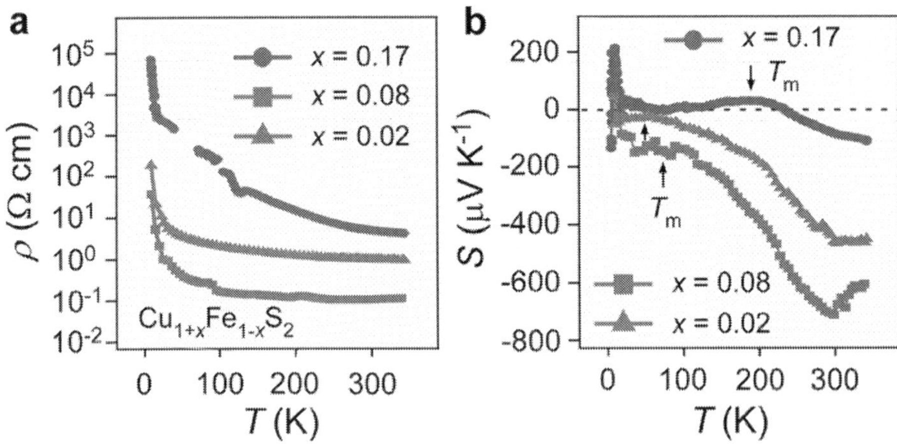

Figure 12. Formation of thermoelectricity by $Cu_{1+x}Fe_{1-x}S_2$. (a) Temperature dependence of resistivity ρ in three natural samples of $Cu_{1+x}Fe_{1-x}S_2$. (b) Temperature dependence of thermoelectric power S for three samples.

In order to track the evolution of electronic states, we carried out thermoelectric power (S) measurement (**Figure 12b**), where the sign of S changes. For $x = 0.17$, the sign of S switches from negative to positive around 235 K with decreasing temperature (**Figure 12b**). It is amazing to observe two unusual peaks: a broad peak (T_m; 32 µV/K, 186 K) and a sharper peak (T_p; 215 µV/K, 11 K), indicating the majority of hole carriers (p-type). It is of particular interest that, for $x = 0.08$ and 0.02, T_p peak utterly disappears, while T_m peak becomes wider and rapidly shifts to a lower temperature, where S presents very large negative values, demonstrating the majority of electron carriers (n-type), in line with negative Hall coefficient R_H (**Figure 13**). Large S_{300K} reached a remarkable value of −713 and −457 µV/K for $x = 0.08$ and 0.02, respectively. Namely, more electrons are activated at room temperature with increasing Fe concentration. For $x = 0.08$, charge carriers' mobility μ_{300K} and density n_{300K} are 1.8 cm² V⁻¹ s⁻¹ and 3.5 × 10¹⁹ cm⁻³, obtained from $R_H = 1/ne$ and $\mu = 1/ne\rho$. In addition, Fe magnetic moment may also play an key role to induce large S, indicative of strong coupling between magnetic ions and doped charge carriers because synthetic $CuFeS_2$ presents AFM ordering at 823 K [15].

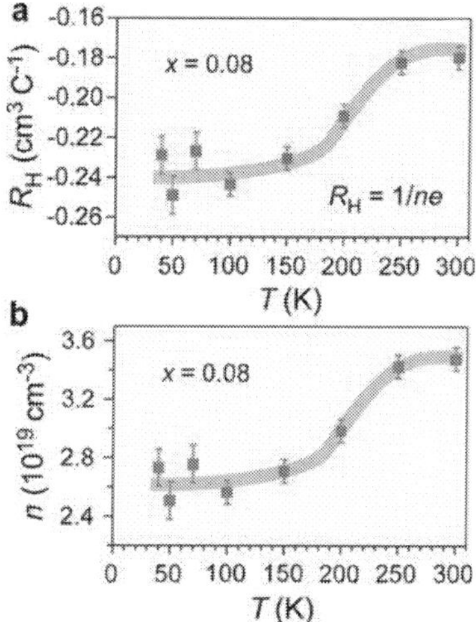

Figure 13. Hall effect of natural sample of $Cu_{1+x}Fe_{1-x}S_2$ ($x = 0.08$). (a) Temperature dependence of Hall coefficient R_H. (b) Temperature dependence of charge carriers' density n. Value of R_H (cm^3 C^{-1}) is determined by n (cm^{-3}) and electron charge e, that is, $R_H = 1/ne$, where $e = 1.602176 \times 10^{-19}$ C. The shadow in bold is guide to the eyes.

4.3. Electron-magnon scattering and large effective mass

The matter of imperative concern is how to understand the origin of T_m peak and conduction mechanism. According to Mott's formula, S can be qualitatively expressed as $S = -\pi^2 k_B^2 T/3e$ $[\sigma'(E_F)/\sigma(E_F)]$, where k_B is Boltzmann constant, $\sigma(E_F)$ is electrical conductivity at Fermi level E_F, and σ' denotes $d[\sigma(E)]/dE$ [35]. If one assumes σ' is a constant accompanied by isotropic electrical transport properties, namely, $\sigma^{-1} = \rho$, then $\Delta S/S_0$ $\Delta\rho/\rho_0$ can be derived. Plot of $\Delta S/S_0$ versus $\Delta\rho/\rho_0$ for $x = 0.17$ (**Figure 14**) shows that all experimental data near T_m at T_0 from 155 to 300 K deviate from the theoretical calculation, the linearity. These results verify that exotic mechanism of $S(T)$ in natural sample is beyond the framework of conventional thermoelectric picture [50].

To better discern intrinsic transport mechanism of $Cu_{1+x}Fe_{1-x}S_2$, we incorporate spin-wave theory to analyze temperature dependence of S. For $x = 0.08$ and 0.02, field-cooling magnetization and loop hysteresis indicate the localized ferromagnetism (FM) at low temperatures because of additional Fe moments (**Figure 15**). However, strong AFM interaction at high temperatures dominates for three natural samples. Generally speaking, spin waves can scatter electrons for AFM or FM materials, resulting in magnon-drag effect [12]. To check this issue, we developed magnon-drag model, $S=S_0+S_{3/2}T^{3/2}+S_4T^4$, where S_0 is value of S at $T = 0$, $S_{3/2}T^{3/2}$ term stems from electron-magnon scattering, and S_4T^4 term is related to spin-wave fluctuation in AFM phase. Using this model of magnon drag, the predicted values for three samples closely matched $S(T)$ data (**Figure 16a** and **b**). As the absolute value of $S_{3/2}$ is nearly six orders of magnitude larger than that of S_4 (**Table 1**), electron-magnon scattering dominates $S(T)$ curve. Thus, T_m peak is predicted to originate from magnon drag due to the strong electron-magnon interaction.

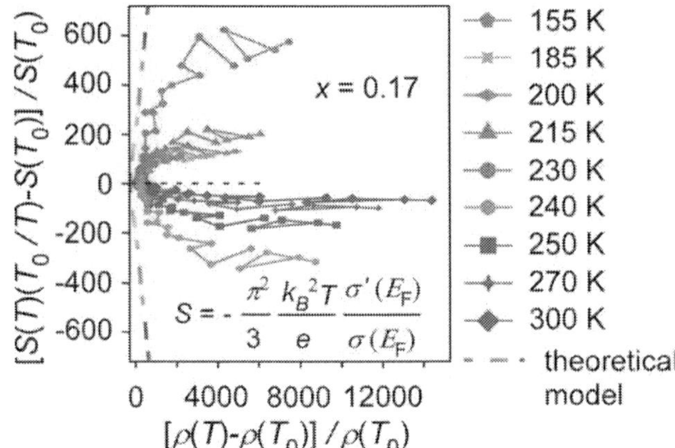

Figure 14. Correlation between thermoelectric power $S(T)$ and resistivity $\rho(T)$. Relative changes of $\Delta S/S_0$ versus $\Delta\rho/\rho_0$ in natural sample with $x = 0.17$ at various temperatures ($T_0 = 155, 185, 200, 215, 230, 240, 250, 270$, and 300 K). The present experimental data substantially deviates from the linear relationship predicted by Mott's formula, which is indicated by dotted line.

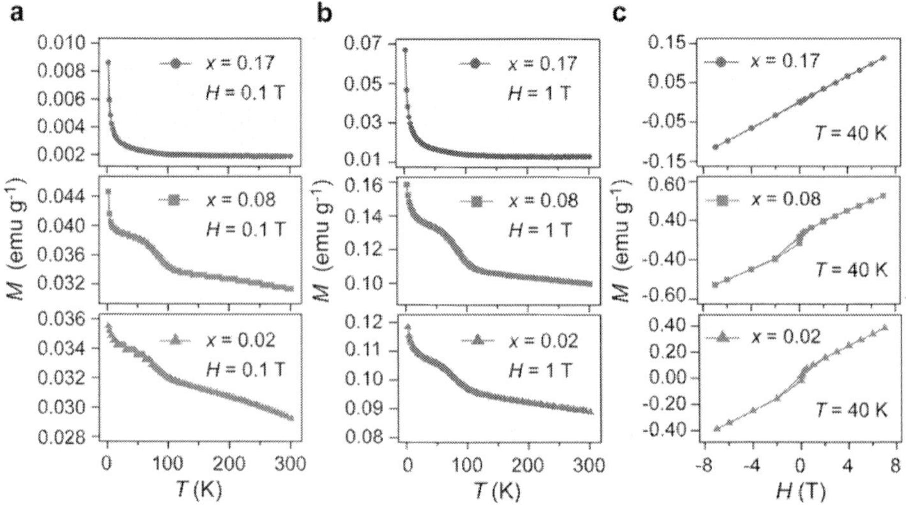

Figure 15. Magnetic properties of natural $Cu_{1+x}Fe_{1-x}S_2$. (**a**, **b**) Temperature dependence of field-cooling (FC) magnetization, M, in three natural samples of $Cu_{1+x}Fe_{1-x}S_2$, measured in applied magnetic field of $H = 0.1$ T (**a**) and $H = 1$ T (**b**). (**c**) Magnetic field dependence of magnetization, M, for three samples, measured at 40 K.

To gain more insight into the correlation between magnon drag, doped carriers, and S, we plotted parameters S_0, $S_{3/2}$, and S_4 as a function of x (**Table 1**). S_0, $S_{3/2}$, and S_4 for $x = 0.08$ has largest absolute values among three natural samples, in agreement with the largest S, smallest ρ, and highest power factor. Unlike S_0 and S_4, dependence of $S_{3/2}$ is quite unique (**Figure 16c**). The sign of $S_{3/2}$ varies from positive to negative with increasing Fe concentration, suggesting the alternation of p-type and n-type charge carriers and orbital degree of freedom of Fe $3d$ band with AFM ordering. Additionally, electron-magnon scattering occupies thermoelectric properties, indicating strong coupling between doped charge carriers and AFM spins. Furthermore, $\rho(T)$ follows TAC model $\rho(T) = \rho_0 \exp(\Delta E / k_B T)$, where ΔE is activation energy [35]. Notably, the fitted energy gap of ΔE (60.1, 4.9, and 11.8 meV for x = 0.17, 0.08, and 0.02, respectively), which verifies the existence of localized Fe spins, is markedly smaller than that of artificial chalcopyrite [21, 29–31]. It is noted that experimental $S(T)$ result is well described by electron-magnon scattering up to ~200 K, while it deviates from theoretical lines for higher temperatures. In particular, power factor S^2/ρ shows an abrupt enhancement above 200 K for x = 0.08 (**Figure 16d**), in agreement with that of R_H and n (**Figure 13**). Above 200 K, large effective mass (m^*) leads to high power factor and large S due to low μ and high n. For x = 0.08, it exhibits the largest m^* value (1.6 m_0) at room temperature, where m_0 is free electron mass. Therefore, we can conclude that robust electron-magnon scattering and large m^* induce unexpected thermoelectricity generation in natural chalcopyrite mineral.

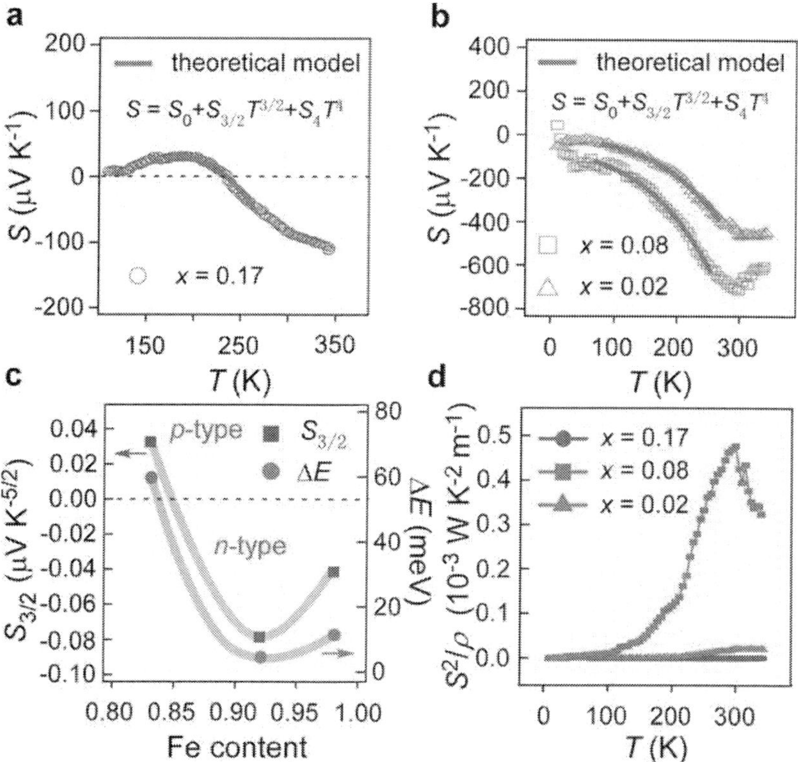

Figure 16. Temperature dependence of S for $Cu_{1+x}Fe_{1-x}S_2$ samples with x = 0.17 (**a**) and x = 0.08 and 0.02 (**b**). Symbols represent experimental data and solid lines correspond to theoretical simulation based on the model of magnon drag, $S = S_0 + S_{3/2}T^{3/2} + S_4 T^4$. (**c**) Obtained parameters $S_{3/2}$ and ΔE are plotted as function of Fe content, where $S_{3/2}$ represents the electron-magnon scattering process and ΔE is activation energy. (**d**) Temperature dependence of power factor, S^2/ρ, for three samples.

Parameter	T_m(K)	$S_0(\mu VK^{-1})$	$S_{3/2}(\mu VK^{-5/2})$	$S_4\ (\mu VK^{-5})$	ΔE(meV)
$x = 0.17$	186	−6.21	0.03	−3.84×10^{-8}	60.1
$x = 0.08$	68	−75.45	−0.08	−5.47×10^{-8}	4.9
$x = 0.02$	38	−10.61	−0.04	−3.95×10^{-8}	11.8

The parameter T_m represents the peak of magnon drag, which stems from the experimental $S(T)$ curve. The parameters S_0, $S_{3/2}$, and S_4 stem from the model of magnon drag, $S = S_0 + S_{3/2}T^{3/2} + S_4T^4$. The parameter ΔE is the activation energy, which stems from the TAC model, $\rho(T) = \rho_0 \exp (\Delta E/k_B T)$.

Table 1. Obtained parameters based on theoretical simulation.

In terms of thermal conductivity κ, phononic component κ_{ph} dominates for three natural samples owing to negligible electronic component κ_e (**Figure 17**). For the optimal sample with $x = 0.08$, value of ZT can reach 0.03 at room temperature (**Figure 17**), thus indicating that natural chalcopyrite semiconductor is a promising candidate for thermoelectric energy materials. It is quite striking that the spontaneous doping process during deep-sea hydrothermal vent mineral precipitations led to natural thermoelectric improvement, which is similar to natural mineral tetrahedrites [51].

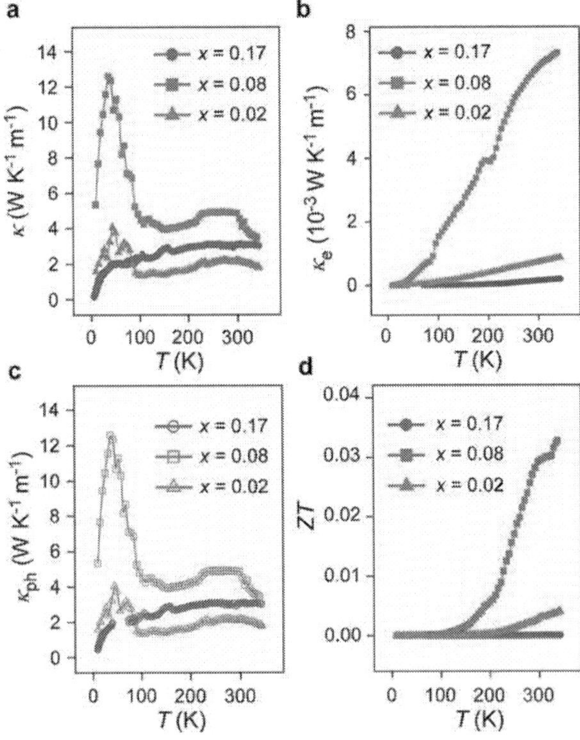

Figure 17. Thermal conductivity and phonon scattering of natural $Cu_{1+x}Fe_{1-x}S_2$. (a) Temperature dependence of total thermal conductivity κ. (b) Temperature dependence of electronic component κ_e. (c) Temperature dependence of phononic component κ_{ph}. (d) Temperature dependence of dimensionless figure of merit ZT.

5. Conclusions

Our results of layered cobaltites $Bi_2Sr_2Co_2O_y$ system based on narrow band model are not only helpful to understand large S and transport mechanism but also differentiate other systems based on a broad band model. In particular, we give the experimental evidence by Hall effect and $C(T)$ measurements, demonstrating that Sommerfeld coefficient γ is dependent on charge carriers' density n, and also as a function of DOS at E_F, which induces exotic enhancement of large S in $Bi_2Sr_{2-x}Ca_xCo_2O_y$. Especially for $Bi_2Sr_{1.9}Ca_{0.1}Co_2O_y$, it may provide an excellent platform to be regarded as potential candidates for thermoelectric materials.

In addition, we demonstrated direct thermoelectricity generation in natural chalcogenides, $Cu_{1+x}Fe_{1-x}S_2$, which was shown to have large S value and high power factor in the low x region, in which electron-magnon scattering and large m^* values were detected. Since doped charge carriers exist in strong coupling with localized spins, the unusual alternation of p- and n-type carriers should be of paramount importance in understanding charge dynamics arising from $3d$ orbital degrees of freedom. Such a finding of exotic thermoelectric properties in natural but not synthetic chalcopyrite opens a novel research field for manipulating low-cost thermoelectricity or even electron/hole carriers, providing therefore a new perspective on technical feasibility for designing and pinpointing the surface-morphology-engineered devices via the naturally abundant materials.

Acknowledgements

The author gratefully thanks L. H. Yin, W. H. Song, Y. P. Sun, A. U. Khan, N. Tsujii, K. Takai, R. Nakamura, and T. Mori for their fruitful collaboration in the study of layered cobaltites and natural chalcogenides for thermoelectrics. This work was supported by the National Natural Science Foundation of China under Contract No. 10904151, the Fund of Chinese Academy of Sciences for Excellent Graduates, and the NIMS Open Innovation Center (NOIC) of Japan. The author thanks the Sichuan University Talent Introduction Research Funding (grant No. YJ201537) and Sichuan University Outstanding Young Scholars Research Funding (grant No. 2015SCU04A20) of China for financial support.

Author details

Ran Ang

Address all correspondence to: rang@scu.edu.cn

1 Key Laboratory of Radiation Physics and Technology, Ministry of Education, Institute of Nuclear Science and Technology, Sichuan University, Chengdu, China

2 Institute of New Energy and Low-Carbon Technology, Sichuan University, Chengdu, China

References

[1] Tarascon J M, Ramesh R, Barboux P, Hedge M S, Hull G W, Greene L H, et al.: New non-superconducting layered Bi-oxide phases of formula $Bi_2M_3Co_2O_y$ containing Co instead of Cu. Solid State Communications. 1989;**71**:663–668. DOI: 10.1016/0038-1098(89)91813-9

[2] Funahashi R, Matsubara I, Sodeoka S.: Thermoelectric properties of $Bi_2Sr_2Co_2O_x$ polycrystalline materials. Applied Physics Letters. 2000;**76**:2385–2387. DOI: 10.1063/1.126354

[3] Koumoto K, Terasaki I, Funahashi R.:Complex oxide materials for potential thermoelectric applications. MRS Bulletin. 2006;**31**:206–210. DOI: 10.1557/mrs2006.46

[4] Terasaki I, Sasago Y, Uchinokura K.: Large thermoelectric power in $NaCo_2O_4$ single crystals. Physical Review B. 1997;**56**:R12685–12687. DOI: 10.1103/PhysRevB.56.R12685

[5] Masset A C, Michel C, Maignan A, Hervieu M, Toulemonde O, Studer F, et al.: Misfit-layered cobaltite with an anisotropic giant magnetoresistance: $Ca_3Co_4O_9$. Physical Review B. 2000;**62**:166–175. DOI: 10.1103/PhysRevB.62.166

[6] Itoh T, Terasaki I.: Thermoelectric Properties of $Bi_{2.3-x}Pb_xSr_{2.6}Co_2O_y$ single crystals. Japanese Journal of Applied Physics. 2000;**39**:6658–6660. DOI: 10.1143/JJAP.39.6658

[7] Heikes R R, Ure R W.: Thermoelectricity: Science and Engineering. New York: Interscience; 1961. 576 p.

[8] Yin L H, Ang R, Huang Y N, Jiang H B, Zhao B C, Zhu X B, et al.: The contribution of narrow band and modulation of thermoelectric performance in doped layered cobaltites $Bi_2Sr_2Co_2O_y$. Applied Physics Letters. 2012;**100**:173503. DOI: 10.1063/1.4705429

[9] Yin L H, Ang R, Huang Z H, Liu Y, Tan S G, Huang Y N, et al.: Exotic reinforcement of thermoelectric power driven by Ca doping in layered $Bi_2Sr_{2-x}Ca_xCo_2O_y$. Applied Physics Letters. 2013;**102**:141907. DOI: 10.1063/1.4801644

[10] Sootsman J R, Chung D Y, Kanatzidis M G.: New and old concepts in thermoelectric materials. Angewandte Chemie International Edition. 2009;**48**:8616–8639. DOI: 10.1002/anie.200900598

[11] Chen Y L, Liu Z K, Analytis J G, Chu J H, Zhang H J, Yan B H, et al.: Single Dirac cone topological surface state and unusual thermoelectric property of compounds from a new topological insulator family. Physical Review Letters. 2010;**105**:266401. DOI: 10.1103/PhysRevLett.105.266401

[12] Costache MV, Bridoux G, Neumann I, Valenzuela SO.: Magnon-drag thermopile. Nature Materials. 2011;**11**:199–202. DOI: 10.1038/nmat3201

[13] Ekwo P I, Okeke C E.: Thermoelectric properties of the PbS ZnS alloy semiconductor and its application to solar energy conversion. Energy Conversion and Management. 1992;**33**:159–164. DOI: 10.1016/0196-8904(92)90121-C

[14] Donnay G, Corliss L M, Donnay J D H, Elliott N, Hastings J M.: Symmetry of magnetic structures: Magnetic structure of chalcopyrite. Physical Review. 1958;**112**:1917–1923. DOI: 10.1103/PhysRev.112.1917

[15] Teranishi T.: Magnetic and electric properties of chalcopyrite. Journal of the Physical Society of Japan. 1961;**16**:1881–1887. DOI: 10.1143/JPSJ.16.1881

[16] Tossell J A, Urch D S, Vaughan D J, Wiech G.: The electronic structure of $CuFeS_2$, chalcopyrite, from x-ray emission and x-ray photoelectron spectroscopy and $X\alpha$ calculations. The Journal of Chemical Physics. 1982;**77**:77–82. DOI: 10.1063/1.443603

[17] Fujisawa M, Suga S, Mizokawa T, Fujimori A, Sato K.: Electronic structures of $CuFeS_2$ and $CuAl_{0.9}Fe_{0.1}S_2$ studied by electron and optical spectroscopies. Physical Review B. 1994;**49**:7155–7164. DOI: 10.1103/PhysRevB.49.7155

[18] Nakamura R, Takashima T, Kato S, Takai K, Yamamoto M, Hashimoto K.: Electrical current generation across a Black Smoker Chimney. Angewandte Chemie International Edition. 2010;**49**:7692–7694. DOI: 10.1002/anie.201003311

[19] Lovesey S W, Knight K S, Detlefs C, Huang S W, Scagnoli V, Staub U.: Acentric magnetic and optical properties of chalcopyrite ($CuFeS_2$). Journal of Physics: Condensed Matter. 2012;**24**:216001. DOI: 10.1088/0953-8984/24/21/216001

[20] Lyubutin I S, Lin C R, Starchikov S S, Siao Y J, Shaikh M O, Funtov K O, et al.: Synthesis, structural and magnetic properties of self-organized single-crystalline nanobricks of chalcopyrite $CuFeS_2$. Acta Materialia. 2013;**61**:3956–3962. DOI: 10.1016/j.actamat.2013. 03.009

[21] Tsujii N, Mori T.: High thermoelectric power factor in a carrier-doped magnetic semiconductor $CuFeS_2$. Applied Physics Express. 2013;**6**:043001. DOI: 10.7567/APEX.6.043001

[22] Goodman C H L, Douglas R W.: New semiconducting compounds of diamond type structure. Physica. 1954;**20**:1107–1109. DOI: 10.1016/S0031-8914(54)80247-3

[23] Austin I G, Goodman C H L, Pengelly A E.: New semiconductors with the chalcopyrite structure. Journal of The Electrochemical Society. 1956;**103**:609–610. DOI: 10.1149/ 1.2430171

[24] Nikiforov K G.: Magnetically ordered multinary semiconductors. Progress in Crystal Growth and Characterization of Materials. 1999;**39**:1–104. DOI: 10.1016/S0960-8974(99) 00016-9

[25] Koschel W H, Sorger F, Baars J.: Optical phonons in I-III-VI$_2$ compounds. Le Journal De Physique Colloques. 1975;**36**:C3:177–181. DOI: http://dx.doi.org/10.1051/jphyscol:1975332

[26] Koschel W H, Bettini M.: Zone-centered phonons in $A^IB^{III}S_2$ chalcopyrites. Physica Status Solidi B. 1975;**72**:729–737. DOI: 10.1002/pssb.2220720233

[27] Sato K, Harada Y, Taguchi M, Shin S, Fujimori A.: Characterization of Fe 3d states in $CuFeS_2$ by resonant X-ray emission spectroscopy. Physica Status Solidi A. 2009;**206**:1096–1100. DOI: 10.1002/pssa.200881196

[28] Woolley J C, Lamarche A M, Lamarche G, Quintero M, Swainson I P, Holden T M.: Low temperature magnetic behaviour of $CuFeS_2$ from neutron diffraction data. Journal of Magnetism and Magnetic Materials. 1996;**162**:347–354. DOI: 10.1016/S0304-8853(96)00252-1

[29] Austin I G, Goodman C H L, Pengelly A E.: Semiconductors with chalcopyrite structure. Nature. 1956;**178**:433. DOI: 10.1038/178433a0

[30] Hamajima T, Kambara T, Gondaira K I, Oguchi T.: Self-consistent electronic structures of magnetic semiconductors by a discrete variational $X\alpha$ calculation. III. Chalcopyrite $CuFeS_2$. Physical Review B. 1981;**24**:3349–3353. DOI: 10.1103/PhysRevB.24.3349

[31] Teranishi T, Sato K, Kondo K.:Optical properties of a magnetic semiconductor: Chalcopyrite $CuFeS_2$.: I. Absorption spectra of $CuFeS_2$ and Fe-Doped $CuAlS_2$ and $CuGaS_2$. Journal of the Physical Society of Japan. 1974;**36**:1618–1624. DOI: 10.1143/JPSJ.36.1618

[32] Tsujii N, Mori T, Isoda Y.: Phase stability and thermoelectric properties of $CuFeS_2$-based magnetic semiconductor. Journal of Electronic Materials. 2014;**43**:2371–2375. DOI: 10.1007/s11664-014-3072-y

[33] Ang R, Khan A U, Tsujii N, Takai K, Nakamura R, Mori T.: Thermoelectricity generation and electron-magnon scattering in a natural chalcopyrite mineral from a deep-sea hydrothermal vent. Angewandte Chemie International Edition. 2015;**54**:12909–12913. DOI: 10.1002/anie.201505517

[34] Yamamoto T, Tsukada I, Uchinokura K, Takagi M, Tsubone T, Ichihara M, et al.: Structural phase transition and metallic behavior in misfit layered (Bi,Pb)-Sr-Co-O System. Japanese Journal of Applied Physics. 2000;**39**:L747–750. DOI: 10.1143/JJAP.39.L747

[35] Mott N F, Davis E A.: Electronic Processes in Non-Crystalline Materials. Oxford: Clarendon; 1971. 437 p.

[36] Zvyagin I P.: On the theory of hopping transport in disordered semiconductors. Physica Status Solidi B. 1973;**58**:443–449. DOI: 10.1002/pssb.2220580203

[37] Kittel C.: Introduction to Solid State Physics. Singapore: Wiley; 2001.

[38] MacDonald D K C.: Thermoelectricity: An Introduction to the Principles. New York: Wiley; 1962. 133 p.

[39] Takeuchi T, Kondo T, Takami T, Takahashi H, Ikuta H, Mizutani U, et al.: Contribution of electronic structure to the large thermoelectric power in layered cobalt oxides. Physical Review B. 2004;**69**:125410. DOI: 10.1103/PhysRevB.69.125410

[40] Asahi R, Sugiyama J, Tani T.: Electronic structure of misfit-layered calcium cobaltite. Physical Review B. 2002;**66**:155103. DOI: 10.1103/PhysRevB.66.155103

[41] Yamamoto T, Uchinokura K, Tsukada I.: Physical properties of the misfit-layered (Bi,Pb)-Sr-Co-O system: Effect of hole doping into a triangular lattice formed by low-spin Co ions. Physical Review B. 2002;**65**:184434. DOI: 10.1103/PhysRevB.65.184434

[42] Watanabe Y, Tsui D C, Birmingham J T, Ong N P, Tarascon J M.: Infrared reflectivity of single-crystal $Bi_2M_{m+1}Co_mO_y$ (M=Ca,Sr,Ba; m=1,2), $Bi_2Sr_3Fe_2O_{9.2}$, and $Bi_2Sr_2MnO_{6.25}$, isomorphic to Bi-Cu-based high-Tc oxides. Physical Review B. 1991;**43**:3026–3033. DOI: 10.1103/PhysRevB.43.3026

[43] Terasaki I, Nakahashi T, Maeda A, Uchinokura K.: Optical reflectivity of single-crystal $Bi_2M_3Co_2O_{9+\delta}$ (M=Ca, Sr, and Ba) from the infrared to the vacuum-ultraviolet region. Physical Review B. 1993;**47**:451–456. DOI: 10.1103/PhysRevB.47.451

[44] Yamamoto T, Tsukada I, Takagi M, Tsubone T, Uchinokura K.: Hall effect in a layered magnetoresistive cobalt oxide. Journal of Magnetism and Magnetic Materials. 2001;**226–230**:2031–2032. DOI: 10.1016/S0304-8853(00)00670-3

[45] Limelette P, Hbert S, Hardy V, Frsard R, Simon Ch, Maignan A.: Scaling Behavior in thermoelectric misfit cobalt oxides. Physical Review Letters. 2006;**97**:046601. DOI: 10.1103/PhysRevLett.97.046601

[46] Ang R, Sun Y P, Luo X, Hao C Y, Song W H.: Studies of structural, magnetic, electrical and thermal properties in layered perovskite cobaltite $SrLnCoO_4$ (Ln = La, Ce, Pr, Nd, Eu, Gd and Tb). Journal of Physics D: Applied Physics. 2008;**41**:045404. DOI: 10.1088/0022-3727/41/4/045404

[47] Takai K, Nunoura T, Ishibashi J, Lupton J, Suzuki R, et al. Variability in the microbial communities and hydrothermal fluid chemistry at the newly discovered Mariner hydrothermal field, southern Lau Basin. Journal of Geophysical Research. 2008;**113**:G02031. DOI: 10.1029/2007JG000636

[48] Tivey M K. The influence of hydrothermal fluid composition and advection rates on black smoker chimney mineralogy: Insights from modeling transport and reaction. Geochimica et Cosmochimica Acta. 1995;**59**:1933–1949. DOI: 10.1016/0016-7037(95)00118-2

[49] Goodman C H L. A new group of compounds with diamond type (chalcopyrite) structure. Nature. 1957;**179**:828–829. DOI: 10.1038/179828b0

[50] Asamitsu A, Moritomo Y, Tokura Y. Thermoelectric effect in $La_{1-x}Sr_xMnO_3$. Physical Review B. 1996;**53**:R2952–2955. DOI: 10.1103/PhysRevB.53.R2952

[51] Lu X, Morelli D T, Xia Y, Zhou F, Ozolins V, Chi H, et al. High performance thermoelectricity in earth-abundant compounds based on natural mineral tetrahedrites. Advanced Energy Materials. 2013;**3**:342–348. DOI: 10.1002/aenm.201200650

Electrical Conductivity, Thermoelectric Power and Crystal and Band Structures of EDOB-EDT-TTF Salts Composed of PF_6^-, AsF_6^- and SbF_6^-

Tomoko Inayoshi

Additional information is available at the end of the chapter

http://dx.doi.org/10.5772/65561

Abstract

Novel 2:1 EDOB-EDT-TTF radical salts with different octahedral PF_6^-, AsF_6^-, and SbF_6^- anions were prepared by electrochemical oxidation. AsF_6 salt was found to be isostructural to PF_6 salt and had a triclinic crystal structure, while SbF_6 salt was not isostructural with PF_6 salt and had monoclinic crystal structure. PF_6 salt had higher metal-to-semiconductor (MS) transition temperature, than that of AsF_6 salt, while SbF_6 salt exhibited semiconductive behavior throughout the temperature range of electrical conductivity measurements. To clarify MS transition of these salts, thermoelectric power measurements were also carried out. Thus, thermoelectric power apparatus was constructed and measurements were performed simultaneously with thermoelectric power and electrical resistivity measurements. Crystal structural features for EDOB-EDT-TTF salts at 90, 293, 330 and 350 K, as well as conductivity, thermoelectric power measurements and band structures before and after MS transition are described.

Keywords: EDOB-EDT-TTF radical salts, conductivity, thermoelectric power

1. Introduction

Highly conducting organic TTF·TCNQ complex composed of tetrathiafulvalene (TTF) and tetracyanoquinodimethane (TCNQ) was reported by Heeger et al. in 1973 [1] and has been studied by many physicists and chemists as a one-dimensional conducting system [2, 3]. TTF·TCNQ provides highly anisotropic charge-transfer (CT) complex with metallic properties down to 58 K along b-axis, in which crystal structure [4] has two columns of TTF and TCNQ. Thermoelectric power of TTF·TCNQ along b-axis is negative and proportional to absolute

temperature down to 140 K, but not along a-axis [5]. This is consistent with electrical conductivity measured along different axes. Apparatus for thermoelectric power measurement on organic single crystals has been reported by Chaikin and Kwak [6]. It is designed specifically for small fragile anisotropic samples, such as TCNQ salts. Measurements can be taken with a small (0.5 K) temperature gradient for good temperature resolution.

Bis(ethylenedithio)tetrathiafulvalene (BEDT-TTF) is a good electron donor, and CT complexes and radical salts composed of BEDT-TTF also grow excellent crystals. BEDT-TTF radical salts afford many superconductors as two-dimensional conducting system. In the case of such two-dimensional system, thermoelectric power often exhibits complicated temperature dependence and anisotropy. Mori and Inokuchi have found an agreement between thermoelectric power and calculations for β-(BEDT-TTF)$_2$I$_3$ and κ-(BEDT-TTF)$_2$Cu(NCS)$_2$ [7].

Bis(ethylenedioxy)dibenzotetrathiafulvalene (BEDO-DBTTF) modified with strong electron-donating groups containing ethylenedioxy groups has been synthesized [8]. BEDO-DBTTF CT complexes and salts afforded no metallic compounds. Therefore, we have synthesized a new unsymmetrical EDOB-EDT-TTF donor, which consists of parts of BEDO-DBTTF and BEDT-TTF, and EDOB-EDT-TTF radical salts with octahedral PF$_6^-$, AsF$_6^-$, and SbF$_6^-$ anions [9, 10]. Based on electrical resistivity measurements, PF$_6$ and AsF$_6$ salts underwent a metal-to-semiconductor (MS) transition. X-ray analyses of these salts elucidated their crystal characteristics. Simultaneous measurements of thermoelectric power and electrical resistivity on a single sample were performed for these salts. MS transition temperatures of these PF$_6$ and AsF$_6$ salts were also determined from their thermoelectric power.

2. Preparation of organic conductors

2.1. Synthesis of unsymmetrical EDOB-EDT-TTF donor

As shown in **Figure 1**, synthesis of EDOB-EDT-TTF was carried out using two synthetic methods: cross-coupling (I) and thermal decomposition (II), resulting in 30 and 27% yields, respectively.

2.1.1. Cross-coupling method

To the suspension of **1** (3.0 g, 12 mmol) and **2** (2.5 g, 12 mmol) in dry benzene (40 mL) was added triethyl phosphate (30 mL, 130 mmol). The mixture was refluxed for 5 h at 80°C. The resulting orange precipitate was removed by filtration, and the filtrate was purified by silica gel column chromatography using CHCl$_3$/hexane (5:1) as the eluent. The second fraction was collected, and the resulting product was recrystallized using ethyl acetate to afford EDOB-EDT-TTF as orange crystals (30% yield): mp 260−262. ^{1}H NMR (CDCl$_3$): δ3.30 (4H, s, −SCH$_2$), 4.23 (4H, s, −OCH$_2$), 6.77 (2H, s, ArH). MS (EI) m/z: 402 (M$^+$). IR (KBr, ν_{max} cm^{-1}): 1575 (w), 1540 (w), 1478 (s), 1456 (s), 1300 (s), 1376 (m), 1360 (m), 1100 (m), 1063 (s), 908 (m), 895 (m), 854 (m), 772 (w). UV-vis (CHCl$_3$) λ_{max}, nm: 454, 344, 313, > 260. Anal. calcd for C$_{14}$H$_{10}$O$_2$S$_6$: C, 41.77; H,

2.50; S, 47.78. Found: C, 41.77; H, 2.47; S, 47.74. The product was subsequently subjected to X-ray crystal structure analysis.

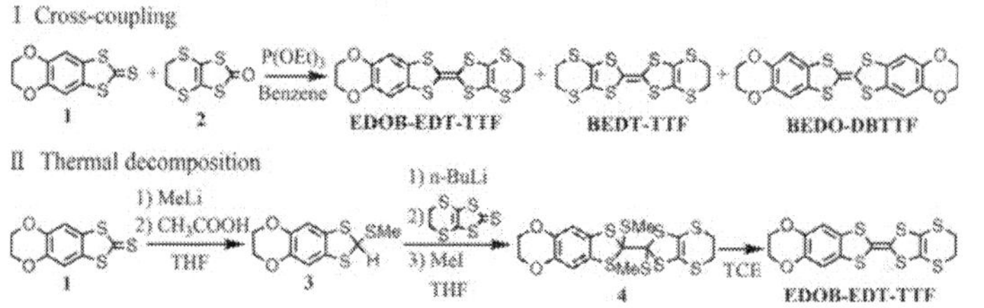

Figure 1. Synthetic routes of EDOB-EDT-TTF.

2.1.2. Thermal decomposition method

2-(Methylthio)-5,6-ethylenedioxy-1,3-benzodithiole (**3**). To a solution of **1** (1.03 g, 4.25 mmol) in dry THF (150 mL) at −78°C under nitrogen was added solution of MeLi (5.1 mL, 6 mmol) dropwise via a syringe. After stirring for 5 h, the mixture was treated with acetic acid (2 mL) and allowed to warm to room temperature. The mixture was combined with water, and the desired compound was extracted using CH_3Cl. The organic layer was dried using $MgSO_4$ and evaporated *in vacuo* to afford the crude product, which was purified using silica gel chromatography with tetrahydrofuran THF/*n*-hexane (1:3) as the eluent to afford **3** as a colorless oil (0.74 g, 66%). ^{1}H NMR ($CDCl_3$): δ2.22 (3H, s, −SCH_3), 4.21 (4H, s, −OCH_2), 5.93 (1H, s, −CH), 6.79 (2H, s, ArH). HRMS (EI) (*m/z*): calcd for $C_{10}H_{10}O_2S_2$, 257.9843; found, 257.9718.

Hexathioorthooxalate (**4**). To a stirred solution of **3** (2.60 g, 10 mmol) in dry THF (120 mL) at −78°C under nitrogen was added *n*-BuLi (6.35 mL, 9.5 mmol) dropwise using a syringe. After stirring for 1 h, 4,5-ethylenedithio-1,3-dithiol-2-thione (2.24 g, 10 mmol) in THF was added dropwise to the mixture, followed by addition of an excess of MeI (1.91 mL, 30 mmol) via a syringe to the mixture after 1 h. After further stirring for 1 h, the solution was treated with portions of 1 mol/dm^3 aq. NH_4Cl. The mixture was allowed to warm to room temperature, and the aqueous layer was extracted with CH_2Cl_2. The organic phase was concentrated and purified using silica gel column chromatography with CH_3Cl/*n*-hexane (1:2) as the eluent to afford **4** as a yellow solid (2.40 g, 45%): mp 207-213 (dec. 150°C);^{1}H NMR ($CDCl_3$): δ2.46 (3H, s, −SCH_3), 2.53 (3H, s, −SCH_3), 3.25 (4H, m, −SCH_2), 4.21 (4H, s, −OCH_2), 6.62 (2H, s, ArH). Anal. calcd for $C_{16}H_{16}O_2S_8$: C, 38.68; H, 3.25. Found: C, 38.56; H, 3.17.

EDOB-EDT-TTF. After refluxing **4** in 1,1,2-trichloroethane (TCE) for 12 h, the crude product was purified by column chromatography with CH_2Cl_2/*n*-hexane (1:1) as the eluent followed by recrystallization from ethyl acetate to yield EDOB-EDT-TTF (27%). ^{1}H NMR and MS data were comparable to those obtained by the cross-coupling method.

The redox potentials of unsymmetrical EDOB-EDT-TTF appeared middle between that of BEDT-TTF and BEDO-DBTTF, as similar to other unsymmetrical TTF derivatives reported in [11]. The difference potential (ΔE) between the first redox potential ($E_{1/2(1)}$) and the second redox potential ($E_{1/2(2)}$) is related to intramolecular on-site Coulomb repulsion energy, U. The ΔE of EDOB-EDT-TTF also showed middle between that of BEDT-TTF and BEDO-BDTTF. The U of EDOB-EDT-TTF decreased compared to that of BEDO-BDTTF.

2.2. Preparation of CT complexes and radical salts

Hot solutions of each donor and acceptor in acetonitrile were mixed. After the reaction mixture was cooled to room temperature (RT), the resulting precipitate was collected by filtration. Complexes were washed with the same organic solvent and dried *in vacuo*. Black octahedral PF_6, AsF_6 and SbF_6 salts were obtained by electrochemical oxidation in distilled 1,2-dichloro-ethane or TCE under constant current of 1 μA in a mixture of the donor and the tetra-*n*-butylammonium salts of the corresponding anions at RT using H-shaped cell with Pt electrodes for 2 weeks. Stoichiometry of CT complexes and radical salts was determined using elemental analysis or X-ray crystallographic analyses.

3. Measurements

3.1. Electrical conductivity measurement

DC conductivities were measured with standard four- or two-probe techniques, using Keithley 220 current source, Keithley 199 voltage/scanner, Keithley 195A voltmeter, and Scientific Instruments 9650 temperature controller. For powder samples, measurements were performed on compressed pellets, which were cut to form orthorhombic shape. Gold wires were glued to the samples with gold paint (Tokuriki Chemical, no. 8560).

3.2. Apparatus for simultaneous measurements of thermoelectric power and resistivity on organic conductors

Simultaneous measurements of thermoelectric power and resistivity by a two-probe method of PF_6, AsF_6, and SbF_6 salts were performed using computer-interfaced system, which schematic diagram is shown in **Figure 2**.

Software program for controlling the system was created in LabVIEW by Computer Automation Co. and System Approach Co. By applying different digital signals using relay and counter timer, we could perform simultaneous measurements of thermoelectric power and two-probe electrical resistivity on a single sample over the entire temperature range [12]. First, by opening the circuit through Keithley 2002 multimeter attached 2001-Scan scanner card as the relay, sample voltage (ΔV) between V^+ and V^- electrodes was measured by Keithley 2182 nanovolt-meter. Temperature gradient, ΔT, between two copper plates was measured by another Keithley 2182 nanovoltmeter. Thermoelectric power was determined. Second, after thermal equilibrium was sufficiently reached between two copper plates, by connecting the circuit through the relay, sample voltage was measured by Keithley 2182 nanovoltmeter, while constant current was supplied to the sample reversing leads to cancel thermal EMFs by Keithley 220 current source. Two-probe resistance measurement was determined by subtracting these two voltages and averaging. Using thermoelectric power stage made by MMR Technologies Inc. radical salts and the reference Cu-constantan or Au/Fe-Chromel thermocouple wire were glued onto copper plates using gold paint. Thermal gradient across the sample was applied by heating chip resistor (110 Ohm), which was applied from loop 2 of LakeShore 331S temperature controller. Thermoelectric power was determined from the slope of the line and was calculated as follows: $S = \Delta V/\Delta T$ [13], where ΔT of 15 total points was typically < 0.5 K. Changing voltage value was measured at specific fixed time intervals by 15 points simultaneously using different nanovoltmeters. Therefore, it is necessary to apply synchronously digital signals to two Keithley 2182 nanovoltmeters. By using National Instruments PCI-6602 counter timer, we could measure 15 points synchronously to determine ΔV and ΔT. It was necessary to subtract the offset drift in order to obtain absolute thermoelectric power of the samples. At each measurement, temperature of the sample holder in the cryostat was controlled using loop 1 of LakeShore 331S temperature controller with sensor (DT-470). System was checked by measurements with Pt standard [14] and TTF·TCNQ complex [5, 15] in temperature range 4–350 K.

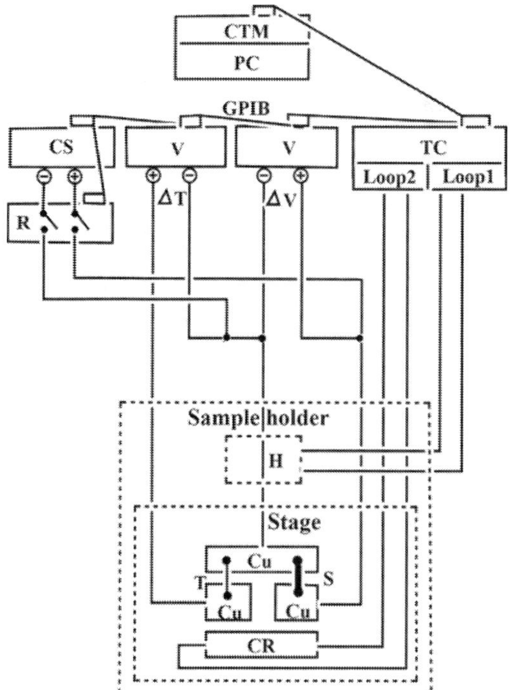

Figure 2. Schematic diagram for simultaneous measurements of thermoelectric power and two-probe electrical resistivity. CTM: counter timer 6602; TC: temperature controller 331S; V: nanovoltmeter 2182; CS: current source 220; R: relay 2002; H: heater; S: sample; T: thermocouple wire; CR: chip resistor.

Simultaneous measurements of thermoelectric power and resistivity of $(EDOB\text{-}EDT\text{-}TTF)_2PF_6$ were also performed on Quantum Design PPMS Model P670 Thermal Transport System (TTO) in temperature range 232-327 K. The crystal, which was glued to two-probe using bar-shaped copper leads, was mounted on a TTO sample puck.

4. Crystal structure

Single crystal structure analyses have been carried out for PF_6 salt at 298 K, AsF_6 salt at 90, 293, 330, and 350 K, SbF_6 salt at 90 and 293 K. Crystallographic data are listed in **Table 1**.

	PF_6 salt at 298 K	AsF_6 salt at 90 K	AsF_6 salt at 293 K	AsF_6 salt at 330 K	AsF_6 salt at 350 K	SbF_6 salt at 90 K	SbF_6 salt at 293 K
Chemical formula	$C_{28}H_{20}F_6O_4PS_{12}$		$C_{28}H_{20}AsF_6O_4S_{12}$			$C_{28}H_{20}F_6O_4S_{12}Sb$	
Formula weight	950.23		994.15			1040.99	
Crystal system	Triclinic		Triclinic			Monoclinic	
Space group	P–1		P–1			$C2/c$	
a/Å	7.003	6.875	7.000	7.036	7.057	37.805	38.105
b/Å	8.074	7.914	8.061	8.092	8.112	8.204	8.340
c/Å	16.326	16.411	16.424	16.411	16.404	11.371	11.429
α/°	76.02	103.40	76.07	76.23	76.34		
β/°	78.07	98.34	78.41	78.46	78.49	103.23	102.77
γ/°	81.50	97.11	81.45	81.07	80.85		
V/Å³	871.7	847.8	876.2	883.4	888.0	3433.0	3542.4
Z	1	1	1	1	1	4	4
T/K	298	90	293	330	350	90	293
CCDC no.		819768	809939	809703	802121	799928	799214

Table 1. Crystallographic data for EDOB-EDT-TTF salts.

4.1. Crystal structure of $(EDOB\text{-}EDT\text{-}TTF)_2PF_6$

Crystal structure of 2:1 $(EDOB\text{-}EDT\text{-}TTF)_2PF_6$ at 298 K is depicted in **Figure 3** and belongs to triclinic P–1 space group. Cation layers of EDOB-EDT-TTF molecules and anion layers of PF_6^- anions are arranged alternately along the direction of a-axis. Donor molecules in the crystal are stacked in alternating orientations along the stacking axis. Average interplanar donor-donor distances in the columns equal to 3.635 and 3.641 Å, respectively. This donor packing arrangement is so-called β-type structure [16] as in β-$(BEDT\text{-}TTF)_2I_3$ [17]. Intermolecular side-by-side short contacts, less than van der Waals (vdW) [18] sum, are observed at S(6)…S(6) (3.55 Å: vdW sum = 3.60 Å) and O(1)…H(11A) (2.63 Å: vdW sum = 2.72 Å). Intermolecular intrastack short contacts O(1)...H(5A) (2.693 Å) and C(8)...H(4A) (2.756 Å: vdW sum = 2.90 Å) are alternatively found along b-axis. Intermolecular S...F short contacts less, than vdW sum (3.27 Å for S...F) between EDOB-EDT-TTF cations and hexafluorophosphate anions, are not observed. Octahedral anion does not show rotational disorder.

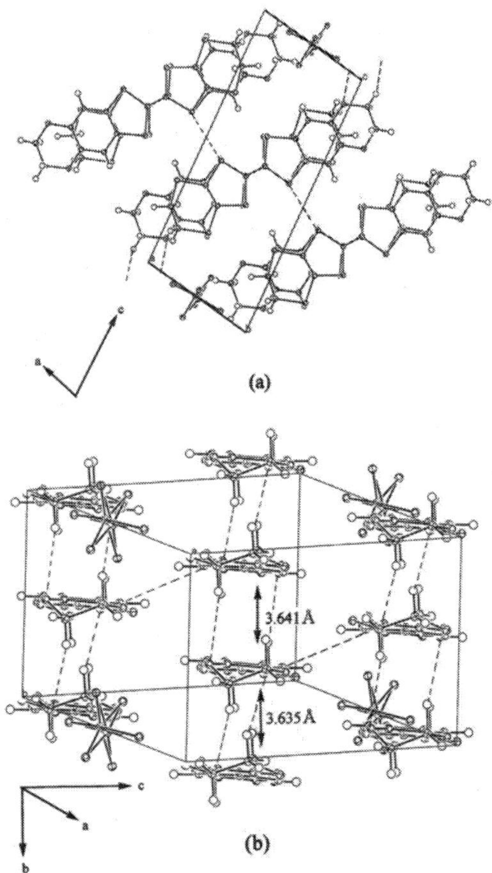

Figure 3. Crystal structure of (EDOB-EDT-TTF)$_2$PF$_6$: (a) viewed along b-axis and (b) viewed along molecular long axis. Broken lines represent intermolecular short contacts.

4.2. Crystal structure of (EDOB-EDT-TTF)$_2$AsF$_6$

Crystal structure of (EDOB-EDT-TTF)$_2$AsF$_6$ is isostructural to PF$_6$ salt and was elucidated at various temperatures (90, 293, 330 and 350 K). Cation layers of donor molecules and anion layers of AsF$_6^-$ anions are arranged alternately along the direction of c-axis as shown in

Figure 4. EDOB-EDT-TTF molecules are packed head-to-tail in face-to-face overlapping manner and alternately stacked with different interplanar along *b*-axis. There are intermolecular short S...S, O...H and S...F contacts, less than vdW sum. Interstack S(1)...S(1) short contacts less, than 3.60 Å, are observed over the entire temperature range (3.530 Å at 90 K, 3.548 Å at 293 K, 3.555 Å at 330 K, and 3.563 Å at 350 K), and S(3)...S(6) short contacts are found at 90 K (3.536 Å), but not at 293 K (3.629 Å), 330 K (3.654 Å), and 350 K (3.669 Å). S...S short contacts are found only between the stacks and not within the stacks. Side-by-side O(2)...H(6B)–C(6) short contacts (2.673 Å at 90 K, 2.647 Å at 293 K, 2.637 Å at 330 K, and 2.634 Å at 350 K), less than vdW sum in intermolecular interstack along *c*-axis, are observed over the entire temperature range as shown in **Figure 4**. O(2)...H(12A)–C(12) short contacts in intermolecular intrastack along *b*-axis (**Figure 4a**) can be seen at 90 K (2.440 Å), 293 K (2.610 Å) and

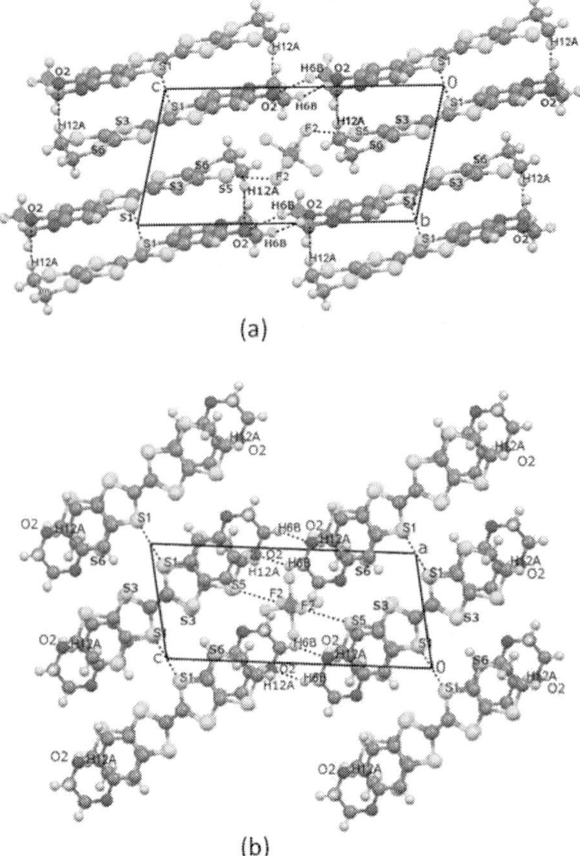

(a)

(b)

Figure 4. Crystal structure of (EDOB-EDT-TTF)$_2$AsF$_6$ at 293 K showing short contacts as dashed lines: (a) projection in *bc* plane and (b) projection in *ac* plane.

330 K (2.721 Å), but not at 350 K (2.757 Å). Only O(2)...H(12A)–C(12) short contacts exist within the intermolecular intrastack dimer. The other O(2)...H(6B), S...S, and S...F short contacts exist in the intermolecular interstack along the transverse direction. Owing to the effect of dimerization, distances of O(2)...H(12A)–C(12) at 90, 293, and 330 K are shorter than those at 350 K. Intermolecular S(5)...F(2) short contacts, less than vdW sum between EDOB-EDT-TTF cations and hexafluoroarsenate anions, are observed at 90 K (3.080 Å), 293 K (3.208 Å), and 330 K (3.255 Å), but not at 350 K (3.279 Å). Octahedral anion does not show rotational disorder.

4.3. Crystal structure of (EDOB-EDT-TTF)$_2$SbF$_6$

(EDOB-EDT-TTF)$_2$SbF$_6$ crystallizes in two forms, plate and needle. Needle form is too much small for X-ray crystal structure analysis. Crystal structure of plate (EDOB-EDT-TTF)$_2$SbF$_6$ is not isostructural to AsF$_6$ and PF$_6$ salts and belongs to monoclinic $C2/c$ space group, which was elucidated at 90 and 293 K. Crystal structure of plate form at 293 K is shown in **Figure 5**, and octahedral SbF$_6^-$ anion does not show a disorder. Molecular arrangement of EDOB-EDT-TTF molecules is quite different from that in (EDOB-EDT-TTF)$_2$AsF$_6$. Two EDOB-EDT-TTF molecules are paired with molecular planes almost parallel, and adjacent pairs are nearly perpendicular to each other. This type of molecular arrangement tends to create two-dimensional networks [19]. However, SbF$_6$ salt is semiconductor. Intermolecular short S...S and O...H contacts are observed, but no S...F contacts are found at 293 and 90 K. S...S short contacts at 293 K are between S(4)...S(5) (3.502 Å) and S(5)...S(6) (3.519 Å) as shown in **Figure 5**. S...S short contacts at 90 K are seen at S(3)...S(6) (3.462 Å) and S(5)...S(6) (3.467 Å) and at S(2)...S(3) (3.563 Å) as shown in **Figure 6**. Dimerization of two EDOB-EDT-TTF molecules becomes stronger at 90 K. O...H short contacts are observed at O(2)...H(13B) (2.607 Å) (along b-axis) and O(2)...H(3) (2.470 Å) (along c-axis) at 293 K and O(1)...H(12A) (2.503 Å) (along b-axis) and O(1)...H(8) (2.425 Å) (along c-axis) at 90 K as shown in **Figures 5** and **6**.

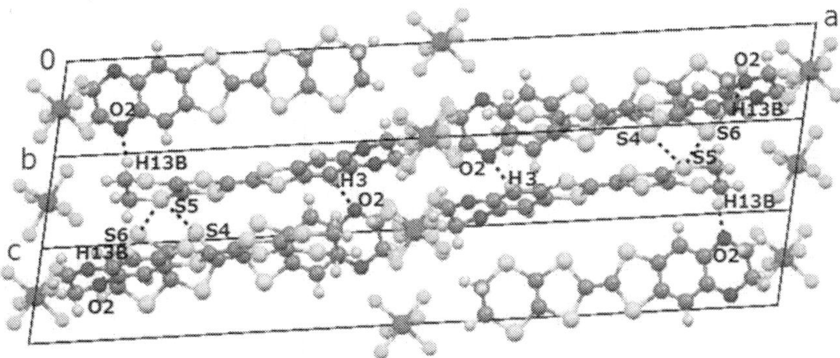

Figure 5. Crystal structure of plate (EDOB-EDT-TTF)$_2$SbF$_6$ at 293 K showing intermolecular short O...H and S...S contacts as dotted lines.

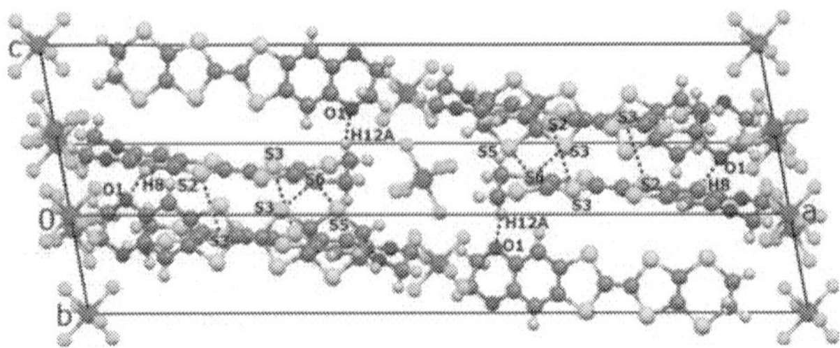

Figure 6. Crystal structure of plate (EDOB-EDT-TTF)$_2$SbF$_6$ at 90 K showing intermolecular short H...H and S...S contacts as dotted lines.

5. Electrical conductivity

Table 2 summarizes appearances, component ratios, metal-to-semiconductor (MS) transition, room-temperature electrical conductivity, and activation energies of EDOB-EDT-TTF complexes and salts. A newly [10] and the previously [9] reported (EDOB-EDT-TTF)$_2$PF$_6$ salts exhibited electrical resistivity decrease with heating (**Figure 7**) and showed resistivity minimum at 340 K, then gradual increase up to 350 K. As shown in **Figure 8**, new black plates (EDOB-EDT-TTF)$_2$AsF$_6$ (σ_{RT} = 2.6 S cm^{-1}) exhibited distinct minimum in resistivity at 315 K confirming MS transition at this temperature. The semiconductive region (< 315 K) showed the activation energy of E_a = 0.13 eV. It was found that T_{MS} decreased with increase in anion size, as well as β-(BEDT-TTF)$_2$X (X = PF$_6$ and AsF$_6$) [20, 21]. Electrical conductivity of new black plate and fine needle SbF$_6$ salts at room temperature was 4.4 × 10^{-2} S cm^{-1} (E_a = 0.13 eV) and 2.9 × 10^{-3} S cm^{-1} (E_a = 0.13 eV), respectively. SbF$_6$ salts behaved as semiconductors from room temperature to 100 K.

Acceptor or anion	Appearance	Stoichiometry D:A	T_{MS}/K	σ_{RT}/S cm^{-1}	E_a/eV	References
M$_2$TCNQ	Black powder	1:1		Insulator[a]		[9]
TCNQ	Black power	1:1		1.9[a]	0.085	[9]
FTCNQ	Dark green powder			9.6 × 10^{-2a}	0.13	[9]
F$_2$TCNQ	Dark green powder	4:1		7.1 × 10^{-2a}	0.13	[9]
PF$_6^-$	Black plate	2:1	340	1.7 × 10^b	0.23	[9]
PF$_6^-$	Black plate	2:1	337	1.0[c]	0.17	[10]
AsF$_6^-$	Black plate	2:1	315	2.6[c]	0.13	[10]
SbF$_6^-$	Black plate	2:1		4.4 × 10^{-2c}	0.13	[10]
SbF$_6^-$	Black fine needle	2:1		2.9 × 10^{-3c}	0.13	[10]

[a]Measured on a compressed pellet by a four-probe method.
[b]by a four-probe method.
[c]by a two-probe method.

Table 2. Electric conductivities of EDOB-EDT-TTF complexes and salts.

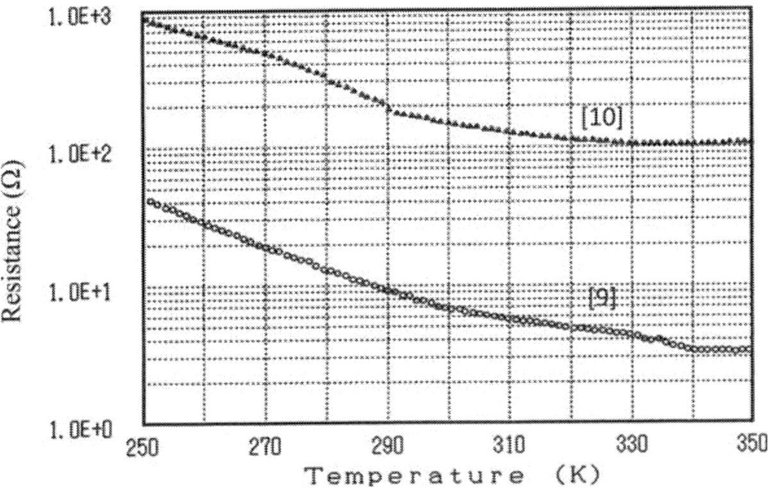

Figure 7. Temperature dependence of resistance of single crystals $(EDOB-EDT-TTF)_2PF_6$ in the heating run. Data for two crystals are plotted.

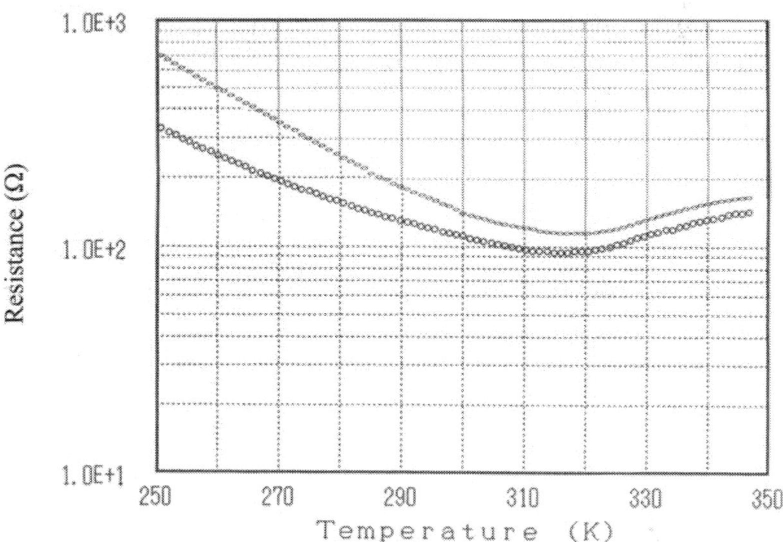

Figure 8. Temperature dependence of resistance of single crystals $(EDOB-EDT-TTF)_2AsF_6$ in the heating run. Data for two crystals are plotted.

6. Thermoelectric power

Chaikin et al. have described in references [15, 22, 23], that thermoelectric power coefficient in metallic state for a single one-dimensional band is given by:

$$S = \frac{-\pi^2 k_B^2 T}{3|e|} \left(\frac{\cos\left(\frac{\pi\rho}{2}\right)}{2|t|\sin^2(\frac{\pi\rho}{2})} + \frac{\tau'(\varepsilon)}{\tau(\varepsilon)_{EF}} \right), \tag{1}$$

where $\tau(\varepsilon)$ is energy-dependent electron scattering time, ρ is amount of charge transfer, t is transfer integral ($4t$ is the bandwidth), k_B is Boltzmann constant, E_F is Fermi energy, and T denotes temperature. When band structure contribution to thermoelectric power dominates in Eq. (1), the sign of thermoelectric power is negative for $\rho < 1$ and positive for $\rho > 1$. S shows linear temperature dependence in metallic region.

Thermoelectric power coefficient for semiconductor is given by:

$$S = \frac{-k_B^2}{|e|} \left(\frac{b-1}{b+1} \frac{E_a}{k_B T} + ln\frac{m_h}{m_e} \right), \tag{2}$$

where b is the ratio of electron-to-hole mobility, and m_h and m_e are, respectively, effective mass of hole and electron. S shows T^{-1} temperature dependence for semiconductor.

Conwell has shown that near-constant S value close to -60 μV/K over wide temperature range is obtained with model, in which there are strong on-site correlations, U [20, 24].

In this study, thermoelectric power measurements were carried out to clarify MS transition of these salts. Thermoelectric powers of PF_6, AsF_6, and SbF_6 salts were measured in temperature range 220–360 K.

6.1. Thermoelectric power of (EDOB-EDT-TTF)$_2$PF$_6$

Positive value of thermoelectric power implies hole-like character of conduction charge carriers. The sign ($\bullet$) of thermoelectric power of PF_6 salt along the crystal growth axis was positive above 235 K as shown in **Figure 9**. Thermoelectric power value of PF_6 salt jumps around 305 and 340 K. Inflection of thermoelectric power curve occurs around 270 K. These jumps and inflection were reproduced for different three samples. Thermoelectric power data ($\blacktriangle$) measured by Quantum Design PPMS also shows jumps around 272 and 305 K. Thermo-electric power jumps of PF_6 salt shown in **Figure 9** is not clear, however, that of PF_6 salt was proportional to absolute temperature down to 320 K, which is characteristic of metallic conduction. Below 320 K, value of thermoelectric power dropped gradually, indicating MS transition around 320 K. These transition temperatures of PF_6 salt corresponded with the results of electrical resistivity (•). Thermoelectric power of PF_6 salt dropped below 305 K, decreasing rapidly below 270 K. Metallic properties denoted both by thermoelectric power and by electrical resistivity measurements.

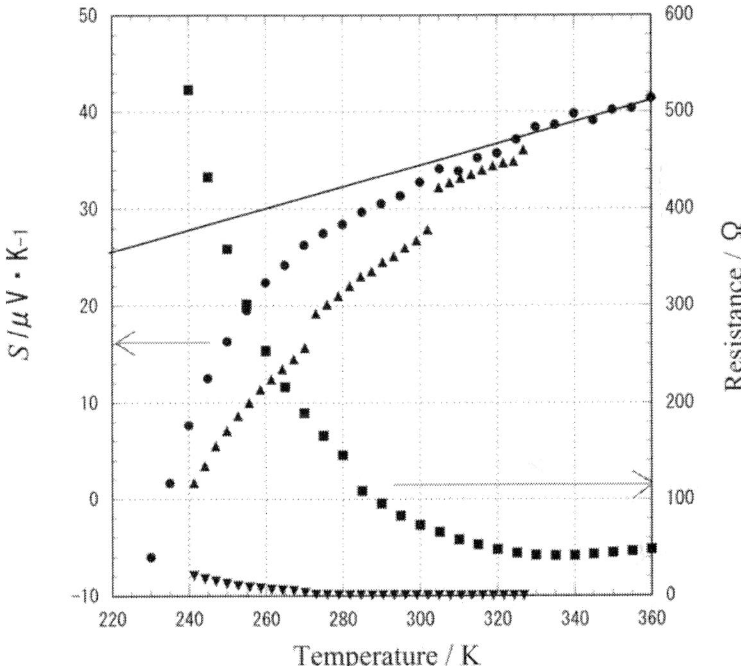

Figure 9. Simultaneous measurements of temperature dependence of thermoelectric power (•) and electrical resistivity (•) of $(EDOB-EDT-TTF)_2PF_6$ salt. Data of thermoelectric power (▲) and resistivity (▼) were measured by Quantum Design PPMS in temperature range 232–327 K. Solid line extrapolates to zero at $T = 0$ K.

6.2. Thermoelectric power of $(EDOB-EDT-TTF)_2AsF_6$

Figure 10 shows simultaneous measurements of temperature dependence of thermoelectric power and electrical resistivity of $(EDOB-EDT-TTF)_2AsF_6$ salt. Thermoelectric power of AsF_6 salt along the crystal growth axis was linear with temperature down to 310 K, which is characteristic of a metal. Thermoelectric power value of AsF_6 salt seems to jump around 305 K and then drops below 300 K, indicating MS transition around 310 K. MS transition temperature observed in thermoelectric power seems to be slightly lower than that of electrical conductivity [25]. Thermoelectric power of AsF_6 salt again showed rapid decrease below 260 K. Inflection of thermoelectric power curve was detected around 260 K. These jumps and inflection were reproduced for different two samples. The sign of thermoelectric power of AsF_6 salt was positive above 220 K. Thermoelectric power of AsF_6 salt became negative below 220 K and decreased with decreasing temperature. MS transition temperatures of PF_6 and AsF_6 salts observed in thermoelectric power decreased in order of 320 and 310 K.

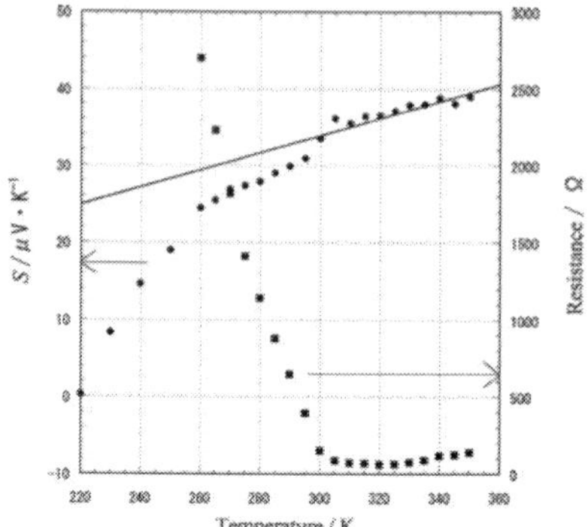

Figure 10. Simultaneous measurements of temperature dependence of thermoelectric power (•) and electrical resistivity (•) of $(EDOB-EDT-TTF)_2AsF_6$ salt. Solid line extrapolates to zero at $T = 0$ K.

6.3. Thermoelectric power of $(EDOB-EDT-TTF)_2SbF_6$

Simultaneous measurements of temperature dependence of thermoelectric power and electrical resistivity of $(EDOB-EDT-TTF)_2SbF_6$ salt are shown in **Figure 11**. Thermoelectric power value of plate SbF_6 salt was negative below 335 K and decreased with decreasing temperature. Negative value of thermoelectric power implies electron-like character of conduction charge carriers. Thermoelectric power exhibits T^{-1}-temperature dependence, which is a characteristic of semiconductor.

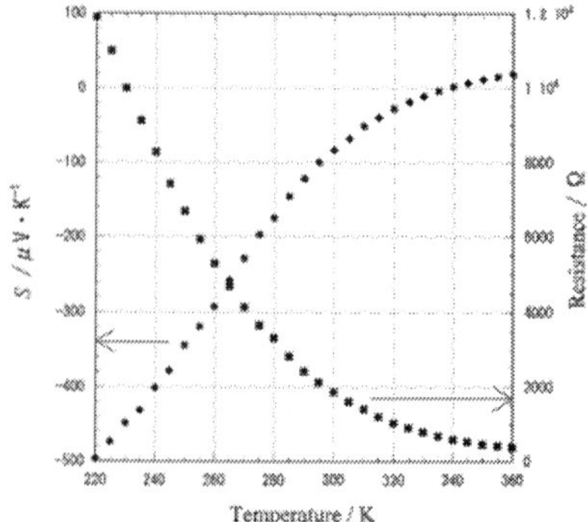

Figure 11. Simultaneous measurements of temperature dependence of thermoelectric power (•) and resistivity (•) of $(EDOB-EDT-TTF)_2SbF_6$ salt.

http://dx.doi.org/10.5772/65561

7. Band structure of AsF_6 salt

Band structures of PF_6 and AsF_6 salts were calculated on the basis of tight-binding approximation using intermolecular overlap integrals of HOMO, which were calculated by extended Hückel method [26]. Calculated intermolecular overlap integrals are listed in **Table 3**. Arrangements of donor centers and intermolecular overlaps in EDOB-EDT-TTF salts viewed onto *ab* plane are depicted in **Figure 12**. **Figure 13** shows energy band structures and Fermi surfaces of AsF_6 salt at 293 and 350 K. Energy bands of these salts are three-quarters filled and metallic. Short S...S contacts are observed between stacks, not within the stacks. Such structural feature provides isotropic two-dimensional electronic structures in Fermi surfaces [27, 28].

Symbol	PF_6 salt at 298 K	AsF_6 salt at 293 K	AsF_6 salt at 350 K
S_a	−3.886	−3.864	−3.896
S_{b1}	16.089	12.819	14.467
S_{b2}	12.464	16.414	13.013
S_{p1}	5.545	0.607	5.922
S_{p2}	0.596	5.489	0.605
S_{q1}	0.058	0.058	0.044
S_{q2}	0.056	0.006	0.052

Symbols S_a- S_{q2} represent the overlap integrals between donors shown in **Figure 12**

Table 3. Intermolecular overlap integrals (× 10^{-3}) of EDOB-EDT-TTF salts.

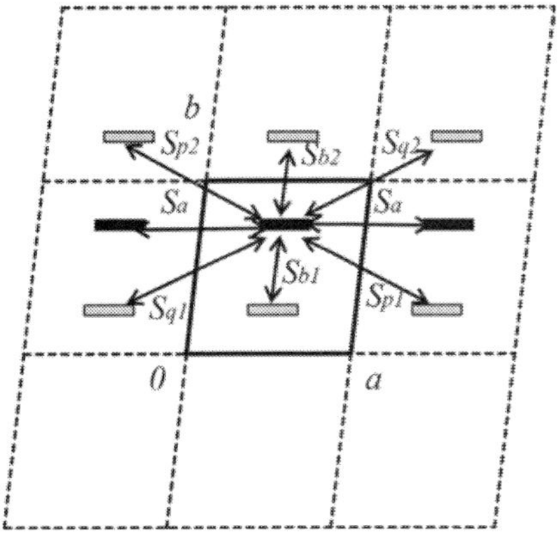

Figure 12. Arrangement of donor centers (▬, ▭) and intermolecular overlaps in PF_6 salt at 298 K, AsF_6 salts at 293 and 350 K viewed onto *ab* plane.

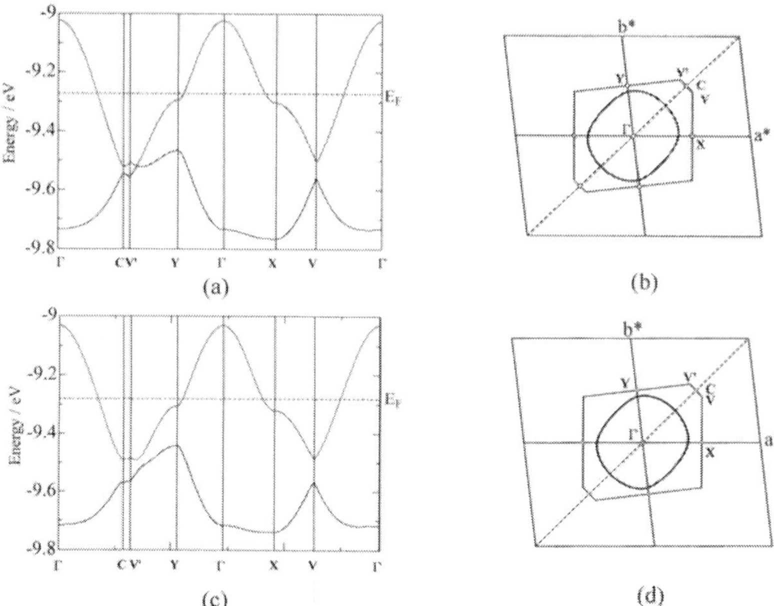

Figure 13. Calculated band structures and corresponding Fermi surfaces of (EDOB-EDT-TTF)$_2$AsF$_6$ salt at 293 K (a and b) and 350 K (c and d).

8. Conclusion

Synthesis of unsymmetrical EDOB-EDT-TTF donors was accomplished by two methods. New radical salts with octahedral PF$_6^-$, AsF$_6^-$ and SbF$_6^-$ anions were prepared by electrochemical oxidation. Crystal structure of AsF$_6$ salt was isostructural to PF$_6$ salt, but crystal structure of plate SbF$_6$ salt was not. According to conventional electrical conductivity measurements, both PF$_6$ and AsF$_6$ salts exhibited MS transitions at 340 and 315 K, respectively. Electrical resistivity and thermoelectric power of PF$_6$, AsF$_6$, and SbF$_6$ salts were measured simultaneously on a single sample. Judging from the results of thermoelectric power measurements, MS transition temperatures of PF$_6$ and AsF$_6$ salts were around 320 and 310 K, respectively. From crystal structure analysis in AsF$_6$ salt follows that intermolecular O…H distance along the stacking b-axis at 350, 330, 293, and 90 K decrease in that order by the dimerization. Formation of dimers results in semiconductor transition. Short S…S contacts were found only between the stacks and not within the stacks. They provided isotropic two-dimensional electronic structure in Fermi surfaces. Two-dimensional electronic structure derived from β-type arrangement is not so stable against packing modification [28]. MS transition is associated with some structural transition.

Acknowledgements

The author wish to thank Drs. Maeda T and Yamashita K of Computer Automation Co. and Dr. Niwa N of System Approach Co. for programming the LabVIEW thermoelectric power measurement system and Emeritus Professor Matsumoto S of Aoyama Gakuin University for giving valuable comments.

Abbreviations

TTF, tetrathiafulvalene; EDT, ethylenedithio; BEDT-TTF, bis(EDT)-TTF; DBTTF, dibenzo-TTF; EDO, ethylenedioxy; BEDO-DBTTF, bis(EDO)-DBTTF; EDOB-EDT-TTF, ethylenedioxybenzo-EDT-TTF; TCNQ, 7,7,8,8-tetracyanoquinodimethane; FTCNQ, 2-fluoro-TCNQ; F_2TCNQ, 2,5-difluoro-TCNQ; Me_2TCNQ, 2,5-dimethyl-TCNQ; TCE, 1,1,2-trichloroethane.

Author details

Tomoko Inayoshi

Address all correspondence to: inayoshi@chem.aoyama.ac.jp

Department of Chemistry and Biological Science, College of Science and Engineering, Aoyama Gakuin University, Sagamihara, Kanagawa, Japan

References

[1] Coleman L B, Cohen M J, Sandman D J, Yamagishi F G, Garito A F, Heeger A J. Superconducting fluctuations and the peierls instability in an organic solid. Solid State Commun. 1973; 12: 1125–732. doi:10.1016/0038-1098(73)90127-0

[2] Heeger A J, Garito A F. In: Keller H J editor. Low-Dimensional Cooperative Phenomena. New York: Plenum Press; 1975. p. 89–124.

[3] Ishiguro T, Kagoshima S, Anzai H. Dc conductivity anomalies at the transition points in TTF-TCNQ. J. Phys. Soc. Jpn. 1976; 41: 351–352. doi:10.1143/JPSJ.41.351

[4] Kistenmacher T J, Phillips T E, Cowan D O. The crystal structure of the 1:1 cation-radical anion salts of 2,2′-bis-1,3-dithiole (TTF) and 7,7,8,8-tetracyanoqunodimethane (TCNQ). Acta Cryst. 1974; B30: 763–768. doi:10.1107/S0567740874003669

[5] Kwak J F, Chaikin P M, Russel A A, Garito A F, Heeger A J. Anisotropic thermoelectric power of TTF-TCNQ. Solid State Commun. 1975; 16: 729–732. doi: 10.1016/0038-1098(75)90062-9

[6] Chaikin P M, Kwak J F. Apparatus for thermopower measurements on organic conductors. Rev. Sci. Instrum. 1975; 46: 218–220. doi:10.1063/1.1134171

[7] Mori T, Inokuchi H. Thermoelectric power of organic superconductors-calculation on the basis of the tight-binding theory. J. Phys. Soc. Jpn. 1988; 57: 3674–3677. doi:10.1143/JPSJ.57.3674

[8] Senga T, Kamoshida K, Kushch L A, Saito G, Inayoshi T, Ono I. Peculiarity of ethylenedioxy group in formation of conductive charge-transfer complexes of bis(ethylenedioxy)-dibenzotetrathiafulvalene (BEDO-DBTTF). Mol. Cryst. Liq. Cryst. 1997; 296: 97–143. doi:10.1080/10587259708032316

[9] Inayoshi T, Matsumoto S, Ono I. Physical properties of CT complexes of a new asymmetric donor: (EDOB)(EDT)TTF. Synth. Met. 2003; 133–134: 345–348. doi:10.1016/S0379-6779(02)00334-X

[10] Inayoshi T, Matsumoto S. Synthesis, electronic properties, thermoelectric power, and crystal and band structures of unsymmetrical EDOB-EDT-TTF salts composed of PF_6^- ASF_6^- SbF_6^- at various temperatures. Mol. Cryst. Liq. Cryst. 2013; 582: 136–153. doi:10.1080/15421406.2013.807156

[11] Mori T, Inokuchi H, Kini A M, Williams J M. Unsymmetrically substituted ethylenedioxytetrathiafulvalenes. Chem. Lett. 1990; 19: 1279–1282. doi:10.1246/cl.1990.1279

[12] Kim G T, Park J G, Lee J Y, Yu H Y, Choi E S, Suh D S, Ha Y S, Park Y W. Simple technique for the simultaneous measurements of the four-probe resistivity and the thermoelectric power. Rev. Sci. Instrum. 1998; 69: 3705–3706. doi:10.1063/1.1149164

[13] Park Y W. Structure and morphology: relation to thermopower properties of conductive polymers. Synth. Met. 1991; 45: 173–182. doi:10.1016/0379-6779(91)91801-G

[14] Barnard R D. Thermoelectricity in Metals and Alloys. London: Taylor & Francis; 1972. p. 49.

[15] Chaikin P M, Kwak J F, Jones T E, Garito A F, Heeger A J. Thermoelectric power tetrathiofulvalinium tetracyanoquinodimethane. Phys. Rev. Lett. 1973; 31: 601–604. doi:10.1103/PhysRevLett.31.601

[16] Mori H. Materials viewpoint of organic superconductors. J. Phys. Soc. Jap. 2006; 75: 051003/1–14. doi:10.1143/JPST.75.051003

[17] Mori T, Kobayashi A, Sasaki Y, Kobayashi H, Saito G, Inokuchi H. Band structure of two types of $(BEDT-TTF)_2I_3$. Chem. Lett. 1984; 13: 957–960. doi:10.1246cl.1984.957

[18] Bondi A. Van der Waals volumes and radii. J. Phys. Chem. 1964; 68: 441–451. doi:10.1021/j100785a001

[19] Ishiguro T, Yamaji K, Saito G. Organic Superconductors. 2nd ed. Berlin: Springer; 1998. p. 125–219.

[20] Kobayashi H, Mori T, Kato R, Kobayashi A, Sasaki Y, Saito G, Inokuchi H. Transverse conduction and metalinsulator transition in β-(BEDT-TTF)$_2$PF$_6$. Chem. Lett. 1983; 12: 581–584. doi:10.1246/cl.1983.581

[21] Senadeera G K R, Kawamoto T, Mori T., Yamaura J, Enoki T. 2 k_F CDW transition in β-(BEDT-TTF)$_2$PF$_6$ family salts. J. Phys. Soc. Jap. 1998; 67: 4193–4197. doi:10.1143/JPSJ.67.4193

[22] Chaikin P M, Greene R L, Etemad S, Engler E M. Thermopower of an isostructural series of organic conductors. Phys. Rev. 1976; B13, 1627–1632. doi:10.1103/PhysRevB.13.1627

[23] Chaikin P M, Kwak J F, Greene R L, Etemad S, Engler E M. Two band transport and disorder effects in a series of organic alloys: (TTF$_{1-x}$TSeF$_x$)-TCNQ (tetrathiafulvalene-tetraselenafulvalene-tetracyanoquinodimethane). Solid State Commun. 1976; 19: 1201–1204. doi:10.1016/0038-1098(76)90819-X

[24] Conwell E M. Thermoelectric power of 1:2 tetracyanoquinodimethanide (TCNQ) salts. Phys. Rev. 1978; B18: 1818–1823. doi:10.1103/PhysRevB.18.1818

[25] Kikuchi K, Saito K, Ikemoto I, Murata K, Ishiguro T, Kobayashi K. Physical properties of (DMET)$_2$X. Synth. Met. 1988; 27: 269–274. doi:10.1016/0379-6779(88)90155-5

[26] Mori T, Kobayashi A, Sasaki Y, Kobayashi H, Saito G, Inokuchi H. The intermolecular interaction of tetrathiafulvalene and bis(ethylenedithio)tetrathiafulvalene in organic metals. Calculation of orbital overlaps and models of energy-band structures. Bull. Chem. Soc. Jpn. 1984; 57: 627–633. doi:10.1246/bcsj.57.627

[27] Wosnitza J. Fermi Surfaces of Low-Dimensional Organic Metals and Superconductors. Berlin: Springer; 1996. p. 34.

[28] Kagoshima S, Kato R, Fukuyama H, Seo H, Kino H. Chapter 4 Interplay of structural and electronic properties. In: Bernier P, Lefrant S, Bidan G, editors. Advances in Synthetic Metals—Twenty Years of Progress in Science and Technology. Amsterdam: Elsevier; 1999. p. 262–316.

Chapter 3

Progress in Polymer Thermoelectrics

Lukas Stepien, Aljoscha Roch, Roman Tkachov and
Tomasz Gedrange

Additional information is available at the end of the chapter

http://dx.doi.org/10.5772/66196

Abstract

This chapter addresses recent progress in the field of polymer thermoelectric materials. It covers a brief introduction to intrinsically conductive polymers and its motivation for thermoelectric utilization. A review about important and recent literature in the field of p-type and n-type polymers for thermoelectric applications is summarized here. For a better understanding of material development issues, doping mechanisms for intrinsically conducting polymers are discussed. Special emphasis is given to n-type polymers, since this group of polymers is often neglected due to unavailability or poor stability during processing. Different possibilities in terms of generator design and fabrication are presented. Recent challenges in this scientific field are discussed in respect to current material development, uncertainty during the measurement of thermoelectric properties as well as temperature stability for the most prominent p-type polymer used for thermoelectric, PEDOT:PSS.

Keywords: printing, coating, intrinsically conductive polymers, PEDOT:PSS, flexible, thermoelectric generator

1. Introduction

Historically, the most famous intrinsically conducting polymer is polyacetylene, which was discovered in 1976 by Alan Heeger, Alan MacDiarmid, and Hideki Shirakawa, who were jointly awarded with Nobel Prize in the year 2000. After the discovery that doping (in the case of polyacetylene it is chemical oxidation with iodine) of polymer chains can increase electrical conductivity of the polymer dramatically, a lot of effort was put in investigating the doping process as well as polymer synthesis itself [1].

In contrast to known solid state semiconductors, polymers can be doped by different approaches following the concept of MacDiarmid on primary and secondary doping. Primary doping includes chemical doping mechanism, which can be oxidizing/reducing or protonating/deprotonating. Secondary doping addresses the change in the polymer morphology like chain alignment, chain orientation, crystallinity, and so on.

However, for a reasonable utilization of these electrical conducting polymers, good processability and stability are the major requirements. This leads to considerable amount of different intrinsically conducting polymers, e.g., polyaniline, polycarbazole, and polythiopene.

In recent years, significant progress in the development of new types of conductive polymers has been achieved. Plenty of scientific publications can be found in scientific journals for polymer synthesis, polymer modification, or characterization.

For thermoelectric applications, both types of conducting polymers were investigated, on the one hand, pure organic polymers such as PANI [2], PPV [3], PPy:Tos [4], PEDOT:PSS [5], or PEDOT: PEDOT:TOS [6, 7], but on the other hand, metal organic complexes [8] as well as composites with nanostructures. The pure organic polymers are generally semiconductors and need to be primary doped in order to become electrically conductive. The oxidation leads to p-type conductivity. The degree of oxidation determines charge carriers' concentration and, therefore, affects the electrical conductivity directly. Electrical conductivity of p-type polymer like PEDOT:PSS with high degree of oxidation can reach up to 2000 S/cm [9] and even 3300 S/cm [10]. It was reported recently [11] about reaching electrical conductivity up to 4600 S/cm by active control of the deposition procedure of polymer PEDOT:PSS. It was possible to achieve conductivity up to 5400 S/cm by changing the counter ions of PEDOT [12]. Vapor-phase-grown single crystal PEDOT nanowires showed electrical conductivity of 8797 S/cm [13], which is the highest value known for this group of polymers. In spite of this outstanding electrical conductivity, Seebeck coefficient is usually relatively small for polymers.

Among p-type polymers, PEDOT, modified in PEDOT:TOS and PEDOT:PSS, respectively, is the most investigated polymer for thermoelectric utilization. Reported ZT values are in the range 0.2–1.02 for PEDOT:TOS [5–7]. Unfortunately, the reported performances often lack of reproducibility by other working groups.

The synthesis of n-type polymers involves other challenges compared to the development of p-type polymers. Stability of n-type polymer under ambient conditions is often critical, and electrical conductivity is in general not as high as for p-type polymers. However, a synthesis approach with metal organic polymers is promising and it was shown, that ZT value of this material can reach 0.2 [8].

Figure 1 shows overview about last years' progress in ZT values of conductive polymers.

2. Intrinsically conductive polymers for thermoelectric application

This section gives an overview of the recent advances made in polymer thermoelectric materials, which covers different material classes. First, considerations regarding the benefits of polymers for thermoelectric applications are discussed. Second, short introduction to doping of intrinsically conductive polymers is given. Subsequently, overview of p-type and n-type polymers is presented.

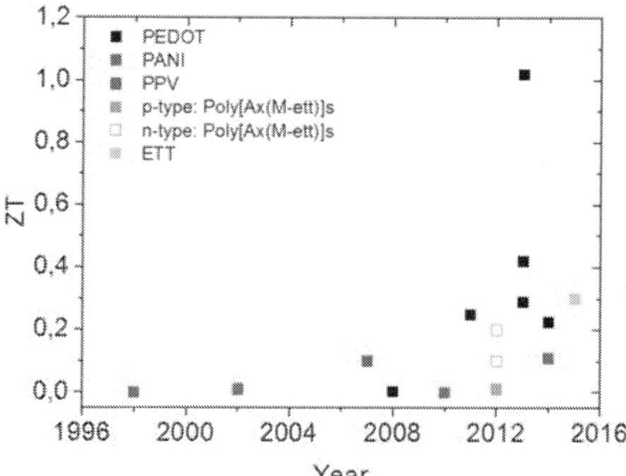

Figure 1. Overview of ZT values of conductive polymers, which were published recently [2, 3, 5–8, 14–16].

The main requirement for a polymer to be electrically conductive is the formation of a conjugated pi-electron system. With the increase of the delocalized pi-electron system, the polymer changes its nature from an isolator to a semiconductor [17]. The band gap of intrinsically conductive polymers can vary from 1.36 eV (for polyacetylene) to 5.96 eV (for polypyrrole) [18]. In order to increase electrical conductivity of these semiconductive polymers, one has to increase the free charge carriers concentration (analogy to silicon-based semiconductors). This doping leads to the formation of additional energy levels (allowed energy levels) within the band gap. Due to these new energy levels, free charge carriers can arise, which can be dislocated along the polymer chain or hop/tunnel between individual polymer chains. A brief introduction can be found in reference [19].

In the beginning, conductive polymers were investigated for thermoelectric applications, especially because of their alleged low thermal conductivity and high electrical conductivity. Low thermal conductivity would make them attractive for producing advanced thermoelectric materials. For technical polymers like PVC or PMMA, thermal conductivity is in the range 0.13–0.3 W/(m K) [20]. In the case of intrinsically conductive polymers, thermal conductivity is higher due to the contribution to heat transport by charge carriers. In the case of drop casted films of PEDOT:PSS, thermal conductivity can vary from 0.29 to 1 W/(m K) for cross-plane and in-plane measurements, respectively. It was also shown, that in-plane thermal conductivity of these films increases with electrical conductivity [21]. For comparison, spark plasma sintered BiTe compounds also show a relatively low thermal conductivity of 0.8 W/(m K) for cross-plane and 1.2 W/(m K) for in-plane measurements [22]. For nanostructured thermoelectric materials, thermal conductivity can be even lower [23]. This example shows also, the importance of distinguishing between in-plane and cross-plane measurements. This should also be applied to electrical properties measurements.

True arguments regarding advantages, comparing to other (brittle) thermoelectric materials, are possible flexibility and mostly nontoxic properties. Abundance, independence on volatile raw material costs, feasible scale-up for material production, as well as, efficient processability of polymer material can be considered as advantages for polymer thermoelectric materials. Polymers are processable by different (industrial) established and sophisticated technologies and the final building up of thermoelectric modules based on polymers is possible with printing techniques. This offers cost-effective production of thermoelectric generators (TEG) at industrial scale.

To facilitate the usage of intrinsically conductive polymers for thermoelectric applications, one has to consider several requirements besides, undoubtedly important, thermoelectric properties. These requirements imply costs, abundance, recyclability, environmental stability, thermal stability, processability, and more.

Basic factors that determine thermoelectric properties of polymers are as follows:

– nature of polymer structure itself;

– kind of counterions;

– polymer chain length, structural order; and

– environmental factors like temperature and humidity.

Of course, other factors like charge carriers density and morphology play an important role in terms of thermoelectric properties and will be discussed below.

2.1. Doping mechanism for intrinsically conductive polymers

Thermoelectric properties of polymers are predominantly governed by doping. While undoped polymers are considered to be bad conductive semiconductors, doping can change this state and transfer into metallic-like behavior. Following doping concept of MacDiarmid [24], doping of polymers can be subdivided into primary and secondary doping.

In general, **primary doping** affects charge carriers density in the polymer material, which influences Seebeck coefficient and electrical conductivity directly. The extent of this effect is related to the used doping agent, which chemically affects polymer backbone, hence creating/reducing number of free charge carriers.

Changes in thermoelectric properties from **secondary doping** are related to morphological modification of polymer chains (atomic scale) or grains/crystals (nano-to-microscale). Since this modification does not change the chemical environment of the polymer, thus not changing of charge carriers density, no influence on Seebeck coefficient is expected.

Mobile charge carriers can be dislocated over the polymer backbone. However, not only movement of charge carriers along polymer chain, but also from chain to chain can occur. Charge transfer between chains is considered in many models (e.g., variable-range-hopping, Sheng's model). Notably, charge carriers' mobility along the chain and interchain is different. Typically, charge carriers' mobility along chain is higher than for hopping events. From this follows that better chain alignment, or crystallinity, will directly affect macroscopic electrical conductivity. Yet, increased orientation will also increase anisotropy of the material.

For instance, electrical conductivity of iodine doped polyacetylene could be increased by one order of magnitude after stretching [25].

2.2. P-type conductive polymers

In spite of high electrical conductivity and power factor, polyacetylene plays no role in thermoelectric usage due to poor processability, unstable doping (with iodine), and high reactivity of solitons (charge carriers), leading to decrease in electrical conductivity over time.

The most prominent representatives of p-type polymers are aromatic polymers with polaron conduction mechanism. This can be found in the family of polythiopenes and others. The most promising polymer is currently PEDOT. Other polymers like P3HT or PANI also showed good improvement, however, they cannot reach the outstanding performance of PEDOT yet.

Review of recently reported material properties are given in **Table 1**. Further data can be found in references [26–32].

Material, reference	Conductivity, S/cm	Seebeck coefficient, μV/K	Power Factor, μW/(mK^2)
PEDOT: TOS [6]	74	210	324
PEDOT:PSS [5]	890	72	469
PP-PEDOT: TOS [7]	1320	98	1270
PEDOT:PSS [33]	677	28	55
PEDOT:PSS [33]	788	32	83
PEDOT:PSS [34]	1400	19	48
PEDOT:PSS [11]	4800	–	–
PEDOT:Sulf-NMP [12]	5400	–	–
PEDOT:PSS/rGO [35]	637	26	45
PEDOT:PSS-Tellurim [36]	215	115	284
P3HT +50 wt. % CNT [37]	345	97	325
P3HT [38]	225	53	62
Poly[Cu$_x$(Cu-ett)] [8]	10	83	7
Poly[Cu$_x$(Cu-ett)] [39]	1	45	0.1
Polyacetylene [40]	700	17	–
PANI [41]	10	12	0.2
PANI + 50 wt. % GN sheets [41]	60	30	8
PANI + SWCNT [42]	1440	–	217
PANI + graphene [43]	814	26	55

Table 1. Thermoelectric properties of selected p-type polymer semiconductors.

2.3. N-type conductive polymers

To make complete thermoelectric (TE) module, an n-type TE material is also required. However, the number of n-type organic materials with good thermoelectric properties is much smaller in comparison to that of p-type materials.

The main reasons are difficulties in n-type doping, because, typically, dopants providing one-electron transfer must have low ionization energies, which leads to instability in air. The next reason is the absence of a large variety of intrinsically conductive polymers. But, in the past 5 years, the number of reports, devoted to n-type organic materials, is steadily growing. It was found that promising materials are n-type fullerenes K_xC70 [44], fullerenes C60 doped with $Cr_2(hpp)_4$ (hpp = 1,3,4,6,7,8-hexahydro-2 H-pyrimido[1, 2-a]pyrimidine) [45], poly[K_x(Ni-ett)] (ett = 1,1,2,2-ethenetetrathiolate) [8], poly{N,N'-bis(2-octyl-dodecyl)-1,4,5,8-napthalene dicarboximide-2,6-diyl]-alt-5,5'-(2,2'-bithiophene)} P(NDIOD-T2) doped by dihydro-1H-benzimidazole-2-yl (N-DBI) derivatives [46], self-doped perylene diimides (PDI) [47], polyethylenimine (PEI)/diethylenetriamine (DETA)-doped CNT, that were further reduced by NaBH4 [48], poly(p-phenylene vinylene) derivatives (FBDPPV) doped with (4-(1,3-dimethyl-2,3-dihydro-1H-benzoimidazol-2-yl)phenyl)dimethylamine (N-DMBI) [49], CoCp2@SWNTs [50], three-dimensional copper 7,7,8,8-tetracyanoquinodimethane (CuTCNQ) [51], isoindigo-based conjugated polymers (IIDT) [52], nanostructured tetrathiotetracene (TCNQ)2 (TTT(TCNQ)2) [53], and polyaniline doped with aprotic ionic liquid [54] (**Table 2**).

Material, reference	Conduc-tivity, S/cm	Seebeck coefficient, μV/K	Power Factor, μW/mK^2
K_xC_{70} [44]	550	−22.5	28
C_60 doped with $Cr_2(hpp)_4$ [45]	4	−175	12
poly[K_x(Ni-ett)] [8]	40	−122	60
P(NDIOD-T2)) doped with N-DBI [46]	0.008	−850	0.6
PDI [47]	0.5	−168	1.4
PEI/DETA-doped CNT [48]	52	−86	38
FBDPPV doped with N-DMBI [49]	14	−141	28
CoCp2@SWNTs [50]	432	−41.8	75.4
CuTCNQ [51]	0.037	−632	1.5
IIDT [52]	7×10^{-8}	−898	
nanostructured TTT(TCNQ)2 [53]	12000 (asses.)	−150 (asses.)	
polyaniline doped with aprotic ionic liquid [54]	0.0023	−138.8	4.43×10^{-3}
poly[K_x(Ni-ett)], prepared by an electrochemical method [55]	200 to 400	−90 to −140	453

Table 2. Thermoelectric properties of selected n-type organic semiconductors.

Taking into account the very recent research of poly[K_x(Ni-ett)] [55], it is one of the most promising n-type materials due to its excellent thermoelectric properties. Despite the very simple synthesis (**Figure 2**), chemical structure of this polymer is still not exhaustively clear, and is currently the research object. However, prototypes of generators based on this polymer have already shown very promising results. So, provided achieving good processability of this polymer, it can become one of main n-type materials for manufacture of TEGs. Below, we describe in a little more detailed manner, the present state-of-the-art in this topic.

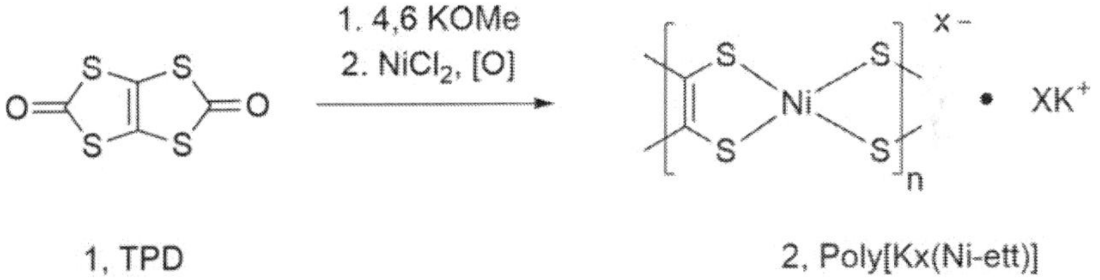

Figure 2. Synthesis of poly[K_x(Ni-ett)].

Poly(nickel1,1,2,2-ethenetetrathiolate), poly[K_x(Ni-ett)] (K is an alkali metal) (**Figure 2**) as an n-type organic polymer has been already known for a long time [56–64], its thermoelectric properties were carefully evaluated by Zhu's group [8] and an "all-organic" TEG device was fabricated.

Thanks to ambient stability and exciting thermoelectric characteristics, ZT value was equal to 0.2 at 400 K. This inspired other groups to work with this polymer for thermoelectrical applications [65] or to develop its composites with carbon nanotubes [66–68]. One major drawback of poly[K_x(Ni-ett)] is its poor processability due to its completely insoluble nature leading to suspensions with broad particle size distributions.

Since this polymer is totally insoluble, commonly used structure characterization methods cannot be applied for this polymer, so these compounds have no clearly established structure. Its thermoelectric properties are not completely reproducible, even by the same personnel in the same laboratory. Due to the same reason, it is not possible to control polymerization reaction, for example, to control the degree of polymerization and have an influence on polydispersity of the resulting product. The exact structures of the terminal groups are also not known. Also, because of insolubility, it is not possible to improve properties of polymer after its production—methods like Soxhlet extraction, reprecipitation, or separation of products with the help of column chromatography cannot be applied. The most significant disadvantage of such insolubility and, as a result, bad processability is difficulty for utilization in devices like thermoelectric generators.

To solve the above-mentioned problem, many researchers tried to obtain different composite materials on the basis of these polymers and another compound—poly(vinylidene fluoride) (PVDF) solution [69], 1-butyl(or decyl)-3-methylimidazolium tetrafluoroborate [70], and dodecyltrimethylammonium bromide [71]. It results in great improvement of its processability, but TE properties of such material decrease dramatically, especially the electrical conductivity.

But, there is also another approach for making paste suitable for printing, which allows avoiding the degradation in thermoelectric performance. It is possible not to modify the polymer itself, but to change the procedure of its preparation (temperature of reaction mixture, speed of rotation, different system of solvents, and the ratio of the solvents, access of oxygen, and so on). Usually, such procedure consists of two sequential stages. First one (preparation of the monomer in fact) is reaction of 1,3,4,6-tetrathiapentalene-2,5-dione (TPD) with potassium methoxide. The second stage (namely, polymerization) is the addition of a metal salt to the solution. Usually [8, 56], both steps are carried out in methanol, and both with reflux. After that, the final product is modified to obtain flexible composites [69–71]. But carrying out the polymerization reaction in N-methylformamide (NMF)-methanol medium under certain temperature conditions allows obtaining of the product as a paste with controllable viscosity. Further evaporation of solvent and washing sequentially with water and methanol provide the thermoelectric material with characteristics similar to the powder.

The product of polymerization reaction poly[K_x(Ni-ett)] (**Figure 2**) is an alternating copolymer with organic and inorganic monomeric units. The important feature of this reaction is that it occurs without a catalyst. Also, at the moment, it is not possible for the quantitative description of the role of atmospheric oxygen as an oxidizing agent in the polymerization process. Therefore, it is not suitable to describe the rate of reaction with patterns of chain growth or step growth mechanism.

By creating the proper conditions of this synthesis, the produced gel has no fluidity and is stable indefinitely under inert conditions. Even 1 month after the formation and storage under an inert atmosphere, it shows no signs of aging. However, when the gel is placed in the ambient atmosphere, it starts to slowly delaminate, by forming two phases—liquid and precipitate. Obviously, the reason of delamination is oxidation of paste components, leading to further polymerization and formation of insoluble product. It is possible to see very clearly the formation of a gel-like phase and its further separation, accompanied by reduction in viscosity which can be shown by rheological experiments. The diagram (**Figure 3**) reflects the dependence of the shear viscosity over time in a methanol-N-methylformamide system. During gel formation, a sharp increase in shear viscosity can be noticed and a subsequent reduction of shear viscosity related obviously to delamination of gel during the oxidation process.

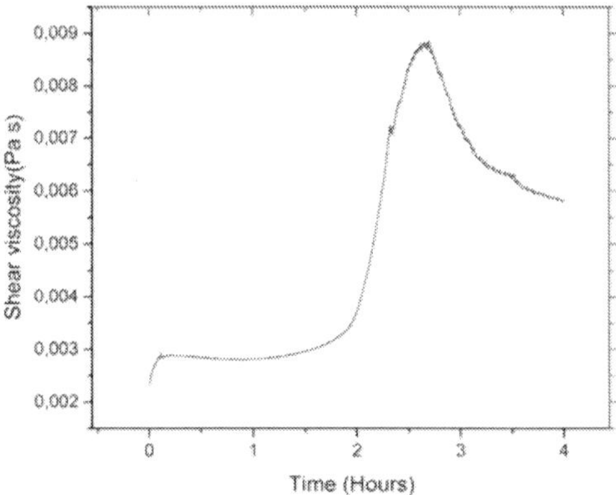

Figure 3. Measurement of shear-viscosity in the ambient atmosphere.

The film, produced by airbrush-gun, is inhomogeneous and thin (~763 nm). But it still has a relatively good conductivity (0,13 S/cm (make units used equal)). More homogeneous films are formed by using dispenser printer. Prepared films on Kapton substrate have shown poor adhesion and could be easily removed in water. However, gentle immersion in water kept the film intact and improved its performance ($S = -53.6$ μV/K, $\sigma = 1.60$ S/cm). Interestingly, it has almost the same thermoelectric characteristics like the powder produced by drying of solvent from paste (see **Table 3**). Bending test shows, that any appreciable change in conductivity of this film occurs only when winding the tube diameter of less than 4 mm. It is important to note, that in an inert atmosphere thermoelectric properties of polymer are stable.

Probe, reference	Conductivity, S/cm	Seebeck coefficient, μV/K	Power Factor, μW/(mK^2)
Cuboid, pressed from powder [8]	40	−122	60
Film [69]	8.31	−67.4	3.71
Film [71]	1.14	−49.0	0.28
Tablet pressed from powder [this work]	1.81	−44.5	0.36
Film made from suspension [this work]	0.13	−34.5	0.02
Film made from paste [this work]	1.60	−53.6	0.46

Table 3. Thermoelectric properties of selected poly[K$_x$(Ni-ett)].

During first day of exposure in the ambient atmosphere, electrical conductivity of the film, as well as, the pure powder tends to deteriorate. After the fifth day, the thermoelectric properties remained stable for several months (shown in **Figure 4**).

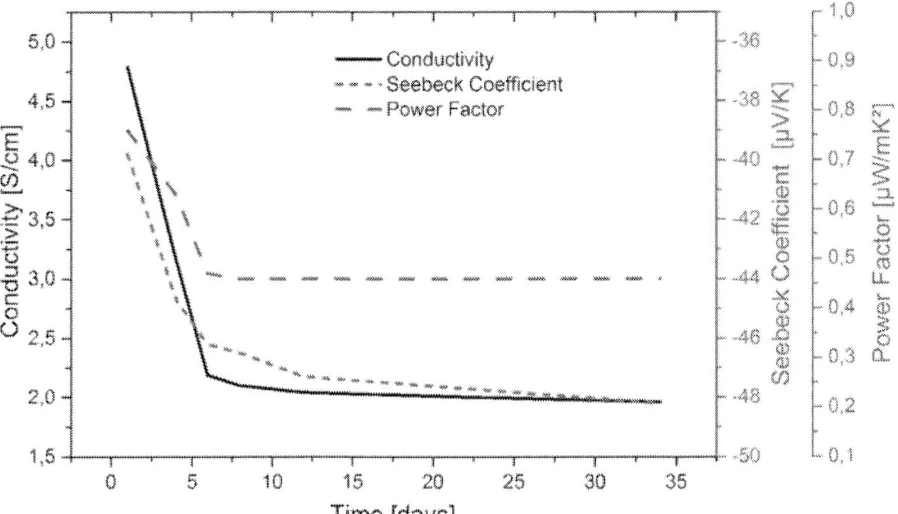

Figure 4. Evolution of thermoelectric properties of Poly [K$_x$(Ni-ett)] under ambient atmosphere.

To conclude, a simple procedure has been proposed to obtain printable paste based on insoluble conductive polymer poly[K_x(Ni-ett)]. This method opens a way for various applications, such as aerosol and dispenser printing.

3. Thermoelectric modules

The performance of polymer TEG is in general lower in comparison to TEGs made by inorganic materials like BiTe. The main reason for this is, of course, the higher thermoelectric performance of BiTe. The ZT value of BiTe is around 1 [22]. ZT values of p-type and n-type polymers are generally clearly below 1 or even below 0.1.

However, polymers offer other advantages like flexibility or easy processing and fabrication techniques. Most polymers can be printed without losing dramatically electrical and thermoelectric performance. For other materials like BiTe that performance is generally dependent on hot pressing or spark plasma sintering (SPS) techniques followed by cutting which may include manual work or other different processing steps. This advantage of polymers opens opportunities for industrial and economical production of thermoelectric devices.

A well-known technique for printing polymer TEG is the dispense technique. With dispenser, both types of TEGs, mono-leg-TEG (only one polymer), as well as, TEG with p- and n- type polymers can be easily printed, as shown in different publications. **Figure 5** shows mono-leg TEG designs.

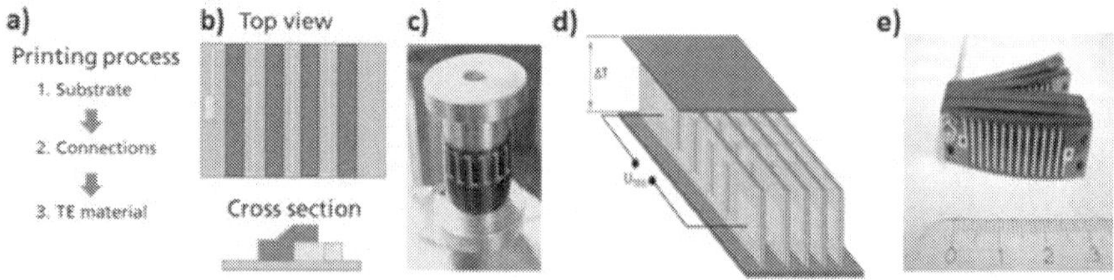

Figure 5. Typical design for printed polymer TEG as mono-leg with one polymer (lateral design). The printing sequence is shown (a). A top view and cross section is displayed (b). In (c), a mono-leg TEG printed with PEDOT:PSS and Ag-paste is shown. TEG is wrapped around holder. (d) Scheme of a stacked mono-leg TEG design. (e) Printed sheets of a mono-leg TEG before assembly.

The same concept is also used for polymer TEG with both p-type and n-type materials. In this case, conductive paste (e.g., silver paste) is replaced by second polymer (see **Figure 6**). The design concepts shown in **Figures 5** and **6** have a disadvantage consisting in the fact that heat flow is parallel to the substrate and, therefore, parasitic heat flows exist through the substrate.

Furthermore, thermal connection to a heat source and a heat sink is relatively complex and difficult. **Figure 7** shows another mono-leg TEG concept orientated on classical TEG design.

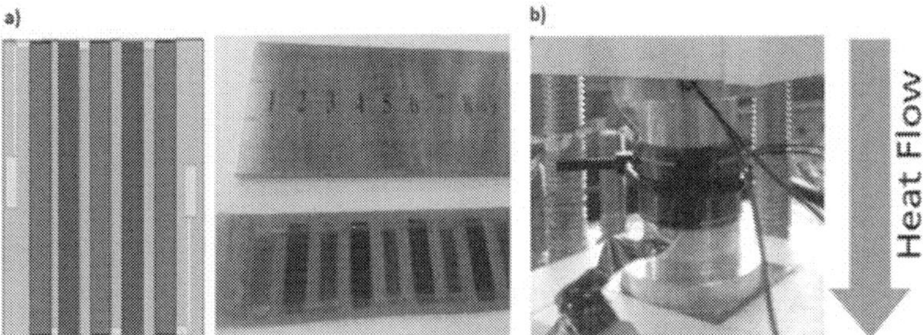

Figure 6. (a) Scheme of a TEG design with p-type (red) and n-type (blue) polymers and the actually printed polymer TEG with PEDOT:PSS and Poly[K$_x$(Ni-ett)] on Polyimide (PI) substrate. (b) TEG wrapped around a holder for measuring power output by applying defined temperature difference. The heat flow is parallel to the substrate.

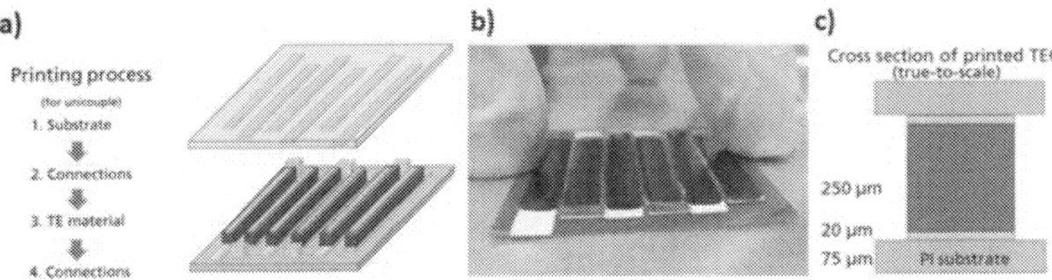

Figure 7. (a) Printing sequence and scheme of a vertical TEG design. (b) Vertical polymer TEG printed with dispenser, substrate is PI, polymer is PEDOT:PSS with 5 % DMSO. (c) Schematic cross section and dimensions of printed vertical TEG.

This design has the advantage consisting in the fact that heat flow goes only thorough the active polymer, preventing parasitic heat flows. Furthermore, fill factor and contact resistances can be reduced by larger contact areas in comparison to the other design. The challenge in printing polymer TEG is, however, that the pastes/ink which are used for printing TEGs have very high amount of solvents, which makes it difficult to quickly build up thick layers. PEDOT:PSS paste used for printing TEG in **Figure 7** contained ~ 2 wt.% PEDOT:PSS. This ratio of solvent and solid component is necessary for avoiding agglomeration and, thus, for providing stable printable polymer paste. Therefore, printing of thick layers is time consuming, because after deposition of each layer it is necessary to heat up polymer to evaporate solvent. Typical thickness of one layer after removing the solvent is in the range of 10–20 µm. Therefore, printing time of TEG shown in **Figure 7** was relatively long about 5 hours (in a lab environment).

Thus, polymer TEGs designs and configurations are still not satisfying and need to be improved. Simulations by Oshima et al. [71] based on the work of Aranguren et al. [72] have shown theoretically the possibility to reach power output in the range of MWh/year by polymer TEGs with a new optimized design based on porous substrate. In this publication, relatively high temperatures of >100°C were assumed. However, experimental realization is still missing.

But, potential application of polymer TEG is strongly limited by physical and chemical properties of polymers. Thermal stability of polymers is very critical for applications, as was shown recently by Stepien et al. It was shown for PEDOT:PSS, that electrical conductivity drops irreversibly down for temperatures >55°C (further details are given in Subsection 4.2.). It was assumed, that the reason might be degradation of the polymer. This behavior was observed under ambient conditions as well as in glove box under an inert atmosphere. That means, degradation of polymer properties starts at temperatures around 50°C and limits possible application dramatically. Thus, investigation by Stepien et al. has shown that the challenge for polymers is, on the one hand, to increase ZT values, but, on the other hand, to increase thermal stability also.

Power output and efficiency of polymer TEG are dependent on both the ZT value and maximum possible temperature difference.

Theoretical expression of the maximum efficiency of TEG is:

$$\eta_{max} = \frac{T_H - T_C}{T_H} \frac{\sqrt{1 + Z\bar{T}} - 1}{\sqrt{1 + Z\bar{T}} + \frac{T_c}{T_H}} \tag{1}$$

with ZT value for TE module according to Eq. (2):

$$Z\bar{T} = \frac{\left(S_p - S_n\right)^2 \bar{T}}{\left[\left(\rho_n \kappa_n\right)^{1/2} + \left(\rho_p \kappa_p\right)^{1/2}\right]^2}, \tag{2}$$

where T_C, T_H, and $\bar{T}$ are the temperature at the cold side, hot side, and average ambient absolute temperature, respectively; S_n, S_p, ρ_n, ρ_p, κ_n, and κ_p are Seebeck coefficient, electrical resistivity, and thermal conductivity of n- and p-type polymers, respectively.

The temperature dependence of TEG efficiency calculated according to Eq. (1) is shown in **Figure 8**. Calculations performed with ZT values equals 1 [7] and 0.2 [8] for p-type and n-type polymers, respectively, efficiency below 0.25% can be reached by applying maximum temperature difference of 60 K (333–273 K).

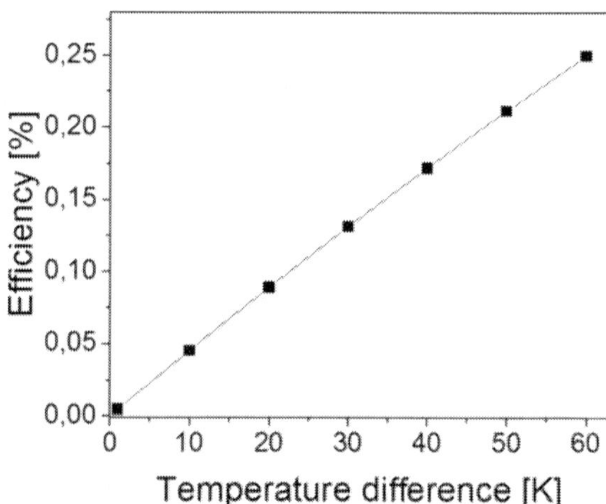

Figure 8. Theoretical temperature dependence of maximum efficiency for polymer TEG with p- and n-type polymers from references [7, 8]. Cold side temperature equals to 273 K. Internal contact resistance was assumed as ideal.

Polymer TEGs based on metal organic materials with 35 thermocouples (p- and n-type legs) have delivered around 1 $\mu W/cm^2$ at temperature difference of 25 K [8]. Another TEG made of PEDOT:TOS in a different design delivered 0.27 $\mu W/cm^2$ at a temperature difference of 30 K.

If ZT value or working temperature difference between hot and cold sides of polymer TEG cannot be increased, then polymer TEG can be used for low power application only. Therefore, there is a big challenge to avoid degradation of polymers at higher temperatures.

Another reason for relatively low power output is high contact resistance between polymer and metal contacts. The measurement of contact resistance between PEDOT:PSS and Silver paste reached 5 x 10^{-2} Ohm x cm^2. This leads to relatively high internal resistance values for printed polymer TEG. Measured internal resistance of printed TEGs was equal to a few kOhm. Internal resistance of commercial TEG with areas of, for example, 4 cm × 4 cm is <10 Ohm and efficiency for energy conversion is indicated around 5% by TEG suppliers. Conventional thermoelectric modules and materials were developed, of course, over decades. Design of commercial TEG was developed and simulated carefully. Leg-length and leg-cross sections were studied in order to get high fill factor, low internal resistances, and optimized power output. Such optimization processes are still missing for polymer TEG.

Estimated costs per Watt ($/W) of conventional TEG are in the range of 4 $/W. More than 50% of the costs are for system itself and manufacturing and just small part for thermoelectric material itself [73].

On the one hand, printing technology is, of course, advantageous for polymer thermoelectrics, because it has the potential for reducing system costs in general. However, on the other hand, if we compare polymer material costs to, for example, today costs of BiTe-based material, we find a large discrepancy. Assuming that costs for producing 1 kg BiTe-based material equals to 1000 €, we have costs for 500 ml of PEDOT:PSS dispersion of ~400 € today. Thus, cost for 500 ml PEDOT:PSS or more precisely 1 kg pure PEDOT:PSS is much higher than that of the BiTe-based material, considering the fact that PEDOT:PSS dispersion is purchased with a solid content of only around 2 wt%. Using cost-saving manufacturing technique like printing is an advantage for polymer thermoelectrics of course; however, this advantage is not the key for economical applications without much lower polymer material costs.

The applications of polymer thermoelectrics seem to be for the medium–long term in the low-power area. The amortization of TEG costs by energy harvesting with polymer TEG is today still difficult and ambitiously.

First applications for polymer TEGs are imaginable if the flexibility and the low weight of the polymer devices are the criteria for choosing the polymer material for thermoelectrics.

4. Challenges

4.1. Material optimization

Concerning optimization of thermoelectric material, one advantage over nonpolymer materials is the possibility to decouple partially chemistry from morphology. Indeed, changes in chemistry also have effect on the conformation of polymer chains, thus the morphology is affected; however, it can be stated that Seebeck coefficient is mainly governed by chemical factors, for example, the kind of counterions used or degree of oxidation, while electrical conductivity is additionally influenced by morphological aspects like crystal orientation, crystal size, or namely charge carrier mobility. Therefore, it is favorable to optimize these two aspects separately. Therefore, parameters of polymer thermoelectric materials can be improved according to the following concept, which can be applied to almost all thermoelectric polymer materials; here special emphasis is given to PEDOT.

4.1.1. Maximizing electrical conductivity

Morphology should be optimized by using a proper deposition technique and/or using cosolvents or other additives. The aim of this step is to maximize electrical conductivity. The solution-sheared deposition technique seems to be a promising approach. In this case, Seebeck coefficient should not be influenced significantly.

4.1.2. Optimization of Seebeck coefficient

Continuing theme of high electrically conductive material, increase in Seebeck coefficient with proper reduction/oxidation (deprotonating/protonating) methods will be required to obtain good thermoelectric material.

Reduction can be done by electrochemical methods due to its good control of the degree of oxidation. Reduction can also be done by other methods like dipping or immersion in corresponding agents. The aim here is to find the optimum between electrical conductivity and Seebeck coefficient, because change in charge carrier density will affect Seebeck coefficient and electrical conductivity to the contrary.

It should be emphasized, that possible formation of crystals, because of the reducing agents used, should be avoided in order to maintain the crystal structure prepared in step one. This is also true for the use of composite materials. Composites with beneficial thermoelectric properties contain often high quantity of nanoparticles like nanowires, graphene sheets, or others (often over 50 wt%), which as a result is nanoparticle-matrix filled with polymers. For these improvements, it is crucial to have absolutely robust processing techniques, as well as, characterization routines.

If these factors cannot be controlled properly, then poor reproducibility of material parameters will be the result. Since thermoelectric properties of polymers are highly dependent on morphological aspects, it is necessary to have ultimate control of film formation and conformation of polymer chains during this processing step.

Reported results often lack comparability because of different characterization methods. For instance, Seebeck coefficient can vary significantly, when electrodes with different geometries are used during the measurement. It was shown, that variation in electrode geometry can influence Seebeck coefficient by a factor of 3 or more [74]. Regarding electrical conductivity, it is known that different methods, for example, van der Pauw or linear 4-point-probe, can result in different sheet resistances because of possible anisotropies in the samples. The same applies for measuring film thickness. Regarding the method used, variations of more than 30% for films in nanometer range can occur. These systematic uncertainties, as well as, non-systematic ones, will have significant impact on power factor and, hence, the figure of merit. **Figure 9** shows the range of uncertainty for power factor over electrical conductivity.

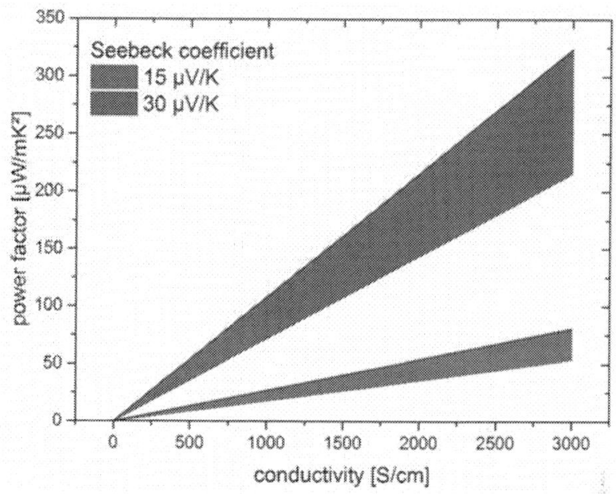

Figure 9. Range of uncertainty for power factor over electrical conductivity. Total uncertainties for Seebeck coefficient and electrical conductivity are 10% and 5%, respectively.

It can be seen, that overestimated Seebeck coefficient (maybe due to not proper electrode contacts) can lead to dramatic differences. This is especially true for high conducting materials.

Note that more efforts focus recently on measurements of thermal conductivities, as well as, anisotropy of the overall material properties and should be further encouraged.

4.2. Thermal stability

Thermoelectric energy conversion is considered to be a robust and sustainable process. This claim can only be fulfilled if material degradation can be avoided. One drawback, compared to other low temperature materials like BiTe, is besides thermoelectric performance, low thermal stability. Usually, glass transition and crystal melting for polymers occurs at much lower temperatures as for BiTe. This can lead to unfavorable change in nano- and microstructure of the polymer material.

In the case of DMSO doped PEDOT:PSS films, it was shown, that during 50 hours of thermal stress at 75°C ambient temperature, electrical conductivity decreased irreversibly by 17% (**Figure 10**). For higher temperatures, decrease in electrical conductivity was even more pronounced. Significant chemical degradation is found to start between 140 and 160°C, which is highly undesirable for real long-term applications. This leads to the conclusion that thermo-electric generators made of currently most promising p-type polymer PEDOT:PSS should be deployed only for moderate temperatures, which limits possible applications.

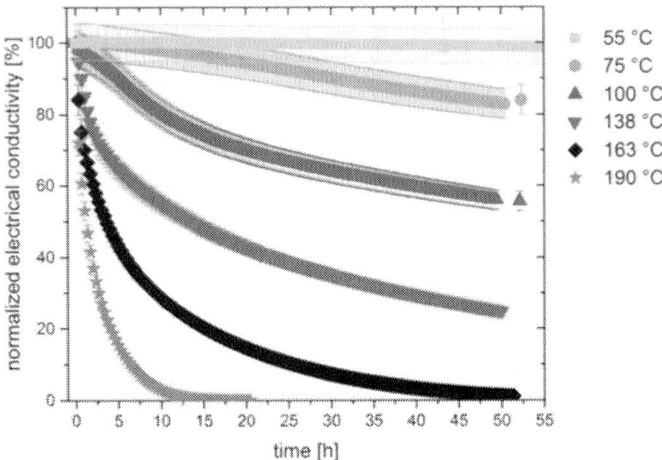

Figure 10. Decrease of normalized electrical conductivity of PEDOT:PSS (+DMSO) films at thermal stress for over 50 hours. Stepien et al. not accepted yet.

Comparing intrinsically conductive polymers with BiTe in terms of thermoelectric performance and thermal stability still a lot of effort has to be put in polymer materials in order to compete with BiTe.

Acknowledgement

This work has received partial funding from the European Unions' Seventh Framework Programme for research, technological development, and demonstration under the grant agreement No 604647.

Author details

Lukas Stepien[1,*], Aljoscha Roch[1], Roman Tkachov[1] and Tomasz Gedrange[2]

*Address all correspondence to: lukas.stepien@iws.fraunhofer.de

1 Fraunhofer Institute for Material and Beam Technology IWS, Dresden, Germany

2 University Hospital Carl Gustav Carus at the TU Dresden, Dresden, Germany

References

[1] Tsai T-S, Cheng H-C, Chen C-H, Whang W-T.: Widely variable Seebeck coefficient and enhanced thermoelectric power of PEDOT:PSS films by blending thermal decomposable ammonium formate. Organic Electronics. 2012;**12**(12):2159–2164. DOI: 10.1016/j. orgel.2011.09.004

[2] Toshima N.: Conductive polymers as a new type of thermoelectric material. Macromolecular Symposia. 2002;**186**(1):81–86 . DOI: 10.1002/1521-3900(200208)

[3] Hiroshige Y, Ookawa M, Toshima N.: Thermoelectric figure-of-merit of iodine-doped copolymer of phenylenevinylene with dialkoxyphenylenevinylene. Synthetic Metals. 2007;**157**(10–12):467–474. DOI: 10.1016/j.synthmet.2007.05.003

[4] Yan H, Ishida T, Toshima N, China S.: Thermoelectric properties of electrically conductive polypyrrole film. Thermoelectrics, 2001. Proceedings ICT 2001. XX International Conference. 2001;310–313. DOI: 10.1109/ICT.2001.979894

[5] Kim G-H, Shao L, Zhang K, Pipe KP.: Engineered doping of organic semiconductors for enhanced thermoelectric efficiency. Nature Materials. 2013;**12**:719–723. DOI: 10.1038/ nmat3635

[6] Bubnova O, Khan ZU, Malti A, Braun S, Fahlman M, Berggren M, Crispin X.: Optimization of the thermoelectric figure of merit in the conducting polymer poly(3,4-ethylenedioxythiophene). Nature Materials. 2011;**10**:429–433. DOI: 10.1038/nmat3012

[7] Park T, Park C, Kim B, Shin H, Kim E.: Flexible PEDOT electrodes with large thermoelectric power factors to generate electricity by the touch of fingertips. Energy & Environmental Science. 2013;**6**:788–792. DOI: 10.1039/c3ee23729j

[8] Sun Y, Sheng P, Di C, Jiao F, Xu W, Qiu D, Zhu D.: Organic thermoelectric materials and devices based on p- and n-Type Poly(metal 1,1,2,2-ethenetetrathiolate)s. Advanced Materials. 2012;**24**:932–937. DOI: 10.1002/adma.201104305

[9] Mengistie DA, Ibrahem MA, Wang P-C, Chu C-W.: Highly conductive PEDOT:PSS treated with formic acid for ITO-Free polymer solar cells. ACS Applied Material Interfaces. 2014;**6**(4):pp. 2292–2299. DOI: 10.1021/am405024d

[10] Tan L, Zhou H, Ji T, Huang L, Chen Y.: High conductive PEDOT via post-treatment by halobenzoic for high-efficiency ITO-free and transporting layer-free organic solar cells. Organic Electronics. 2016;**33**:316–323. DOI: 10.1016/j.orgel.2016.03.037

[11] Worfolk BJ, Andrews SC, Park S, Reinspach J, Liu N, Toney MF, Mannsfeld SCB, Bao Z.: Ultrahigh electrical conductivity in solution-sheared polymeric transparent films. PNAS. 2015;**112**(46):14138–14143. DOI: 10.1073/pnas.1509958112

[12] Gueye MN, Carella A, Massonnet N, Yvenou E, Brenet S, Faure-Vincent J, Pouget S, Rieutord F, Okuno H, Benayad A, Demadrille R, Simonato J-P.: Structure and dopant

engineering in PEDOT thin films: Practical tools for a dramatic conductivity enhancement. Chemistry of Materials. 2016;**28**(10):3462–3468. DOI: 10.1021/acs.chemmater.6b01035

[13] Cho B, Park KS, Baek J, Oh HS, Koo Lee Y-E, Sung MM.: Single-Crystal Poly(3,4-ethylenedioxythiophene) Nanowires with Ultrahigh Conductivity. Nano Letters. 2014;**14**(6):3321–3327. DOI: 10.1021/nl500748y

[14] Mateeva N, Niculescu H, Schlenoff J, Testardi LR.: Correlation of Seebeck coefficient and electric conductivity in polyaniline and polypyrrole. Journal of Applied Physics. 1998;**83**(6):3111–3117 . DOI: 10.1063/1.367119

[15] Li J, Tang X, Li H, Yan Y, Zhang Q.: Synthesis and thermoelectric properties of hydrochloric acid-doped polyaniline. Synthetic Metals. 2010;**160**(11-12):1153–1158. DOI: 10.1016/j.synthmet.2010.03.001

[16] Feng-Xing J, Jing-Kun X, Bao-Yang L, Yu X, Rong-Jin H, Lai-Feng L.: Thermoelectric performance of Poly(3,4-ethylenedioxythiophene): Poly(styrenesulfonate). Chinese Physics Letters. 2008;**25**(6):2202–2205.

[17] Atkins P, d'Sousa J. Atkins' Physical Chemistry. 10th ed. Oxford: Oxford University Press; 2014. 1008 p. DOI: ISBN 978-0199697403

[18] Botelho AI, Shin Y, Liu J, Li X.: Structure and optical bandgap relationship of π-conjugated systems. PLoS One. 2014;**9**(1):e86370. DOI: 10.1371/journal.pone.0086370

[19] Rehan M.: Elektrisch leitfähige Kunststoffe. Chemie unserer Zeit. 2003;**37**(1):18–29.

[20] Thyssen Krupp Plastics. Technische Kunststoffe im Überblick [Internet]. Available from: http://www.thyssenkrupp-plastics.de/fileadmin/inhalte/07_Publikationen/06_Prospekte/0750_Techn_Kunststoffe_150dpi.pdf [Accessed: 18.08.2016].

[21] Liu J, Wang X, Li D, Coates NE, Segalman RA, Cahill DG.: Thermal conductivity and elastic constants of PEDOT:PSS with high electrical conductivity. Macromolecules. 2015;**48**:585–591. DOI: 10.1021/ma502099t

[22] Han L, Spangsdorf SH, Nong NV, Hung LT, Zhang YB, Pham HN, Chen YZ, Roch A, Stepien L, Pryds N.: Effects of spark plasma sintering conditions on the anisotropic thermoelectric properties of bismuth antimony telluride. RSC Advances. 2016;**6**(64):59565–59573. DOI: 10.1039/C6RA06688G

[23] Kim SI, Lee KH, Mun HA, Kim HS, Hwang SW, Roh JW, Yang DJ, Shin WH, Li XS, Lee YH, Snyder GJ, Kim SW.: Dense dislocation arrays embedded in grain boundaries for high-performance bulk thermoelectrics. Science. 2015;**348**(6230):109–114. DOI: 10.1126/science.aaa4166

[24] MacDiarmid AG, Epstein AJ.: The concept of secondary doping as applied to polyaniline. Synthetic Metals. 1994;**65**(2–3):103–116. DOI: 10.1016/0379-6779(94)90171-6

[25] Tsukamoto J, Takahashi A, Kawasaki K.: Structure and electrical properties of poly-acetylene yielding a conductivity of 105 S/cm. Japanese Journal of Applied Physics. 1990;**29**(1):125 –130.

[26] He M, Qiu F, Lin Z.: Towards high-performance polymer-based thermoelectric materials. Energy & Environmental Science. 2013;**6**(5):1341–1642. DOI: 10.1039/c3ee24193a

[27] Kamarudin MA, Sahamir SR, Datta RS.: A Review on the Fabrication of Polymer-Based Thermoelectric Materials and Fabrication Methods. The Scientific World Journal. 2013;1–17. DOI: 10.1155/2013/713640

[28] Culebras M, Gómez CM, Cantarero A.: Review on polymers for thermoelectric applications. Materials. 2014;**7**:6701–6732. DOI: 10.3390/ma7096701

[29] Sun K, Zhang S, Li P, Xia Y, Zhang X, Du D, Isikgor FH, Ouyang J.: Review on application of PEDOTs and PEDOT:PSS in energy conversion and storage devices. Journal of Material Sciences. 2015;**26**:4438–4462. DOI: 10.1007/s10854-015-2895-5

[30] Wei Q, Mukaida M, Kirihara K, Naitoh Y, Ishida T.: Recent progress on PEDOT-based thermoelectric materials. Materials. 2015;**8**:732–750. DOI: 10.3390/ma8020732

[31] Katz HE, Poehler TO, editors.: Innovative Thermoelectric Materials. 1st ed. London: Imperial College Press; 2016. 292 p. DOI: ISBN 978-1783266050

[32] Russ B, Glaudell A, Urban JJ, Chabinyc ML, Segalman RA.: Organic thermoelectric materials for energy harvesting and temperature control. Nature Reviews Materials.2016;**1**:16050. DOI: 10.1038/natrevmats.2016.50

[33] Lee SH, Park H, Son W, Choi HH.: Novel solution-processable, dedoped semiconductors for application in thermoelectric devices. Journal of Materials Chemistry A. 2014;**2**:13380. DOI: 10.1039/c4ta01839g

[34] Xiong J, Jiang F, Zhou W, Liu C, Xu J.: Highly electrical and thermoelectric properties of a PEDOT:PSS thin-film via direct dilution–filtration. RSC Advances. 2015;**5**:60708. DOI: 10.1039/c5ra07820b

[35] Yoo D, Kim J, Kim JH.: Direct synthesis of highly conductive poly(3,4-ethylenedioxythiophene):poly(4-styrenesulfonate) (PEDOT:PSS)/graphene composites and their applications in energy harvesting systems. Nano Research. 2014;**7**(5):717–730. DOI: 10.1007/s12274-014-0433-z

[36] Bae EJ, Kang YH, Jang K-S, Cho SY.: Enhancement of thermoelectric properties of PEDOT:PSS and tellurium-PEDOT:PSS hybrid composites by simple chemical treatment. Scientific Reports. 2016;**6**:18805. DOI: 10.1038/srep18805

[37] Hong CT, Kang YH, Ryu J, Cho SY, Jang K-S.: Spray-printed CNT/P3HT organic thermoelectric films and power generators. Journal of Materials Chemistry A. 2015;**3**:21428–21433. DOI: 10.1039/C5TA06096F

[38] Qu S, Yao Q, Wang L, Chen Z, Xu K, Zeng H, Shi W, Zhang T, Uher C, Chen L.: Highly anisotropic P3HT films with enhanced thermoelectric performance via organic small molecule epitaxy. NPG Asia Materials. 2016;8:e292. DOI: doi:10.1038/am.2016.97

[39] Jiao F, Di CA, Sun Y, Sheng P, Xu W, Zhu D.: Inkjet-printed flexible organic thin-film thermoelectric devices based on p- and n-type poly(metal 1,1,2,2-ethenetetrathiolate) s/polymer composites through ball-milling. Philosophical Transactions of the Royal Society A: Mathematical, Physical and Engineering Sciences. 2013;10:20130008. DOI: 10.1098/rsta.2013.0008

[40] Zuzok R, Kaiser AB, Pukacki W, Roth S.: Thermoelectric power and conductivity of iodine-doped "new" polyacetylene. Journal of Chemical Physics. 1991;95:1270. DOI: 10.1063/1.461107

[41] Du Y, Shen SZ, Yang W, Donelson R, Cai K, Casey PS.: Simultaneous increase in conductivity and Seebeck coefficient in a polyaniline/graphene nanosheets thermoelectric nanocomposite. Synthetic Metals. 2012;161(23–24):2688–2692. DOI: 10.1016/j.synthmet.2011.09.044

[42] Wang L, Yao Q, Xiao J, Zeng K, Qu S, Shi W, Wang Q, Chen L.: Engineered molecular chain ordering in single-walled carbon nanotubes/polyaniline composite films for high-performance organic thermoelectric materials. Chemistry - An Asian Journal. 216;11(12) DOI: 10.1002/asia.201600212

[43] Wang L, Yao Q, Bi H, Huang F, Wang Q, Chen L.: PANI/graphene nanocomposite films with high thermoelectric properties by enhanced molecular ordering. Journal of Material Chemistry A. 2015;3:7086–7092. DOI: 10.1039/C4TA06422D

[44] Wang ZH, Ichimura K, Dresselhaus MS, Dresselhaus G, Lee WT, Wang KA, Eklund PC.: Electronic transport properties of KxC70 thin films. Physical Review B. 1993;48(14):10657–10660. DOI: 10.1103/PhysRevB.48.10657

[45] Menke T, Ray D, Meiss J, Leo K, Riede M.: In-situ conductivity and Seebeck measurements of highly efficient n-dopants in fullerene C60. Applied Physics Letters. 2012;100:093304. DOI: 10.1063/1.3689778

[46] Schlitz RA, Brunetti FG, Glaudell AM, Miller PL, Brady MA, Takacs CJ, Hawker CJ, Chabinyc ML.: Solubility-limited extrinsic n-Type doping of a high electron mobility polymer for thermoelectric applications. Advanced Materials. 2014;26:2825–2830. DOI: 10.1002/adma.201304866

[47] Russ B, Robb MJ, Brunetti FG, Miller PL, Perry EE, Patel SN, Ho V, Chang WB, Urban JJ, Chabinyc ML, Hawker CJ, Segalman RA.: Power factor enhancement in solution-processed organic n-Type thermoelectrics through molecular design. Advanced Materials. 2014;26:3473–3477. DOI: 10.1002/adma.201306116

[48] Kim SL, Choi K, Tazebay A, Yu C.: Flexible power fabrics made of carbon nanotubes for harvesting thermoelectricity. ACS Nano. 2014;8:2377–2386. DOI: 10.1021/nn405893t

[49] Shi K, Zhang F, Di C-A, Yan T-W, Zou Y, Zhou X, Zhu D, Wang J-Y, Pei J.: Toward high performance n-type thermoelectric materials by rational modification of BDPPV backbones. Journal of the American Chemical Society. 2015;**137**:6979–6982. DOI: 10.1021/jacs.5b00945

[50] Fukumaru T, Fujigaya T, Nakashima N.: Development of n-type cobaltocene-encapsulated carbon nanotubes with remarkable thermoelectric property. Science Reports. 2015;**5**:7951. DOI: 10.1038/srep07951

[51] Sun Y, Zhang F, Sun Y, Di C, Xu W, Zhu D.: n-Type thermoelectric materials based on CuTCNQ nanocrystals and CuTCNQ nanorod arrays. Journal of Materials Chemistry A. 2015;**3**:2677. DOI: 10.1039/c4ta06475e

[52] Hu D, Liu Q, Tisdale J, Lei T, Pei J, Wang H, Urbas A, Hu B.: Seebeck Effects in n-Type and p-Type polymers driven simultaneously by surface polarization and entropy differences based on conductor/polymer/conductor thin-film devices. ACS Nano. 2015;**9**:5208. DOI: 10.1021/acsnano.5b00589

[53] Sanduleac I, Casian A.: Nanostructured TTT(TCNQ)2 organic crystals as promising thermoelectric n-Type materials: 3D modeling . Journal of Electronic Materials. 2016;**45**:1316–1320. DOI: 10.1007/s11664-015-4018-8

[54] Yoo D, Lee JJ, Park C, Choi HH, Kim JH.: N-type organic thermoelectric materials based on polyaniline doped with the aprotic ionic liquid 1-ethyl-3-methylimidazolium ethyl sulfate. RSC Advances. 2016;**6**:37130–37135. DOI: 10.1039/c6ra02334g

[55] Sun Y, Qiu L, Tang L, Geng H, Wang H, Zhang F, Huang D, Xu W, Yue P, Guan Y, Jiao F, Sun Y, Tang D, Di C, Yi Y, Zhu D.: Flexible n-Type high-performance thermoelectric thin films of Poly(nickel-ethylenetetrathiolate) prepared by an electrochemical method. Advanced Materials. 2016;**28**:3351–3358. DOI: 10.1002/adma.201505922

[56] Poleschner H, John W, Hoppe F, Fanghänel E, Roth S.: Synthese und Eigenschaften elektronenleitender Poly-Dithiolenkomplexe mit Ethylentetrathiolat und Tetrathiafulvalentetrathiolat als Bruckenliganden. J. Praktische Chem. 1983;**325**:957–975. DOI: 10.1002/prac.19833250612

[57] Holdcroft GE, Underhill AE.: Preparation and electrical conduction properties of polymeric transition metal complexes of 1,1,2,2-ethenetetrathiolate ligand. Synthetic Metals. 1985;**10**:427. DOI: 10.1016/0379-6779(85)90200-0

[58] Vicente R, Ribas J, Cassoux P, Valade L.: Synthesis, characterization and properties of highly conducting organometallic polymers derived from the ethylene tetrathiolate anion. Synthetic Metals. 1986;**13**:265. DOI: 10.1016/0379-6779(86)90188-8

[59] Vogt T, Faulmann C, Soules R, Lecante P, Mosset A, Castan P, Cassoux P, Galy J.: A LAXS (Large Angle X-ray Scattering) and EXAFS (Extended X-ray Absorption Fine Structure) investigation of conductive amorphous nickel tetrathiolato polymers. Journal of American Chemical Society. 1988;**110**:1833–1840. DOI: 10.1021/ja00214a028

[60] Bellitto C, Bonamico M, Fares V, Imperatori P, Patrizio S.: Tetrathiafulvalenium salts of planar PtII, PdII, and CuII 1,2-dithio-oxalato-S,S' anions. Synthesis, chemistry and molecular structures of bis(tetrathiafulvalenium) bis(1,2-dithio-oxalato-S,S')palladate(II), [ttf]2[Pd(S2C2O2)2], and of bis(tetrathiafulvalenium)tetrathiafulvalene bis(1,2-dithio-oxalato-S,S')platinate(II), [ttf]3[Pt(S2C2O2)2]. Journal of the Chemical Society, Dalton Transactions. 1989;719–727. DOI: 10.1039/DT9890000719

[61] Faulmann C, Cassoux P, Vicente R, Ribas J, Jolly CA, Reynolds JR.: Conductive amorphous metal-tetrathiolato polymers: Synthesis of a new precursor C6O2S8 and its derived polymers and laxs structural studies. Synthetic Metals. 1989;29:557–562. DOI: 10.1016/0379-6779(89)90349-4

[62] Yoshioka N, Nishide H, Inagaki K, Tsuchida E.: Electrical conductive and magnetic properties of conjugated tetrathiolate nickel polymers. Polymer Bulletin. 1990;23:631–636. DOI: 10.1007/BF01033109

[63] Sarangi R, George SD, Rudd DJ, Szilagyi RK, Ribas X, Rovira C, Almeida M, Hodgson KO, Hedman B, Solomon EI.: Sulfur K-Edge X-ray absorption spectroscopy as a probe of Ligand–Metal bond covalency: Metal vs Ligand oxidation in copper and nickel dithiolene complexes. Journal of the American Chemical Society. 2007;129:2316–2326. DOI: 10.1021/ja0665949

[64] Yamamoto T.: Assignment of pre-edge peaks in K-edge x-ray absorption spectra of 3d transition metal compounds: electric dipole or quadrupole? X-Ray Spectrometry. 2008;37:572–584. DOI: 10.1002/xrs.1103

[65] Massonnet N, Carella A, Jaudouin O, Rannou P, Laval G, Celle C, Simonato J.: Improvement of the Seebeck coefficient of PEDOT:PSS by chemical reduction combined with a novel method for its transfer using free-standing thin films. Journal of Materials Chemistry C. 2014;2:1278–1283. DOI: 10.1039/C3TC31674B

[66] Toshima N, Oshima K, Anno H, Nishinaka T, Ichikawa S, Iwata A, Shiraishi Y.: Novel hybrid organic thermoelectric materials: three-component hybrid films consisting of a nanoparticle polymer complex, carbon nanotubes, and vinyl polymer. Advanced Materials. 2015;27:2246–2251. DOI: 10.1002/adma.201405463

[67] Asano H, Sakura N, Oshima K, Shiraishi Y, Toshima N.: Development of ethenetetrathiolate hybrid thermoelectric materials consisting of cellulose acetate and semiconductor nanomaterials. Japanese Journal of Applied Physics. 2016;55:2S. DOI: 10.7567/JJAP.55.02BB02

[68] Oshima K, Asano H, Shiraishi Y, Toshima N.: Dispersion of carbon nanotubes by poly(Ni-ethenetetrathiolate) for organic thermoelectric hybrid materials. Japanese Journal of Applied Physics. 2016;55:2S. DOI: 10.7567/JJAP.55.02BB07

[69] Jiao F, Di C, Sun Y, Sheng P, Xu W, Zhu D.: Inkjet-printed flexible organic thin-film thermoelectric devices based on p- and n-type poly(metal 1,1,2,2-ethenetetrathiolate)s/polymer composites through ball-milling. Philosophical Transactions of the Royal Society of London. 2014;372DOI: 10.1098/rsta.2013.0008

[70] Faulmann C, Chahine J, Jacob K, Coppel Y, Valade L, Caro D.: Nickel ethylene tetra-thiolate polymers as nanoparticles: a new synthesis for future applications?. Journal of Nanoparticle Research. 2013;**15**:1. DOI: 10.1007/s11051-013-1586-5

[71] Oshima K, Shiraishi Y, Toshima N.: Novel nanodispersed polymer complex, Poly(nickel 1,1,2,2-ethenetetrathiolate): preparation and hybridization for n-Type of organic ther-moelectric materials. Chemistry Letters. 2015;**44**:1185–1187. DOI: 10.1246/cl.150328

[72] Aranguren P, Roch A, Stepien L, Abt M, von Lukowicz M, Dani I, Astrain D.: Optimized design for flexible polymer thermoelectric generators. Applied Thermal Engineering. 2016;**102**:402–411. DOI: 10.1016/j.applthermaleng.2016.03.037

[73] LeBlanc S, Yee SK, Scullin ML, Dames C, Goodson KE.: Material and manufacturing cost considerations for thermoelectrics. Renewable and Sustainable Energy Reviews. 2014;**32**:313–327. DOI: 10.1016/j.rser.2013.12.030

[74] van Reenen S, Kemerink M.: Correcting for contact geometry in Seebeck coefficient measurements of thin film devices. Organic electronics. 2014;**15**(10):2250–2255. DOI: 10.1016/j.orgel.2014.06.018

Chapter 4

Tetrahedrites: Prospective Novel Thermoelectric Materials

Christophe Candolfi, Yohan Bouyrie, Selma Sassi,
Anne Dauscher and Bertrand Lenoir

Additional information is available at the end of the chapter

http://dx.doi.org/10.5772/65638

Abstract

Since their discovery in 1845, tetrahedrites, a class of minerals composed of relatively earth-abundant and nontoxic elements, have been extensively studied in mineralogy and geology. Despite a large body of publications on this subject, their transport properties had not been explored in detail. The discovery of their interesting high-temperature thermoelectric properties and peculiar thermal transport has led to numerous experimental and theoretical studies over the last 4 years with the aim of better understanding the relationships between the crystal, electronic, and thermal properties. Tetrahedrites provide a remarkable example of anharmonic system giving rise to a temperature dependence of the lattice thermal conductivity that mirrors that of amorphous compounds. Here, we review the progress of research on the transport properties of tetrahedrites, highlighting the main experimental and theoretical results that have been obtained so far and the important issues and questions that remain to be investigated.

Keywords: tetrahedrites, thermoelectric, thermal conductivity, exsolution, composite

1. Introduction

Thermoelectric effects provide a reliable way for converting waste heat into useful electricity and vice versa [1, 2]. This solid-state conversion process is realized without hazardous gas emissions and moving parts, ranking this technology among clean and sustainable energy sources. Thermoelectric generators have been successfully used to reliably power deep-space probes and rovers over several decades, and have been used as solid-state coolers for electronic devices [1, 2]. Yet, a widespread use of this versatile technology is hampered by the rather low conversion efficiency achieved. The thermoelectric efficiency, with which a

thermoelectric device converts heat into electricity and vice versa is directly dependent of the dimensionless figure of merit ZT of the active thermoelectric legs [1, 2]. This parameter is defined at an absolute temperature T as $ZT = \alpha^2 T / \rho(\kappa_L + \kappa_e)$, where α is the thermopower (or Seebeck coefficient), ρ is the electrical resistivity, and κ_L and κ_e are the lattice and electronic thermal conductivities, respectively. Thus, a good thermoelectric material should possess a combination of high thermopower to produce a sizeable thermoelectric effect, low electrical resistivity to avoid Joule effect, and low thermal conductivity to maintain a large thermal gradient across the device [3, 4].

While quite simple at first sight, this expression however underlies a formidable material challenge since the ideal thermoelectric material should be concomitantly a thermal insulator and an electrical conductor. The question is therefore how far these two seemingly conflicting aspects can be reconciled within the same material. The quest for this long-sought ideal material has led to thousands of experimental and theoretical investigations on a large number of material's families and on the possibilities to optimize their thermoelectric performances through various strategies such as the optimization of the carrier concentration by doping or the reduction of the lattice thermal conductivity via substitutions or nanostructuring [1–4]. These studies have increased the number of known crystalline compounds that show the remarkable ability to conduct heat akin to glassy systems [5–14]. In addition to being ideal systems for improving our understanding of the physical mechanisms leading to this behavior, these materials provide interesting playgrounds to achieve high thermoelectric performances. When the lattice thermal conductivity is intrinsically lowered to a value close to the theoretical minimum value, the electrical resistivity and thermopower remain the only key properties to be optimized to reach high ZT values.

This approach has led to the identification of several new families of thermoelectric materials, some of which exhibiting thermoelectric performances that surpass those of the state-of-the-art thermoelectric materials such as PbTe or $Si_{1-x}Ge_x$ alloys at moderate and high temperatures, respectively [2, 5–14]. Among these new families, tetrahedrites have recently draw attention due to the relatively nontoxic, earth-abundant elements that enter their chemical composition [15]. These compounds are a class of copper antimony sulfosalt minerals geologically formed in hydrothermal veins at low-to-moderate temperatures making them abundant in the Earth's crust. Tetrahedrites are minor ores of copper that were first discovered in 1845 in Germany. While they were the subject of a large number of experimental and theoretical studies in geology and mineralogy, it is not until recently, however, that their transport properties have been investigated in detail [15, 16]. Both natural and synthetic tetrahedrites, i.e., synthesized in laboratory environment, have been recently studied indicating that these compounds are interesting candidates for thermoelectric applications in power generation.

This chapter provides an updated review on the experimental and theoretical results obtained and an overview of the status of the research activities on the thermoelectric properties of these materials. Our goal is to highlight the important structural and chemical aspects that influence their transport properties and thus play a role in their thermoelectric performances. This review will also cover the first experimental attempts at scaling-up the synthesis process via chemical or metallurgical approaches.

2. Crystal structure and chemical composition

The general chemical formula of tetrahedrites can be written as $A_{12}X_4Y_{13}$, where A is mainly Cu that can be partially substituted by transition metals (Ag, Zn, Fe, Ni, Co, Mn, and Hg), X is a pnictide (Sb or As with possible partial substitution by Te or Bi), and Y is sulfur (S can be substituted in small quantities by Se). All these elements can be found in the chemical composition of natural tetrahedrites and numerous experimental and theoretical studies in geology and mineralogy focused on the link between their composition and the geological place where they were discovered.

With no exceptions, all tetrahedrites, should they be natural or synthetic, crystallize within a cubic crystal structure described in the $I\bar{4}3m$ space group with 58 atoms per unit cell (**Figure 1**) [17, 18]. The different A, X, and Y elements are distributed over five distinct crystallographic sites. Taking the ternary compound $Cu_{12}Sb_4S_{13}$ as an example, the copper atoms possess two different chemical environments. The Cu1 atoms show a tetrahedral coordination with three S and one surrounding Sb atoms. The Cu2 atoms lie within a triangular environment formed by three sulfur atoms in a nearly coplanar coordination. Twelve of the 13 S atoms exhibit a tetrahedral environment while the remaining S atom is surrounded by six Cu atoms forming an octahedron. The main peculiarity of this crystal structure is related to the Cu2 atoms that show large and anisotropic atomic thermal displacement parameters (ADPs), i.e., a strong ability to vibrate about their equilibrium position (see **Figure 1**) [19]. In addition, the tetrahedral environment of the Sb atoms lacks one sulfur atom to be complete. The presence of only three S atoms thus gives rise to 5s lone pair electrons on the Sb atoms. These "free" electrons can be oriented along the missing vertex of the tetrahedron according to the valence shell electron pair repulsion theory. Several studies have pointed out the decisive influence of lone pair electrons on the thermal conductivity, the delocalization of the lone pair away from the Sb nucleus yielding anharmonic forces in the lattice [20–22]. As we shall see below, a similar situation is probably at play in tetrahedrites, explaining their extremely low lattice thermal conductivity values.

While the crystal structure of tetrahedrites is simple to describe, their chemical composition displays some subtleties that make these compounds particularly interesting. Specifically, synthetic tetrahedrites often show deviations from stoichiometry, a characteristic usually absent in natural specimens that possess the exact 12–4–13 composition to within the detection limits of the instruments used [23–26]. The most prominent example of such behavior is provided by the ternary compound $Cu_{12}Sb_4S_{13}$, which has been the subject of thorough experimental studies in the 1970s [23–26]. These investigations have shown that the chemical composition of this compound is best described by the general chemical formula $Cu_{12+x}Sb_{4+y}S_{13}$ with $0.08 \leq x \leq 1.72$ and $0.06 \leq y \leq 0.30$ [23, 24]. This formula indicates that an excess of Cu is systematically observed together with a possible excess on the Sb sites, both of which depend on the synthesis conditions. These deviations are clearly correlated to the lattice parameter: an increase in either the Cu or Sb content always expands the unit cell from 10.327 Å for $(x, y) = (0.08, 0.06)$ up to 10.448 Å for $(x, y) = (1.72, 0.09)$. In addition to these deviations from the ideal stoichiometry, the ternary compositions undergo an exsolution process, i.e., a separation of the main tetrahedrite phase into two tetrahedrite phases of close compositions below the so-called exsolution temperature [23, 24]. Such phase separation has been widely observed in minerals and often results in lamellar microstructures. In $Cu_{12+x}Sb_{4+y}S_{13}$, this process occurs below about 120°C, the exact value slightly varying with the chemical composition [23, 24]. Below this temperature, Cu-rich and Cu-poor phases coexist, the lattice parameters of both phases differing significantly from each other. This mechanism is reversible and disappears upon heating above 120°C to show up again upon cooling back below this temperature.

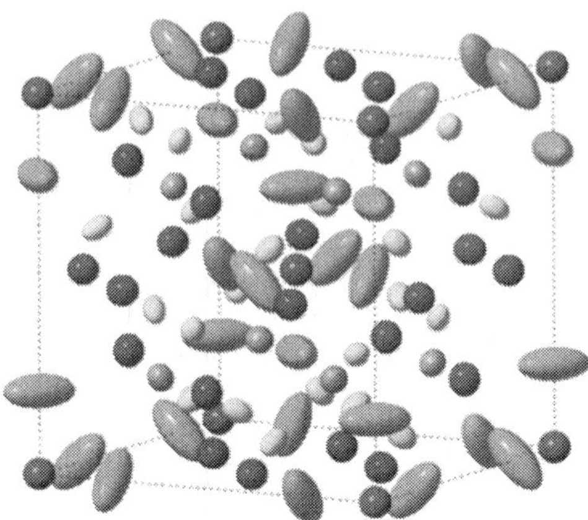

Figure 1. Perspective view of the crystal structure of $Cu_{12}Sb_4S_{13}$ in the ellipsoidal representation (drawn at the 95% probability level). The Cu1 and Cu2 atoms are in blue and green, respectively. Sb atoms are shown in brown while S1 and S2 atoms are in yellow and red, respectively.

3. Electronic properties

While the chemical and structural trends in natural and synthetic tetrahedrites are rather well understood, their transport properties have been investigated in detail only very recently [15, 16]. In order to better understand why these materials may be interesting for thermoelectric applications, it is helpful to assume that the atomic bonds are purely ionic. Within this assumption, the general chemical formula may be rewritten as $(Cu^+)_{10}(Cu^{2+})_2(Sb^{3+})_4(S^{2-})_{13}$ which corresponds to 204 valence electrons per chemical formula [27]. From an electronic point of view, these valence electrons do not entirely fill the valence bands leaving two holes per formula unit. The ternary compound is thus predicted to behave as a p-type metal. This metallic state can nevertheless be driven toward a semiconducting state when two electrons per chemical formula are added (resulting in a total of 208 valence electrons per chemical formula) [27]. This progressive filling thus leads to highly doped semiconducting states more favorable to achieve high thermoelectric performances. Further, based on this simple electronic structure model, it was argued that natural tetrahedrites crystallize preferentially with a composition that corresponds to 208 valence electrons in agreement with the results based on a large survey of the literature data [27]. In particular, this model predicts that metallic compositions are energetically less favorable which might explain why $Cu_{12}Sb_4S_{13}$ tends to exsolve into Cu-poor and Cu-rich phases.

The first experimental results on the transport properties of the ternary tetrahedrite $Cu_{12}Sb_4S_{13}$ have been reported by Suekuni et al. [15] who measured the temperature dependence of the magnetic susceptibility, electrical resistivity, thermopower, and thermal conductivity between 5 and 350 K (**Figure 2**). The results have shown that this compound exhibits several interesting features. A first one is a metal-insulator transition that sets in near 85 K and leaves clear signatures on the transport and magnetic properties. Below this temperature, the electrical resistivity significantly increases by approximately two orders of magnitude upon cooling from 85 to 5 K. A concomitant strong increase in the thermopower values from 25 μV K^{-1} at 85 K to 100 μV K^{-1} at 60 K further corroborates a semiconducting-like state of the low-temperature phase. The thermal conductivity drops below the transition temperature due to both a reduced electronic contribution as a result of the increase in ρ and to the influence of this transition on the lattice thermal conductivity. The magnetic susceptibility, indicative of paramagnetic behavior across the entire temperature range, suddenly drops below about 100 K.

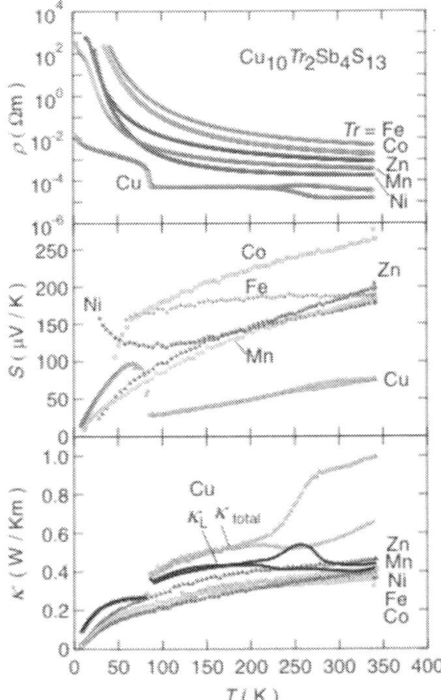

Figure 2. Temperature dependence of the electrical resistivity, thermopower, and total thermal conductivity of the tetrahedrite $Cu_{10}Tr_2Sb_4S_{13}$ (Tr = Cu, Fe, Co, Ni, Zn, and Mn). Copyright 2012 by the Japan Society of Applied Physics.

Using low-temperature powder X-ray diffraction measurements, May et al. [28] demonstrated that this transition is accompanied by a cubic-to-tetragonal lattice distortion characterized by an in-plane ordering that doubles the unit cell volume (**Figure 3**). The low-temperature crystal structure has been described in the $P\bar{4}$ space group using a supercell model $\sqrt{2}a \times \sqrt{2}a \times c$. Owing to the high number of atomic position parameters to be refined (83), the crystal structure could not be entirely solved and remains to be determined in future studies.

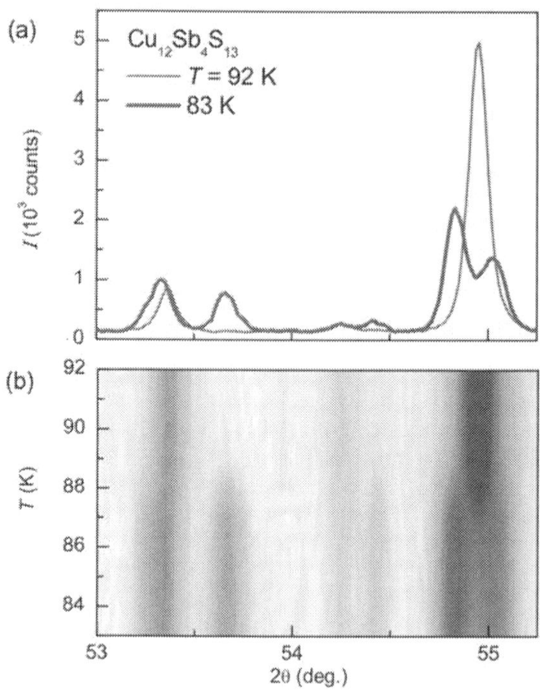

Figure 3. (Panel a) Powder X-ray diffraction pattern collected upon cooling across the 85 K phase transition in Cu$_{12}$Sb$_4$S$_{13}$. The appearance of additional Bragg reflections (Panel b) is indicative of the cubic-to-tetragonal structural transition that accompanies the metal-insulator transition. Reproduced **Figure 1** with permission from May et al. [28]. Copyright 2016 by the American Physical Society. DOI: 10.1103/PhysRevB.93.064104.

Kitagawa et al. [29] further investigated the metal-to-insulator transition through measurements of the electrical resistivity and magnetic susceptibility at high pressures (up to 4.06 GPa) and of the ^{63}Cu-NMR spectra. The results evidence a transition from a paramagnetic bad metal to a nonmagnetic insulating state below 85 K at ambient pressure. The nonmagnetic ground state evolves toward a metallic state under pressure. Tanaka et al. [30] further investigated the pressure dependence of this transition as well as its evolution upon substituting As for Sb (Cu$_{12}$Sb$_{4-x}$As$_x$S$_{13}$). In agreement with the study of Kitagawa et al., the transition is suppressed under pressure. The substitution of As for Sb that decreases the unit cell volume acts similarly and supresses the transition for relatively low substitution levels ($x \leq 0.5$).

In their initial study on Cu$_{12}$Sb$_4$S$_{13}$, Suekuni et al. [15] have also reported the transport properties of several tetrahedrites Cu$_{10}Tr_2$Sb$_4$S$_{13}$ with Cu substituted by various transition metals

(Tr = Ni, Zn, Co, Fe, or Mn) (see **Figure 2**). In agreement with the aforementioned simple ionic model, all these quaternary tetrahedrites exhibit semiconducting-like properties characterized by an activated-like temperature dependence of the electrical resistivity and high thermopower values. Further, all samples display extremely low thermal conductivity values of the order of 0.4 W m^{-1} K^{-1} at 300 K regardless of the nature of the transition metal. However, although this study has demonstrated that a semiconducting state can be achieved thanks to substitution on the Cu site, the too high electrical resistivity values measured preclude achieving high ZT values. In order to optimize the thermoelectric properties, it thus appears necessary to adjust the concentration of the transition metal to optimize the hole concentration and hence, the power factor. This strategy was at the core of the study reported by Lu et al. [16] who reported high-temperature transport properties measurements on the tetrahedrites $Cu_{12-x}Fe_xSb_4S_{13}$ and $Cu_{12-x}Zn_xSb_4S_{13}$. This investigation has demonstrated for the first time that high thermoelectric performances could be achieved around 700 K with maximum ZT values of 0.8 for the composition $Cu_{11.5}Fe_{0.5}Sb_4S_{13}$.

These encouraging results led other groups to investigate in detail the influence of several transition metals on the crystal structure and the high-temperature thermoelectric properties [31–41]. All these studies have confirmed the main traits of these compounds, i.e., a favorable combination of intrinsically extremely low thermal conductivity values and semiconducting-like electrical properties that can be tuned by varying the concentration of the substituting element. Peak ZT values ranging between 0.7 and 0.9 around 700 K were achieved in Ni-, Co-, or Mn-substituted tetrahedrites. The highest ZT value of 1.1 at 575 K has been reported by Heo et al. [33] in Mn-substituted tetrahedrites. This value, not reproduced independently so far to the best of our knowledge, mainly originates from thermal conductivity values twice lower than usually measured in these materials. Further investigations seem necessary before drawing a definitive conclusion on the validity of this high value.

Lu et al. [39] explored the possibility to substitute Te for Sb and showed that Te also provides additional electrons that enable optimizing the power factor. As a result, a maximum ZT value of 0.92 at 723 K has been achieved for the composition $Cu_{12}Sb_3TeS_{13}$. Bouyrie et al. [38] further extended these investigations and synthesized Te-containing tetrahedrites by considering two different synthetic routes: using precursor compounds (CuS, Sb_2S_3, and Te) and from direct reaction of pure elemental powders, both syntheses being performed in evacuated and sealed silica tubes. Surprisingly, the results have evidenced that both routes are not strictly equivalent indicating that the final chemical compositions could be sensitive to the synthesis conditions used. The differences between the two series of samples was revealed by significantly higher lattice parameters for Te-concentrations below x = 1.5 in the series of samples prepared from precursors (**Figure 4**).

In the series of samples synthesized from direct reaction of the elements, the lattice parameter monotonically increases in a linear manner with increasing the Te-content. While this difference does not seem to affect the thermoelectric performances at high temperatures, the measurements of the low-temperature transport properties showed that these quaternary tetrahedrites undergo an exsolution process at 250 K [42]. This phenomenon has a drastic influence on the transport properties and more particularly on the thermal transport. Below the exsolution temperature, the lattice thermal conductivity drops significantly by 40% reaching values as low as 0.25 W m^{-1} K^{-1} around 200 K (**Figure 5**). This behavior, which seems to be tied to the large lattice parameters of these samples, is not present in the series of samples prepared by direct reaction of the elements. The exact origin of these differences is not yet settled and requires further investigations. In addition, low-temperature transmission electron microscopy experiments on the Te-containing tetrahedrites would be helpful in determining the microstructure and perhaps the chemical composition of the two exsolved phases.

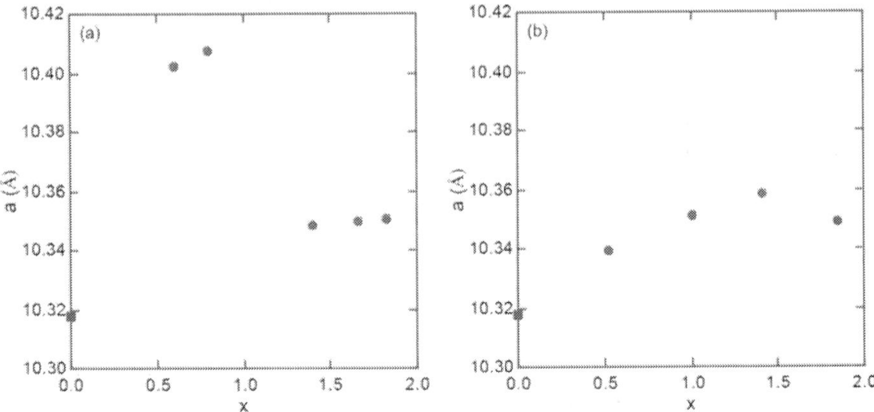

Figure 4. Lattice parameter a as a function of the actual Te content x for the series of samples prepared from precursors (a) and directly from reaction of elemental powders (b). As a reference, the lattice parameter of the ternary compound $Cu_{12}Sb_4S_{13}$ (filled blue square) has been added. Reproduced from Ref. [38] with permission from The Royal Society of Chemistry.

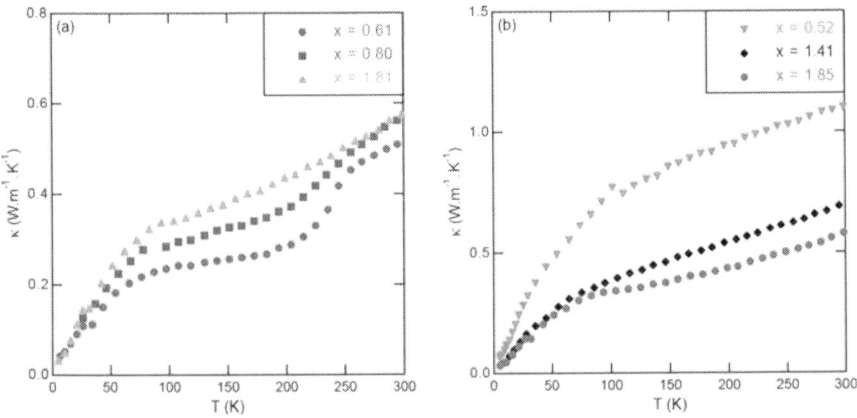

Figure 5. Temperature dependence of the total thermal conductivity for the samples prepared from precursors (a) and from reaction of elemental powders (b). For the first series, the exsolution process results in a significant drop in the thermal conductivity values near 250 K (panel a). Reprinted (adapted) with permission from Bouyrie et al. [42]. Copyright 2015 by the American Chemical Society.

The possibility to substitute on the S site has only been recently considered by Lu et al. [43] who reported a detailed study on the $Cu_{12}Sb_4S_{13-x}Se_x$ tetrahedrites. Although a maximum ZT value of 0.9 was achieved for $x = 1$, these authors have shown that the presence of selenium tends to result in phase separation yielding samples with rather poor chemical homogeneity.

While all these studies focused on the influence of a single isovalent or aliovalent substitution on the thermoelectric properties, only few works have been devoted so far to double substitutions. Lu et al. [40] have extended their investigations to double-substituted tetrahedrites with Ni and Zn substituting for Cu. These authors have shown that this combination of elements results in higher thermoelectric performances with a peak ZT value of 1 at 700 K. Of note, this increase was mainly due to increased thermopower values while maintaining the electrical resistivity to relatively low values. These results suggest that judicious combinations of elements substituting for Cu can lead to improved thermoelectric properties. Following these ideas, Bouyrie et al. [41, 44] have investigated double substitutions on both the Cu and Sb sites with Co and Te, respectively. The presence of Co and Te did not result in enhanced thermoelectric performances with respect to the single-substituted compounds with a maximum ZT value of 0.8 at 673 K achieved in $Cu_{11.47}Co_{0.82}Sb_{3.78}Te_{0.41}S_{13}$.

4. Thermal properties

In addition to being one of the key ingredients that leads to high ZT values, the extremely low lattice thermal conductivity of tetrahedrites is a remarkable property on its own. Both the values measured and the temperature dependence of the lattice thermal conductivity are reminiscent to those observed in amorphous systems (**Figure 6**) [45].

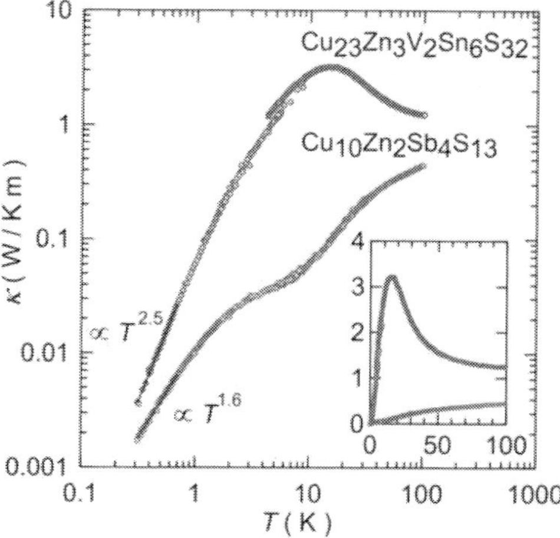

Figure 6. Temperature dependence of the total thermal conductivity of the tetrahedrite $Cu_{10}Zn_2Sb_4S_{13}$ and of the colusite $Cu_{23}Zn_3V_2Sn_6S_{32}$. The temperature dependence observed in the tetrahedrite below 1 K is similar to that observed in glassy SiO_2. Copyright 2015 by the Physical Society of Japan (Suekuni et al. [45]).

The large and anisotropic thermal displacement parameters of the Cu2 atoms had been thought to play a major role in disrupting efficiently the heat-carrying acoustic waves. A detailed study of the lattice dynamics of tetrahedrites has been undertaken recently using a combination of inelastic neutron scattering on poly- and single-crystalline tetrahedrites [46]. The conventional temperature dependence of the lattice thermal conductivity in the Cu-deficient tetrahedrite $Cu_{10}Te_4S_{13}$ offered an interesting experimental platform to unveil the microscopic mechanisms responsible for the low, glass-like thermal conductivity of tetrahedrites (**Figure 7**).

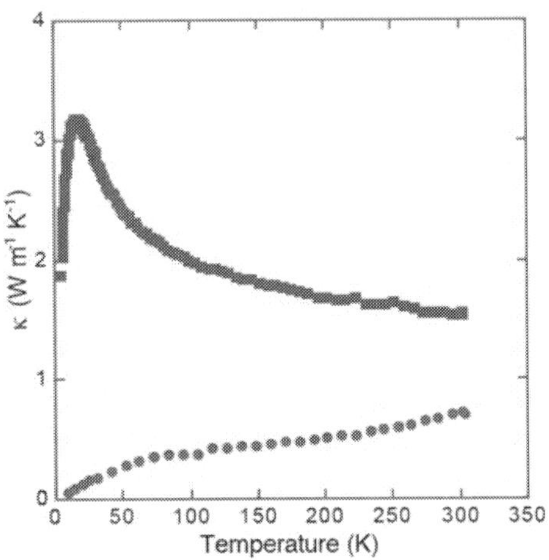

Figure 7. Temperature dependence of the total thermal conductivity of the tetrahedrites $Cu_{12}Sb_2Te_2S_{13}$ (red filled circles) and $Cu_{10}Te_4S_{13}$ (blue filled squares). Adapted from Ref. [46] with permission from the PCCP Owner Societies.

Bouyrie et al. [46] carried out a comparison study between this compound and the tetrahedrite $Cu_{12}Sb_2Te_2S_{13}$ that behaves as a glassy system. Despite adopting the same crystal structure, the contrast between the thermal transports in these compounds suggests that distinct microscopic mechanisms are at play.

A first important difference between these two compounds was found in the temperature dependence of the ADPs of the Cu2 atoms investigated by laboratory X-ray diffraction on single crystals. The ADPs inferred in $Cu_{10}Te_4S_{13}$ were nearly three times smaller than those observed in $Cu_{12}Sb_2Te_2S_{13}$ providing a first experimental hint of the direct link between the thermal vibrations of the Cu2 atoms and the lattice thermal conductivity. Further decisive evidences were delivered by inelastic neutron scattering and Raman spectroscopy performed on polycrystalline samples of $Cu_{10}Te_4S_{13}$ and $Cu_{12}Sb_2Te_2S_{13}$ between 2 and 500 K. The results showed the presence of an excess of vibrational density of states at low energies in $Cu_{12}Sb_2Te_2S_{13}$, which is clearly absent in the isostructural compound $Cu_{10}Te_4S_{13}$ (**Figure 8**). This finding is consistent with recent INS measurements performed by May et al. [28] on the ternary compound $Cu_{12}Sb_4S_{13}$. The temperature dependence of this low-energy excess of vibrational states further indicates that this excess can be unambiguously attributed to the thermal vibrations of the

Cu2 atoms. Upon cooling, this excess experiences a strong renormalization of its characteristic energy, which shifts significantly toward lower energies. This dependence, at odds with a conventional quasi-harmonic behavior, indicates a strongly anharmonic character of this excess.

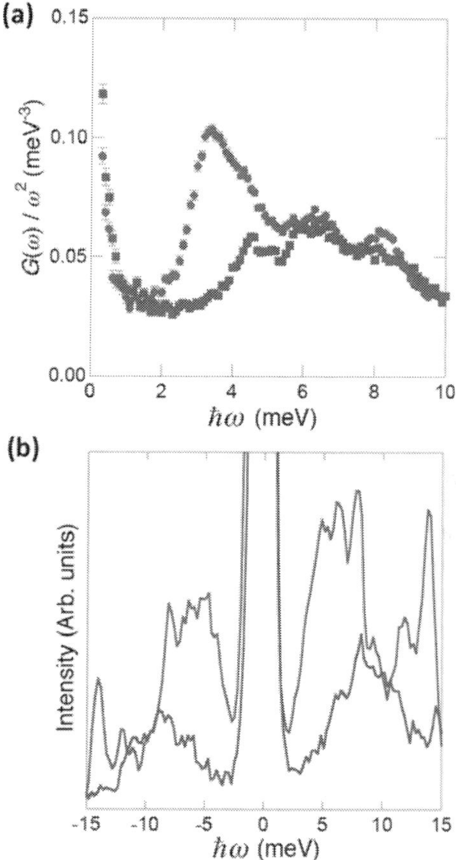

Figure 8. (a) Generalized phonon density of states measured at room temperature on the tetrahedrites $Cu_{12}Sb_2Te_2S_{13}$ (red filled circles) and $Cu_{10}Te_4S_{13}$ (blue filled squares). A clear excess at low energies is present in $Cu_{12}Sb_2Te_2S_{13}$. (b) Raman spectra of $Cu_{12}Sb_2Te_2S_{13}$ (red) and $Cu_{10}Te_4S_{13}$ (blue) measured at the Stokes and anti-Stokes line. Reproduced from Ref. [46] with permission from the PCCP Owner Societies.

INS measurements performed on natural, single-crystalline specimen further shed light on the role played by this excess on the lattice thermal conductivity [46]. These experiments enable to directly probe the dispersion of transverse acoustic phonons and the optical branch associated with the thermal vibrations of the Cu2 atoms. The low-energy optical branch strongly limits the phase space over which the acoustic phonon branch disperses (**Figure 9**). This strong limitation is accompanied by a drastic suppression of their intensity. The presence of this

low-energy optical mode has two main consequences: (i) the suppression of the acoustic phonon states that are the main heat carriers and (ii) the presence of a novel channel of Umklapp
processes that remain active even at low temperatures. The first of these two consequences
naturally explains the very low lattice thermal conductivity values measured in tetrahedrites,
while the second consequence explains the absence of an Umklapp peak at low temperatures
in $Cu_{12}Sb_2Te_2S_{13}$ suppressed by active Umklapp processes. The lack of this excess in $Cu_{10}Te_4S_{13}$
is thus at the origin of its higher lattice thermal conductivity values and the presence of the
Umklapp peak centered at 25 K.

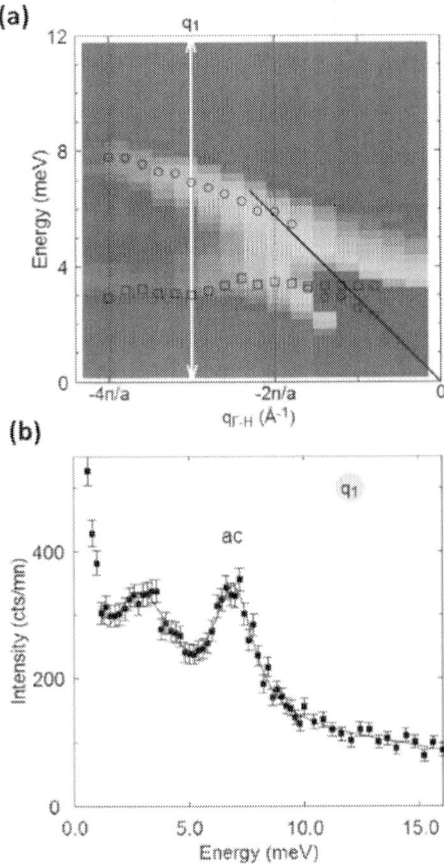

Figure 9. (a) Mapping of the transverse acoustic phonon propagating along the (011) direction and polarized along
the (100) direction measured in a natural specimen. (b) Raw data of constant energy scan performed at the constant
wave vector q_1 = 0.91 Å⁻¹. The black solid line corresponds to the fit of the measured scattering profile using damped
harmonic oscillator for acoustic phonons (ac) and Gaussian functions for the low-lying optical excitations. Reproduced
from Ref. [46] with permission from the PCCP Owner Societies.

The origin of the strong anharmonicity has been attributed to the active $5s$ lone pair electrons of the Sb atoms located at either side of the Cu2 atoms. Electronic band structure calculations have suggested that the lone pair electrons are inactive in $Cu_{10}Te_4S_{13}$. A study of Wei et al. [47] based on a combination of theoretical calculations with synchrotron X-ray diffraction has further corroborated that the strong anharmonic potential felt by the Cu2 is linked to the lone pair electrons of the Sb atoms.

5. Scaling up tetrahedrite synthesis

Owing to their interesting thermoelectric properties, tetrahedrites hold promise to be used as p-type legs in thermoelectric generators. Yet, any widespread thermoelectric application would require fast, easily scalable, and cost-efficient synthetic methods to produce these materials in high yield. Several studies have dealt with these issues and focused either on the direct use of natural ore to decrease the cost and the time of preparation or on different synthetic routes faster than conventional powder metallurgy techniques. In this regard, James et al. [48] have recently developed a solvothermal route to produce synthetic tetrahedrites without using long-chain ligands that usually leads to a strong increase in the electrical resistivity due to difficulties in separating them from the targeted compound. Upon optimizing their process, these authors successfully synthesized phase-pure tetrahedrites within only one day at moderate temperatures (around 150°C). Of note, this fast synthesis process was not at the expense of the thermoelectric performances since the tetrahedrites prepared by solvothermal and solid-state synthesis methods showed similar ZT values.

Barbier et al. [49] have investigated another processing technique based on a combination of high-energy ball milling of stoichiometric mixtures of elemental powders and spark plasma sintering. This study, performed on the composition $Cu_{10.4}Ni_{1.6}Sb_4S_{13}$, has demonstrated that pure, highly dense samples of this tetrahedrite can be synthesized over a reduced period of time (estimated to eight times shorter by the authors) with respect to conventional solid-state synthesis. Similarly to the study of James et al. [48], the thermoelectric performances were not adversely affected and were comparable to those prepared by conventional methods in prior studies with a peak ZT value of 0.8 at 700 K.

Besides these two direct processes, Gonçalves et al. [50] used a different approach to synthesize within less than one day a tetrahedrite phase. This approach relies on the preparation of a glass of composition $Cu_{12}Sb_{3.6}Bi_{0.4}S_{10}Se_3$ by melt-spinning, i.e., fast quenching of the melt on a fast-rotating copper wheel, and subjected to controlled heat treatments to crystallize the targeted tetrahedrite phase. Within this strategy, both Se and Bi were introduced as vitrifying and nucleation agents to produce a homogeneous glassy sample. The authors further showed that annealing treatments at temperatures close to the crystallization peaks (around 200°C) leads to the crystallization of a tetrahedrite phase. Analysis of the chemical composition revealed the presence of Se partially substituting S, while Bi could not be detected within the experimental uncertainty of the instruments used. Thermopower and electrical resistivity measurements resulted in room-temperature power factors that are close to those measured in the ternary $Cu_{12}Sb_4S_{13}$ tetrahedrite (~400 $\mu W\ m^{-1}\ K^{-2}$). Further investigations on the thermal conductivity of these samples will be essential in determining whether high thermoelectric performances can be equally achieved by this technique.

Finally, a radically different approach has been used by Lu et al. [31] who synthesized "composite" tetrahedrites from a mixture of synthetic and natural samples. Achieving high ZT values in such composite system would significantly reduce the time and cost required to synthesize synthetic specimens. While tetrahedrite minerals exhibit too high electrical resistivity to be viable thermoelectric materials, mixing ore with a fraction of synthetic tetrahedrite exhibiting metallic-like properties were shown to result in maximum ZT values of 1.0 at 700 K. The authors used two ores of composition $Cu_{10.5}Fe_{1.5}As_{3.6}Sb_{0.4}S_{13}$ and $Cu_{9.7}Zn_{1.9}Fe_{0.4}As_4S_{13}$, i.e., two As-rich tetrahedrites which are then named tennantites. These ores were mixed with the ternary tetrahedrite $Cu_{12}Sb_4S_{13}$ that behaves as a metal by ball milling. Remarkably, the powder X-ray diffraction pattern collected after ball-milling showed only one single tetrahedrite phase. This process thus provides an interesting and time-efficient way of producing single-phased tetrahedrite displaying high thermoelectric performances. Further investigations aiming at determining the influence on the chemical composition of the mineral on the thermoelectric properties will be interesting to undertake. In this regard, a preliminary study carried out on minerals from various geographical origins has shown that all of them show semiconducting-like properties despite differences in their chemical composition [51].

6. Conclusion

Progress in synthesizing tetrahedrites in laboratory environment and in understanding their transport properties have significantly advanced over the last 4 years, thanks to both experimental and theoretical efforts. Tetrahedrites possess transport properties not only interesting for thermoelectric applications but also for fundamental reasons. Subtle differences in their chemical compositions have a sizeable influence on their transport properties thereby adding another degree of freedom to study the interplay between their crystallographic, chemical, and transport properties. This is one of the main reasons why these materials have attracted attention in thermoelectricity yielding several p-type materials with ZT values ranging between 0.7 and 1.0 at 700 K. These high values originate from a favorable combination of semiconducting-like electronic properties and extremely low lattice thermal conductivity. The electronic properties can be tuned by substituting on the three possible sites, all of them resulting in either a control of the hole concentration or a favorable modification of the electronic band structure. Spectroscopic tools used to investigate the lattice dynamics of these materials have unveiled the presence of strong anharmonicity whose exact origin seems to be tied to the lone pair electrons revolving around the Sb atoms.

On the application side, several studies have successfully speed up the synthetic procedures used to obtain phase-pure tetrahedrites. Combined with the possibility to mix synthetic tetrahedrites with natural ores, these techniques may lead to the production in high yield of low-cost efficient tetrahedrites for thermoelectric applications. Despite proved to be feasible, these "composite" tetrahedrites have so far received little attention and future research will lead to an improved knowledge of their transport properties and of the influence on the chemical composition of the ore on the thermoelectric properties of the composite.

Author details

Christophe Candolfi, Yohan Bouyrie, Selma Sassi, Anne Dauscher and Bertrand Lenoir*

*Address all correspondence to: bertrand.lenoir@univ-lorraine.fr

Institut Jean Lamour, University of Lorraine, Nancy, France

References

[1] Goldsmid H J. Thermoelectric Refrigeration. Temple Press Books Ltd: London; 1964. DOI:10.1007/978-1-4899-5723-8.

[2] Rowe D M, editor. Thermoelectrics and its Energy Harvesting, Boca Raton, Florida (USA), CRC Press; 2012.

[3] Snyder G J, Toberer E S. Complex thermoelectric materials. Nat. Mater. 2008; 7:105–114. DOI:10.1038/nmat2090.

[4] Sootsman J R, Chung D Y, Kanatzidis M G. New and old concepts in thermoelectric materials. Angew. Chem. Int. Ed. 2009;48:8616–8639. DOI:10.1002/anie.200900598.

[5] Brown S R, Kauzlarich S M, Gascoin F, Snyder G J. $Yb_{14}MnSb_{11}$: New high efficiency thermoelectric material for power generation. Chem. Mater. 2006;18:1873-1877. DOI:10.1021/cm060261t.

[6] Grebenkemper J H, Hu Y, Barrett D, Gogna P, Huang C K, Bux S K, Kauzlarich S M. High temperature thermoelectric properties of $Yb_{14}MnSb_{11}$ prepared from reaction of MnSb with the elements. Chem. Mater. 2015;27:5791-5798. DOI:10.1021/acs.chemmater.5b02446.

[7] Bux S K, Zevalkink A, Janka O, Uhl D, Kauzlarich S M, Snyder G J, Fleurial J P. Glass-like lattice thermal conductivity and high thermoelectric efficiency in $Yb_9Mn_{4.2}Sb_9$. J. Mater. Chem. A. 2014;2:215-220. DOI:10.1039/c3ta14021k.

[8] Toberer E S, Zevalkink A, Crisosto N, Snyder G J. The Zintl compound $Ca_5Al_2Sb_6$ for low-cost thermoelectric power generation. Adv. Funct. Mater. 2010;20:4375-4380. DOI:10.1002/adfm.201000970.

[9] Zhou T, Lenoir B, Colin M, Dauscher A, Al Rahal Al Orabi R, Gougeon P, Potel M, Guilmeau E. Promising thermoelectric properties in $Ag_xMo_9Se_{11}$ compounds ($3.4 \leq x \leq 3.9$). Appl. Phys. Lett. 2011;98:162106. DOI:10.1063/1.3579261.

[10] Gougeon P, Gall P, Al Rahal Al Orabi R, Fontaine B, Gautier R, Potel M, Zhou T, Lenoir B, Colin M, Candolfi C, Dauscher A. Synthesis, crystal and electronic structures, and thermoelectric properties of the novel cluster compound $Ag_3In_2Mo_{15}Se_{19}$. Chem. Mater. 2012;24:2899-2908. DOI:10.1021/cm3009557.

[11] Al Rahal Al Orabi R, Gougeon P, Gall P, Fontaine B, Gautier R, Colin M, Candolfi C, Dauscher A, Hejtmanek J, Malaman B, Lenoir B. X-ray characterization, electronic band structure, and thermoelectric properties of the cluster compound $Ag_2Tl_2Mo_9Se_{11}$. Inorg. Chem. 2014;**53**:11699-11709. DOI:10.1021/ic501939k.

[12] Kurosaki K, Yamanaka S. Low-thermal-conductivity group 13 chalcogenides as high-efficiency thermoelectric materials. Phys. Status Solidi A. 2013;**210**:82-88. DOI:10.1002/pssa.201228680.

[13] Kurosaki K, Kosuga A, Muta H, Uno M, Yamanaka S. Ag_9TlTe_5: A high-performance thermoelectric bulk material with extremely low thermal conductivity. Appl. Phys. Lett. 2005;**87**:061919. DOI:10.1063/1.2009828.

[14] Guo Q, Assoud A, Kleinke H. Improved bulk materials with thermoelectric figure-of-merit greater than 1: $Tl_{10-x}Sn_xTe_6$ and $Tl_{10-x}Pb_xTe_6$. Adv. Energy Mater. 2014;**4**:1400348. DOI:10.1002/aenm.201400348.

[15] Suekuni K, Tsuruta K, Ariga T, Koyano M. Thermoelectric properties of mineral tetrahedrites $Cu_{10}Tr_2Sb_4S_{13}$ with low thermal conductivity. Appl. Phys. Express. 2012;**5**:051201. DOI:10.1143/APEX.5.051201.

[16] Lu X, Morelli D T, Xia Y, Zhou F, Ozolins V, Chi H, Zhou X, Uher C. High performance thermoelectricity in earth-abundant compounds based on natural mineral tetrahedrites. Adv. Energy Mater. 2013;**3**:342–348. DOI:10.1002/aenm.201200650.

[17] Wuensch B J. The crystal structure of tetrahedrite, $Cu_{12}Sb_4S_{13}$. Z. Kristallogr. 1964;**119**:437–453. DOI:10.1524/zkri.1964.119.5-6.437.

[18] Pauling L, Neuman E W. The crystal structure of binnite, $(Cu, Fe)_{12}As_4S_{13}$, and the chemical composition and structure of minerals of the tetrahedrite group. Z. Kristallogr. 1934;**88**:54–62. DOI:10.1524/zkri.1934.88.1.54.

[19] Pfitzner A, Evain M, Petricek V. $Cu_{12}Sb_4S_{13}$: A temperature-dependent structure investigation. Acta Crystallogr. 1997;**B53**:337–345. DOI:10.1107/S0108768196014024.

[20] Skoug E J, Morelli D T. Role of lone-pair electrons in producing minimum thermal conductivity in nitrogen-group chalcogenide compounds. Phys. Rev. Lett. 2011;**107**:235901. DOI:10.1103/PhysRevLett.107.235901.

[21] Nielsen M D, Ozolins V, Heremans J P. Lone pair electrons minimize lattice thermal conductivity. Energy Environ. Sci. 2013;**6**:570–578. DOI:10.1039/c2ee23391f.

[22] Walsh A, Payne D J, Egdell R G, Watson G W. Stereochemistry of post-transition metal oxides: Revision of the classical lone pair model. Chem. Soc. Rev. 2011;**40**:4455–4463. DOI:10.1039/c1cs15098g.

[23] Skinner B J, Luce F D, Makovicky E. Studies of the sulfosalts of copper III. Phases and phase relations in the system Cu-Sb-S. Econ. Geol. 1972;**67**:924–938. DOI:10.2113/gsecongeo.67.7.924.

[24] Makovicky E, Skinner B J. Studies of the sulfosalts of copper. VI. Low-temperature exsolution in synthetic tetrahedrite solid solution, $Cu_{12+x}Sb_{4+y}S_{13}$. Can. Mineral. 1978;**16**: 611–623.

[25] Tatsuka K, Morimoto N. Composition variation and polymorphism of tetrahedrite in the Cu-Sb-S system below 400°C. Am. Mineral. 1973;**58**:425–434.

[26] Tatsuka K, Morimoto N. Tetrahedrite stability relations in the Cu-Sb-S system. Econ. Geol. 1977;**72**:258–270. DOI:10.2113/gsecongeo.72.2.258.

[27] Johnson M L, Jeanloz R. A brillouin-zone model for compositional variation in tetrahedrite. Am. Mineral. 1983;**68**:220–226.

[28] May A F, Delaire O, Niedziela J L, Lara-Curzio E, Susner M A, Abernathy D L, Kirkham M, McGuire M A. Structural phase transition and phonon instability in $Cu_{12}Sb_4S_{13}$. Phys. Rev. B. 2016;**93**:064104. DOI:10.1103/PhysRevB.93.064104.

[29] Kitagawa S, Sekiya T, Araki S, Kobayashi T C, Ishida K, Kambe T, Kimura T, Nishimoto N, Kudo K, Nohara M. Suppression of nonmagnetic insulating state by application of pressure in mineral tetrahedrite $Cu_{12}Sb_4S_{13}$. J. Phys. Soc. Jpn. 2015;**84**:093701. DOI: 10.7566/JPSJ.84.093701.

[30] Tanaka H I, Suekuni K, Umeo K, Nagasaki T, Sato H, Kutluk G, Nishibori E, Kasai H, Takabatake T. Metal-semiconductor transition concomitant with a structural transformation in tetrahedrite $Cu_{12}Sb_4S_{13}$. J. Phys. Soc. Jpn. 2016;**85**:014703. DOI: 10.7566/JPSJ.85.014703.

[31] Lu X, Morelli D T. Natural mineral tetrahedrite as a direct source of thermoelectric materials. Phys. Chem. Chem. Phys. 2013;**15**:5762–5766. DOI:10.1039/c3cp50920f.

[32] Suekuni K, Tsuruta K, Kunii M, Nishiate H, Nishibori E, Maki S, Ohta M, Yamamoto A, Koyano M. High-performance thermoelectric mineral $Cu_{12-x}Ni_xSb_4S_{13}$ tetrahedrite. J. Appl. Phys. 2013;**113**:043712. DOI:10.1063/1.4789389.

[33] Heo J, Laurita G, Muir S, Subramanian M A, Keszler D A. Enhanced thermoelectric performance of synthetic tetrahedrites. Chem. Mater. 2014;**26**:2047–2052. DOI:10.1021/cm404026k.

[34] Chetty R, Prem Kumar D S, Rogl G, Rogl P, Bauer E, Michor H, Suwas S, Puchegger S, Giester G, Mallik R C. Thermoelectric properties of a Mn substituted synthetic tetrahedrite. Phys. Chem. Chem. Phys. 2015;**17**:1716–1727. DOI:10.1039/c4cp04039b.

[35] Barbier T, Lemoine P, Gascoin S, Lebedev O I, Kaltzoglou A, Vaqueiro P, Powell A V, Smith R I, Guilmeau E. Structural stability of the synthetic thermoelectric ternary and nickel-substituted tetrahedrite phases. J. Alloys Compd. 2015;**634**:253–262. DOI:10.1016/j.jallcom.2015.02.045.

[36] Lara-Curzio E, May A F, Delaire O, McGuire M A, Lu X, Liu C Y, Case E D, Morelli D T. Low-temperature heat capacity and localized vibrational modes in natural and synthetic tetrahedrites. J. Appl. Phys. 2014;**115**:193515. DOI:10.1063/1.4878676.

[37]   Chetty R, Bali A, Naik M H, Rogl G, Rogl P, Jain M, Suwas S, Mallik R C. Thermoelectric properties of co-substituted synthetic tetrahedrite. Acta Mater. 2015;**100**:266–274. DOI:10.1016/j.actamat.2015.08.040.

[38]   Bouyrie Y, Candolfi C, Ohorodniichuk V, Malaman B, Dauscher A, Tobola J, Lenoir B. Crystal structure, electronic band structure and high-temperature thermoelectric properties of Te-substituted tetrahedrites $Cu_{12}Sb_{4-x}Te_xS_{13}$ (0.5 ≤ x ≤ 2.0). J. Mater. Chem. C. 2015;**3**:10476–10487. DOI:10.1039/c5tc01636c.

[39]   Lu X, Morelli D T. The effect of Te substitution for Sb on thermoelectric properties of tetrahedrite. J. Electron. Mater. 2014;**43**:1983–1987. DOI:10.1007/s11664-013-2931-2.

[40]   Lu X, Morelli D T, Xia Y, Ozolins V. Increasing the thermoelectric figure of merit of tetrahedrites by co-doping with nickel and zinc. Chem. Mater. 2015;**27**:408–413. DOI:10.1021/cm502570b.

[41]   Bouyrie Y, Candolfi C, Vaney J B, Dauscher A, Lenoir B. High temperature transport properties of tetrahedrite $Cu_{12-x}M_xSb_{4-y}Te_yS_{13}$ (M = Zn, Ni) compounds. J. Electron. Mater. 2016;**45**:1601–1605. DOI:10.1007/s11664-015-4128-3.

[42]   Bouyrie Y, Candolfi C, Dauscher A, Malaman B, Lenoir B. Exsolution process as a route toward extremely low thermal conductivity in $Cu_{12}Sb_{4-x}Te_xS_{13}$ tetrahedrites. Chem. Mater. 2015;**27**:8354–8361. DOI:10.1021/acs.chemmater.5b03785.

[43]   Lu X, Morelli D T, Wang Y, Lai W, Xia Y, Ozolins V. Phase stability, crystal structure, and thermoelectric properties of $Cu_{12}Sb_4S_{13-x}Se_x$ solid solutions. Chem. Mater. 2016;**28**:1781–1786. DOI:10.1021/acs.chemmater.5b04796.

[44]   Bouyrie Y, Sassi S, Candolfi C, Vaney J B, Dauscher A, Lenoir B. Thermoelectric properties of double-substituted tetrahedrites $Cu_{12-x}Co_xSb_{4-y}Te_yS_{13}$. Dalton Trans. 2016;**45**:7294–7302. DOI:10.1039/c6dt00564k.

[45]   Suekuni K, Tanaka H I, Kim F S, Umeo K, Takabatake T. Glasslike versus crystalline thermophysical properties of the Cu-S based minerals: Tetrahedrite and colusite. J. Phys. Soc. Jpn. 2015;**84**:103601. DOI:10.7566/JPSJ.84.103601.

[46]   Bouyrie Y, Candolfi C, Pailhès S, Koza M M, Malaman B, Dauscher A, Tobola J, Boisron O, Saviot L, Lenoir B. From crystal to glass-like thermal conductivity in crystalline minerals. Phys. Chem. Chem. Phys. 2015;**17**:19751–19758. DOI:10.1039/c5cp02900g.

[47]   Lai W, Wang Y, Morelli D T, Lu X. From bonding asymmetry to anharmonic rattling in $Cu_{12}Sb_4S_{13}$ tetrahedrites: When lone-pair electrons are not so lonely. Adv. Funct. Mater. 2015;**25**:3648–3657. DOI:10.1002/adfm.201500766.

[48]   James D J, Lu X, Morelli D T, Brock S L. Solvothermal synthesis of tetrahedrite: Speeding up the process of thermoelectric material generation. ACS Appl. Mater. Interfaces. 2015;**7**:23623–23632. DOI:10.1021/acsami.5b07141.

[49] Barbier T, Rollin-Martinet S, Lemoine P, Gascoin F, Kaltzoglou A, Vaqueiro P, Powell A. V, Guilmeau E. Thermoelectric materials: A new rapid synthesis process for non-toxic and high-performance tetrahedrite compounds. J. Am. Ceram. Soc. 2016;**99**:51–56. DOI:10.1111/jace.13838.

[50] Gonçalves A P, Lopes E B, Monnier J, Bourgon J, Vaney J B, Piarristeguy A, Pradel A, Lenoir B, Delaizir G, Pereira M. F. C, Alleno E, Godart C. Fast and scalable preparation of tetrahedrite for thermoelectrics via glass crystallization. J. Alloys Compd. 2016;**664**:209–217. DOI:10.1016/j.jallcom.2015.12.213.

[51] Levinsky P, Vaney J B, Candolfi C, Dauscher A, Lenoir B, Hejtmánek J. Electrical, thermal, and magnetic characterization of natural tetrahedrites-tennantites of different origin. J. Electron. Mater. 2016;**45**:1351–1357. DOI:10.1007/s11664-015-4032-x.

Chapter 5

Thermoelectric Effect and Application of Organic Semiconductors

Nianduan Lu, Ling Li and Ming Liu

Additional information is available at the end of the chapter

http://dx.doi.org/10.5772/65872

Abstract

Human development and society progress require solving many pressing issues, including sustainable energy production and environmental conservation. Thermoelectric power generation looks like promising opportunity converting huge heat from the sun and waste heat from industrial sector, housing appliances and infrastructure and automobile and other fuel combustion exhaust directly to electrical energy. Thermoelectric power generation will be of high demand, when technology will be affordable, providing low price, high conversion efficiency, reliability, easy applicability and advanced ecological properties of end products. In this context, organic thermoelectric materials attract great interest caused by non-scarcity of raw materials, non-toxicity, potentially low costs in high-scale production, low thermal conductivity and wide capabilities to control thermoelectric properties. In this chapter, we focus mainly on thermoelectric effect in several organic semiconductors, both crystalline and disordered. We present theory of some transport phenomena determining thermoelectric properties of organic semiconductors, including general expression of thermoelectric effect, percolation theory of Seebeck coefficient, hybrid model of Seebeck coefficient, Monte Carlo simulation and first-principle theory. Finally, a future outlook of this field is briefly discussed.

Keywords: organic semiconductor, thermoelectric effect, theoretical model

1. Introduction

1.1. Organic semiconductors

1.1.1. History

Organic semiconductors have revealed promising performance and received considerable attentions due to large area, low-end, lightweight and flexible electronics applications [1]. Currently, organic semiconductors have appealed for a broad range of devices including sensors, solar cells, light-emitting devices and thermoelectric application [2, 3]. Historically, organic materials were

viewed as insulators with applications commonly seen in inactive packaging, coating, containers, moldings and so on. The earliest research on electrical behavior of organic materials dates back to the 1960s [4]. In the 1970s, photoconductive organic materials were recognized and were used in solar cells and xerographic sensors, etc. [5]. In about the same age, organic thermoelectric materials, such as conducting polymers, had been investigated [6], although First European Conference on Thermoelectric in 1988 had no mention of organic materials in their proceeding. Since proof of concept for organic semiconductors occurred in the 1980s, remarkable development of organic semiconductors has promoted improvement of performance that maybe competitive with amorphous silicon (a-Si), increasing their suitability for commercial applications [7]. Appearance of conductive polymers in the late 1970s, and of conjugated semiconductors and photoemission polymers in the 1980s, greatly accelerated development in the field of organic electronics [8]. Polyacetylene was one of the earliest polymer materials known to be potential as conducting electricity [9], and one could find that oxidative dopant with iodine could greatly increase conductivity by 12 orders of magnitude [10]. This discovery and development of highly conductive organic materials were attributed to three scientists: Alan J. Heeger, Alan G. MacDiarmid and Hideki Shirakawa, who were jointly awarded the Nobel Prize in Chemistry in 2000 for their discovery in 1977 and development of oxidized, iodine-doped polyacetylene. After that, plenty of organic semiconductor materials were synthesized, and research field of organic electronics matured over the years from proof-of-principle phase into major interdisciplinary research area, involving physics, chemistry and other disciplines. As important branch of organic semiconductors, in the past decade, organic thermoelectric effect has received much attention. **Figure 1** shows Thomson Reuters Web of Science publication report for the topic "organic thermoelectric Seebeck effect" for the last 16 years [3]. Research interest in organic thermoeletrics Seebeck effect has been growing remarkably over the last 5 years.

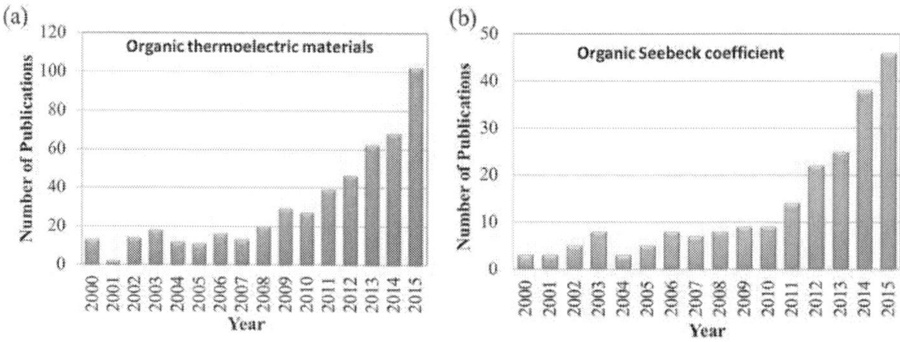

Figure 1. Thomson Reuters web of science publication report for the topic "organic thermoelectric Seebeck effect" from 2000 to 2015.

1.1.2. Structure

Organic semiconductors can be usually classified as two types: crystalline and amorphous materials, in terms of crystalline fraction (also static disorder) [3, 11].

All organic semiconductor materials are generally characterized by weak van der Waals bonding, which leads to weak intermolecular interactions. This weak coupling of molecules would induce weak interaction energy to give narrow electronic bandwidths. Otherwise, existing narrow electronic bands will be eliminated by statistical variation of width in energy level distribution of molecules, which, hence, creates Anderson charge localization. For crystalline organic semiconductor materials, localization of charge carriers is attributed to intermolecular thermal fluctuations (dynamic disorder), at which size of localized wave function is expected to be on the order of molecular spacing and charge carrier transport in this weakly localized field is treated as "intermediate hopping transport regime" [12]. Because of high density of crystal imperfections in disordered organic semiconductors, such as impurities, grain boundaries, dangling bonds and periodicity loss of crystal, localization of charge carriers is attributed to spatial and energetic disorder due to weak intermolecular interactions [13]. Disorder in organic semiconductors results in basic charge transport mechanism, common for very rich variety of such materials [14], incoherent tunneling (hopping) of charge carriers between localized states [15].

1.2. Charge transport mechanism in organic semiconductors

1.2.1. Dispersive transport

Charge carriers are always slowing down during conduction processes in dispersive transport regime. It happens only when charge carriers' distribution is thermally nonequilibrium. In dispersive transport regime, energy fluctuations allow release (emission) of charge carrier captured by localized state (trap), when state (trap) becomes temporarily shallow during fluctuations [16]. Otherwise, charge carrier release time is strongly dependent on both temperature and energy. Charge carriers localized on shallow states are usually released before the states can change their energies noticeably. In contrast, charge carriers, always being trapped in energetically deeper and deeper states, have to perform longer and longer tunneling transitions to hop to the next destination site. As a result, temperature-dependent distribution of effective activation energies just follows the density of states (DOS) function within domain of shallow states, while these distributions appear to be very different for deep traps.

1.2.2. Nearest-neighbor hopping

Transport in disordered organic semiconductors is generally characterized by charge carrier's localization and hopping transport mechanism [17]. Localized states, randomly distributed in energy and space, form discrete array of sites in hopping space. The most probable hop for charge carrier on site with particular energy is to the closest empty site, that is, to its nearest-neighbor site in hopping space. In conjunction with hopping probability rate, it gives mobility for carriers at this energy. In a word, nearest-neighbor hopping describes hopping regime, in which tunneling part of hopping rates in Eq. (1) referred as hopping rates of Miller-Abrahams is so much slower than energy contribution, that only the nearest neighbors are addressed in hops [18]:

$$\gamma_{ij} = v_0 \times \exp\left(-R\right) = v_0 \times \begin{cases} \exp\left(-2 \times \alpha \times R_{ij} - \dfrac{E_j - E_i}{k_B T}\right), & E_j - E_i > 0, \\ \exp\left(-2 \times \alpha \times R_{ij}\right), & E_j - E_i < 0 \end{cases} \qquad (1)$$

where v_0 is attempt-to-jump frequency, R is hopping range, α is inverse localized length describing extension of wave function of localized state, R_{ij} is distance between site i and site j, E_i and E_j are energies of sites i and j, respectively. As long as charge carrier can find shallow and empty sites with energies below its current state, it will perform nearest-neighbor hopping to energetically lower sites, since in this case rates are limited by spatial tunneling distances only.

1.2.3. Variable-range hopping

Variable-range hopping (VRH) theory was first proposed by Neville Mott in 1971 [19] and hence was called Mott VRH, which is model describing low temperature conduction in strongly disordered systems with localized states. VRH transport has characteristic temperature dependence of 3D electrical conductance:

$$\sigma = \sigma_0 \times \exp\left[-\left(\frac{T_0}{T}\right)^{1/4}\right], \tag{2}$$

here $k_B T_0 = \frac{\beta}{g(E_f)\times a^3}$, σ_0 is prefactor, α^{-1} is localization length, k_B is Boltzmann constant, $g(E_f)$ is DOS function at Fermi energy E_f, and β is constant coefficient with value in interval 10.0–37.2 according to different theories.

For 3D electrical conductance, and in general for d-dimensions, VRH transport is expressed as:

$$\sigma = \sigma_0 \times \exp\left[-\left(\frac{T_0}{T}\right)^{1/(d+1)}\right], \tag{3}$$

here d is dimensionality.

1.2.4. Multiple trapping and release theory

Multiple trapping and release theory assume that charge carrier transport occurs in extended states, and that most of charge carriers are trapped in localized states. Energy of localized state is separated from mobility edge energy. When localized state energy is slightly lower mobility edge, then localized state acts as shallow trap, from which charge carrier can be released (emitted) by thermal excitations. But, if that energy is far below mobility edge energy, then charge carrier cannot be thermally excited (emitted). In multiple trapping and release theory, total charge carriers' concentration n_{total} is equal to sum of concentrations n_e in extended states and in localized states as in Ref. [20]:

$$n_{total} = n_e + \int_{-\infty}^{0} g(E)f(E)dE, \tag{4}$$

where $f(E)$ is Fermi-Dirac distribution function.

2. Organic thermoelectric materials

2.1. Polymer-based thermoelectric materials

Polymers as thermoelectric materials recently have attracted much attention due to easy fabrication processes and low material cost [21, 22], as well as, low thermal conductivity, which is highly desirable for thermoelectric applications. Different types of polymers have been used in thermoelectric devices [23–25], such as polyaniline (PANI), poly(p-phenylenevinylene) (PPV), poly(3,4-ethylenedioxythiophene) (PEDOT), tosylate(tos), poly(styrenesulfonate) (PSS), and poly(2,5-dimethoxy phenylenevinylene) (PMeOPV), poly[2-methoxy-5-(2-ethylhexyloxy)-1,4-phenylenevinylene] (MEHPPV) and poly(3-hexylthiophene-2,5-diyl) (P3HT). **Figure 2** shows chemical structure of some polymers. These polymers are chosen as thermoelectric materials for their conductive nature. Takao Ishida's group has reported high power factor (PF) values of over 100 μW/(m K^2) on PEDOT films through chemical reduction with different chemicals in their review [25], as shown in **Table 1**.

Thermoelectric properties of 1,1,2,2-ethenetetrathiolate(ett)–metal coordination polymers poly [Ax(M–ett)] (A = Na, K; M = Ni, Cu) also have been studied, as shown in **Figure 3a** [26]. P-type poly[Cux(Cu–ett)] exhibited best ZT of 0.014 at 380 K with electrical conductivity of ~15 S/cm, Seebeck coefficient of 80 μV/K and thermal conductivity of 0.45 W/(m K); n-type poly[Kx(Ni–ett)] showed best ZT of 0.2 at 440 K with electrical conductivity of ~60 S/cm, Seebeck coefficient of 150 μV/K and thermal conductivity of 0.25 W/(m K). Otherwise, thermoelectric module based on p-type poly[Cux(Cu–ett)] and n-type poly[Nax(Ni–ett)] (i.e., ZT of 0.1 at 440 K) was built (**Figure 3b** and **c**). More recently, Pipe et al. reported thermoelectric measurements of PEDOT: PSS with 5% of dimethylsulfoxide (DMSO) and ethylene glycol (EG) after submerging films in EG several times (from 0 to 450 min) [27]. Insulating polymer (PSS) is removed, and consequently, electrical conductivity and Seebeck coefficient increase simultaneously. ZT value of 0.42 reported in Pipe et al.'s work is the highest ever obtained for polymer until date.

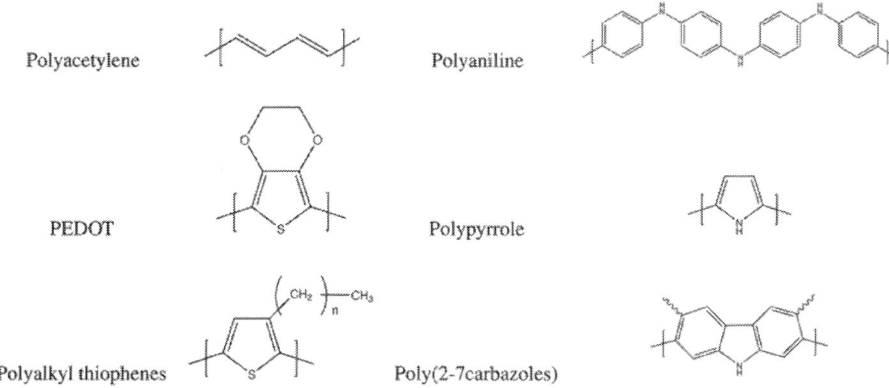

Figure 2. Molecular structures of some typical polymers [23].

Materials	S (μV/K)	PF (μW/m K^2)	ZT
PEDOT:PSS	22	47	0.1
PEDOT:tos (dedoped)	200	324	0.25
PEDOT:PSS	73	469	0.42
PEDOT:tos	~85	12,900	–
PEDOT:tos	55	453	–
PEDOT:BTFMS	~40	147	0.22
PEDOT:PSS (dedoped)	~50	112	0.093
PEDOT:PSS (dedoped)	43	116	0.2
PEDOT:PSS	65	355	~0.3

Table 1. Thermoelectric properties of polymer-based materials (tos: tosylate, and PSS: poly(stryrenesulfonate) [25].

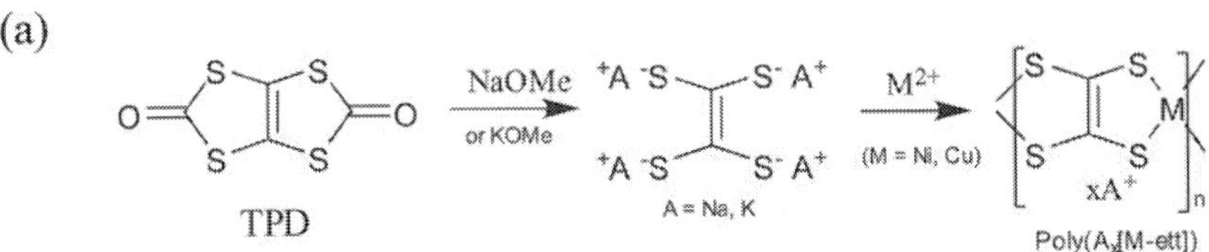

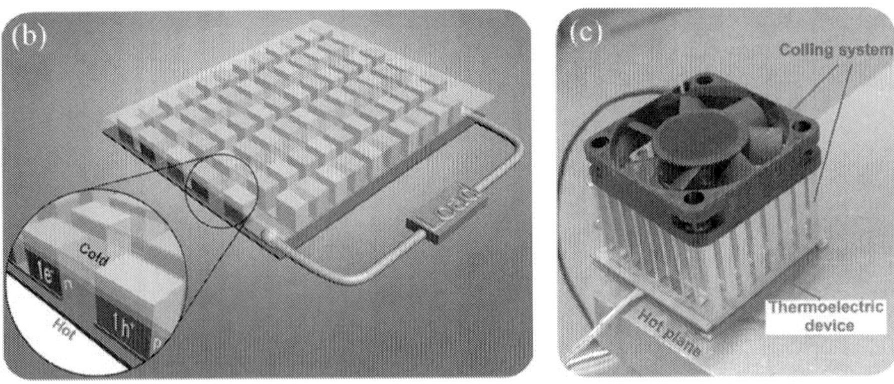

Figure 3. (a) Synthetic route of poly[Ax(M-ett)]s, (b) module structure and (c) photograph of module and measurement system with a hot plane and cooling fan.

Although thermoelectric performance of organic semiconductors has been increased by using different fabricated methods or dopant, as compared with inorganic thermoelectric materials, organic thermoelectric materials still exhibit lower ZT so far. Nevertheless, researchers are making their great efforts in organic semiconductors instead of inorganic materials, due to several more advantages in organic thermoelectric materials, for example, the non-scarcity of raw materials, non-toxicity and large area applications.

2.2. Small molecule-based thermoelectric materials

Currently, investigation of small molecule-based organic thermoelectric materials is lagging behind that of polymers-based organic materials. However, small molecule-based organic thermoelectric materials exhibit plenty of attractions. For example, small molecules are easier to be purified and crystallized and may be more feasible to achieve n-type conduction. **Figure 4** shows chemical structures of some small molecules studied as organic thermoelectric materials [24].

As the benchmark of p-type organic semiconductors, pentacene is the best-known small molecule studied for organic thermoelectric applications. A remarkable attraction for pentacene is attributed to its high mobility up to 3 cm^2/V s in thin film transistors (TFT). Due to low charge carriers' concentration of pentacene in neutral state, appropriate doping treatment is critical to optimization of its thermoelectric properties. So far, F$_4$TCNQ and iodine are efficient dopant for pentacene [24]. F$_4$TCNQ is strong electron acceptor that is frequently used as p-type dopant in organic electronics [28]. Harada et al. have achieved maximum electrical conductivity of 4.1 × 10^{-2} S/cm and maximum PF of 0.16 μW/(m K^2) by using dopant of 2.0 mol.%, as shown in **Figure 5a**. Furthermore, when thickness of pentacene layer was varied, electrical conductivity could be optimized, while Seebeck coefficient was unaffected (around 200 μV/K). Finally, maximum PF of 2.0 μW/(m K^2) was obtained in 6-nm-thick pentacene sample, as shown in **Figure 5b**. For using iodine dopant, Minakata et al. have achieved highest electrical conductivity of 60 S/cm with Seebeck coefficient in the range of 40–60 μV/K [29]. As a result, the highest PF is 13 μW/(m K^2), which is more than six times that of pentacene/F$_4$TCNQ bilayer sample. In a word, as compared with polymer-based thermoelectric materials, small molecule-based materials are relatively less explored and need to put in more effort in the future.

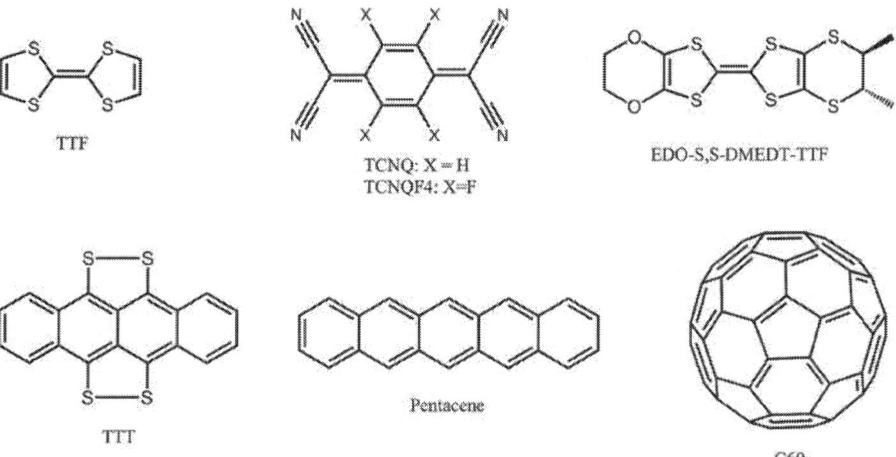

Figure 4. Chemical structures of some small molecules studied as organic thermoelectric materials.

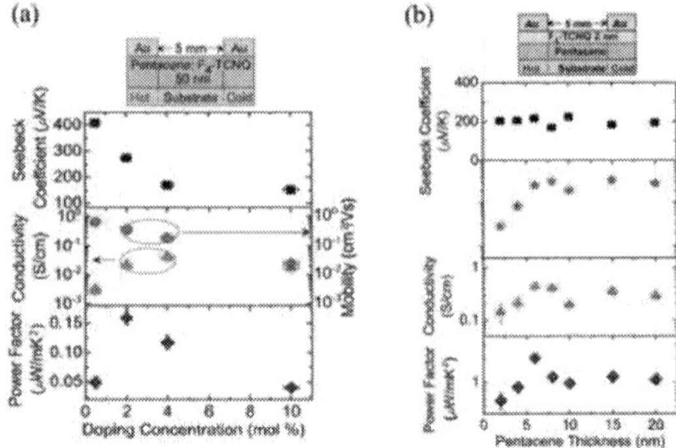

Figure 5. Thermoelectric properties of pentacene samples doped by (a) co-evaporation with F$_4$TCNQ and (b) forming pentacene/F$_4$TCNQ bilayer structures.

3. Thermoelectric transport theory of organic semiconductors

3.1. General expression of thermoelectric effect

Since organic semiconductors consist of amorphous and crystal structure, theoretical model of charge carrier thermoelectric transport should be more general. Derivation of present expression of thermoelectric effect was inspired by work of Cutler and Mott [30]. Basic expression was in terms of electrical conductivity σ. Based on definition of Cutler and Mott, for hopping system in disordered lattice at zero and finite temperature, σ was expressed as:

$$\sigma = -\int \left[\sigma_E \frac{\partial f}{\partial E} \right] dE. \tag{5}$$

Otherwise, one can start out by writing electrical conductivity as integral over single states neglecting electron correlation effects [31]:

$$\sigma = e \int g(E)\mu(E)f(E)[1-f(E)]dE. \tag{6}$$

Then, energy dependence of electrical conductivity is written as:

$$\sigma(E) = eg(E)\mu(E)f(E)[1-f(E)], \tag{7}$$

here $g(E)$ is density of states, $\mu(E)$ is mobility, and $f(E)$ is Fermi-Dirac distribution function. Seebeck coefficient S is related to Peltier coefficient Π as follows:

$$S = \frac{\Pi}{T}. \tag{8}$$

Peltier coefficient Π is energy carried by electrons per unit charge. Carried energy is characterized connected with Fermi energy E_f. Contribution to Π of each electron is in proportion to its correlative contribution to total conduction. Weighting factor for electrons in interval dE at energy E is thus $\frac{\sigma(E)}{\sigma}dE$, with energy dependence of conductivity $\sigma(E)$ as Eq. (7). Therefore, one can obtain general expression of Seebeck coefficient as [32]:

$$S = \frac{-k_B}{e}\int\left(\frac{E-E_f}{k_B T}\right)\frac{\sigma(E)}{\sigma}dE. \tag{9}$$

To distinguish crystalline solids, general Seebeck coefficient also can be defined as shape of transport energy with mean energy of conducting charge carriers as in Ref. [33]:

$$S = -\frac{1}{eT}\times(E_{trans}-E_f), \tag{10}$$

where transport energy E_{trans} is defined as averaged energy weighted by electrical conductivity distribution:

$$E_{trans} = \int E\frac{\sigma(E)}{\sigma}dE. \tag{11}$$

Usual expression of Seebeck coefficient was used early in doped organic materials by Roland Schmechel in 2003 [33]. In Schmechel's article, a detailed method to complex hopping transport in doped system (p-doped zinc-phthalocyanine) was proposed and used to discuss experimental data on effect of doping on conductivity, mobility and Seebeck coefficient.

3.2. Percolation theory of Seebeck coefficient

Percolation theory is considered as the best way known to analytically investigate charge carriers hopping transport characteristics. Percolation problem for charge carriers transport properties in disordered semiconductors has been argued early by Ziman and a number of researchers [34, 35], at which charge transport should be in proportion to percolation probability $P(p)$. A simple definition is that approximates firstly electrical conductivity as [36]:

$$\sigma(E) = \sigma_0\times P(p(E)), \tag{12}$$

where $P(p)$ percolation probability is fraction of the volume allowed, but not isolated, and σ_0 denotes a large allowed value of the material. $P(p)$ is known to vanish for p less than critical value B_c, but drops sharply to zero as $p \to B_c$. To make percolation question simple, researchers have put forward two kinds of standpoints for critical B_c, that is, $B_c = 1$, and $B_c = 2.8$ or $B_c = 2.7$. Although researchers have not achieved unified agreement, percolation theory in hopping system was generally established to explain charge carrier transport characteristics.

Generally speaking, charge carrier transport is described by a four-dimensional (4D) hopping space, including three spatial coordinates and one energy coordinate, at which probability of charge carrier hopping between localized sites associated with these four coordinates. Therefore, charge carrier transport would be more complex based on percolation approach addressing, if all of positions, that is, spatial positions of sites and their energies and occupation of sites, must be included. In Eq. (12), to simulate electrical conductivity $\sigma(E)$, the key is to seek out percolation path in hopping space. Thus, a random resistor network linking all of molecular sites under percolation model is essential. **Figure 6** shows schematic diagram of charge carrier transport in hopping system and corresponding percolation current through polymer matrix for charge carrier to travel through [37].

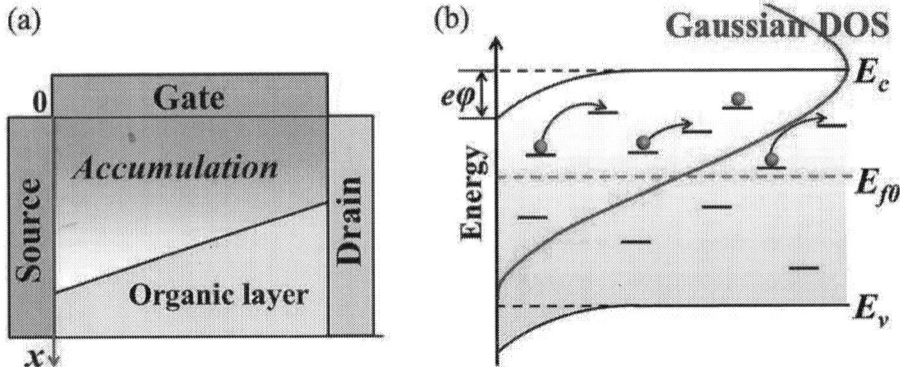

Figure 6. (a) Schematic diagram of charge carrier transport in hopping space with density of states and (b) corresponding percolation current in disordered organic semiconductor.

Based on the following general definition through Kelvin-Onsager relation to Peltier coefficient Π, percolation theory to calculate Seebeck coefficient S in hopping transport is expressed as in Eq. (8), where Π is generally identified with average site energy on percolation cluster and can be written as:

$$\Pi = \int E_i P(E_i) dE_i, \tag{13}$$

where $P(E_i)$ is probability that site of energy E_i is on current-carrying percolation cluster and was further given by:

$$P(E_i) = \frac{g(E_i) P_1(Z_m|E_i)}{\int_{-E_m}^{E_m} g(E_i) P_1(Z_m|E_i) dE}, \tag{14}$$

where $g(E_i)$ is density of states per unit volume, E_m is maximum site energy, and $P_1(Z_m|E_i)$ is probability that second smallest resistance emanating from site with energy E_i is not larger than maximum resistance on percolation cluster Z_m. Expression of probability $P_1(Z_m|E_i)$ is written as:

$$P_1(Z_m|E_i) = 1 - \exp\left[-P(Z_m|E_i)\right] \times \left[1 + P(Z_m|E_i)\right], \tag{15}$$

where $P(Z_m|E_i)$ is bonds' density, which means average number of resistance of Z_m or less connected to site energy E_i. To calculate Peltier coefficient (or Seebeck coefficient), an expression for $P(Z_m|E_i)$ is necessary.

Based on percolation theory, disordered organic material is regarded as a random resistor network (see **Figure 6b**). To address total electrical conductivity in disordered system, the initial is to obtain reference conductance Z and eliminate all conductive pathways between sites i and j with $Z_{ij} < Z$. Conductance between sites i and j is given by $Z_{ij} \propto \exp(-S_{ij})$ with [38]:

$$S_{ij} = 2\alpha R_{ij} + \frac{|E_i - E_f| + |E_j - E_f| + |E_i - E_j|}{2k_B T}. \tag{16}$$

The density of bonds $P(Z_m|E_i)$ then can be written as:

$$P(Z_m|E_i) = \int 4\pi R_{ij}^2 \, g(E_i)g(E_j)dR_{ij}dE_i dE_j \theta(S_c - S_{ij}). \tag{17}$$

If the density of participating sites is P_s, then critical parameter S_c is found by solving equation:

$$P(Z_m|E_i) = B_c P_s = B_c \int g(E)dE\theta(S_c k_B T - |E - E_f|). \tag{18}$$

Based on numerical studies for three-dimensional amorphous system, the formation of infinite cluster corresponds to $B_c = 2.7$ [38]. By connecting Eqs. (16)–(18), bonds' density can be formulated as:

$$P(Z_m|E_i) = \frac{4\pi R^3}{3(2\alpha)^3} \times$$

$$\begin{cases} \int_{\epsilon_f}^{\epsilon_i}(S_c - \epsilon_j + \epsilon_f)^3 g(\epsilon_j)d\epsilon_j + \int_{\epsilon_i}^{S_c + \epsilon_f}(S_c - \epsilon_j + \epsilon_f)^3 g(\epsilon_j)d\epsilon_j + \int_{\epsilon_i - S_c}^{\epsilon_f}(S_c - \epsilon_i + \epsilon_j)^3 g(\epsilon_j)d\epsilon_j \\ \int_{\epsilon_i}^{\epsilon_f}(S_c + \epsilon_i - \epsilon_f)^3 g(\epsilon_j)d\epsilon_j + \int_{\epsilon_f - S_c}^{\epsilon_i}(S_c + \epsilon_i - \epsilon_f)^3 g(\epsilon_j)dE_j + \int_{\epsilon_f}^{\epsilon_i + S_c}(S_c - \epsilon_j + \epsilon_i)^3 g(\epsilon_j)d\epsilon_j \end{cases} \quad \begin{cases} \epsilon_i > \epsilon_f \\ \epsilon_i < \epsilon_f \end{cases}$$

$$\tag{19}$$

here ϵ is normalized energy as $\epsilon = \frac{E}{k_B T}$. This expression has been split into two regimes of $\epsilon_i > \epsilon_f$ and $\epsilon_i < \epsilon_f$, which are corresponding to contributions of ϵ_i above or below Fermi energy to $P(Z_m|E_i)$ and, therefore, Seebeck coefficient S, respectively. Above Fermi energy, charge carriers in shallow states will move by hopping to other shallow states. While below Fermi energy, charge carriers in deep states will move by thermal excitation to shallower states. By substituting Eqs. (13)–(15) and Eq. (19) into Eq. (8), one can obtain the final result of Seebeck coefficient.

Figure 7 shows simulated and experimental dependences of Seebeck coefficient on charge carriers' density; simulation is based on percolation theory, experimental data measured by using field-effect transistor (FET) from three kinds of conjugated polymers, that is, IDTBT, PBTTT and PSeDPPBT [39]. Model of percolation theory can reasonably reproduce experimental data under the whole range of charge carriers' density for different conjugated polymers.

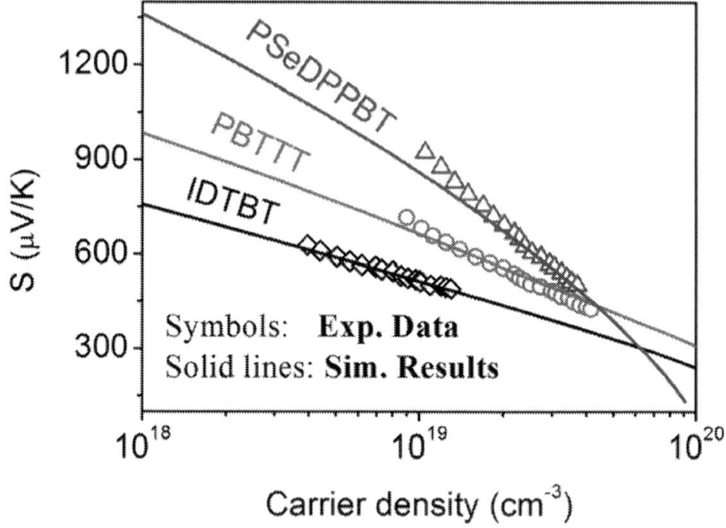

Figure 7. Charge carrier's density dependence of Seebeck coefficient for different materials at room temperature [37]. Symbols and solid lines are experimental and simulated results, respectively.

3.3. Hybrid model of Seebeck coefficient

Usual behavior of Seebeck coefficient is to decrease with increasing charge carriers' density. However, Germs et al. have observed unusual thermoelectric behavior for pentacene in TFT [40], indeed, at room temperature, increasing charge carriers' density results in expected decrease in S, while with decreasing temperature to values below room temperature, S demonstrates growth with increasing charge carriers' density at $T = 250$ K and even more pronounced at $T = 200$ K, as shown in **Figure 8**.

To explain this unusual thermoelectric behavior, Germs et al. developed simplified hybrid model that incorporates both variable-range hopping (VRH) and mobility edge (ME) transport [40]. Charge carrier and energy transport can be described independently by two processes: VRH-type process that occurs within exponential tail of localized states and band-like type transport that occurs within band-like states above mobility edge. Then, Seebeck coefficient of hybrid model is expressed as conductivity-weighted average of two contributing transport channels:

$$S = \frac{S_{ME}\sigma_{ME} + S_{VRH}\sigma_{VRH}}{\sigma_{ME} + \sigma_{VRH}}, \tag{20}$$

where S_{ME} and σ_{ME} are Seebeck coefficient and electrical conductivity in ME part, and S_{VRH} and σ_{VRH} are Seebeck coefficient and electrical conductivity in VRH part, respectively.

Then, general expression of Seebeck coefficient in Eq. (9) reduces:

$$S_{ME} = \frac{(E_c - E_f)}{eT} + A, \tag{21}$$

with

$$A = \frac{\displaystyle\int_0^\infty \frac{\varepsilon}{eT}\sigma(\varepsilon)d\varepsilon}{\displaystyle\int_0^\infty \sigma(\varepsilon)d\varepsilon}, \tag{22}$$

where $\varepsilon = E - E_c$, E_c is energy value at mobility edge.

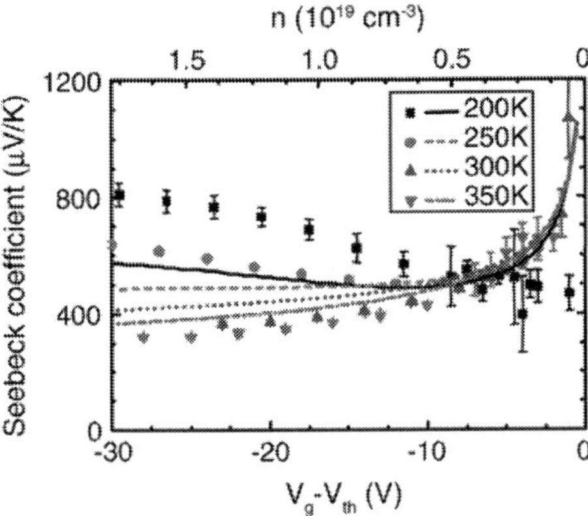

Figure 8. Measurements (symbols) and calculation (lines) of Seebeck coefficient as function of gate bias in pentacene TFT. Gate voltage V_g is corrected for threshold voltage V_{th} of TFT.

In Eq. (21), A accounts for excitations beyond the band edge with 1–20% of S_{ME}. Similarly, within VRH model, it is assumed that transport is determined by hopping event from equilibrium energy state to relatively narrow transport energy E^* [41], and Eq. (9) becomes:

$$S_{VRH} = \frac{(E^* - E_f)}{eT}, \tag{23}$$

Electrical conductivity in ME part is calculated as:

$$\sigma_{ME} = en_{free}\mu_{free}(T), \tag{24}$$

with power law dependence on temperature, $\mu_{free}(T) = \mu_0 T^{-m}$.

VRH part is described by Mott-Martens model that assumes transport to be dominated by hops from Fermi energy to transport level E^*. Electrical conductivity in VRH part is subsequently calculated by optimizing Miller-Abrahams expression as:

$$\sigma_{VRH} = \sigma_0 \exp\left[-2\alpha R^* - \frac{(E^* - E_f)}{k_B T}\right], \tag{25}$$

where position of transport level E^* and typical hopping distance R^* is connected via percolation argument:

$$B_c = \frac{4}{3}\pi(R^*)^3 \int_{E_f}^{E^*} g(E)dE, \tag{26}$$

where $B_c = 2.8$ is critical number of bonds, $g(E)$ represents DOS, which here is simplified to single exponential trap tail below mobility edge and constant density of extended states above E_c:

$$g(E) = \begin{cases} \dfrac{n_{trap}}{k_B T_0} \exp\left(-\dfrac{E}{k_B T_0}\right) & for\ E < E_c \\ \dfrac{n_0}{k_B T_0} & for\ E \geq E_c \end{cases} \quad (E_c = 0), \tag{27}$$

where n_0 is divided by $k_B T_0$ for dimensional reasons. The number of charge carriers above E_c, n_{free} follows from Fermi-Dirac distribution.

Figure 9 shows measured and calculated dependences of Seebeck coefficient on difference between gate voltage V_g and threshold voltage V_{th} on TFT at T = 200 K. One can see that at 200 K heat transported at mobility edge $(E_c - E_f)$ and heat transported at transport level $(E^* - E_f)$ both decrease with increasing charge carriers' density, accounting for downward trends in S_{ME} and S_{VRH}. Consequently, weight-averaged Seebeck coefficient S_{Hyb} shifts from S_{VRH} values at small gate bias up toward S_{ME} for large gate bias. Relatively large value of S_{ME} at lower temperatures follows from Eq. (21) and temperature independence of E_C.

3.4. Monte Carlo simulation

As compared with numeric model, analytical thermoelectric transport models exhibit more context and physical property, but they also have inevitable shortcoming due to the use of plenty of free parameters during simulation and calculation. In order to eliminate these hindrances, universal method has been used based on Monte Carlo (MC) simulation for describing hopping transport and insuring validity and accuracy of thermoelectric transport.

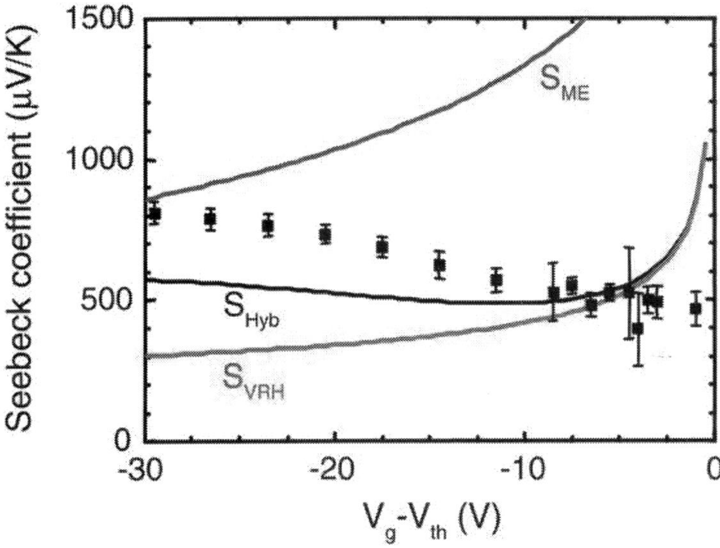

Figure 9. Measured (symbols) and calculated (lines) dependences of Seebeck coefficient on gate bias at T = 200 K [40]. S_{Hyb} is conductivity-weighted average of S_{VRH} and S_{ME}.

Kinetic Monte Carlo simulation generally includes six steps as follows [42]:

i. Initializing site energy E_i. Random energy E_i at site i derives from Gaussian or exponential DOS.

ii. Initial placement of charge carriers. Fermi-Dirac occupation probabilities will be used in randomly placing N_{ch} charge carriers.

iii. Choosing hopping events. If neglecting the polaron effect, hopping rates v_{uo} from site i to site j are based on Miller-Abrahams expression in Eq. (1). Corresponding setting is: (a) hopping rates equal to 0 to prevent hopping into already occupied sites; (b) introducing cut-off distance and set $v_{uo} = 0$ for jumps longer than this distance.

Otherwise, renormalizing hopping rates Γ_{ij} as $p_{ij} = \dfrac{\Gamma_{ij}}{\sum\limits_{i,j}\Gamma_{ij}}$. Sum of rates includes only

jumps from occupied to unoccupied sites, that is, $\Gamma_{ij} = 0$ for occupied site j or unoccupied site i. For every pair ij, index k (i.e., $ij \rightarrow k$ and $p_{ij} = p_k$), where $k \in \{1, ..., k_{max}\}$ with k_{max} being total number of all possible hopping events. Then, partial sum S_k is defined for every index k:

$$S_k = \sum_{k'=1}^{k} p_{k'}. \tag{28}$$

Apparently, for every k, extent from interval $[S_{k-1}, S_k]$ equals to probability p_k for k^{th} jump, and total extent of all intervals equals to 1, for example $S_{kmax} = 1$. Then, determining a random real number r from interval $[0, 1]$ and finding index k, here $S_{k-1} << r << S_k$, and one can find hopping event. After determining hoping event, one would move charge carrier between corresponding sites i and j.

iv. Calculation of waiting time. After finding each hopping transport, total simulation time t and waiting time τ would be added that has passed until the event took place. This time is determined by describing a random number from exponential waiting time distribution $P(\tau) = \nu_i \exp(-\nu_i \tau)$ with $\nu_i = \sum_j \nu_{ij}$ being the total rate for charge carrier hopping from site i. It is, therefore, written as:

$$\tau = \frac{-1}{\nu_i} \ln(x), \tag{29}$$

where random number x is drawn from interval $[0, 1]$.

v. Calculating current density. Every time, when predefined numbers of jumps have occurred, current density $J(t)$ can be expressed as:

$$J(t) = \frac{e(N^+ - N^-)}{t N_y N_z a^2}, \tag{30}$$

where N^+ and N^- are the total number of jumps in and opposite direction of electric field for cross-sectional slice in yz plane, and a is lattice constant.

vi. Calculating Seebeck coefficient. Seebeck coefficient is given by expression as in Eq. (10), where transport energy is defined as averaged energy weighted by electrical conductivity distribution:

$$E_{trans} = \frac{\int E\sigma(E, T)\left(-\frac{\partial f}{\partial E}\right)dE}{\sigma(T)}, \tag{31}$$

with $\sigma(T) = \int \sigma(E, T)\left(-\frac{\partial f}{\partial E}\right)dE$.

Although kinetic MC technique gives a direct simulation of thermoelectric transport in organic semiconductor materials and, therefore, it is accurate for the most description of electronic conductivity, its negative factor is to require extensive computational resources, which leads to difficultly analyze and fit experimental data. **Figure 10** shows comparison of analytical model with MC simulation for Seebeck coefficient [42]. It is seen in **Figure 10**, that results exhibit qualitative agreement for all values of parameter α. For large α in **Figure 10a**, S_{sa} and S_{MC} show not only qualitative, but even relatively good quantitative agreement in energy interval $E < 0$ (corresponding to relative charge carriers' concentration $n/N_0 < 0.5$). For higher energies (and thus for higher concentrations), difference between S_{sa} and S_{MC} increases. As parameter α decreases, functional dependencies S_{sa} and S_{MC} remain very close to each other, but S_{sa} gets shifted with respect to S_{MC} (**Figure 10b**).

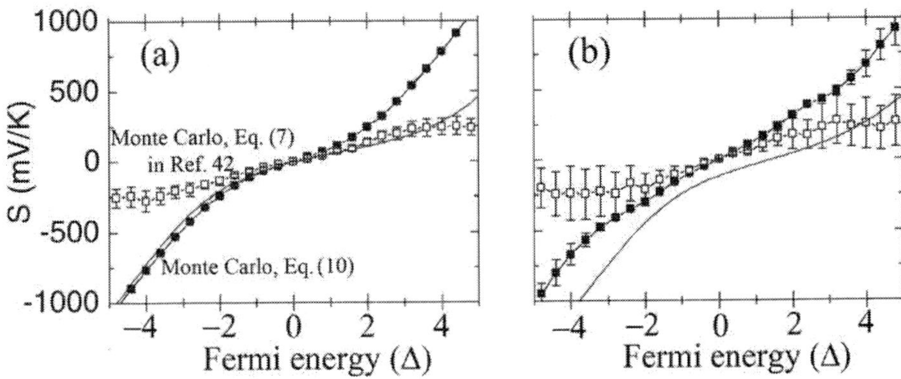

Figure 10. Monte Carlo and semi-analytical calculations of Seebeck coefficient for different values of localization length (a) α^{-1} = 1 nm and (b) α^{-1} = 0.2 nm. E_f is in units of Δ, $\Delta = 4k_BT$ and T = 300 K.

3.5. First-principle calculation theory

First-principle (*ab initio*) theory would be deemed to the best type of theory for hopping charge transport in organic semiconductors, since it starts from particular chemical and geometrical structure of the system, and it starts directly at the level of established science and does not make assumptions, such as empirical model and fitting parameters. So far, a few researches on charge carriers transport properties based on first-principle theory are hardly beyond the scope of crystalline. Otherwise, direct calculation of Seebeck coefficient is hardly realistic. Current method combines generally first-principle calculations with transport theory. For example, Gao et al. have investigated theoretically Seebeck coefficient of narrow bandgap crystalline polymers, including crystalline solids β-Zn$_4$Sb$_3$ and AuIn$_2$ and these polymers, based on muffin-tin orbital and full-potential linearized augmented plane-wave (FLAPW) electronic structure code [43]. In essence, Gao et al.'s method for calculation of transport properties of crystalline solid is firstly based on semiclassical Boltzmann theory, following as:

$$\sigma_0(T) = \frac{e^2}{3} \int \tau(E,T) N(E) v^2(E) \left(-\frac{\partial f(E)}{\partial E} \right) dE, \tag{32}$$

where e, τ, f and v represent free electron charge, electronic relaxation time, Fermi-Dirac distribution function and Fermi velocity, respectively. If relaxation time for electron scattering processes is assumed to be constant, that is, $\tau(E,T) = $ const, which may yield reasonable simulated results in a wide range of materials, then temperature dependence of $\sigma_0(T)$ can be simulated based on constant relaxation time τ:

$$\frac{\sigma_0(T)}{\tau} = \frac{e^2}{3} \int N(E) v^2(E) \left(-\frac{\partial f(E)}{\partial E} \right) dE. \tag{33}$$

Then, Seebeck coefficient is calculated from ratio of the zeroth and first moments of electrical conductivity with respect to energy:

$$S(T) = \frac{1}{eT} \times \frac{I^1}{I^0},$$

(34)

where

$$I^x(T) = \int \tau(E,T)N(E)v^2(E)(E-E_f)^x\left(-\frac{\partial f(E)}{\partial E}\right)dE.$$

(35)

Product of the density of states $N(E)$ and arbitrary quantity g, which is relative to energy and k-vector as in Eqs. (33) and (35), can be calculated by using integration on constant energy surface S in k space. Electronic band structure can be calculated by using WIEN97 and WIEN2K FLAPW [44] or pseudopotential plane-wave code in Vienna *ab initio* simulation package (VASP) [45].

Figure 11a and **b** shows simulated energy band structure of polythiophene polymer. Here, internal structural parameters of the polymer are fully optimized, and electronic band structure is calculated by pseudopotential plane-wave calculations employing ultrasoft Vanderbilt pseudopotential and generalized gradient scheme. Simulated results display that band structure of polythiophene is very simple, which exhibits semiconductor performance with band gap of 0.9 eV. Neutral polythiophene is electrical insulator. Otherwise, by inspecting band structure, one can find that, except very close to zone center, where the density of states is high, band dispersions are considerable. The special band structure, thus, induces very low value of Seebeck coefficient (~20 μV/K), as shown in **Figure 11**.

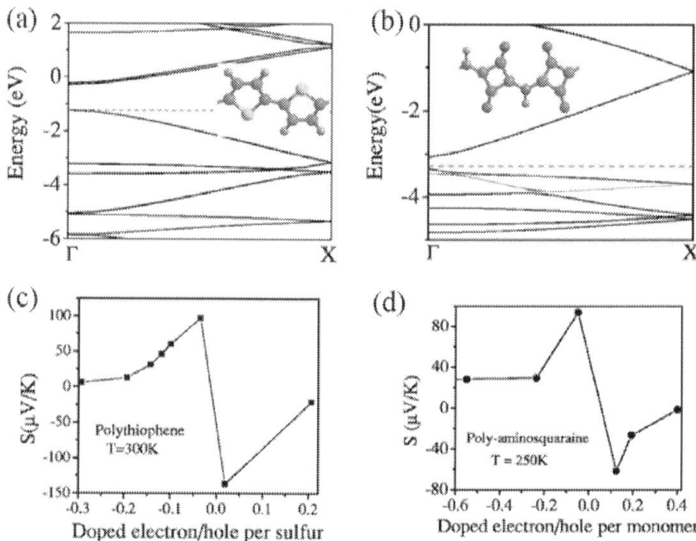

Figure 11. Calculated energy band structure (a), (b), and Seebeck coefficient (c), (d), for polythiophene (left) and polyaminosquarine (right), respectively [43]. Isolated planar polymer chains were used for this calculation.

Similar to calculated method from Gao et al., Shuai et al. have combined first-principles band structure calculations and Boltzmann transport theory to study thermoelectric in pentacene and rubrene crystals [46]. Electronic contribution to Seebeck coefficient is obtained in approximations of constant relaxation time and rigid band, as shown in **Figure 12**. Calculation results exhibited also the similar trend compared with experimental Seebeck coefficient.

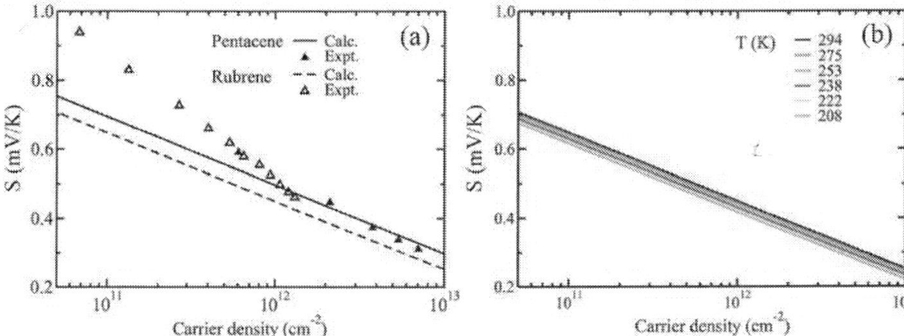

Figure 12. Seebeck coefficient calculated as function of charge carrier's concentration (a) for pentacene and rubrene at room temperature and compared to FET measurements. Calculated Seebeck coefficients have been averaged over three crystal directions (b) for rubrene at temperature in the range between 200 and 300 K [46].

Afterward, Shuai et al. applied also combining method to calculate thermoelectric properties of organic materials, which is used to calculating α-form phthalocyanine crystals H2Pc, CuPc, NiPc and TiOPc [47]. This combining method includes first-principles band structure calculations, Boltzmann transport theory and deformation potential theory. They used first-principles calculations in VASP to calculate Seebeck coefficient. After obtaining band structure, Boltzmann transport theory was performed to calculate properties related to charge carrier transport, as in Eqs. (32) and (33).

Being different from Gao et al.'s and Shuai et al.'s previous works, it is assumed that relaxation time is a constant, which can be estimated by deformation potential theory for treating electron-phonon scattering. In terms of corresponding articles [46, 48], acoustic phonon scattering in both pristine and doped system was simulated by this theory including scattering matrix element for electrons from $\mathbf{k}$ state to $\mathbf{k}'$ state expressing as:

$$|M(\mathbf{k},\mathbf{k}')|^2 = \frac{E_1{}^2}{C_{ii}} k_B T, \tag{36}$$

where E_1 is deformation potential constant that represents energy band shift caused by crystal lattice deformation, and C_{ii} is elastic constant in the direction of lattice wave's propagation. Then, relaxation time can be expressed by scattering probability:

$$\frac{1}{\tau(i,k)} = \sum_{k' \in BZ} \left\{ \frac{2\pi}{\hbar} |M(\mathbf{k},\mathbf{k}')|^2 \delta[\varepsilon(i,k) - \varepsilon(i,k')](1 - \cos\theta) \right\}, \tag{37}$$

where $\delta[\varepsilon(i,k)-\varepsilon(i,k')]$ is Dirac delta function, and θ is angle between $\mathbf{k}$ and $\mathbf{k}'$. In Bardeen and Shockley's method, scattering is assumed to be isotropic, and matrix element of interactions $M(\mathbf{k},\mathbf{k}')$ is not relative with $\mathbf{k}$ and $\mathbf{k}'$.

Calculated dependences of Seebeck coefficient on charge carriers' concentration are shown in **Figure 13**. Calculated dependences of Seebeck coefficient display different properties such as positive S values for holes and negative S values for electrons. Seebeck coefficient value is isotropic at first glance, and it decreases rapidly as charge carriers' concentration increases.

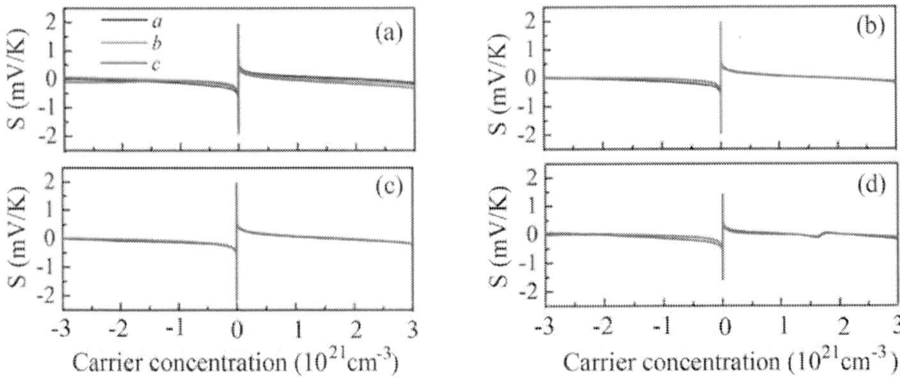

Figure 13. Seebeck coefficient for (a) H2Pc, (b) CuPc, (c) NiPc, (d) TiOPc calculated as a function of the charge carriers concentration at 298 K [47]. a, b and c denote a, b and c crystal axes, respectively.

4. Conclusion and outlook

In the past decades, the research on organic thermoelectric materials has made great progress. Rich variety of novel organic materials has been synthesized and applied in thermoelectric devices, and thermoelectric performances of organic semiconductors have been promoted greatly. However, as compared to inorganic thermoelectric materials, organic thermoelectric materials still exhibit lower ZT so far. However, situation looks like that progress in theoretical study of organic thermoelectric effect lags far behind experimental investigation in the last 30 years, but has been changed remarkably until the recent five years. Here, we have tried to describe organic thermoelectric materials and theoretical approaches, which allow to calculate characteristics of charge carrier transport processes responsible for thermoelectric effect in organic semiconductors. We hope that these contexts can be helpful to improve thermoelectric effect in organic materials and provide motivation for growth of thermoelectric applications of organic semiconductors. We believe that quest for green energy sources will stimulate intensive research and development works in the field of novel organic thermoelectric materials and devices that will result in serious improvement in thermoelectric efficiency of organic thermoelectric materials and enable development of novel high-performance and affordable organic thermoelectric devices.

Abbreviations

DOS	density of states
VRH	variable-range hopping,
PANI	polyaniline,
PPV	poly(p-phenylenevinylene),
PEDOT	poly(3,4-ethylenedioxythiophene),
PSS	poly(styrenesulfonate),
PMeOPV	poly(2,5-dimethoxy phenylenevinylene),
MEHPPV	poly[2-methoxy-5-(2-ethylhexyloxy)-1,4-phenylenevinylene],
P3HT	poly(3-hexylthiophene-2,5-diyl),
DMSO	dimethylsulfoxide,
EG	ethylene glycol,
S	Seebeck coefficient,
Π	Peltier coefficient,
TFT	thin film transistors,
PF	power factor,
FET	field-effect transistor,
ME	mobility edge,
V_g	gate voltage,
V_{th}	threshold voltage,
MC	Monte Carlo,
FLAPW	full-potential linearized augmented plane-wave,
VASP	Vienna ab initio simulation package.

Author details

Nianduan Lu*, Ling Li and Ming Liu

*Address all correspondence to: lunianduan@ime.ac.cn

Key Laboratory of Microelectronic Devices & Integrated Technology, Institute of
Microelectronics of Chinese Academy of Sciences, Beijing, China

References

[1] Reese C, Roberts M, Ling M, Bao Z.: Organic thin film transistors. Mater. Today. 2004;**7**
(9):20–27. doi:10.1016/S1369-7021(04)00398-0

[2] Elkington D, Cooling N, Belcher W, Dastoor PC, Zhou XJ.: Organic thin-film transistor
(OTFT)-based sensors. Electronics. 2014;**3**:234–254. doi:10.3390/electronics3020234

[3] Lu ND, Li L, Liu M.: A review of carrier thermoelectric-transport in organic semiconductors. Phys. Chem. Chem. Phys. 2016;**18**:19503. doi:10.1039/C6CP02830F

[4] Owens RM, Malliaras GG.: Organic electronics at the interface with biology. MRS Bull. 2010;**35**(06):449–456. doi:10.1557/mrs2010.583

[5] Law KK.: Organic photoconductive materials: recent trends and developments. Chem. Rev. 1993;**93**(1):449–486. doi:10.1021/cr00017a020

[6] Shirakawa H, Louis EJ, MacDiarmid AG, Chiang CK, Heeger AJ.: Synthesis of electrically conducting organic polymers: halogen derivatives of polyacetylene, (CH)x. J. Chem. Soc. Chem. Commun. 1977;(16):578–580. doi:10.1039/C39770000578

[7] Chason M, Brzis PW, Zhang J, Kalyanasundaram K, Gamota DR.: Printed organic semiconductors devices. Proc. IEEE. 2006;**93**(7):1348–1356. doi:10.1109/JPROC.2005. 850306

[8] Li FM, Nathan F, Wu YL, Ong BS.: Organic thin film transistor integration: a hybrid approach. Wiley-Vch Verlag GmbH & Co. KGaA, Boshstr. Weinheim Germany; 2011. 270 pp. doi:10.1002/9783527634446

[9] Ito T. Shirakawa H, Ikeda S.: Simultaneous polymerization and formation of polyacetylene film on the surface of concentrated soluble Ziegler-type catalyst solution. J. Polym. Sci.: Polym. Chem. Ed. 1974;**12**(1):11–20. doi:10.1002/pol.1974.170120102

[10] Chiang CK, Fincher CB, Park JYW, Heeger AJ.: Electrical conductivity in doped polyacetylene. Phys. Rev. Lett. 1977;**39**(17):1098–1011.

[11] Minder NA, Ono S, Chen ZH, Facchetti A, Morpurgo AF.: Band-like electron transport in organic transistors and implication of the molecular structure for performance optimization. Adv. Mater. 2012;**24**:503–508. doi:10.1002/adma.201103960

[12] Novikov SV, Tyutnev AP.: Charge carrier transport in molecularly doped polycarbonate as a test case for the dipolar glass model. J. Chem. Phys. 2013;**138**:104120. doi:10.1063/1.4794791

[13] Tessler N, Preezant Y, Rappaport N, Roichman Y.: Charge transport in disordered organic materials and its relevance to thin-film devices: a tutorial review. Adv. Mater. 2009;**21**:2741–2761. doi:10.1002/adma.200803541

[14] Coropceanu V, Cornil J, Filho DS, Olivier Y, Silbey R, Brédas JL.: Charge transport in organic semiconductors. Chem. Rev. 2007;**107**:926–952. doi:10.1021/cr050140x

[15] Li L, Meller G, Kosina H.: Carrier concentration dependence of the mobility in organic semiconductors. Syn. Metals. 2007;**157**:243–246. doi:10.1016/j.synthmet.2007.03. 002

[16] Arkhipov VI, Adriaenssens GJ.: Trap-controlled dispersive transport in systems of randomly fluctuating localized states. Philos. Mag. B. 1997;**76**(1):11–22. doi:10.1080/ 01418639708241075

[17] Lu ND, Li L, Banerjee W, Sun PX, Gao N, Liu M.: Charge carrier hopping transport based on Marcus theory and variable-range hopping theory in organic semiconductors. J. Appl. Phys. 2015;**118**:045701. doi:10.1063/1.4927334

[18] Miller A, Abrahams E.: Impurity conduction at low concentrations. Phys. Rev. 1960;**120**:745. doi:10.1103/PhysRev.120.745

[19] Mott NF, Davis EA. Electronic Processes in Non-Crystalline Materials. 2nd ed. Oxford: Oxford University Press; 1971. 32 p.

[20] Li L, Lu ND, Liu M.: Field effect mobility model in oxide semiconductor thin film transistors with arbitrary energy distribution of traps. IEEE Electron Dev. Lett. 2014;**35** (2):226. doi:10.1109/LED.2013.2291782

[21] Bubnova O, Khan ZU, Malti A, Braun S, Fahlman M, Berggren M, Crispin X.: Optimization of the thermoelectric figure of merit in the conducting polymer poly(3,4-ethylenedioxythiophene). Nat. Mater. 2011;**10**:429–433. doi:10.1038/nmat3012

[22] Weathers A, Khan ZU, Brooke R, Evans D, Pettes MT, Andreasen JW, Crispin X, Shi L.: Significant electronic thermal transport in the conducting polymer poly(3,4-ethylenedioxythiophene). Adv. Mater. 2015;**27**:2101–2106. doi:10.1002/adma. 201404738

[23] Culebras M, Gómez CM, Cantarero A.: Review on polymers for thermoelectric applications. Materials. 2014;**7**:6701–6732. doi:10.3390/ma7096701

[24] Zhang Q, Sun YM, Xu W, Zhu DB.: Organic thermoelectric materials: emerging green energy materials converting heat to electricity directly and efficiently. Adv. Mater. 2014;**26**:6829–6851. doi:10.1002/adma.201305371

[25] Wei QS, Mukaida M, Kirihara K, Naitoh Y, Ishida T.: Recent progress on PEDOT-based thermoelectric materials. Materials. 2015;**8**:732–750. doi:10.3390/ma8020732

[26] Sun YM, Sheng P, Di CG, Jiao F, Xu W, Qiu D, Zhu DB.: Organic thermoelectric materials and devices based on p- and n-type poly(metal 1,1,2,2-ethenetetrathiolate)s. Adv. Mater. 2012;**24**:932. doi:10.1002/adma.201104305

[27] Kim GH, Shao L, Zhang K, Pipe KP.: Engineered doping of organic semiconductors for enhanced thermoelectric efficiency. Nat. Mater. 2013;**12**:719–723. doi:10.1038/nmat3635

[28] Harada K, Sumino M, Adachi C, Tanaka S, Miyazaki K.: Improved thermoelectric performance of organic thin-film elements utilizing a bilayer structure of pentacene and 2,3,5,6-tetrafluoro-7,7,8,8 -tetracyanoquinodimethane(F4-TCNQ). Appl. Phys. 2010;**96**:253304. doi:10.1063/1.3456394

[29] Minakata T, Nagoya I, Ozaki M.: Highly ordered and conducting thin film of pentacene doped with iodine vapor. J. Appl. Phys. 1991;**69**:7354. doi:10.1063/1.347594

[30] Cutler M, Mott NF.: Observation of Anderson localization in an electron Gas. Phys. Rev. 1969;**181**:1336.

[31] Nagel P. Electronic transport in amorphous semiconductors. In: Brodsky MH, editor. Amorphous Semiconductors. Springer-Verlag: Springer Berlin Heidelberg; 1979. pp. 113–158. doi:10.1007/3-540-16008-6_159

[32] Fritzsche H.: A general expression for the thermoelectric power. Solid State Commun. 1971;**9**:1813–1815.

[33] Schmechel R.: Hopping transport in doped organic semiconductors: a theoretical approach and its application to p-doped zinc-phthalocyanine. J. Appl. Phys. 2003;**93**:4653–4660. doi:10.1063/1.1560571

[34] Ziman JM.: The localization of electrons in ordered and disordered systems I. Percolation of classical particles. J. Phys. C. 1968;**1**:1532–1538.

[35] Shante VKS, Kirkpatrick S.: An introduction to percolation theory. Adv. Phys. 1971;**20**:325–357.

[36] Kirkpatrick S.: Classical transport in disordered media: scaling and effective-medium theories. Phys. Rev. Lett. 1971;**27**:1722–1725.

[37] Lu ND, Li L, M. Liu.: Universal carrier thermoelectric-transport model based on percolation theory in organic semiconductors. Phys. Rev. B. 2015;**91**:195205. doi:10. 1103/ PhysRevB.91.195205

[38] Limketkai BN, Jadhav P, Baldo MA.: Electric-field-dependent percolation model of charge-carrier mobility in amorphous organic semiconductors. Phys. Rev. B. 2007;**75**:113203. doi:10.1103/PhysRevB.75.113203

[39] Venkateshvaran D, Nikolka M, Sadhanala A, Lemaur V, Zelazny M, Kepa M, Hurhangee M, Kronemeijer AJ, Pecunia V, Nasrallah I, Romanov I, Broch K, McCulloch I, Emin D, Olivier Y, Cornil J, Beljonne D, Sirringhuas H.: Approaching disorder-free transport in high-mobility conjugated polymers. Nature. 2014;**515**:384–388. doi:10.1038/nature13854

[40] Germs WC, Guo K, Janssen RAJ, Kemerink M.: Unusual thermoelectric behavior indicating a hopping to bandlike transport transition in pentacene. Phys. Rev. Lett. 2012;**109**:016601. doi:10.1103/PhysRevLett.109.016601

[41] Baranovskii SD, Faber T, Hensel F, Thomas P.: The applicability of the transport-energy concept to various disordered materials. J. Phys. Condens. Matter. 1997;**9**:2699–2706.

[42] Ihnatsenka S, Crispin X, Zozoulenko IV.: Understanding hopping transport and thermoelectric properties of conducting polymers. Phys. Rev. B. 2015;**92**:035201. doi:10.1103/ PhysRevB.92.035201

[43] Gao X, Uehara K, Klug DD, Patchkovskii S, Tse JS, Tritt TM.: Theoretical studies on the thermopower of semiconductors and low- band-gap crystalline polymers. Phys. Rev. B. 2005;**72**:125202. doi:10.1103/PhysRevB.72.125202

[44] Perdew JP, Burke K, Ernzerhof M.: Generalized gradient approximation made simple. Phys. Rev. Lett. 1996;**77**:3865.

[45] Kresse G, Furthmuller J.: Efficiency of ab-initio total energy calculations for metals and semiconductors using a plane-wave basis set. Comput. Mater. Sci. 1996;**6**:16–50. doi:10.1016/0927-0256(96)00008-0

[46] Wang D, Tang L, Long MQ, Shuai ZG.: First-principles investigation of organic semiconductors for thermoelectric applications. J. Chem. Phys. 2009;**131**:224704. doi:10.1063/ 1.3270161

[47] Chen JM, Wang D, Shuai ZG.: First-principles predictions of thermoelectric figure of merit for organic materials: deformation potential approximation. J. Chem. Theory Comput. 2012;**8**:3338–3347. doi:10.1021/ct3004436

[48] Shi W, Zhao TQ, Xi JY, Wang D, Shuai ZG.: Unravelling doping effects on PEDOT at the molecular level: from geometry to thermoelectric transport properties. J. Am. Chem. Soc. 2015;**137**:12929–12938. doi:10.1021/jacs.5b06584

Chapter 6

Review of Research on the Thermoelectric Material ZnSb

Xin Song and Terje G. Finstad

Additional information is available at the end of the chapter

http://dx.doi.org/10.5772/65661

Abstract

The thermoelectric material ZnSb has been studied intensively in recent years and has shown promising features. The other zinc-antimonide compound, Zn_4Sb_3 has remarkable low thermal conductivity, but it is accompanied with phase transitions at moderate temperature and has inherent stability problems. Compared to that, ZnSb is relatively phase stable and has a relative high charge carrier mobility and Seebeck coefficient, thus yielding a decent power factor. Meanwhile, its thermal conductivity can be reduced by means of nanostructuring, thus giving a good figure of merit at moderate temperatures, 400–600 K. Many researchers have dedicated their efforts to study and improve ZnSb properties, and the figure of merit has been reported to be above one. Still, ZnSb as a thermoelectric material has features and behaviours that are not well-understood. The behaviour and properties of its intrinsic defects are not understood, but have interested researchers in recent years. This chapter intends to offer a comprehensive review on ZnSb to the readers. By combining own experiences from research on thermoelectric materials, the authors address the prospect for improving the thermoelectric properties of ZnSb and the concerns of transferring lab results to manufacturing.

Keywords: zinc antimonide, impurity band conduction, intrinsic defects, vacancies, p-type

1. Introduction

The thermoelectric effect in ZnSb has been known for almost two centuries. The first documented encounter can be traced back to the original work of Seebeck on thermoelectric current generation on different materials and alloy pairs in 1819–1827 [1, 2]. Quantitative measurements of the Seebeck voltage of ZnSb have been carried out by Becquerel in 1866 [3],

and thermoelectric generators using Zn-Sb alloy were fabricated for practical purposes since 1870 [4]. From the early twentieth century, many attempted to solve the crystal structure, but were barely successful [5–7], until Almin, finally, in 1948 determined the crystal structure of ZnSb together with CdSb [8]. A large interest in ZnSb followed and was benefitted by the progress on semiconductors since the 1950s. **Figure 1** shows a thermoelectric device made by ZnSb in the 1950s. It shows a solar thermoelectric generator prototype built by p-ZnSb and n-Bi$_{91}$Sb$_9$ with an overall efficiency of 0.63% [9]. Around the same time, a thermoelectric refrigerator made of n-PbTe and p-ZnSb was demonstrated in the USSR in the 1950s [10]. However, research reporting on ZnSb faded in the 1970s, and there was little improvement in terms of its efficiency. In recent years, renewed attention appeared due to a nanostructuring boom in material science, and the thermoelectric properties of ZnSb have been intensively studied and improved. There are also other energy-related applications of ZnSb actively being explored, such as electrodes for rechargeable Li-ion batteries [11], or phase change memory cells [12].

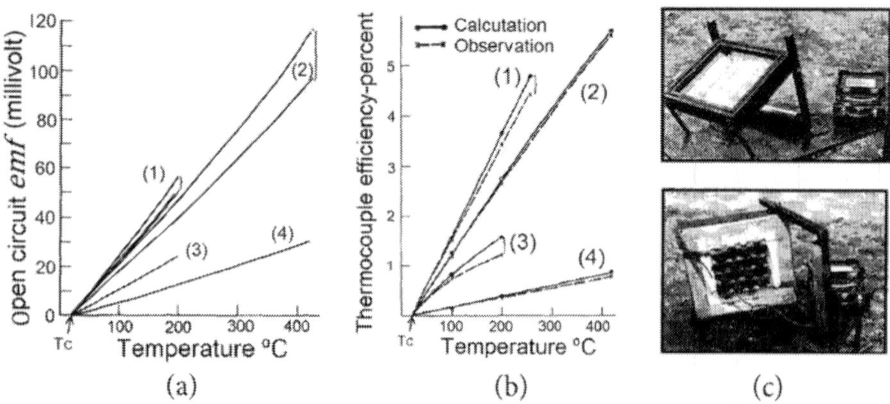

Figure 1. Solar thermoelectric generator. (Reprinted from [9], with the permission of AIP Publishing.) (a) Open circuit *emf* (Seebeck coefficient) as a function of temperature on ZnSb-alloys comparing with other thermal couples. Materials: (1) ZnSb (Sn, Ag, Bi)-Bi$_{91}$Sb$_9$, (2) ZnSb (Sn, Ag, Bi)-constantan, (3) Bi$_{91}$Sb$_9$Sn$_5$-Bi$_{91}$Sb$_9$, (4) Chromel P-constantan. (b) Thermoelectric efficiency on aforesaid materials; (c) Solar radiation thermopile contained 25 junctions built by p-ZnSb in plot (a) and (b) and n-Bi$_{91}$Sb$_9$. Above: front surface exposed to the sun; below: rear view shows the thermoelectric junctions.

Table 1 lists some reported figure of merit zT. The performance of ZnSb strongly depends on doping concentration and operation temperature. In addition, preparation-induced defects (phase impurity, oxidation and grain size) and intrinsic defects (vacancies, interstitials, clusters) influence strongly the material's performance. These need to be controlled in order to bring further progress, and in general, handling of the material has a learning curve and a detailed understanding of the material, per se, is needed.

Report year	zT	At temperature (K)	Sample	Reference
(1961)	0.6	460	Single crystal	[13]
(1964)	0.42	300	Single crystal	[14]
(1964)	0.3	300	Polycrystalline	[14]
(1966)	0.2	273	Single crystal	[15]
(2010)	0.07	373	Polycrystalline	[16]
(2010)	0.8	573	Polycrystalline	[17]
(2012)	0.9	659	Polycrystalline with Cu	[18]
(2013)	1.15	670	Polycrystalline with Ag	[19]
(2014)	0.9	635	Polycrystalline with Ag	[20]
(2014)	0.8	600	Polycrystalline with Cu	[20]
(2014)	1	630	Polycrystalline with Sn+Cd	[21]
(2014)	1.5	673	Polycrystalline with Cu	[22]
(2015)	0.8	700	Polycrystalline with Zn vacancies	[23]
(2015)	0.55	525	Polycrystalline	[24]

Table 1. List of reported zT of ZnSb.

2. Crystal structure of ZnSb

2.1. Crystallographic structure and covalent bonds

The crystallographic structure of ZnSb has been determined by Almin [8]. According to this determination, ZnSb has orthorhombic crystal structure, oP16 and belongs to the space group Pbca no. 61. The structure of ZnSb has been studied and confirmed by different techniques, albeit gave slightly different interatomic distances [25–28]. **Figure 2** shows the crystal structure of ZnSb that was generated by the structural data from Mozharivskyj [27].

The crystal structure can be viewed as a deformed zinc blende structure. The distorted edge-sharing $ZnSb_4$ tetrahedra generate a peculiar five-fold coordination of each atom, as seen in **Figure 2a**: one of the same kind and four of the other kind. In the context of bonding and its relation to conduction, all atoms in the crystal structure are tied together in a network—a point, we will return to below.

Another way to systematize the structure is to group the atoms together in planar rhomboid rings of Zn_2Sb_2 that have short Zn-Zn bonds connected to two different longer Zn-Sb bonds, as seen in **Figure 2a**. These motifs are also tied together in a network that completes the crystal structure. The crystal structure can be recognized in atomic scale by scanning transmission electron microscopy (STEM), as shown in **Figure 2b**.

The interatomic distances for each bond are also annotated in **Figure 2a**. The bonds in ZnSb have been categorized in three groups, shown in **Figure 2c** and **d**: the bonds *i–vi* are covalent bonds for building up the tetrahedron; the bonds *vii* and *viii* form the Zn_2Sb_2-ring; and the bonds ix_{1-6} are the dimers that connect the Zn_2Sb_2-rings. The angles between each Sb-Zn-Sb bond are also annotated in **Figure 2d**. Notice that the angles are close to those in regular tetrahedron, 109.5°. Therefore, even though the structure has five-fold coordination, one can still expect that ZnSb satisfies the tetrahedron rules to some extent. One of them is the electron count per bond. ZnSb has seven valence electrons per formula unit, whereas a regular tetrahedral binary semiconductor has eight valence electrons. Such electron-poor valence often indicates a metallic bond. Yet, ZnSb behaves as a semiconductor. The clue about its semiconducting property is probably due to the sp^3-hybrid orbitals in the tetrahedron, that often represents a semiconducting bond [30]. The counting rules cannot be applied on a per-bond-basis without considering the complete network. Therefore, all the atoms in the crystal structure are bonded together in a network. ZnSb, thereby, has been classified as an electron-poor framework semiconductor (EPFS) [26, 31]. A quite similar interpretation for the structure has been applied to the related compounds CdSb [30, 32–34] and ZnAs [35].

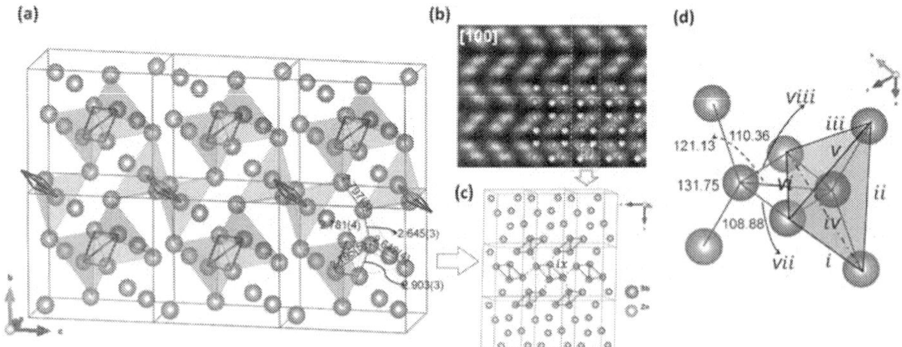

Figure 2. Crystal structure, bonding and coordination environment of ZnSb. (Reprinted with author's permission from [29]. ©X. Song, 2016.) (a) Structure in a 1×2×3 cell generated by Mozharivskyj's data [27]. Both the deformed tetrahedra and the Zn_2Sb_2-ring, as well as the interatomic distances between the nearest neighbours are annotated. (b) High-resolution scanning transmission electron microscopic (HR-STEM) imaging reveals atomic scale structure that is recognized corresponding to (a) and (c). (c) 3×1×3 cell shows the Zn_2Sb_2-ring network. Each ring has six nearest neighbours, connected by bonds ix_{1-6}. (d) The angles between Sb-Zn-Sb are similar to that of the standard tetrahedron, 109.5°. The bonds *i-vi* are covalent bonds for building up the tetrahedron, while *vii* and *viii* are the bonds to form Zn_2Sb_2-ring; bonds ix_{1-6} in (c) are the dimers, that connect the Zn_2Sb_2-rings.

Valence electrons in ZnSb have a certain distribution. **Figure 3** shows the theoretical *ab-initio* calculations by GGA-PBE (Generalized Gradient Approximation with Perdew-Burke-Ernzerhof exchange functionals) of the so-called deformation charge [36].

The results show that a Zn atom transfers a small average fraction of 0.26 electrons to Sb along Zn-Sb bond [37], which is expected due to the difference in Pauling electronegativity between Zn (1.65 eV) and Sb (1.96 eV). Essentially, the same transfer was found in the calculations of Benson *et al.* [38]. On the other hand, X-ray photoelectron spectroscopy (XPS) and electron energy loss spectroscopy (EELS) measurements indicated a shift of a Zn Auger peak and softening of EELS fine structure could be caused by a small net charge transfer of 0.1 electron from Zn to Sb [39]. The apparent difference between reported experiments and calculations can be related to experimental uncertainties and different volumes that were chosen for the calculation.

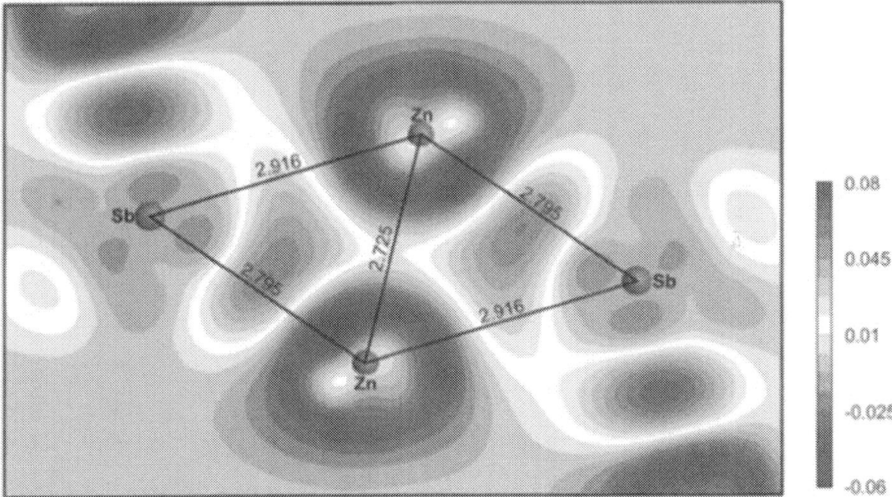

Figure 3. Deformation charge density distribution in electron per Å^2 in the plane of a Zn_2Sb_2-ring calculated by GGA–PBE. The colour map indicates the isocharge density lines: red indicates accumulation of electrons, whereas blue shows loss of electrons in the relaxed structure of the compound compared with the number of electrons in the free atoms [36], (DOI:10.1088/0953-8984/26/36/365401. © IOP Publishing. Reproduced with permission. All rights reserved.)

2.2. Electronic structure

2.2.1. Band calculations

The band structure of ZnSb has been calculated by *ab-initio* methods by many groups in recent years [2, 26, 31, 36–42]. The calculations are based upon density functional theory (DFT), but with varying detailed approximations and trade-offs between computational cost and accuracy. By comparing different calculated band diagrams, one can see some features that are common for most of the calculations and expect to filter out methodological errors in the different reports. **Figure 4** shows calculation of ZnSb band diagram.

The zero energy position corresponds to the largest energy of filled states. Thus, the states below 0 correspond to the valence band, while those above correspond to the conduction band. The value of the band gap is severely underestimated by the computational approximations. By using more accurate methods but at the expense of increased computation time, such as HSE-hybrid functional (Heyd-Scuseria-Ernzerhof), the band gap was calculated to be around 0.5 eV [37]. This value is close to the experimental value for single crystal ZnSb. On the other hand, the shape of the bands may be less affected by computational approximations. The band diagram in **Figure 4** shares some of the features found in most of the calculations listed. The band gap is indirect with the maximum of the valence band along the symmetry line Γ-X and the minimum of the conduction band along Γ-Z. One can see that the conduction band contains more satellites within 0.5eV of the minimum than pockets of the maximum in the valence band. Thus, one can expect ZnSb would have acted better as an *n*-type material than a *p*-type material, if stable *n*-type doping could have been achieved.

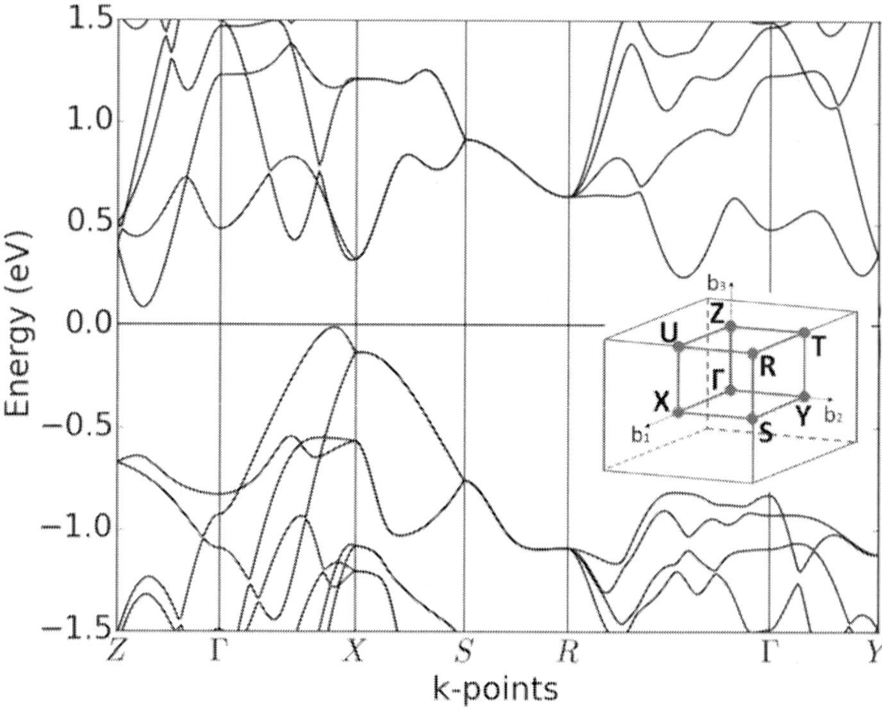

Figure 4. Band diagram of ZnSb showing energy states along high symmetry directions in *k*-space calculated by Berland *et al.* [2] using GGA−PBE. The Brillouin zone for orthorhombic lattice with the high symmetry symbols are shown in inset for indication. (Reprinted with author's permission from [29]. ©X. Song, 2016.)

It is worth mentioning the band diagram that was calculated by Yamada in 1978 by using pseudopotentials [43]. The maximum of the valence band was found to be on the line from Γ-X at k =(0.93 π/*a*, 0.0), which is close to the DFT values, and the band gap was 0.6eV. However, a minimum of the conduction band is located at k=(0.47 π/*a*, 0.0) along Γ-X, which is not in agreement with most DFT calculations, and it is hard to determine experimentally due to the difficulty in preparing *n*-type ZnSb.

2.2.2. Experimental band gap

The band gap has been measured by different methods. Values of 0.5–0.53eV for single crystal ZnSb, which were measured by optical absorption, have been reported [44, 45]. These values are essentially identical when considering the uncertainties, and 0.5 eV has been considered to be a reference value for single crystal ZnSb. Many research have also determined the thermo-dynamic band gap from the temperature dependence of the charge carrier concentration in the intrinsic regime. The charge carrier concentration is determined from the Hall coefficient, R_H. A linear fit of $\ln R_H T^{\frac{3}{2}}$ vs. $\frac{1}{T}$ will yield the activation energy, which is half of the thermodynamic energy gap. (This assumes an effective density of states proportional to $T^{\frac{3}{2}}$.). Values of 0.47–0.65 eV have been reported in the literature for single crystal ZnSb based upon Hall effect measurements [46, 47].

For polycrystalline ZnSb, the band gap is often reported to be around 0.31–0.35 eV [13, 48], which is smaller than that of single crystal material. It is justified to discuss whether the band gap really is different, or if it is due to that the idealizations of the measurement methods do not hold for polycrystalline material. One of the methods that has been used for estimating the band gap is to measure the temperature dependence of the Seebeck coefficient. By determining the temperature, T_{max}, for the maximum of the absolute value of the Seebeck coefficient, α_{max}, the band gap can be estimated from the Goldsmid formula, $E_g = 2q\alpha_{max}T_{max}$ [49]. This method works well for many semiconductors, for instance Zr-NiSn half-Heusler [50]. However, there have been reports on some systems that the Goldsmid formula does not apply [51, 52]. ZnSb may be one of them. Guo and Luo have reported that the Goldsmid formula returned a value of E_g=0.3 eV for their polycrystalline ZnSb samples [23]. However, Böttger *et al.* justified that for the samples that have a band gap of 0.3 eV obtained by Goldsmid formula, the tempera-ture-dependent resistivity was fitted better with a value of 0.44 eV [53]. It is fair to state that it depends upon the context whether it is best to reconsider the band gap value for polycrystal-line or reconsider the idealizations used in the measurement methods.

We should also keep in mind that heavily doping may influence the density of states near the band edges, forming tails that are extending into the band gap. The defect states in the band gap also influence the E_g values. This is a well-documented phenomenon (for Si) [54], even if the precision in a detailed quantitative understanding may be lacking. We have reported that the maximum of the Seebeck coefficient of ZnSb could be varied considerably by the presence of defect states in the band gap [52].

2.2.3. Effective mass and density of states

The idealized single parabolic band (SPB) model is convenient for analysing experimental results. The model has been successful for finding the optimum doping concentration of many thermoelectric materials [55–57]. When applied to Seebeck measurements on ZnSb with differ-ent doping concentrations, it has been observed that Pisarenko plot (the Seebeck coefficient vs. the charge carrier concentration) does not follow the SPB model with a single density of states effective mass m^*, which was determined to be $(0.42–0.49)\times m_0$, where m_0 is free electron mass [15, 53], but rather a different mass fit for different ranges of doping concentrations. A previous study by Böttger *et al.* has suggested that deviations from idealized SPB behaviour could be induced by impurity band states [53]. We have showed that deviations from simple SPB behaviour at varying doping concentration could be modelled by introducing an impurity band [52]. One may comment that the best fit with varying m^* is not necessary to imply that the energy band curvature varies. It suffices that the temperature-dependent position of the Fermi level varies differently than that in an idealized SPB case.

3. Electrical properties and doping effect

3.1. Electrical properties and scattering mechanisms

Table 2 summarizes the band gap, the charge carrier concentration and other electrical and thermal properties reported for ZnSb in the literature.

Given the crystal structure, single crystal ZnSb exhibits anisotropic conduction, which is strongly dependent on the anisotropic effective mass [58]. Böttger *et al.* determined each mass tensor component from band diagram calculations [53]. Different components were given by $m_a^* = 0.1811m_0$; $m_b^* = 0.4913m_0$; $m_c^* = 0.0837m_0$. The results agree with the anisotropic Hall mobility and the highest electrical conductivity is measured along the *c*-axis [15, 45]. Aniso-tropic energy surface has also been observed by optical absorption indicating that the constant energy surface in *k*-space in ZnSb is a spheroid [59].

There is much literature devoted to scattering mechanisms in semiconductors [65], while there are fewer reports dealing with that topic specifically for ZnSb. It is expected that ZnSb has similar behaviour to those semiconductors which have been much studied and follows similar trends. ZnSb appears to have favourably small polarity due to small electronegativity differ-ence between Zn and Sb. This would in turn lead to a negligible polar optical phonon scatter-ing compared to III–V and II–VI compounds where polar optical phonon scattering may be dominant. Roughly, transport in ZnSb is dominated by impurity scattering at low tempera-ture, while at higher temperature, when lattice vibrations are stronger, longitudinal acoustic phonon scattering dominates (deformation potential scattering). The hole mobility varies with temperature as $\mu \propto T^r$, thus a plot of lnμ vs. T will give the scattering factor r that implies the scattering mechanism. A value of -1 to -1.5 typically indicates that longitudinal acoustic phonon scattering dominates. In **Figure 5a**, the slopes of the Hall mobility approaches -1.5 as the doping concentration decreases. It indicates that acoustic phonon scattering dominates within the temperature range. The deviation from -1.5 for each individual curve is attributed to additional ionized impurity scattering [66]. It is also seen that the higher hole concentration corresponds to an increase in ionized impurity scattering, which limits the hole mobility.

Not only the intentional dopants, but also defects, that are ionized or neutralized, screened or unscreened, contribute to scattering. Likely scattering centres are Zn vacancies, interstitials, internal strain, grain boundaries and dislocations. **Figure 5b** shows the resistivity for an unprocessed ZnSb ingot that was obtained directly from solidification and hot-pressed ZnSb pellets with different dopant concentrations. The temperature coefficient, $\frac{d\rho}{dT}$, for the unprocessed ingot is positive, which is commonly observed for metals and semiconductors in a certain temperature range where phonon scattering dominates. For the processed samples, either with or without Ag, the temperature coefficient $\frac{d\rho}{dT}$ is negative. The reason for the negative temperature coefficient could be the dominant Coulomb scattering due to charged defects or impurities.

Reference	Report year	Structure	Dopant	σ (S)	ρ ($\Omega\cdot$cm)	μ (cm^2/(V·s))	p (cm^{-3})	α (μV/K)	E_g(eV)	κ (W/(m·K))
[60]	(1947)	–	–	–	0.0072	–	–	250	–	1.4
[9]	(1954)	–	2at.% Sn+0.1at.% Ag	–	0.0019	–	–	210	–	–
[61]	(1960)	Single crystal	–	2.8	3.57	350	5.1×10^{16}	440	0.52–0.57	–
[46]	(1961)	Single crystal	–	3	0.33	480	3×10^{16}	550	0.49	–
[62]	(1963)	Single crystal	–	1.6	0.63	248	4.1×10^{16}	–	–	–
[62]	(1963)	Single crystal	–	5.6	0.18	1150	3.1×10^{16}	–	–	–
[45]	(1964)	Single crystal	–	6.4	0.16	10	4×10^{18}	110	0.53	1.1
[14]	(1964)	Single crystal	–	19.7	0.05	384	3.2×10^{17}	490	–	–
[14]	(1964)	Polycrystalline	–	20	0.05	95.2	1.3×10^{18}	375	–	1.3
[14]	(1964)	Polycrystalline	0.1at.% Ga	2.04	0.49	210	6×10^{16}	675	–	–
[14]	(1964)	Polycrystalline	0.1at.% Cu	110	–	680	1.1×10^{18}	–	–	–
[63]	(1964)	Single crystal		–	–	–	3×10^{17}	–	0.48	–
[15]	(1966)	Single crystal		825	0.001	800	2×10^{16}	182	–	3.7
[15]	(1966)	Single crystal	Cu	42	0.02	660	4×10^{17}	–	–	–
[15]	(1966)	Single crystal	Cu	898	0.001	510	1×10^{19}	–	–	–
[47]	(1967)	Single crystal	Au	26.9	0.04	233	4.2×10^{17}	404	0.47–0.65	–
[47]	(1967)	Single crystal	In	37.5	0.03	326	7.7×10^{17}	288	0.47–0.65	–
[47]	(1967)	Single crystal	Te	51	0.02	379	8.3×10^{17}	400	0.47–0.65	–
[16]	(2010)	Polycrystalline	2wt.% Ag	–	0.01	–	–	185	–	2.3
[17]	(2010)	Polycrystalline		28.2–32.1	–	–	–	400–500	–	1.41
[19]	(2013)	Polycrystalline	0.002at.% Ag	–	0.0016	111	2.6×10^{19}	181	–	1.5
[20]	(2014)	Polycrystalline	Sn	180	–	–	4.9×10^{18}	250	–	1.8
[64]	(2014)	Polycrystalline	Cu	640	–	–	2.5×10^{19}	134	–	–
[21]	(2014)	Polycrystalline	0.15at.% Cu	935	–	–	2.4×10^{19}	142	–	–
[23]	(2015)	Polycrystalline	3at.% V$_{Zn}$	800	0.0012	66	7.9×10^{17}	275	0.29	1.75

Table 2. Summary of electrical properties of ZnSb.

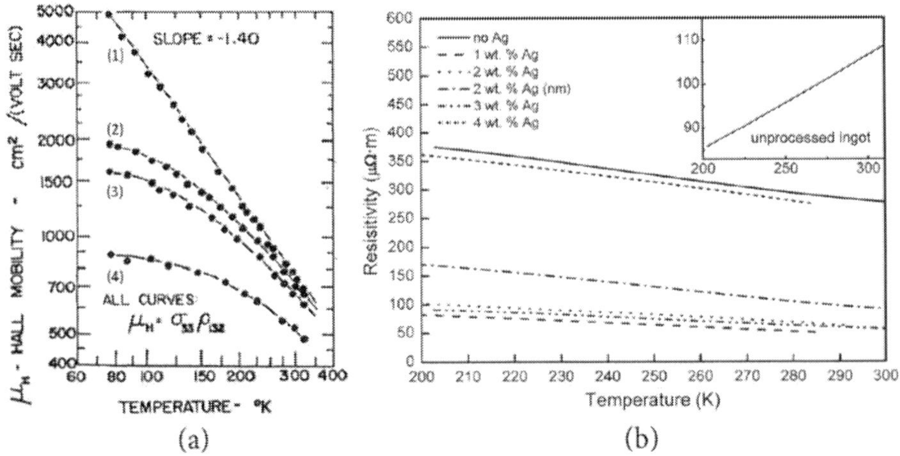

Figure 5. (a) Hall mobility along the *c*-axis as a function of temperature for *p*-type ZnSb at various hole concentrations (1) 3×10^{16} cm^{-3}; (2) 4×10^{17} cm^{-3}; (3) 5.5×10^{17} cm^{-3}; (4) 1×10^{19} cm^{-3}. (Reprinted with permission from [15]. Copyright (1966) by the American Physical Society.) (b) Resistivity vs. temperature of ball-milled and hot-pressed samples with varying silver content. Inset: unprocessed sample (Ingot) [16]. (Reprinted with permission of Springer.)

3.2. Doping effects

The most direct purpose of doping is to vary the charge carrier concentration. A broad range of dopant elements has been reported for ZnSb. The selection of dopant is often rationalized based on the same valence electron counting scheme that is applied to the elemental group IV, III–V or II–VI tetrahedrally bonded semiconductors. These considerations are applied for acceptors, while donors are challenging. One will have acceptors by replacing group I elements for Zn or group IV elements for Sb. It is expected that donors can be substitutes for group III elements on Zn sites and group VI elements on Sb sites. In all cases, there may be an issue with doping efficiency, i.e. not all the added dopant atoms will be electrically active. It is common and qualitatively well-understood for other semiconductors that this inefficiency involves segregation (solid solubility limit), clustering of dopant atoms and/or agglomeration of complexes of dopant atoms and point defects. A theoretical calculation predicts that the optimum hole concentration for the thermoelectric efficiency of ZnSb is around 2×10^{19} cm^{-3} [53], which is achievable in *p*-type ZnSb. For donors in ZnSb, there may also be additional issues to what has been mentioned above.

3.2.1. Acceptors

3.2.1.1. I_{Zn}—acceptors as elements of group I

Cu$_{Zn}$, Au$_{Zn}$, Ag$_{Zn}$ all yield *p*-type conduction [47, 67]. Most of the reported charge carrier concentrations are below the optimum value, and probably depend upon the details of sample preparation. However, the highest hole concentration that have been reported for Ag doping is $4 \times 10^{19}\,cm^{-3}$ (for 0.02 at.% Ag) [19] while that for Cu is $2 \times 10^{19}\,cm^{-3}$ (for 0.1 at.% Cu) [15, 21, 64]. There are several interesting behaviours that involve additional states in the band gap when doping with Cu. Some details can be seen in [52].

3.2.1.2. IV_{Sb} — acceptors as the elements of group IV

Hole concentrations around $(4–14) \times 10^{18}\,cm^{-3}$ were obtained in materials with a content of (0.06–3) at.% Sn [53, 68]. The hole concentration variations with Sn doping concentrations where apparently opposite for these two studies. The highest hole concentration of 14×10^{18} cm^{-3} was obtained for content 0.1 at.% Sn [68] and yielded the highest mobility. It was suggested that two different doping mechanisms are effective in different temperature ranges involving different intrinsic defects, Sn on different lattice sites and their variation with temperature [68].

3.2.1.3. $I_{Zn}IV_{Sb}$ — co-doping

Hole concentrations of $(2–2.5) \times 10^{19}\,cm^{-3}$ have been reported by co-doping of group I (0.15 at.% Cu or Ag)/IV (0.6 at.% Pb, Sn, or Ge)/Cd [20, 21]. The measured transport coefficients at different regions indicate two types of impurity acceptor: one embedded into Zn sites, and another into Sb sites. Here, Cd is not expected to act as an acceptor, but for increasing the phonon scattering and thereby reducing the thermal conductivity. A similar intended function has been applied by adding P to increase alloy phonon scattering in the Cu doped ZnSb [18, 69].

3.2.2. Donors

n-Type ZnSb is desired because (*i*) thermoelectric modules are preferably built of parallel legs of n-type and p-type materials, and it is preferable to use the same material (ZnSb) for minimizing the thermal stress; (*ii*) theoretically, n-type ZnSb is believed to be a much better thermoelectric material than p-type [36, 40, 42]. However, no real successful stable n-type doping has been achieved. But, temporary n-type behaviours have been reported by doping with group III and group VI elements.

3.2.2.1. III_{Zn} — donors as elements of group III

Group III elements have been used as donors to yield n-type conduction in CdSb [70]. It was later reported that ZnSb could also be made n-type by In doping, probably substituting Zn as In_{Zn}. Al_{Zn} and Ga_{Zn} also exhibited temporary n-type behaviour [71, 72]. However, the n-type conduction did not always occur. Justi *et al.* did not achieve n-type ZnSb with Ga despite several attempts with single and polycrystalline ZnSb [14]. Niedziolka *et al.* predicted theoretically by DFT calculations that boron would electronically be a good candidate for n-type ZnSb, but did not succeed to synthesize the material and ascribed it to the high formation energy of a boron atom on a Zn site [36].

3.2.2.2. VI_{Sb} — donors as elements of group VI

Some success with Te doping has been reported. Ueda *et al.* reported that a Te content in very narrow window around 2 at.% yielded *n*-type, possibly by forming substitutions of Te atoms on Sb sites, Te_{Sb}; while at lower concentration (<1 at.%) and higher concentrations (>3 at.%), the samples are always *p*-type. Excess doping with Te results in precipitation of the ZnTe phase and a change in conduction from *n*- to *p*-type [73]. No *n*-type doping was observed by S doping [29].

3.2.2.3. n-Type to p-type transition

A temporary *n*-type behaviour with a transition to *p*-type has been reported. Explanations for these behaviours are related to such factors as oxygen migration on internal surfaces or grain boundaries [71]. Schneider has reported an *n*-type to *p*-type transition in *n*-type $Zn_xCd_{1-x}Sb$ every time when an oxygen gas was flushed into the sample container, and relaxed back to *n*-type after a certain time [71]. Another factor entering into explanations are Zn vacancies acting as acceptors and their migration [36, 42, 74]. A similar transition from *n*-type to *p*-type also occurs in the related compound CdSb, which has been attributed to Cd-vacancies [75].

4. Zn vacancies and intrinsic defects

The theoretical intrinsic charge carrier concentration of perfect ZnSb at room temperature is approximately $2 \times 10^{14} cm^{-3}$ given by $p_i = 2\left(\frac{2\pi m^* k_B T}{h^2}\right)^{\frac{3}{2}} e^{\eta}$, where $\eta = -\frac{E_g}{2k_B T}$, and taking $E_g = 0.53$ eV and $m^* = 0.42 \times m_0$ [15]. However, the experimental measurements show that the charge carrier concentration of the best single crystals at room temperature is around $(1-2) \times 10^{16} cm^{-3}$ [15, 46, 61, 62] and up to $\sim 10^{18} cm^{-3}$ for polycrystalline samples without intended dopants [14, 45]. This deviation is considered due to the intrinsic defects, giving a net hole concentration. The most favoured intrinsic defects in ZnSb are Zn vacancies, which are believed to yield *p*-type conduction [42, 74]. The intrinsic defects in ZnSb have been calculated on by DFT methods [42, 74]. The calculations gave much lower formation energies for Zn vacancies than other intrinsic defects, meaning that Zn vacancies will out-number other intrinsic defects by orders of magnitude. Discrete vacancy defect states are considered to accept electrons from (or donate electrons to) the bands, if they are negatively (or positively) charged. The charge state of the defect will depend on the Fermi level for electrons. This can, in principle, be calculated from the condition of electrical charge conservation (charge neutrality). **Figure 6** shows a conceptual schematic model of a Zn vacancy in different charge states. It is based upon a combination of interpretation of Bjerg *et al.*'s work [74] and a popularization of Fairs Vacancy model for silicon [76].

The vacancy can in principle have any charge states, but only -2, -1 and 0 seem readily accessible by doping and temperature variation. The formation energy for V_{Zn}^- in this config-uration was calculated to be 0.32 eV. The net hole concentration in ZnSb without any doping was then calculated from the requirement of charge neutrality and assuming equilibrium number of vacancies in different charge states (-1 and -2 dominated). The net hole concentration in this configuration was calculated to be 8.8×10^{17} cm^{-3} at room temperature, which is in the range of the experimental Hall concentrations measured by Böttger *et al.* $(4–10) \times 10^{17}$ cm^{-3} [16]. One should also compare the calculated net hole concentration to the hole concentration of undoped single crystal ZnSb that is significantly lower than 8.8×10^{17} cm^{-3}. This apparent discrepancy indicates that exact values of formation energies should be used with caution. However, the concepts and the idea of V_{Zn} as a very important defect seem valid.

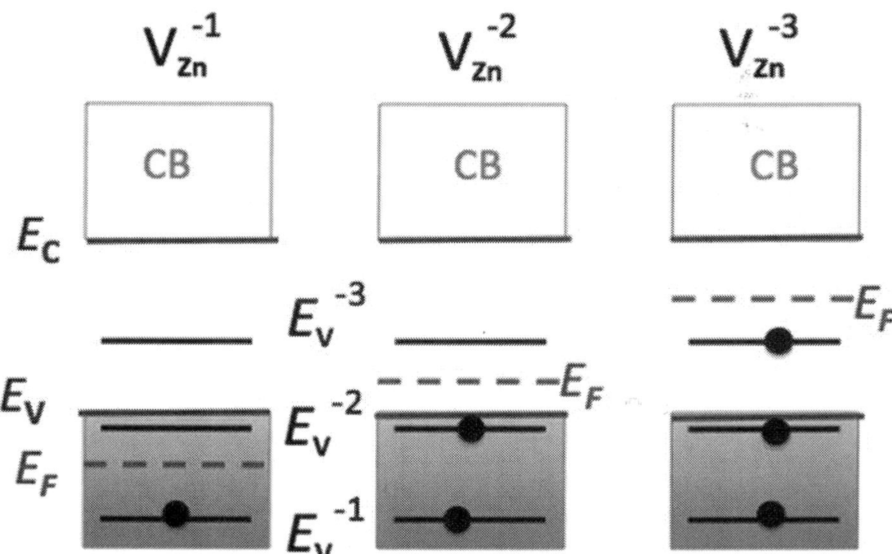

Figure 6. Schematic drawing of occupancy of localized states associated with the Zn vacancy in ZnSb. (Reprinted with author's permission from [29]. ©X. Song, 2016.) The vacancy will have different charge states -1, -2, -3 dependent upon the position of the Fermi level E_F with respect to the levels E_V^{-1}, E_V^{-2} and E_V^{-3}. E_V is the valence band onset and E_C is the conduction band (CB) onset.

There have been many reports on changes in charge carrier concentration after heat treatments, both for single crystal and polycrystalline ZnSb. Many observed a slow recovery to the initial values of the charge carrier concentration [15, 20, 44, 45, 47, 61, 77]. Andronik *et al.* specifically attributed this change to Frenkel defects [77]: by Zn atoms leaving their lattice sites and becoming vacancy-interstitial pairs, V_{Zn}-Zn$_I$. By assuming that all of the measured changes in charge carrier concentration were due to Frenkel defect formation, the activation energy of the process could be determined. The Frenkel defect concentration n_φ was assumed to be given by $n_\varphi = \sqrt{NN'}\,e^{-\frac{E_\varphi}{2k_BT}}$, where $n_\varphi = n_i - n_0$ is the concentration of Frenkel defect, n_0 and n_i are the concentration before and after heating, respectively, N is the concentration of atoms, N' is the concentration of interstitials (there are more than one interstitial position in the lattice), E_φ is the formation energy of a Frenkel pair. The formation energy was determined from experiments to be 0.5 eV. This experimental value seems in the range of formation energies calculated by Bjerg *et al.* [74]. On the other hand, the formation energy for a Zn vacancy calculated by Jund *et al.* gave a different value of 0.8 eV [42]. This calculation used a $Zn_{64}Sb_{64}$ supercell with 128 lattice sites and compared it to the energy of a $Zn_{63}Sb_{64}$ supercell at 0 K, which corresponds to a vacancy concentration of 1.5 at.%.

Recently, we have studied evaporation of Zn by thermogravimetry and Zn vacancy created during vaporization [29]. The net hole concentration was measured to be about $6 \times 10^{18} cm^{-3}$, corresponding to a vacancy concentration of about 0.03 at.%. Schematically, two processes in series were considered as:

$$ZnSb(s) \rightleftharpoons Zn_{1-x}Sb(s) + x \cdot Zn(g), \tag{1}$$

$$ZnSb(s) \rightleftharpoons Zn(g) + Sb(s). \tag{2}$$

Reaction (1) occurs when the V_{Zn} was created but within the dilute limit, while reaction (2) applies to a situation where the Zn vacancy concentration is beyond the solubility limit and ZnSb decomposes into Sb phase and Zn vapour.

5. Impurity band conduction

In an idealized semiconductor, which is well-approximated by pure Si crystals [78], the charge carrier concentration shows the so-called freeze out at low (cryogenic) temperatures: the dopant atoms are not ionized and the charge carrier concentration goes towards zero, characterized by an infinite Hall coefficient in Hall measurements. However, the Hall coefficient in undoped ZnSb has shown a turning point at low temperatures, typically below 50 K, as seen in **Figure 7**, and then a decrease with further cooling, which is explained by impurity band conduction [79]. Here, the term impurity band is most likely tied to defects, but observed phenomena are similar to what can be observed for high doping concentration in semiconductors. Impurity band conduction has been reported in many materials [80–82]. It was also observed for ZnSb by Justi *et al.* in the 1960s [14]. Recently, several others have reported on impurity bands in ZnSb [53, 79, 83].

Let us here analyse sketchily the conditions for observing the characteristics shown in **Figure 7**. The specifics of this observation are given in [79]. The changes in hole concentration as the sample was cooled is in principle similar to that of a textbook low doping concentration semiconductor, where the charge carriers are frozen out of the valence band. The valence band of the sample in **Figure 7** will be nearly empty (for holes) at the lowest temperature. The holes are transferred to the acceptor-based impurity band. However, the holes are mobile in the impurity band and contribute to conduction and Hall effect. Thus, the Hall coefficient decreases with cooling to the lowest temperatures. In order to get conduction in the impurity band, some donor compensating centres are needed. (Without donor level, the impurity band would be full of holes, i.e. empty for electrons and there is no conduction).

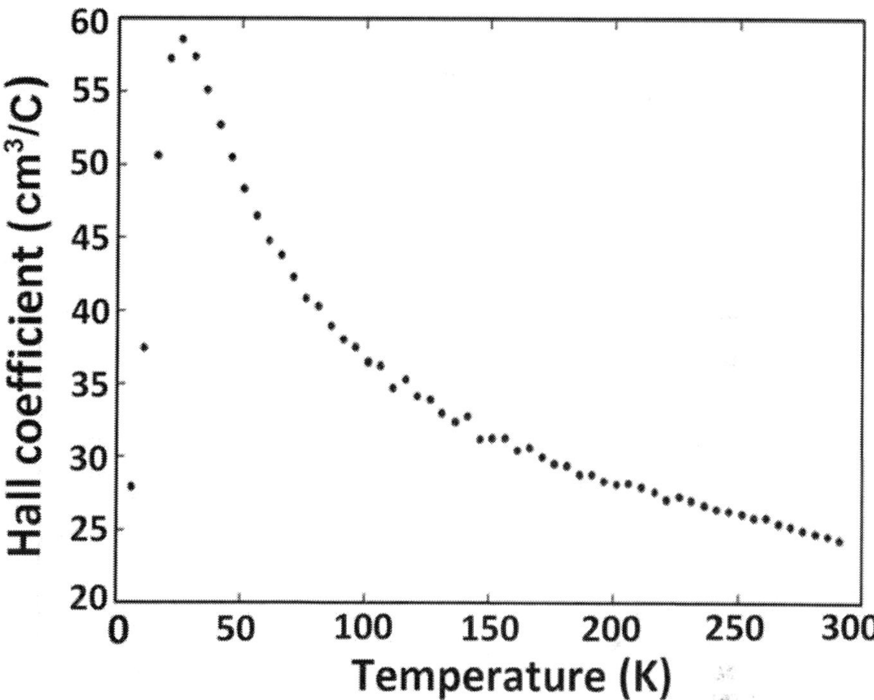

Figure 7. The Hall coefficient of an undoped hot-pressed ZnSb sample. (Reprinted with author's permission from [29]. ©X. Song, 2016.) The Hall coefficient in ideal cases is inversely proportional to the charge carrier concentration. If there was regular freeze-out of the charge carriers, the Hall coefficient goes to infinity at low temperatures. However, in the presented case of undoped ZnSb, there is a turning point in the Hall coefficient, which was interpreted as a signature of an impurity band.

We turn to a situation where the doping concentration is much higher than that of the sample in **Figure 7**, and first consider the large difference in characteristic features of the change in the charge carrier concentration with temperature. **Figure 8** compares the Hall concentration of the charge carrier and the Hall mobility (inset) at low temperature for high (0.3 at.% Cu) and lower concentration (no Cu), respectively (data in Refs. [52, 79]). One can see that the characteristic feature of impurity band conduction vanished in the highly doped sample at low temperature. The situation is qualitatively as follows: A highly doped sample is equivalent to a degenerate semiconductor, where the hole concentration is high and the Fermi level is located in the valence band. The native impurity band would be full of holes and so would the top of the valence band. At the lowest temperatures, the conduction will occur within the valence band. Therefore, one cannot have a similar change with temperature as in the case of undoped ZnSb where there was a vanishing conduction in the valence band at the lowest temperature.

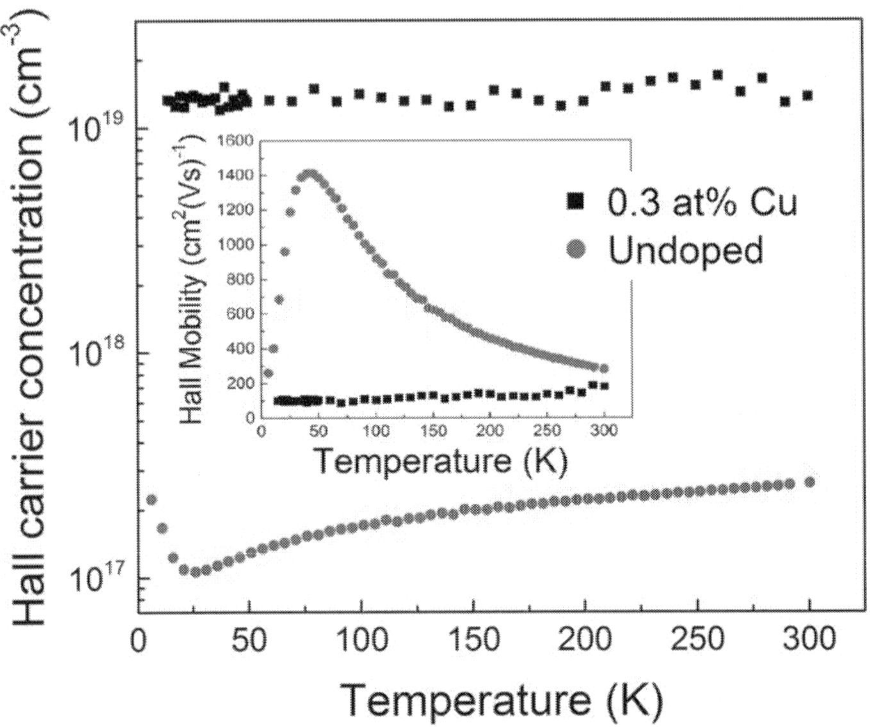

Figure 8. Temperature dependence of the Hall concentration of charge carriers at the temperature lower than 300 K of highly doped ZnSb (0.3 at.% Cu content) and undoped ZnSb. (Reprinted with author's permission from [29]. ©X. Song, 2016.). Inset: the Hall mobility.

5.1. The impurity band in ZnSb — its nature, origin and specific points of interest

The nature and theoretical treatment of general impurity band can be found in textbooks [84]. The band states are considered to come from interactions, which set in for concentrations above a certain value of defect species, such as dopant atoms, impurities or intrinsic crystal defects.

The formation of impurity band is illustrated in **Figure 9** for an n-type semiconductor. Electrons at a donor level can be transferred to a neighbour donor by thermal activation and tunnelling, but without entering into states in the conduction band. One would then have the hopping regime for transport in the material as illustrated in **Figure 9a**. When the donor concentration increases further, the wave functions of the donor states overlap and can form a band, where one has impurity band conduction, as illustrated in **Figure 9b**.

The mobility in the impurity band is typically small because the band is relatively narrow, and as a consequence, one would have a small dispersion curvature in $E(\vec{k})$ and large effective mass. In a semiconductor, at low temperature, conduction can occur dominantly in the impurity band if there are some compensating levels, such that the impurity band is not fully occupied. At a particular higher concentration, there may be a smear of the impurity band and the conduction band, as illustrated in **Figure 9c**. Beyond this concentration, the Mott transition, which is a sharp transition from insulator to metal behaviour, can take place.

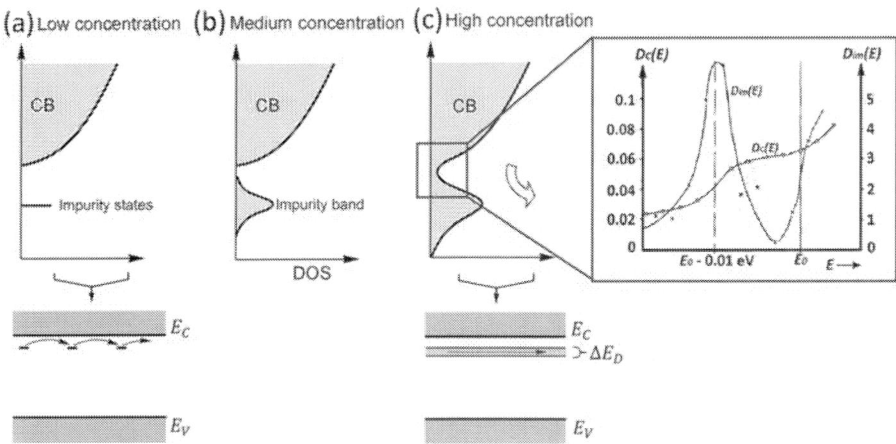

Figure 9. Schematics of impurity state/band and conduction band (CB) for n-type semiconductor at low impurity concentration, medium concentration and high concentration at random impurity distribution. Band diagrams illustrate the hopping conduction and impurity band conduction. (Reprint with author's permission from [84]. ©E. Fred Schubert, 2015). Inset: dimensional schematic calculated density of states for high impurity concentration. E_0 is assumed to be the onset of the conduction band; $D_C(E)$ is the number of conduction band and $D_{im}(E)$ is the density of states of impurities in periodically arrangement. (Reprinted with permission from [85]. Copyright (1953) American Chemical Society.)

Usually, the Mott criterion for when the transition occurs is $N_I \geq 0.014 a_B^{-3}$, where N_I is the atomic concentration of impurities and a_B is the Bohr radius of the impurity. There are several different treatments yielding essentially the same numbers [84]. The most considered situations are those of high concentrations of shallow donors or acceptors. Here, the Mott criterion gives the insulator-metal transition. However, one can create impurity bands with a certain energy anywhere in the band gap. Some of the concentrations for the Mott transition then approximately correspond to the concentration to have an impurity band. Rawat *et al.* have suggested a Yb mid-gap impurity band in PbTe affecting the thermoelectric properties [86].

With the doping concentrations for optimum thermoelectric performance that typically is $\sim10^{19}$ cm^{-3}, it is reasonable to expect that the formation of impurity bands is rather common in thermoelectric materials. Also, impurity band formation is just one of several high doping effects one should expect, such as the Mott transition, band tailing and band gap renormalization [84]. Thus, one should discuss thermoelectric material in the framework of the theory related to heavily doped semiconductors.

When it comes to ZnSb specifically, it has been estimated that the impurity band may exist for the impurity concentrations that are well within the observed doping concentrations. We have determined the critical impurity concentration in ZnSb to be about $6 \times 10^{17}\,cm^{-3}$ in previous study, which is around the acceptor concentrations used in the model to fit the low temperature measurements [79]. For the doping concentration of 10^{18}–$10^{19}\,cm^{-3}$ in ZnSb, one can expect impurity bands to form. For undoped or very lightly doped material, we suggested that point defects, especially Zn vacancies can be expected to be involved in impurity bands. From the discussion of vacancy formation in Section 4, it appears likely that a high concentration of vacancies can be created by annealing, even to some extent they could combine with any other possible impurities or defects in impurity band. The impurities will likely be dependent on the specific dopants and the preparation technique. For example, oxygen is expected to be an impurity in ball-milled material and the amount introduced will depend upon the atmosphere during processing. Presently, one cannot make a conclusion about the importance of oxygen in this context and the solid solubility of oxygen is unknown. Fedorov *et al.* observed evidence for impurity levels in the band gap associated with different combinations of group I (Ag or Cu) and group IV (Sn or Ge) acceptor elements, which all were similar to each other, but different to those of the single acceptor element [21]. Temperature-dependent transport coefficients were also measured that were interpreted as a temperature-dependent energy state being present with a level in the valence band. The energy states were considered as hybridized states formed by mixing characteristics of the valence band and the impurity band. In a general case, a temperature-dependent level can have a similar effect as a defect chemistry reaction involving growth and decrease with temperature of the population of two defect species having different energy, as suggested in Ref. [52].

5.2. Impact of impurity band on thermoelectric properties

An impurity band will have an effect on the transport properties and the thermoelectric device properties. It may not be immediately transparent how. The conduction in the impurity band is perhaps a minor effect in this context. The most important effect may be on the Seebeck coefficient.

5.2.1. Effect on conduction

The effect of impurity bands on the electrical conductivity is expected to be largest when holes in the valence band (for *p*-type) do not contribute to the conduction [53]. This is expected to have a strong effect for samples where the doping is below degenerate, but sufficiently close to the Mott criterion, and in addition at low temperatures. It is expected, that in a thermoelectric material, the mobility of the charge carriers in an impurity band is much lower than that in the valence band, thus the impurity band should only have a modest effect on the electrical conductivity when valence band conduction is strong.

5.2.2. Effect upon Seebeck—effective density of states mass

The density of states effective mass may be affected by an impurity band. The density of states may be changed by several high doping effects including the formation of impurity bands. Qualitatively a smear of the impurity band and valence band is expected. Thus, even though the conduction of impurity band is often only observable at low temperature, the Seebeck coefficient can be affected above room temperature. The details to calculate the transport coefficients can be found in Ref. [79]. The Seebeck coefficient is sensitive to the position of the Fermi level and how the density of states varies with energy. Both these factors can be affected by an impurity band. There have been reports on change of density of states effective mass with varying doping concentration [53]. From a Pisarenko plot, one can find the density of states effective mass by fitting measured data. In Ref. [52], we obtained the best fitting by assuming an impurity band. Further, it was shown that camel-shaped curves of Seebeck coefficient with temperature, which is unusual in ZnSb, could be modelled by a temperature-dependent impurity band. It has been suggested that if one could engineer the energy structure of impurity band, then one could have a tool to enhance the thermoelectric performance [87].

Another impact of the impurity band may be on the n-type to p-type transition. This was suggested by Schneider by the observation of a sharp increase in the electrical conductivity on a temporary n-type and following n-type to p-type transition [71]. On a similar topic, though with different statements, Fedorov *et al.* rationalized the difficulty in doping ZnSb n-type by the formation of an impurity band close to the conduction band [21].

6. ZnSb synthesis techniques

Common synthesis methods for most of bulk thermoelectric materials can be categorized into three groups according to different processing steps, namely stoichiometric melts (SM), powder metallurgic method (PM), pseudo-pulverized and intermixed elements sintering method (Pseudo-PIES), as shown in **Figure 10**.

Polycrystalline ZnSb ingots can be synthesized by the so-called SM methods, i.e. melting of the elemental zinc and antimony followed by solidification in air or quenching in cold water. The purity of the starting elemental zinc and antimony materials has a significant impact on the resulting electrical properties. For example, starting materials with purities of 99.99% and 99.9999% allows to obtain the charge carrier concentration of $\sim10^{19}$ and $\sim10^{16}\,cm^{-3}$, respectively, on undoped polycrystalline ZnSb [14]. Since ZnSb does not melt congruently, solidification will result in a mix of the phases, Zn_4Sb_3, ZnSb and Sb. This mix can be homogenized by sufficient heat treatments to reach the thermodynamic equilibrium state with a single uniform ZnSb phase [14, 15]. Another problem with the solidification is that the sample contains cracks that significantly influence on the electron transport [13].

The solidified ingots are often milled into fine powder and pressed to pellets, which is a procedure that includes the basic ingredients of standard powder metallurgy (PM). Milling offers access to nanosized grains, thus providing possibility to enhance the thermoelectric properties. Earlier studies show that different milling techniques led to a trend of grain size as 80.0, 44.6 and 32.4nm for manually grinding, dry-milling and wet-milling [16], as well as 10 nm for cryo-milling [88], respectively.

The PIES method (pulverized and intermixed elements sintering method) has been introduced into preparation of thermoelectric $(Bi/Sb)_2(Te/Se)_3$ materials, where all the elements are initially mixed and milled to fine powder before hot-pressing (no melting process) [89]. The electrical conductivity of the sample that was synthesized by this method has been reported about 5 times higher than that for the SM sample [90]. Distinguished from a typical PIES method, we often used pseudo-PIES for ZnSb, which partially mixed the dopant element with SM ingot in ball-milling, and then processed hot-pressing.

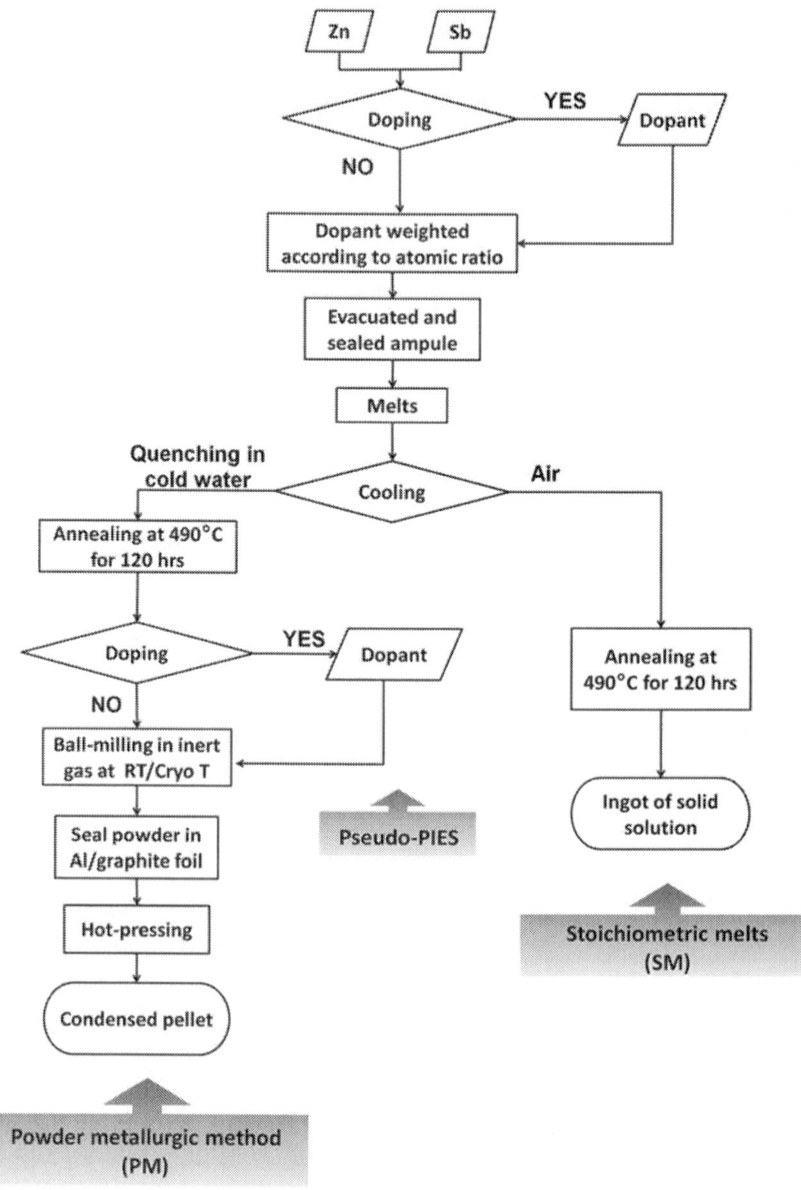

Figure 10. Flow chart of synthesis procedures. (Reprinted with author's permission from [29]. ©X. Song, 2016.)

There are different kinds of compaction techniques that follows powder metallurgy and have been used for fabrication of ZnSb samples. The most common ones are cold-pressing (at room temperature with ultra-pressure 2–10 GPa [83, 91]), hot-pressing (>450°C with pressure of 20–300 MPa [16, 19, 20, 35]), and spark plasma sintering (SPS) (the electrical current is passed through the sample with 5 min reaction time at 350–450°C [83, 92]). One important difference among hot-presses is the manipulation of secondary phases; both removal and proportioning are possible, and obviously depends upon the temperature and duration, but also on details of the instrument design and the environment of the ZnSb powder. Xiong *et al.* have reported that the volume fraction of Sb phase was estimated to be ~2 wt.% by Rietveld refinement in a hot-pressed sample [19] at 673 K in vacuum followed by an evacuated quartz ampoule and annealed at 673 K for 80 h. A recent study on SPS-samples showed also that Sb phase domains were distributed along the samples, accompanied with Zn_4Sb_3 on the surface [92]. Another difference is the final grain size of pellets. We have reported that rapid hot-press helped to minimize the grain size due to shorter cooling time [88].

Another important consideration of the synthesis technique is the ability to produce large amounts of thermoelectric materials in a cost-effective way. Considering that one of the favourable aspects of ZnSb from a commercial point of view is the low materials cost, there have been several efforts where the cost efficiency of the synthesis technique is important [92–94].

7. From laboratory to fabrication

ZnSb practical devices have been produced [9, 95], and there has been a promising achievement on thermoelectric performance of ZnSb in the laboratory. However, it is still challenging to transfer the achievement from laboratory to modern manufacturing. Progress in synthesis from different points of view have also to go through many tests regarding machinability, mechanical stability, thermal stability, thermal cycling and long-term stability, as well as compatibility with targeted fabrication techniques. Several of these issues are expected to contribute to—as well as benefit from—a further fundamental understanding of ZnSb, when practical solutions on short and long timescales are targeted. On a short to medium timescale, ZnSb can take advantage of new fabrication technologies that has been developed, but using traditional approaches for the device functionality. On a longer perspective, ZnSb may also be brought further into the explorations of new nanotechnology approaches to improve the performance of possible future generations of thermoelectrics.

One hindrance towards an ideal thermoelectric module made entirely of ZnSb is the inability to synthesis of stable *n*-type ZnSb. Although theoretical modelling shows favourable electronic structure of *n*-type ZnSb, there seems to be no promising paved routes to success. The direct synthesis of *n*-type ZnSb by doping would need a breakthrough. From an optimistic point of view that may arise indirectly from various other investigations on ZnSb, perhaps through defect engineering or a combination of different approaches, for example, modulation doping by embedded higher band gap materials with the appropriate band offsets for supply of electrons combined with compensation of Zn vacancy acceptors. A practical compromising route towards module-making is using another semiconductor than ZnSb for the n-type leg, for instance $Mg_2Sn_{1-x}Si_x$ that matches ZnSb well in expected operation temperature, has the same environmentally friendly profile [20], as well as a low cost on raw materials. There might also be other suitable material candidates. Any practical problems with thermal expansion mismatch of materials in a module would have to be solved. Previous experiences with ZnSb modules have made it necessary to dope or add elements to ZnSb in order to achieve suitable mechanical properties. Fortunately, there has been a large development in packaging technology for electronic devices in the last couple of decades. New options for substrates and bonding techniques may be offered and meet the requirements on thermal expansion of the semiconductors.

The thermal stability of the synthesized ZnSb needs to be tested and probably be improved. This is one area where both fundamental studies and practical solutions may enter. ZnSb samples are subjected to Zn evaporation at high temperatures. The evaporation depends naturally very much on the ambient and the surface conditions. It is possible that protective layers can be applied to minimize evaporation. The situation has some similarities to that of several binary electronic materials, such as III–V materials, where one of the elements have a much higher vapour pressure, than what can be tolerated at the desired processing temperature. For GaAs, dielectric films SiO_2 and Si_3N_4 have been used for the purpose of preventing arsenic evaporation. A similar approach with a conformal deposition of a protective dielectric layer may be advantageous for ZnSb. Thermal stability also has to do with the thermal generation of point defects and their diffusion at elevated temperatures. The understanding of the phenomena is unsatisfactory from an academic point of view, in particular on the level of defect chemistry and electronic structure, but there are many experimental observations of the simple electrical parameters. Several authors have reported that after a heat treatment of ZnSb, the electrical conductivity and the charge carrier concentration increased, while the Seebeck coefficient decreased. The change was followed by a slow recovery towards the initial values at room temperature [15, 20, 44, 45, 47, 61, 77]. The characteristics can be related it to the V_{Zn}-Zn_I Frenkel pair formation at elevated temperatures, and the recovery caused by their slower recombination at lower temperature. The vacancy concentration was linked to hole concentration. It was rationalized that these hysteresis effects would not be significant at high doping concentrations [15]. The doping effect on the vacancy concentration was then not considered. A detailed understanding of the vacancies, interstitials and their energy levels, ionization and formation energies is needed to understand the influence for higher doping concentration. The influence of more complicated defects can also be a large challenge. There are also reports on various temperature-cycling phenomena [52, 96, 97], involving the doping atoms and energy levels of these. Some of these effects may differ for different synthesis details.

The thermal stability is referring to all properties of the material, including the thermal conductivity. We have reported grain growth induced by heating, particularly in nanostructured bulk samples [88]. The grain growth will naturally induce a change of thermal conductivity due to the dependence on phonon scattering. To which extent, it constitutes a practical problem depends upon the targeted operation temperature. Approaches to minimize grain growth usually consist of adding atoms that segregate in grain boundaries, thereby preventing grain growth. This is an area that needs further study.

There are various routes that can make ZnSb a part of long-term exploration to improve thermoelectrics by nanostructures (not just nanograins). Some of them may use ways of depositing films of ZnSb, such as by MOCVD, sputtering [98, 99] and electroplating [11, 100], *etc.* By these techniques, it could be feasible to make composites and layered structures in a controlled way with different materials with suitable band offsets for energy filtering [101]. Possibilities of making high quality epitaxial films may also be attractive for fundamental material property studies. One might also do band engineering in the material by introducing misfit stress. Thin-film deposition may also offer a possibility to grow template nanostructures and exploiting the possibilities of quantum confinement in ZnSb and study the conduction band properties of ZnSb by injection into nanostructures.

Acknowledgements

The authors acknowledge support by the Norwegian Research Council under contract NFR11-40-6321 (NanoThermo) and the University of Oslo. The authors also thank to Dr. Patricia Almeida Carvalho for her contribution to transmission electron microscopic imaging.

Author details

Xin Song* and Terje G. Finstad

*Address all correspondence to: xins@fys.uio.no

Department of Physics, University of Oslo, Oslo, Norway

References

[1] Seebeck, T.J., Magnetische Polarisation der Metalle und Erze durch Temperatur Differenz. A. J. v. Oettingen, 1823.

[2] Berland, K., et al.: Enhancement of thermoelectric properties by energy filtering: Theoretical potential and experimental reality in nanostructured ZnSb. Journal of Applied Physics. 2016; **119**(12):125103.

[3] Becquerel, E.: Mempire sur les pouvoirs thermo-electroiques des corps et sur les piles thermo-electroques. Annales de chimie et de physique, 1866; **4**(8):389.

[4] in La Nature. 1874. p. 19.

[5] Halla F, Adler J.:, Röntgenographische Untersuchungen im System Cadmium-Antimon. Zeitschrift für anorganische und allgemeine Chemie. 1929; **185**(1):184-192.

[6]   Halla, F, Nowotny H, Tompa H.: Röntgenographische Untersuchungen im System (Zn, Cd)–Sb.II. Zeitschrift für anorganische und allgemeine Chemie. 1933; **214**(2):196-200.

[7]   Olander A.: The Crystal Structure of CdSb. Zeitschrift für Kristallographie. 1935; **91**(3/4):243-247.

[8]   Almin KE.: The Crystal Structure of CdSb and ZnSb. Acta Chemica Scandinavica. 1948; **3**(3-4):400-407.

[9]   Telkes M.: Solar Thermoelectric Generators. Journal of Applied Physics. 1954; **25**(6):765-777.

[10]   Vedernikov MV, Iordanishvili EK. A.F. Ioffe and origin of modern semiconductor thermoelectric energy conversion. In: Proceedings of The XVII International Conference on Thermoelectrics (ICT 98); 1998. p. 37-42.

[11]   Saadat S, et al.: Template-Free Electrochemical Deposition of Interconnected ZnSb Nanoflakes for Li-Ion Battery Anodes. Chemistry of Materials. 2011; **23**(4):1032-1038.

[12]   Wang G, et al.: Investigation on pseudo-binary ZnSb-Sb$_2$Te$_3$ material for phase change memory application. Journal of Alloys and Compounds, 2015:341-346.

[13]   Miller RC. Survey of Known Thermoelectric Materials: ZnSb, In: Heikes RR, Ure Jr. RW, editors. Thermoelectricity: Science and engeneering. New York: Interscience Publishers; 1961. p. 405.

[14]   Justi E, Rasch W, Schneider G.: Untersuchungen an zonengeschmolzenen ZnSb-einkristallen. Advanced Energy Conversion. 1964; **4**(1):27-38.

[15]   Shaver PJ, Blair J.: Thermal and Electronic Transport Properties of p-Type ZnSb. Physical Review. 1966; **141**(2):649-663.

[16]   Böttger PHM, et al.: Influence of Ball-Milling, Nanostructuring, and Ag Inclusions on Thermoelectric Properties of ZnSb. Journal of Electronic Materials, 2010. **39**(9): p. 1583-1588.

[17]   Okamura C, Ueda T, Hasezaki K.: Preparation of Single-Phase ZnSb Thermoelectric Materials Using a Mechanical Grinding Process. Materials Transactions. 2010; **51**(5):860-862.

[18]   Valset K, et al.: Thermoelectric properties of Cu doped ZnSb containing Zn$_3$P$_2$ particles. Journal of Applied Physics. 2012; **111**(2):023703.

[19]   Xiong D-B, Okamoto NL, Inui H.: Enhanced thermoelectric figure of merit in p-type Ag-doped ZnSb nanostructured with Ag$_3$Sb. Scripta Materialia. 2013; **69**(5):397-400.

[20]   Fedorov MI, et al.: New Interest in Intermetallic Compound ZnSb. Journal of Electronic Materials. 2014; **43**(6):2314-2319.

[21]   Fedorov MI, et al.: Thermoelectric efficiency of intermetallic compound ZnSb. Semiconductors. 2014; **48**(4):432-437.

[22]   Valset K, Song X, Finstad TG. Stability and thermoelectric properties of Cu doped ZnSb. In: Proceedings of The European Conference on Thermoelectrics (ECT 2014); 2014. Madrid, Spain.

[23] Guo Q, Luo S.: Improved thermoelectric efficiency in p-type ZnSb through Zn deficiency. Functional Materials Letters. 2015; **08**(02):1550028.

[24] Valset K. A Technique to Measure Thermoelectric Figure of Merit and Heat Flow at High Temperatures by Cancelling Heat Losses. In: Proceedings of The 12th European Conference on Thermoelectrics; Materials Today. 2015; **2**(2):721-728.

[25] Carter FL, Mazelsky R.: The ZnSb structure; A further enquiry. Journal of Physics and Chemistry of Solids. 1964; **25**(6):571-581.

[26] Mikhaylushkin AS, Nylén J, Häussermann U.: Structure and Bonding of Zinc Antimonides: Complex Frameworks and Narrow Band Gaps. Chemistry – A European Journal. 2005; **11**(17):4912-4920.

[27] Mozharivskyj Y, et al.: A Promising Thermoelectric Material: Zn_4Sb_3 or $Zn_{6-\delta}Sb_5$. Its Composition, Structure, Stability, and Polymorphs. Structure and Stability of $Zn_{1-\delta}Sb$. Chemistry of Materials. 2004; **16**(8):1580-1589.

[28] Toman K.: On the structure of ZnSb. Journal of Physics and Chemistry of Solids. 1960; **16** (1): 160-161.

[29] Song X. Thermoelectric Transport and Microctructure of ZnSb, In: Department of Physics. 2016, University of Oslo: Oslo, Norway. p. 138.

[30] Mooser E, Pearson WB.: Chemical Bond in Semiconductors. Physical Review. 1956; **101** (5): 1608-1609.

[31] Haussermann U, Mikhaylushkin AS.: Electron-poor antimonides: complex framework structures with narrow band gaps and low thermal conductivity. Dalton Transactions. 2010. **39**(4):1036-1045.

[32] Arushanov EK.: Crystal growth, characterization and application of II-V compounds. Progress in Crystal Growth and Characterization. 1986. **13**(1):1-38.

[33] Justi E, Lautz G.: Über die Störstellen- und Eigenhalbleitung intermetallischer Verbindungen. In: Zeitschrift für Naturforschung A. 1952. p. 191.

[34] Velický B, Frei V.: The chemical bond in CdSb. Cechoslovackij fiziceskij zurnal B. 1963; **13** (8):594-598.

[35] Fischer A, et al.: Synthesis, Structure, and Properties of the Electron-Poor II–V Semiconductor ZnAs. Inorganic Chemistry. 2014; **53**(16):8691-8699.

[36] Niedziolka K, et al.: Theoretical and experimental search for ZnSb-based thermoelectric materials. Journal of Physics: Condensed Matter., 2014; **26**(36):365401.

[37] Niedziółka K, Jund P.: Influence of the Exchange-Correlation Functional on the Electronic Properties of ZnSb as a Promising Thermoelectric Material. Journal of Electronic Materials. 2015; **44**(6):1540-1546.

[38] Benson D, Sankey OF, Häussermann U.: Electronic structure and chemical bonding of the electron-poor II-V semiconductors ZnSb and ZnAs. Physical Review B. 2011; **84** (12):125211.

[39] Böttger PHM, et al.: Electronic structure of thermoelectric Zn-Sb. Journal of Physics: Condensed Matter. 2011; **23**(26):265502.

[40] Bjerg L, Madsen GKH, Iversen BB.: Enhanced Thermoelectric Properties in Zinc Antimonides. Chemistry of Materials 2011; **23**(17):3907-3914.

[41] Zhao J-H, et al.: First Principles Study on the Electronic Properties of $Zn_{64}Sb_{64-x}Te_x$ Solid Solution (x = 0, 2, 3, 4). International Journal of Molecular Sciences. 2011; **12**(5):3162-3169.

[42] Jund P, et al.: Physical properties of thermoelectric zinc antimonide using first-principles calculations. Physical Review B. 2012; **85**(22):224105.

[43] Yamada Y.: Band structure calculation of ZnSb. physica status solidi (b). 1978; **85**(2):723-732.

[44] Turner WJ, Fischler AS, Reese WE.: Physical Properties of Several II-V Semiconductors. Physical Review. 1961; **121**(3):759-767.

[45] Komiya H, Masumoto K, Fan HY.: Optical and Electrical Properties and Energy Band Structure of ZnSb. Physical Review. 1964; **133**(6A):A1679-A1684.

[46] Eisner RL, Mazelsky R, Tiller WA.: Growth of ZnSb Single Crystals. Journal of Applied Physics 1961; **32**(10):1833-1834.

[47] Kostur NL, Psarev VI.: Electrical properties of doped single crystals of ZnSb. Soviet Physics Journal. 1967; **10**(2):21-23.

[48] Zhang LT, et al.: Effects of ZnSb and Zn inclusions on the thermoelectric properties of β-Zn_4Sb_3. Journal of Alloys and Compounds. 2003; **358**(1-2):252-256.

[49] Goldsmid HJ, Sharp JW.: Estimation of the thermal band gap of a semiconductor from Seebeck measurements. Journal of Electronic Materials. 1999; **28**(7):869-872.

[50] Schmitt J, et al.: Resolving the true band gap of ZrNiSn half-Heusler thermoelectric materials. Materials Horizons. 2015; **2**(1):68-75.

[51] Gibbs ZM, et al.: Band gap estimation from temperature dependent Seebeck measurement-Deviations from the $2e|S|_{max}T_{max}$ relation. Applied Physics Letters. 2015; **106**(2):022112.

[52] Valset K, Song X, Finstad TG.: A study of transport properties in Cu and P doped ZnSb. Journal of Applied Physics. 2015; **117**(4):045709.

[53] Böttger PHM, et al.: Doping of p-type ZnSb: Single parabolic band model and impurity band conduction. physica status solidi (a). 2011; **208**(12):2753-2759.

[54] Van Overstraeten RJ, Mertens RP.: Heavy doping effects in silicon. Solid-State Electronics. 1987; **30**(11):1077-1087.

[55] Toberer ES, et al.: Traversing the Metal-Insulator Transition in a Zintl Phase: Rational Enhancement of Thermoelectric Efficiency in $Yb_{14}Mn_{1-x}Al_xSb_{11}$. Advanced Functional Materials. 2008; **18**(18):2795-2800.

[56] May AF, et al.: Characterization and analysis of thermoelectric transport in n-type $Ba_8Ga_{16-x}Ge_{30+x}$. Physical Review B. 2009; **80**(12):125205.

[57] Toberer ES, May AF, Snyder GJ.: Zintl Chemistry for Designing High Efficiency Thermoelectric Materials. Chemistry of Materials. 2010; **22**(3):624-634.

[58] Mlnaříková L, Tříska A, Štourač L.: The transport phenomena of pure and doped p-type ZnSb. Czechoslovak Journal of Physics B. 1970; **20**(1):63-72.

[59] Stevenson MJ, Cyclotron Resonance in II-V Semiconductcors. In: Proceedings of The International Conference on Semiconductor Physics. 1961. Prague: Czechoslovakian Academy of Science.

[60] Telkes M.: The Efficiency of Thermoelectric Generators. I. Journal of Applied Physics. 1947; **18**(12):1116-1127.

[61] Kot MV, Kretsu IV.: Anisotropy of certain electrical properties of single crystals of Zinc Antimony. Fiz.Tverd.Tela. 1960; 2:1250 (Soviet Physics-Solid State. 1960; 2:1134).

[62] Hrubý A, Beránková J, Míšková V.: Growth of ZnSb Single Crystals. physica status solidi (b). 1963; **3**(2):289-293.

[63] Závêtová M.: Absorption Edge of ZnSb. physica status solidi (b). 1964; **5**(1):K19-K21.

[64] Prokofieva LV, et al.: Doping and defect formation in thermoelectric ZnSb doped with copper. Semiconductors. 2014; **48**(12):1571-1580.

[65] Ridley BK. Quantum Processes in Semiconductors. 4th ed. Oxford: Oxford Science Publications; 1999.

[66] Goldsmid HJ. Introduction to Thermoelectricity. Oxford: Oxford University Press; 2009.

[67] Ito M, Ohishi Y, Muta H, Kurosaki K, Yamanaka S.: Thermoelectric properties of Zn-Sn-Sb based alloys. In: MRS Proceedings Symposium LL – Thermoelectric Materials for Solid-State Power Generation and Refrigeration. 2011.

[68] Prokofieva LV, Konstantinov PP, Shabaldin AA.: On the tin impurity in the thermoelectric compound ZnSb: Charge-carrier generation and compensation. Semiconductors. 2016; **50** (6): 741-750.

[69] Sottmann J, et al.: Synthesis and Measurement of the Thermoelectric Properties of Multiphase Composites: ZnSb Matrix with Zn_4Sb_3, Zn_3P_2, and Cu_5Zn_8. Journal of Electronic Materials. 2013;. **42**(7):1820-1826.

[70] Šmirous K, Hrubý A, Štourač L.: The influence of impurities on the electric and thermoelectric properties of CdSb single crystals. Cechoslovackij fiziceskij zurnal B. 1963; **13** (5):350-357.

[71] Schneider G.: Preparation and Properties of n-Type ZnSb. physica status solidi (b). 1969; **33**(2):K133-K136.

[72] Zeid AA, Schneider G.: Various Donors in n-ZnSb and the Influence of Sample Treatment. Zeitschrift für Naturforschung A. 1975; **30**(3):381.

[73] Ueda T, et al.: Effect of Tellurium Doping on the Thermoelectric Properties of ZnSb. Journal of the Japan Institute of Metals and Materials. 2010; **74**(2):110-113.

[74] Bjerg L, Madsen GKH, Iversen BB.: Ab initio Calculations of Intrinsic Point Defects in ZnSb. Chemistry of Materials. 2012; **24**(11):2111-2116.

[75] Silvey GA, Lyons VJ, Silvestri VJ.: The Preparation and Properties of Some II – V Semiconducting Compounds. Journal of The Electrochemical Society. 1961; **108**(7):653-658.

[76] Campbell SA. The Science and Engineering of Microelectronic Fabrication. Oxford: University Press; 1996.

[77] Kretsu IV, Kot MV.: Thermal Dissociation of Cadmium and Zinc Antimonide Crystals. Uchenye Zapiski Kazanskogo Universiteta (Proceedings of Kazan University), 1961; **49**: 105-111.

[78] Singleton J, ed. Band Theory and Electronic Properties of Solids. Oxford Master Series in Condensed Matter Physics. 2001.

[79] Song X, et al.: Impurity band conduction in the thermoelectric material ZnSb. Physica Scripta. 2012; **2012**(T148):014001.

[80] Hung CS.: Theory of Resistivity and Hall Effect at Very Low Temperatures. Physical Review. 1950; **79**(4):727-728.

[81] Hung CS, Gliessman JR.: The Resistivity and Hall Effect of Germanium at Low Temperatures. Physical Review. 1950; **79**(4):726-727.

[82] Hung CS, Gliessman JR.: Resistivity and Hall Effect of Germanium at Low Temperatures. Physical Review. 1954; **96**(5):1226-1236.

[83] Eklof D, et al.: Transport properties of the II-V semiconductor ZnSb. Journal of Materials Chemistry A. 2013; **1**(4):1407-1414.

[84] Schubert EF. Physical Foundations of Solid-State Devices. 2006.

[85] James HM, Ginzbarg AS.: Band Structure in Disordered Alloys and Impurity Semiconductors. The Journal of Physical Chemistry. 1953; **57**(8):840-848.

[86] Rawat PK, Paul B, Banerji P.: Impurity-band induced transport phenomenon and thermoelectric properties in Yb doped $PbTe_{1-x}I_x$. Physical Chemistry Chemical Physics. 2013; **15**(39):16686-16692.

[87] Goldsmid HJ.: Impurity Band Effects in Thermoelectric Materials. Journal of Electronic Materials. 2012; **41**(8):2126-2129.

[88] Song X, et al.: Nanostructuring of Undoped ZnSb by Cryo-Milling. Journal of Electronic Materials. 2015; **44**(8):2578-2584.

[89] Ohta T, et al.: Pulverized and Intermixed Elements Sintering Method for (Bi, Sb)$_2$(Te, Se)$_3$ based n-type Thermoelectric Devices. IEEJ Transactions on Power and Energy. 1991; **111** (6): 670-674.

[90] Toshitaka O, Takenobu K.: PIES Method of Preparing Bismuth Alloys, In: CRC Handbook of Thermoelectrics. CRC Press; 1995.

[91] Zhao WY, Wang Y, Zhai P, Tang X, Zhang Q.: Method for forming ZnSb-based block thermoelectric material at ultra-high pressure and cold pressure. 2007, Google Patents.

[92] Blichfeld AB, Iversen BB.: Fast direct synthesis and compaction of phase pure thermoelectric ZnSb. Journal of Materials Chemistry C. 2015; **3**(40):10543-10553.

[93] Lee HB, et al.: Thermoelectric properties of screen-printed ZnSb film. Thin Solid Films. 2011; **519**(16):5441-5443.

[94] Pothin R, et al.: Preparation and properties of ZnSb thermoelectric material through mechanical-alloying and Spark Plasma Sintering. Chemical Engineering Journal. 2016; **299**: 126-134.

[95] Farmer M.G. Improvement in thermo-electric batteries. 1870, Google Patents.

[96] Shabaldin AA, et al.: The Influence of Weak Tin Doping on the Thermoelectric Properties of Zinc Antimonide. Journal of Electronic Materials, 2016. **45**(3): p. 1871-1874.

[97] Shabaldin AA, et al.: Acceptor Impurity of Copper in ZnSb Thermoelectric. Materials Today: Proceedings of The 12th European Conference on Thermoelectrics. 2015; **2**(2):699-704.

[98] Fan P, et al.: Thermoelectric properties of zinc antimonide thin film deposited on flexible polyimide substrate by RF magnetron sputtering. Journal of Materials Science: Materials in Electronics. 2014; **25**(11):5060-5065.

[99] Zheng Z-H, et al.: Enhanced thermoelectric properties of Cu doped ZnSb based thin films. Journal of Alloys and Compounds. 2016; **668**:8-12.

[100] Mann O, Freyland W.: Mechanism of formation and electronic structure of semiconducting ZnSb nanoclusters electrodeposited from an ionic liquid. Electrochimica Acta. 2007; **53**(2):518-524.

[101] Flage-Larsen EL, Martin O. Band structure guidelines for higher figure-of-merit; analytic band generation and energy filtering. In: Rowe DM, editor. Thermoelectrics and its Energy Harvesting: Materials, Preparation, and Characterization in Thermoelectrics. CRC Press; 2012.

Chapter 7

Silver-Antimony-Telluride: From First-Principles Calculations to Thermoelectric Applications

Yaron Amouyal

Additional information is available at the end of the chapter

http://dx.doi.org/10.5772/66086

Abstract

Silver-antimony-telluride (AgSbTe$_2$) based compounds have emerged as a promising class of materials for thermoelectric (TE) power generation at the mid-temperature range. This Chapter demonstrates utilization of first-principles calculations for predicting TE properties of AgSbTe$_2$-based compounds and experimental validations. Predictive calculations of the effects of La-doping on vibrational and electronic properties of AgSbTe$_2$ compounds are performed applying the density functional theory (DFT), and temperature-dependent TE transport coefficients are evaluated applying the Boltzmann transport theory (BTE). Experimentally, model ternary (AgSbTe$_2$) and quaternary (3 at. % La-AgSbTe$_2$) compounds were synthesized, for which TE transport coefficients were measured, indicating that thermal conductivity decreases due to La-alloying. The latter also reduces electrical conductivity and increases Seebeck coefficients. All trends correspond with those predicted from first-principles. Thermal stability issues are essential for TE device operation at service conditions, e.g. changes of matrix composition and second-phase precipitation, and are also addressed in this study on both computational and experimental aspects. It is shown that La-alloying affects TE figure-of-merit positively, e.g., improving from 0.35 up to 0.50 at 260 °C. We highlight the universal aspects of this approach that can be applied for other TE compounds. This enables us screening their performance prior to synthesis in laboratory.

Keywords: silver-antimony-telluride, first-principles calculations, thermoelectric transport properties, Boltzmann transport theory, lattice dynamics, thermal stability

1. Introduction

It is of utmost technological importance to develop predictive tools that will provide us with information about design of materials' functional properties. In this context, density functional theory (DFT) first-principle calculations offer us such possibilities [1–4], allowing us

calculation of structural, interfacial, vibrational, and electronic properties. Knowledge of these properties and how they depend on temperature and material's composition are essential to assess total thermoelectric (TE) performance of thermoelectric device. Among recently investigated TE materials, silver-antimony-telluride ($AgSbTe_2$)-based alloys have emerged as a promising class of materials for TE power generation in low- to mid-temperature range. These compounds are derivatives of lead-antimony-silver-telluride (LAST)-based alloys of $AgPb_mSbTe_{2+m}$ form [5–9], which exhibit large TE figure-of-merit ZT values ranging from 1.3 to 1.7 [5, 10, 11], which are associated with the intrinsically good TE properties of $AgSbTe_2$-phase.

$AgSbTe_2$-based alloys serve not only for TE power conversion or cooling, but also for non-volatile electronic memory, being classified as phase change materials, as demonstrated in thorough investigation by Wuttig and coworkers [12–15]. They have attracted scientific interest owing to their special nature of interatomic bonding and vibrational properties [16–18]. On TE aspect, their superior performance is associated mainly to glass-like, intrinsically low values of lattice thermal conductivities, as low as $0.6\ W\ m^{-1}\ K^{-1}$ [19]. This anomaly is associated either to strong anharmonicity of interatomic forces [20, 21] or to relatively large variance of interatomic forces prevailing between Ag^+ and Sb^{3+} cations, encouraging phonon scattering [22]. Additionally, resonant bonding yields high level of structural instability, that is accompanied, for instance, by spontaneous phase decomposition [23, 24]. This, intriguingly, what makes $AgSbTe_2$-based alloys good materials for both TE and phase change applications. Owing to these peculiarities, these alloys have recently been investigated extensively, either experimentally [19, 25–30] or computationally [18, 22, 31–33].

Despite of relatively high ZT values of $AgSbTe_2$ phase, it is still challenging to increase them to the range of 2–3. Reaching at this limit will enable us employing this material for energy conversion at power levels >500 W [34]. Reduction in lattice thermal conductivity is a conventional way to enhance TE performance and is achieved by either doping with solute elements [35] or formation of second phases to stimulate phonon scattering [36–38]. These lattice defects affect, of course, electronic properties, mainly electrical conductivity and Seebeck coefficient. Attempts to improve TE properties of $AgSbTe_2$-based alloys by doping with different elements [19, 25, 39–47], as well as, by formation of second phase precipitates [46, 48–51], proved to be successful, as reported in the exhaustive studies of Zhang et al., Du et al., and Mohanraman et al. Alloying with second phase forming elements raises imperative question about material's thermal stability, when employed in TE generators under service conditions, with engineering implications [51–53].

Among efforts to improve TE properties of $AgSbTe_2$-based alloys, Min et al. reported on improvement of electron transport properties due to La-doping [27]. Positive effects on PbTe compound due to La-doping were recorded, as well [54, 55]. In their recent study, Min et al. introduce a complete analysis of TE properties of $AgSbTe_2$ doped with La of different concentrations [56].

Notwithstanding the aforementioned successful experimental and computational attempts, a set of experimental routines, that is initiated and directed by predictions from first principles for complete TE performance or any other computational procedure, is missing.

A significant step in this direction is introduced by our previous investigations of $AgSbTe_2$-based phase, involving both computational and experimental aspects [57, 58]. Vibrational properties of both $AgSbTe_2$-based and La-doped-$AgSbTe_2$ alloys, including frequency-dependent vibrational density of states functions (v-DOS), temperature-dependent heat capacity, sound velocities, and Debye temperatures were evaluated employing lattice dynamics first-principles calculations. It was reported that La-doping reduces average sound velocity and varies v-DOS of $AgSbTe_2$-based phase [57, 58]. Quantitatively, lattice thermal conductivity of $La_{0.125}Ag_{0.875}SbTe_2$ alloy was calculated to be lower by ca. 14%, than that of $AgSbTe_2$-based phase at 300 K [58]. Experimental validations of these effects of La-alloying on reducing lattice thermal conductivity were made, as well [58].

This chapter introduces a refined approach of evaluating temperature-dependent lattice thermal conductivity from data obtained *ab-initio*, as well as, calculations of electronic transport coefficients. Most importantly, this chapter presents experimental validations for the entire dataset obtained from first-principles, including thermal and electrical measurements.

2. Chapter outline

This chapter consists primarily of original computational and experimental data along with data, that were reported by us earlier [57, 58], and is aimed at drawing a complete picture depicting the role of lanthanum-alloying in silver-antimony-telluride-based alloys on a broad TE view. Herein, we demonstrate how alloying of $AgSbTe_2$ (P4/mmm) alloy with lanthanum solute atoms brings about significant reduction in thermal conductivity with positive effects on TE power factor, as well; thus, achieving improved ZT values. This is achieved by DFT calculations of structural, interfacial, vibrational, and electronic properties performed for La-free and La-doped alloys, followed by experimental validation implemented by thermal and electronic transport measurements.

Computational procedures are divided into the following steps:

1. Total energy calculations for different polymorphs of $AgSbTe_2$ phase are implemented to evaluate their Helmholtz free energies, indicating which one is the most stable around and above room temperature.

2. Vibrational calculations are performed for both La-free and La-doped lattices, including phonon dispersion and density of states, average sound velocity, and Debye temperature. These values enable us evaluating temperature-dependent lattice thermal conductivity values.

3. Electronic calculations of band structures of both La-free and La-doped lattices are performed, and the resulting transport coefficients are derived applying Boltzmann transport theory.

4. To consider the case in which Sb_2Te_3- and Sb_8Te_3-phases precipitate inside $AgSbTe_2$-matrix, similar DFT transport coefficient calculations are performed for both phases, as well. In

this context, the molar formation energies of both phases and the free energies of their interfaces with $AgSbTe_2$-matrix are simulated to predict their thermal stability and nucleation sequence.

5. Additionally, to address the influence of deviations from $AgSbTe_2$-stoichiometry on electron transport properties, the latter is simulated for two off-stoichiometric model alloys Ag_3SbTe_4 and $AgSb_3Te_4$.

Experimental procedures are divided into the following steps:

1. Model ternary ($AgSbTe_2$) and quaternary (3 at.% La-$AgSbTe_2$) alloys are synthesized by vacuum melting followed by quenching and hot-pressing. The appropriate conditions enabling formation of $AgSbTe_2$-matrix that dissolves La-atoms with no La-rich precipitates are found.

2. Differential scanning calorimetry (DSC) tests are implemented for both La-free and La-doped alloys to address thermal stability issues and how they are influenced by La-additions.

3. Temperature-dependent thermal conductivity of both alloys is determined to realize effects of La-doping and to compare them with those predicted from first-principles.

4. Similarly, both temperature-dependent electrical conductivity and Seebeck coefficients are measured for ternary and quaternary alloys to realize effects of La-doping and to compare them with those predicted from first-principles.

5. Finally, to assess whether La-doping contributes to conversion efficiency, TE power factor and figure-of-merit are evaluated for La-free and La-doped materials.

3. Research methods

This section provides a brief description of computational and experimental methods applied in this research.

3.1. First-principles calculation

The primary calculations are performed for $AgSbTe_2$ stoichiometric phase. To address, however, both optional cases of second-phase nucleation and deviations from stoichiometric composition, as described in Section 2, the following phases are simulated, as well: Sb_2Te_3, Sb_8Te_3, Ag_3SbTe_4, and $AgSb_3Te_4$.

3.1.1. The base AgSbTe₂ phase—structural and vibrational calculations

Silver-antimony-telluride of $AgSbTe_2$ stoichiometry is commonly known to introduce a cubic lattice structure; however, it was suggested, that it may coexist with tetragonal and rhombohedral forms [59]. The following optional space group symmetries: cubic (Pm-3m, No. 221), tetragonal (P4/mmm, No. 123), and rhombohedral (R-3m, No. 166) have been simulated from first-principles [57]. Calculations of temperature-dependent Helmholtz free energy for these three polymorphs indicate that P4/mmm polymorph is the most stable one at temperatures above 400 K, whose energy exhibits close proximity to that of Pm-3m polymorph. Based on calculated v-DOS function for P4/mmm model alloy consisting of 4 atoms per simulation cell, **Figure 1a**, it was decided to test effects of doping with lanthanum atoms (to be discussed further below). To represent effects of La-doping with effective concentration of La atoms, that is close to realistic doping levels, a model compound of $Ag_7LaSb_8Te_{16}$ stoichiometry was constructed having the same P4/mmm space group symmetry as of the original $AgSbTe_2$ lattice. In this compound, consisting of 32 atoms per simulation cell, **Figure 1b**, La-atom substitutes for $^1/_8$ of Ag-atoms, so that, the resulting concentration is 3.125 at.% La. Computational parameters concerning structural relaxation and vibrational properties are provided in detail [57]. The effects of La-doping on vibrational and thermal properties will be discussed further below.

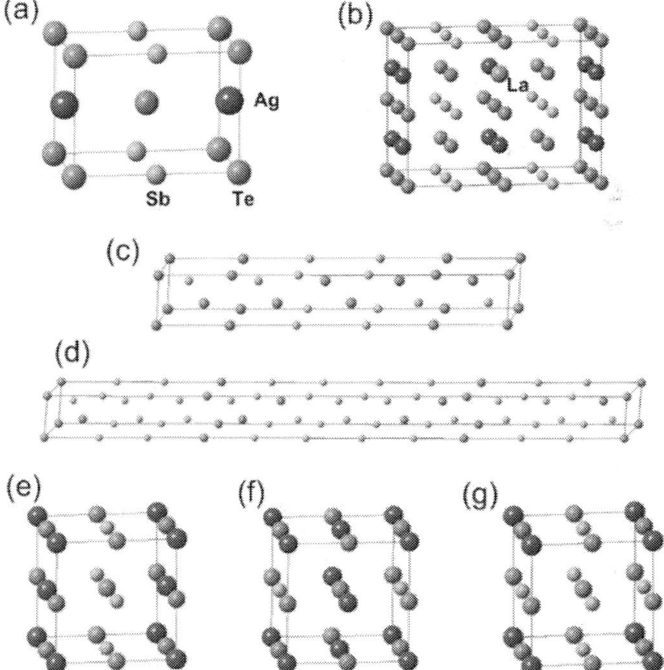

Figure 1. The lattice structures of model alloys discussed in this study and their space group symmetries: (a) $AgSbTe_2$ (P4/mmm); (b) $Ag_7LaSb_8Te_{16}$ (P4/mmm); (c) Sb_2Te_3 (R-3m); (d) Sb_8Te_3 (R-3m); (e) $(AgSbTe_2)_2$ (cubic P1); (f) Ag_3SbTe_4 (cubic P1); and (g) $AgSb_3Te_4$ (cubic P1).

3.1.2. The base AgSbTe₂ phase—electronic calculations

To simulate the effects of La-doping on electrical conductivity and Seebeck coefficient, electronic band structures are calculated for both lattices from first principles. A plane-wave basis set is implemented in Vienna *ab-initio* simulation package (VASP) [60–62] and *MedeA*® software environment [63]. The exchange-correlation electronic energy is expressed by means of generalized gradient approximation (GGA) using PBEsol energy functional [64] and projector augmented wave (PAW) potentials, which are utilized to represent core electron density [65]. Sampling of Brillouin zone is carried out using a set of uniform Monkhorst-Pack k-point mesh with density ranges between 0.14 and 0.17 Å^{-1} and smearing method of linear-tetrahedron with Blöchl corrections [66]. To represent Kohn-Sham electronic wave functions, the plane waves are spanned with 400 eV energy cutoff for the structural relaxation or electronic calculations, respectively. Electronic optimization procedures are performed applying 10^{-5} eV energy convergence threshold.

The calculated 0 K band structures are used for evaluation of temperature-dependent electrical conductivity, electronic component of thermal conductivity, and Seebeck coefficient, applying near-equilibrium Boltzmann transport theory with constant relaxation time approximation, as implemented in BoltzTrap code [67].

The partial electrical conductivity tensor, $\sigma'_{\alpha\beta}(i, \mathbf{k})$, represented for ith energy band and a given $\mathbf{k}$-point, is obtained from Cartesian component of electron group velocity by derivation of ith energy band, $\varepsilon_{i,k}$, with respect to α- and β-components of electron's wave vector [68]. $\sigma'_{\alpha\beta}(i, \mathbf{k})$ is then given by:

$$\sigma'_{\alpha\beta}(i, \mathbf{k}) = e^2 \, \tau_{i,k} \frac{1}{\hbar^2} \frac{\partial^2 \varepsilon_{i,k}}{\partial k_\alpha \partial k_\beta} \, , \tag{1}$$

where e is electron unit charge, $\hbar$ is reduced Planck constant, and $\tau_{i,k}$ is electron relaxation time, which is assumed to be constant. This yields temperature and chemical potential, μ, dependent electrical conductivity tensor with respect to α- and β-components, summed over N-energy bands:

$$\sigma_{\alpha\beta}(T, \mu) = \frac{1}{\Omega} \sum_{i=1}^{N} \int \sigma'_{\alpha\beta}(\varepsilon_i) \left[-\frac{\partial f_0(T, \varepsilon, \mu)}{\partial \varepsilon} \right] \mathrm{d}\,\varepsilon_i \, , \tag{2}$$

where Ω is characteristic unit cell volume and $f_0(T, \varepsilon, \mu)$ is equilibrium Fermi-Dirac distribution function [69]. The electronic component of thermal conductivity tensor, κ^e, is, accordingly, expressed by:

$$\kappa^e_{\alpha\beta}(T, \mu) = \frac{1}{e^2 T\Omega} \sum_{i=1}^{N} \int \sigma'_{\alpha\beta}(\varepsilon_i) \cdot (\varepsilon_i - \mu)^2 \left[-\frac{\partial f_0(T, \varepsilon, \mu)}{\partial \varepsilon} \right] \mathrm{d}\,\varepsilon_i \, . \tag{3}$$

Finally, the explicit expression for Seebeck coefficient tensor, S_{ij}, is given by [70, 71]:

$$S_{ij}(T, \mu) = [\sigma]^{-1}_{\alpha i} \frac{1}{eT\Omega} \sum_{i=1}^{N} \int \sigma'_{j\alpha}(\varepsilon_i) \cdot (\varepsilon_i - \mu) \left[-\frac{\partial f_0(T, \varepsilon, \mu)}{\partial \varepsilon} \right] \mathrm{d}\,\varepsilon_i \, . \tag{4}$$

3.1.3. The Sb_2Te_3 and Sb_8Te_3 phases

Nonmagnetic DFT calculations are performed for Sb_2Te_3 and Sb_8Te_3 crystal structures having (R-3m) space group symmetry, which incorporate 15 and 33 atoms per simulation cell, respectively. Both lattice structures are rendered in **Figure 1c** and **d**, respectively. A computational routine similar to aforementioned one is implemented with several differences. GGA approximation is applied for a set of uniform $9 \times 9 \times 9$ Monkhorst-Pack k-point mesh, and plane waves are spanned with either 400 or 350 eV energy cutoff for structural relaxation or electronic calculations, respectively. Electronic optimization procedures are performed applying 10^{-6} eV energy convergence threshold.

Structural relaxation procedures are first performed, allowing variation of cell volume and atom positions at all degrees of freedom, setting a convergence threshold of 10^{-4} eV Å^{-1} for Hellman-Feynman forces. The resulting lattice parameters obtained for relaxed crystal structures are: a = b = 4.34 Å and c = 31.21 Å for Sb_2Te_3; and: a = b = 4.37 Å and c = 64.93 Å for Sb_8Te_3, which are in good agreement with data reported in the literature [72, 73]. Then, electronic band structure calculations are performed for both relaxed structures.

Band structures are calculated in the same manner as mentioned above, allowing calculations of temperature-dependent electrical conductivity, electronic component of thermal conductivity, and Seebeck coefficient values. These calculations yield p-type behavior for both structures, and we fine-tune the positions of electronic chemical potential to reside at the top of the valence bands. This yields Seebeck coefficient values that are very similar to those measured by us experimentally for pure Sb_2Te_3 standard. Additionally, we set electron relaxation time to be 8 fs, so as to fit electrical conductivity values calculated for Sb_2Te_3 with those measured for the same standards. We, then, apply the same relaxation time for Sb_8Te_3, as well.

To address bulk and interfacial energetic aspects related with nucleation of Sb_2Te_3 and Sb_8Te_3 phases in $AgSbTe_2$ phase, we have simulated formation energies of Sb_2Te_3 and Sb_8Te_3 phases and their interfaces with $AgSbTe_2$ phase. Molar formation energy of model Sb_pTe_q cell, $E^{tot}_{Sb_pTe_q}$, is calculated using the following expression [57]:

$$E^{f}_{Sb_pTe_q} = \frac{E^{tot}_{Sb_pTe_q} - p \times \mu^{o}_{Sb} - q \times \mu^{o}_{Te}}{p+q}, \tag{5}$$

where $E^{tot}_{Sb_pTe_q}$ is cell's molar total energy and μ^{o}_{Sb} and μ^{o}_{Te} are chemical potentials of Sb- and Te-atoms in their standard states, which are evaluated to be −397.72 and −303.12 kJ mol^{-1}, respectively. The free energy of silver-antimony-telluride (AST)/antimony-telluride (SBT) interface is calculated constructing a slab model having AST/SBT generic form, and using the following expression [74]:

$$\gamma = \frac{1}{2A}\left(E^{f}_{AST/SBT} - n_{AST}\,E^{f}_{AST} - n_{SBT}\,E^{f}_{SBT}\right), \tag{6}$$

where A is AST/SBT interface cross-sectional area, $E^{f}_{AST/SBT}$ is calculated formation energy of slab model, E^{f}_{AST} and E^{f}_{SBT} are calculated molar formation energies of AST and SBT sub-cells, and n_{AST} and n_{SBT} are their number of moles in the entire model slab, respectively.

3.1.4. Ag_3SbTe_4 and $AgSb_3Te_4$ model compounds

To simulate the effects of deviations from stoichiometric $AgSbTe_2$ composition, we construct three model alloys based on P4/mmm space group symmetry, which is reduced to cubic P1 symmetry, by setting equal lattice parameter of a = 6.113 Å for all. The resulting structures simulated are: $(AgSbTe_2)_2$, Ag_3SbTe_4, and $AgSb_3Te_4$, which appear in **Figure 1e, f,** and **g,** respectively. All three structures contain 8 atoms per unit cell and Sb/Ag ratios of 1, 1/3, and 3, respectively. To calculate band structures of these three model alloys, spin-orbit (SO) magnetic calculations were performed utilizing a similar GGA/PAW routine as described above for uniform 7 × 7 × 7 Monkhorst-Pack k-point mesh and 400 eV energy cutoff to represent Kohn-Sham electronic wave functions, applying 10^{-6} eV energy convergence threshold. SO coupling is often being considered in band structure calculations [32, 70, 75]. TE transport properties were calculated according to the procedure detailed by Eqs. (1)–(4).

3.2. Experimental procedure

3.2.1. Materials synthesis

Experimental procedures implemented in this study are intended to validate the effects of La-alloying on TE performance, as predicted from first principles. They include synthesis of two model alloys, La-free and La-alloyed, having molar ratios (Ag:Sb:Te:La) of 18:29:53:0 and 15.75:29:53:2.25, respectively. Generally, synthesis procedures comprise vacuum melting and iced-water quenching, followed by uniaxial hot-pressing at two distinct temperatures, 540 and 500°C, yielding two series of 12.7 mm dia. pellets referenced below as Series A and Series B, respectively. The difference between these two series of alloys is manifested by their phase contents and average composition in matrix. These factors significantly affect TE performance, as will be discussed further below. A detailed description of the experimental procedures appears elsewhere [58].

3.2.2. Materials and thermoelectric property characterization

Materials characterization procedures include microstructure, phase identification, and composition analysis employing scanning electron microscopy and X-ray diffraction [58]. Assessment of alloys' thermal stability is investigated using SETARAM 1600 DSC with a scanning rate of 25 K min^{-1} at temperatures ranging from room temperature through 973 K.

Temperature-dependent electrical conductivity, $\sigma(T)$, and Seebeck coefficient, $S(T)$ (thermopower), of these pellets are measured in temperature range from 300 to ~700 K employing *Nemesis*® SBA-458 apparatus (Netzsch GmbH), which is designed for simultaneous measurements of electrical conductivity and thermopower for planar geometry [76–78].

MicroFlash® LFA-457 laser flash analyzer (LFA; Netzsch GmbH) is utilized to measure directly of thermal diffusivity, $\alpha(T)$, of pellets in the same temperature range applying pulse-corrected Cowan approximation to consider heat loss of the samples [79], yielding instrumental accuracy of 2%. Material's density, ρ, is measured at room temperature, and density's dependence on temperature is neglected. Temperature-dependent heat capacity, $C_p(T)$, is simultaneously measured in LFA by comparative method using pure Al_2O_3—reference sample having similar geometry [76]. The resulting accuracy of evaluation of thermal conductivity values is equal to 10%. Pellets' thermal conductivity values, κ, are then determined by measuring their temperature-dependent thermal diffusivity and heat capacity, as well as, density; κ is then expressed by [80]:

$$\kappa(T) = \alpha(T) \cdot \rho \cdot C_p(T) . \tag{7}$$

4. Effects of La-alloying on thermoelectric performance

In this section, we introduce the concept resting behind La-alloying: its origin and implications, predictions from first principles, and experimental validations. Comparative discussion of the results in view of TE performance is provided.

4.1. Predictions from first-principles

4.1.1. The $AgSbTe_2(P4/mmm)$ phase

4.1.1.1. Structural and vibrational properties

P4/mmm form of $AgSbTe_2$ phase is found to be the most stable one compared to all three polymorphs at temperatures larger than 400 K and exhibits Helmholtz free energy values with close proximity to those of cubic polymorph [57]. Frequency-dependent v-DOS, $g_p(\omega)$, calculated for this compound applying Debye approximation exhibits two major peaks at ca. 2.0 and 2.7 THz, and discloses interesting feature. Whereas, 2.7 THz peak comprises equal contributions from lattice vibrations of all sublattice sites, 2.0 THz one is primarily ascribed to vibrations of Ag-sublattice site atoms [57, 58]. This opens up the option of tuning v-DOS pattern by introducing point defects, a discipline for which the term *phonon engineering* has been coined [35]. Particularly, substitutions for Ag-sublattice sites by elements of different mass or atomic radius are expected to modify v-DOS with respect to that of pure $AgSbTe_2$ phase by suppressing its major v-DOS peak. This, consequently, will reduce lattice thermal conductivity. La has been suggested as optional substitution atom due to its relatively large mass and atomic radius compared to average values of $AgSbTe_2$, that is, 138.91 a.m.u. and 187 pm vs. 121.21 a.m.u. and 143.98 pm, respectively, giving rise to enhanced phonon scattering by point defects [81–84]. Furthermore, La-alloying has commercial outcomes, since La is the most inexpensive element compared to constituents of $AgSbTe_2$ alloy and is one of the less inexpensive ones among *energy-critical elements* [85].

Three substitutional options were tested, in which La substitutes for Ag, Sb, or Te, and it was found that substitution at Ag-sublattice sites is the most energetically preferred state for P4/mmm symmetry [57]. Accordingly, La-doped structure was constructed, in which one La-atom substitutes for 1/8. of Ag-atoms, and is shown in **Figure 1b**. First, v-DOS was calculated for La-doped structure and 2.0 THz peak was suppressed, as expected. Second, phonon dispersion curves were calculated for both $AgSbTe_2$ and $LaAg_7Sb_8Te_{16}$ alloys close to

Γ-point along c-crystallographic direction, indicating, that the slopes of the one longitudinal and two transverse acoustic modes of $AgSbTe_2$-lattice are greater, than those of La-alloyed one [57]. Quantitatively, average sound velocities derived for pure and La-alloyed materials are 1727 and 1046 m s^{-1}, respectively. Moreover, temperature-dependent heat capacity functions were determined for both structures, yielding slightly lower values for La-alloyed material. Both values of sound velocity and heat capacity that are found to decrease due to La-alloying imply, that La-alloying should reduce lattice thermal conductivity [57]. Additional calculations employing Debye approximation for low-temperature range of heat capacity yield Debye temperatures and sound velocities for both pure and La-alloyed materials, which are 112 K and 1684 m s^{-1} vs. 104 K and 1563 m s^{-1}, respectively [58]. It is noteworthy that evaluation of sound velocity in this manner is considered to be more physically reliable, since it represents the entire space of lattice directions, rather than individual one. It is, therefore, expected that this way of calculation should yield thermal conductivity values, that fit experimental data better than the former way does.

4.1.1.2. Effects of La-doping on thermal conductivity

Average sound velocity, v_s, and Debye temperature, θ_D, evaluated from first-principles serve as input, that is required to evaluate lattice thermal conductivity, κ_p. To this end, one possibility is to employ Callaway model for lattice thermal conductivity [86, 87], which has become conventional, particularly in the field of TE materials [36–38, 88–93]. In present case, however, there is no need to employ Callaway model for several reasons. First, Callaway model is specified for low temperatures, where contributions of either Normal (N)- or Umklapp (U)-processes are at the same order of magnitude. For temperatures adequately higher than Debye temperature (e.g., $\theta_D \approx 112$ K for $AgSbTe_2$ alloy) [30, 57], only U-processes dominate. Second, Callaway model considers $g_p(\omega)$ and $C_p(T)$ functions that are simplistically approximated based on Debye model [94]. In present case, however, the explicit $g_p(\omega)$ and $C_p(T)$ functions have already been calculated for both pure and La-alloyed materials. Alternatively, the following expression for lattice thermal conductivity is employed [94–96]:

$$\kappa_p = \frac{1}{3} C_v v_s^2 \tau , \tag{8}$$

where τ is phonon relaxation time. To first approximation, it has been assumed that La-doping influences mostly sound velocity and heat capacity and has negligible effect on τ. The ratio of $C_v v_s^2$-products obtained for $LaAg_7Sb_8Te_{16}$ and $AgSbTe_2$ alloys, therefore, reflects the lower limit of relative reduction in thermal conductivity due to La-alloying. Applying dispersion curves close to Γ-point along c-crystallographic direction, it is predicted, that κ_p should decrease by factor of ca. 2.7 due to La-doping. Alternatively, applying sound velocity values derived from Debye approximation, κ_p is expected to decrease by ca. 14% at room temperature due to La-doping [58].

A more thorough and accurate treatment of expression (8) considers the effects of La-alloying on τ, as well. To evaluate τ, contributions of two major scattering mechanisms are taken into account. The first one is phonon-phonon inelastic interactions, i.e., U-processes, that prevail for these alloys above room temperature. Relaxation time for U-processes, τ_U, is represented by [96, 97]:

$$\tau_U^{-1} \approx \frac{\hbar \gamma^2}{M v_s^2 \theta_D} \omega^2 T e^{\left(-\frac{\theta_D}{3T}\right)}, \tag{9}$$

where γ is Grüneisen parameter that reflects the degree of lattice anharmonicity [69], and M is average atomic mass of alloy. Second, to account for internal composition inhomogeneity or compositional modulations at unit-cell length scales, that are typical for such materials [5, 8, 16, 91], the boundary scattering mechanism is employed for characteristic period l, represented by relaxation time τ_B, so that [96]:

$$\tau_B^{-1} \approx \frac{v_s}{l}. \tag{10}$$

To consider dependence of v-DOS on phonon frequency, frequency-averaged expression for τ_U is introduced, so that $g_p(\omega)$ serves as weighting function:

$$\langle \tau_U^{-1} \rangle_\omega = \frac{\hbar \gamma^2}{M v_s^2 \theta_D \omega_D} T e^{\left(-\frac{\theta_D}{3T}\right)} \int_0^{\omega_D} \omega^2 g_p(\omega) d\omega, \tag{11}$$

where ω_D is Debye frequency. Equivalent relaxation time is then expressed as:

$$\tau^{-1} = \langle \tau_U^{-1} \rangle_\omega + \tau_B^{-1}. \tag{12}$$

The resulting values of lattice thermal conductivity for $LaAg_7Sb_8Te_{16}$ and $AgSbTe_2$ alloys are obtained from Eq. (8) by substituting the respective physical magnitudes for both alloys in Eqs. (9)–(11) [19, 20, 30, 57, 58, 92] with $l \approx 1$ nm [5, 8, 16, 91]. Lattice thermal conductivity for $LaAg_7Sb_8Te_{16}$ and $AgSbTe_2$ alloys calculated as function of temperature appear in **Figure 2**.

It is shown that thermal conductivity exhibits realistic values, that correspond with data documented in the literature [19, 20, 25, 98] with marked decrease due to La-doping, ranging between relative values of 11 and 19%, depending on temperature.

4.1.1.3. Effects of La-doping on electrical properties

It was shown that La-alloying reduces lattice thermal conductivity values, which affects TE performance positively. To address, however, the total effects of La-alloying on TE performance, evaluation of electrical conductivity and Seebeck coefficient is essential. This goal was achieved from first-principles applying Boltzmann transport theory as described above for $LaAg_7Sb_8Te_{16}$ and $AgSbTe_2$ alloys. The results are plotted in **Figure 3** in temperature range 50–1000 K.

It is found that La-doping results in reduction in electrical conductivity (e.g., from ca. 1800 down to 250 S cm^{-1} at room temperature) and, at the same time, increase in Seebeck coefficient, e.g., from ca. 4 up to 40 μV K^{-1} at room temperature. For the sake of comparison, Jovovic and Heremans reported on experimental measurements of electrical conductivity and Seebeck coefficients of stoichiometric and doped $AgSbTe_2$ alloys at temperatures up to 400 K [19, 98]. For example, they report on electrical resistivity value of 5×10^{-5} Ohm m at 100 K for stoichiometric $AgSbTe_2$ alloy, which is equivalent to 200 S cm^{-1}.

They report also on electrical resistivity that increases with temperature, indicating charge carriers scattering. Additionally, electrical resistivity may either increase or decrease with doping, depending on dopant's chemical identity. In the present case, electrical conductivity values are significantly larger, e.g., ca. 1900 S cm^{-1} at 100 K, and are decreasing with temperature, where La-doping reduces conductivity. Seebeck coefficient values reported by Jovovic et al. exhibit general trend of increase with temperature, which corresponds to trend calculated in the present case. Also, they report on general trend of increase in Seebeck coefficient values due to doping (except doping with AgTe), in agreement with the present study for La. Complementary trend is reported by Du et al. [25, 99]. Most interestingly, effects of La-doping on electrical properties of $AgSbTe_2$ alloy are reported by Min et al. [27]. They report on trends that are qualitatively similar to those of the present study. First, La-doping was also reported to reduce electrical conductivity, e.g., from ca. 400 S cm^{-1} for undoped $AgSbTe_2$ down to 66 S cm^{-1} for 3 at.% La-doping at room temperature. Second, La-doping increases Seebeck coefficients, e.g., from ca. 90 $\mu V \cdot K^{-1}$ for undoped $AgSbTe_2$ up to ca. 220 μV K^{-1} for 3 at.% La-doping at room temperature. Quantitatively, values of electrical conductivity calculated in this study are considered to be large with respect to the above cited studies. Conversely, Seebeck coefficient values calculated in this study are considered to be smaller than those reported by the above studies. We note, however, that such calculations are most meaningful for comparative purposes, since they rest upon values, that should be calibrated against experimental data, such as electronic chemical potential and relaxation times.

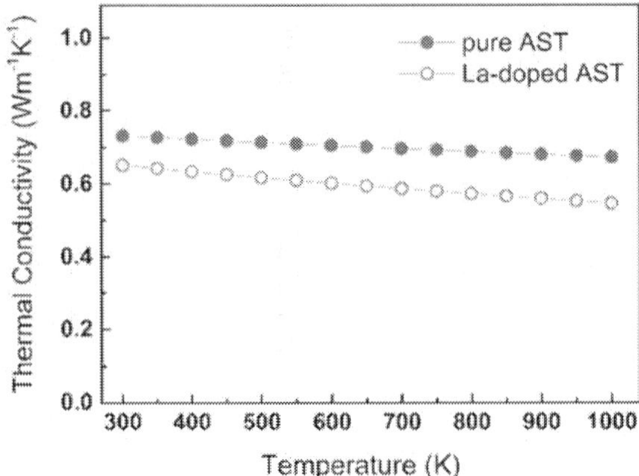

Figure 2. The lattice thermal conductivity values calculated from first-principles for $AgSbTe_2$ (pure AST; filled red circles) and $LaAg_7Sb_8Te_{16}$ (La-doped AST; empty red circles) alloys in temperature range 300–1000 K.

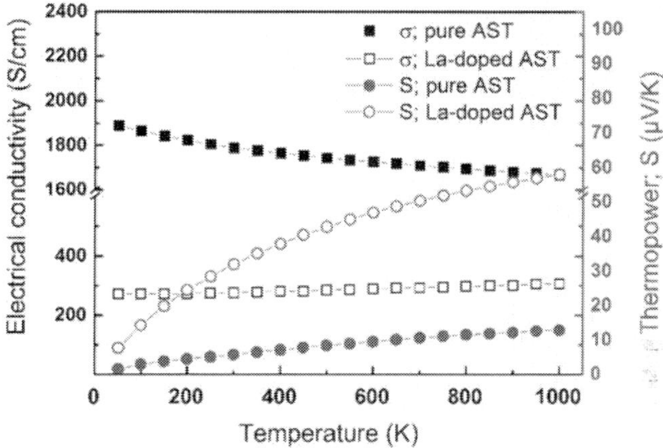

Figure 3. Electrical conductivity and Seebeck coefficient values calculated for $AgSbTe_2$ (pure AST; filled black squares and blue circles, respectively) and $LaAg_7Sb_8Te_{16}$ (La-doped AST; empty black squares and blue circles, respectively) alloys in temperature range 50–1000 K from first-principles applying Boltzmann transport theory.

It is indicated that the effects of La-doping on electrical conductivity and Seebeck coefficient are opposite to each other. Evaluation of TE power factor (PF; $S^2\sigma$) is, therefore, necessary in order to realize how La affects TE power conversion. **Figure 4** displays PF calculated for $LaAg_7Sb_8Te_{16}$ and $AgSbTe_2$ alloys in temperature range 50–1000 K.

It is shown that La-doping has considerably positive effect on PF. This also corresponds with the data reported by Min et al. [27], specifically for low La-concentration regime. We note that, moreover, La-doping reduces lattice thermal conductivity, as shown in **Figure 2**. We conclude that La-alloying should improve energy conversion efficiency of $AgSbTe_2$ (P4/mmm), as reflected by increased TE figure-of-merit.

4.1.2. Formation of Sb_2Te_3 and Sb_8Te_3(R-3m) phases

The single δ-phase is Sb-rich phase based on $AgSbTe_2$ alloy. Since it has limited solubility to Sb with relatively moderate slope of Sb-solvus, it is likely to decompose to δ+Sb_2Te_3 phase mixture [28, 29, 100–103], whereas Sb_2Te_3 is equilibrium phase and may appear as different homologous forms [72, 73, 104–107]. Precipitation of antimony-telluride second phase in δ-matrix is expected to affect TE performance due to contributions from both matrix and precipitate phases or variation of the average matrix composition. In the following sections, we address both aspects. Section 4.1.2.1 introduces the issue of precipitation sequence based on bulk/interfacial energetic considerations, and Section 4.1.2.2 predicts the effects of phase formation on electronic properties. Then, Section 4.1.3 deals with compositional variations in the matrix and their effects on electronic properties.

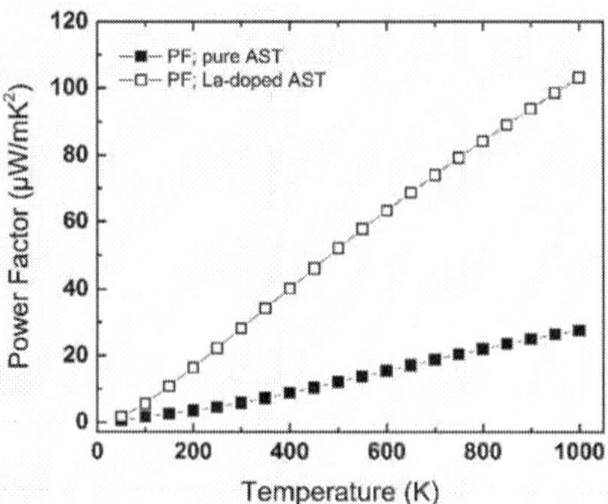

Figure 4. Thermoelectric power factor (PF) calculated for AgSbTe$_2$ (pure AST; filled black squares) and LaAg$_7$Sb$_8$Te$_{16}$ (La-doped AST; empty black squares) alloys in temperature range 50–1000 K from first-principles applying Boltzmann transport theory.

4.1.2.1. The precipitation sequence: energetic aspects

Nucleation of Sb$_2$Te$_3$ and Sb$_8$Te$_3$ phases from Sb-saturated δ-AgSbTe$_2$ matrix has been observed, which can be associated to different experimental conditions. To account for the sequence of phase formation, information on both bulk and interfacial energetics is required. The formation energies of Sb$_2$Te$_3$ and Sb$_8$Te$_3$ (R-3m) phases, calculated according to Eq. (5), are −62 and −56.6 kJ mol^{-1}, respectively. This implies that Sb$_2$Te$_3$ is more energetically favorable, assuming, that Sb$_2$Te$_3$ precipitates are adequately large, so that, interfaces do not play significant role. To address the role of interfaces, free energies of Sb$_2$Te$_3$/AgSbTe$_2$ and Sb$_8$Te$_3$/AgSbTe$_2$ interfaces are evaluated. To this end, two slab models of (Sb$_2$Te$_3$)$_2$/(AgSbTe$_2$)$_5$/(Sb$_2$Te$_3$)$_2$ (40 atoms) and (AgSbTe$_2$)$_3$/(Sb$_8$Te$_3$)$_6$/(AgSbTe$_2$)$_3$ (90 atoms) forms are constructed, respectively, consisting of two interfaces each, exhibiting (111)$_{AgSbTe_2}$ ‖ (0001)$_{Sb_pTe_q}$ and ⟨10$\bar{1}$⟩$_{AgSbTe_2}$ ‖ ⟨$\bar{2}$110⟩$_{Sb_pTe_q}$ orientation relationship, which was observed experimentally [100]. Both structures are displayed in **Figure 5**.

It is noted that two interfaces presented in both slabs shown in **Figure 5a** and **b** consist of different Sb- and Te-terminating planes, so that interfacial free energies calculated according to Eq. (6) represent an average value for both terminations. Correction factor is, therefore, applied to represent interfacial free energy of low-energy Sb-termination. The resulting values for Sb$_2$Te$_3$/AgSbTe$_2$ and Sb$_8$Te$_3$/AgSbTe$_2$ interfaces are γ = 208 and 175 mJ m^{-2}, respectively. These values are considered to be relatively low compared to those of intermetallic compounds and are comparable with those of pure metals [108]. This is, however, not surprising, considering the extremely small atomic misfit between the (111)$_{AgSbTe_2}$ and (0001)$_{Sb_pTe_q}$ crystallographic planes [100], which encourages formation of Sb$_2$Te$_3$ or Sb$_8$Te$_3$ precipitates in the form of long lamellae along these planes [28, 29, 57, 58, 100–103]. These low values of interfacial free energy also initiate fast nucleation, thanks to low activation energy for nucleation, which is proportional to γ^3 [109].

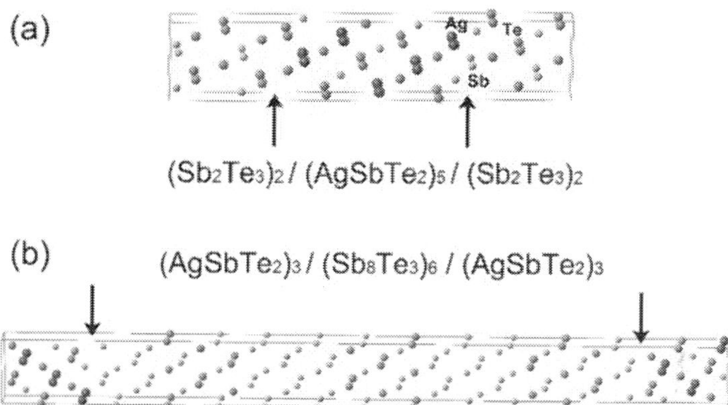

Figure 5. Two slab models of (a) $(Sb_2Te_3)_2/(AgSbTe_2)_5/(Sb_2Te_3)_2$ (40 atoms) and (b) $(AgSbTe_2)_3/(Sb_8Te_3)_6/(AgSbTe_2)_3$ (90 atoms) including two $Sb_2Te_3/AgSbTe_2$ and $Sb_8Te_3/AgSbTe_2$ interfaces each, respectively. All interfaces, marked by arrows, are of $(111)_{AgSbTe2} \parallel (0001)_{SbpTeq}$ and $\langle 10\bar{1}\rangle_{AgSbTe2} \parallel \langle \bar{2}110\rangle_{SbpTeq}$ orientation relationship.

Interestingly, Sb_2Te_3 phase exhibits lower value of formation energy and higher value of interfacial free energy compared to those of Sb_8Te_3 phase. This implies that Sb_8Te_3 is metastable phase that may form prior to the nucleation of Sb_2Te_3 equilibrium phase [28, 29, 100–103]. Suggested nucleation sequence is, therefore, supersaturated-δ → supersaturated-δ + Sb_8Te_3 → equilibrium-δ + Sb_2Te_3.

4.1.2.2. Effects of Sb_2Te_3 and Sb_8Te_3 formation on electronic properties

In view of aforementioned prospect for the presence of either of Sb_2Te_3- or Sb_8Te_3-phases in AgSbTe_2-matrix, calculations of transport coefficients of these phases provide us with predictions of the effects such phase mixture on TE performance. **Figure 6** displays electrical conductivity and Seebeck coefficients calculated in temperature range 50–1000 K.

It is shown that both electrical conductivity and Seebeck coefficient values of Sb_2Te_3 phase are larger than those of Sb_8Te_3 phase in wide temperature range, e.g., ca. 2100 S cm^{-1} and 85 µV K^{-1} for Sb_2Te_3 compared to 1380 S cm^{-1} and 29 µV K^{-1} for Sb_8Te_3 at 300 K, respectively. Moreover, comparison of these results with the data shown in **Figure 3** for AgSbTe_2-matrix implies that precipitation of Sb_2Te_3 phase yields positive influence on TE performance: both electrical conductivity and Seebeck coefficient increase. It is strikingly indicated that the effects of Sb_2Te_3 precipitation are even greater, than those of La-doping. It is, therefore, concluded that the desirable material from TE viewpoint is La-doped, Sb-supersaturated δ-AgSbTe_2-matrix that is aged for a certain duration to form considerable amount of Sb_2Te_3 phase. The effects of Sb_8Te_3 phase on TE performance are, conversely, inferior to those of Sb_2Te_3 phase. Sb_8Te_3 is, however, metastable phase and is not expected to prevail for long durations at elevated temperatures (e.g., under service conditions of TE generator) due to low thermal stability.

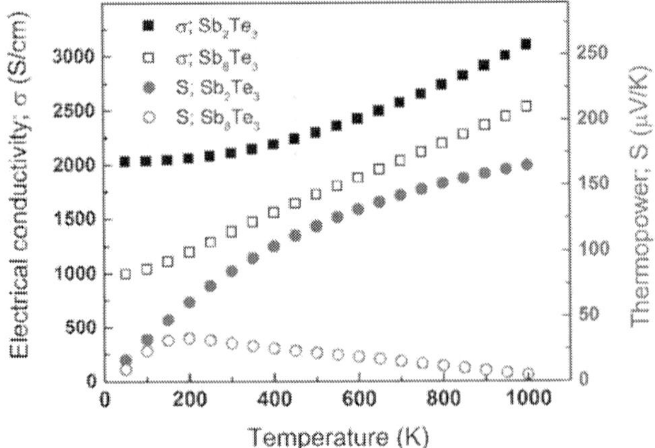

Figure 6. Electrical conductivity and Seebeck coefficient values calculated for Sb_2Te_3 (filled black squares and blue circles, respectively) and Sb_8Te_3 (empty black squares and blue circles, respectively) compounds in temperature range 50–1000 K from first-principles applying Boltzmann transport theory.

4.1.3. Effects of off-stoichiometry on electronic properties of the $AgSbTe_2$ phase

An additional effect taking place during precipitation of any Sb_pTe_q phase from Sb-supersaturated δ-matrix is enrichment of δ-matrix with Ag-atoms and depletion of Sb. To simulate these compositional variations, two off-stoichiometric model alloys are constructed, namely Ag_3SbTe_4 and $AgSb_3Te_4$, in addition to stoichiometric $AgSbTe_2$ phase. These model compounds, appearing in **Figure 1e, g,** and **f,** exhibit Sb/Ag ratios of 1/3, 3, and 1, respectively. **Figure 7** displays electrical conductivity and Seebeck coefficient values calculated in temperature range 50–1000 K.

It is shown, that increase in Sb/Ag ratio results in decrease in electrical conductivity simultaneously with increase in Seebeck coefficient. This trend corresponds well with study of Jovovic and Heremans [19], who reported on decrease in both Seebeck coefficient and electrical resistivity due to additions of 2% AgTe to stoichiometric $AgSbTe_2$-phase, i.e., reducing Sb/Ag ratio. It should be noted that comparison of this trend with data reported in the literature is not straightforward, since compositional changes involve in practice not only Sb/Ag ratio, but also ratio of Te to any of the other species. Additionally, deviations from given stoichiometry often involve formation of second phases, which is not directly simulated here. For instance, Zhang et al. reported on dependence of TE properties on composition for $Ag_{2-y}Sb_yTe_{1+y}$-based alloys and found that electrical conductivity increases, while Seebeck coefficient decreases with y-values increasing from 1.26 up to 1.38 [101].

Most importantly, this predicted effect of Sb/Ag ratio on electrical properties has major implications on the temporal evolution of TE performance of the material during aging heat treatments (below Sb-solvus), or of TE generator during service. Since Sb_pTe_q phases nucleate from Sb-supersaturated δ-matrix during heat treatments, Sb/Ag ratio in δ-matrix decreases. This should be accompanied by increase in electrical conductivity concurrently with decrease in Seebeck coefficient.

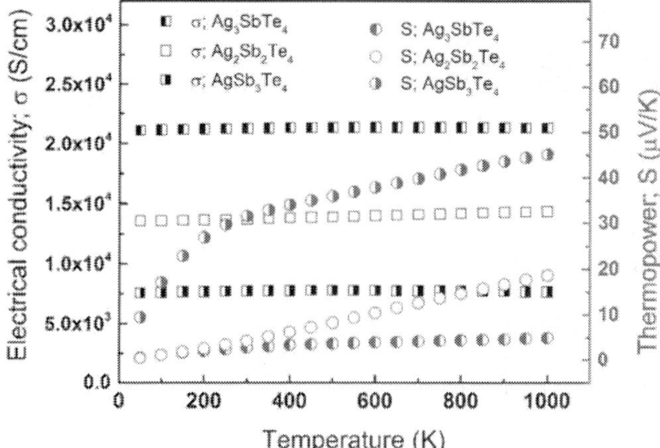

Figure 7. Electrical conductivity and Seebeck coefficient values calculated for Ag_3SbTe_4 (left-half-filled black squares and blue circles, respectively), $(AgSbTe_2)_2$ (empty black squares and blue circles, respectively), and $AgSb_3Te_4$ (right-half-filled black squares and blue circles, respectively) alloys in temperature range 50–1000 K from first-principles applying Boltzmann transport theory.

5. Experimental results

It was shown above how first-principles calculations provide us with information about all aspects concerning TE transport behavior, including thermal and electrical conductivity and Seebeck coefficient. Most importantly, this assists us in tailoring the material by introducing lattice defects to enhance its TE performance. In the following section, we introduce experimental procedures taken for validating the above predictions. Comparing between both aspects is, moreover, very instructive not only on engineering aspects, but also on universal aspect, by realizing how to implement computational tools to predict properties of other materials.

5.1. Microstructure and implications on thermoelectric behavior

As mentioned above, two classes of La-alloyed $AgSbTe_2$-based materials were prepared by uni-axial hot-pressing at 540 or 500°C, and are classified as Series A and Series B, respectively. The ideal case for testing the effects of La-doping is single δ-phase dissolving La homogeneously. This, however, is difficult to achieve. Hot-pressing at 540°C, that is, above Sb-solvus, expected to yield the desirable single δ-phase that does not contain Sb_2Te_3 precipitates [101–103]. These Sb_2Te_3 precipitates, indeed, were not observed in Series A samples; however, La-rich precipi-tates having stoichiometry close to $LaTe_2$ were observed [110]. As a result, δ-matrix was found to be depleted of La [58], which does not allow us comparison between La-free and La-doped samples. Conversely, samples of Series B, that were hot-pressed at 500°C, exhibit considerable amount of Sb_2Te_3 phase, which is expected; however, $LaTe_2$ precipitates are dissolved, so that δ-matrix contains adequately large amount of La, close to its nominal concentration. Series B is, therefore, more suitable to exemplify the effects of La-doping. Moreover, as predicted from first-principles, the presence of Sb_2Te_3 precipitates, in addition to La solute atoms, has positive effects on electronic transport.

5.2. Thermal analysis

Thermal conductivity measurements performed for both Series A and Series B indicate the expected trend. First, all thermal conductivity values lie in the range 0.6–0.8 W m^{-1} K^{-1} [58]. Second, samples of Series A did not exhibit any considerable difference between La-free and La-doped materials [58]. This is associated to depletion of δ-matrix from La solute atoms, so that, matrix composition of La-doped and La-free materials is practically the same. Third, and most importantly, it was found that thermal conductivity of La-doped materials is significantly lower than those of La-free materials of Series B, e.g., 0.8 W m^{-1} K^{-1} for La-free and 0.6 W m^{-1} K^{-1} for La-doped samples at 500 K. This is strikingly corroborated by predictions from first-principles, both qualitatively and quantitatively, as shown in **Figure 2**. It is also noteworthy that both values coincide at temperatures larger than 650 K, which can be associated with phase transition [58, 101–103]. Thorough discussion of thermal conductivity values measured for Series A and Series B materials and their relationship with microstructure appears elsewhere [58]. To address this issue of phase transition, DSC measurements were implemented for both La-free and La-doped samples, **Figure 8**.

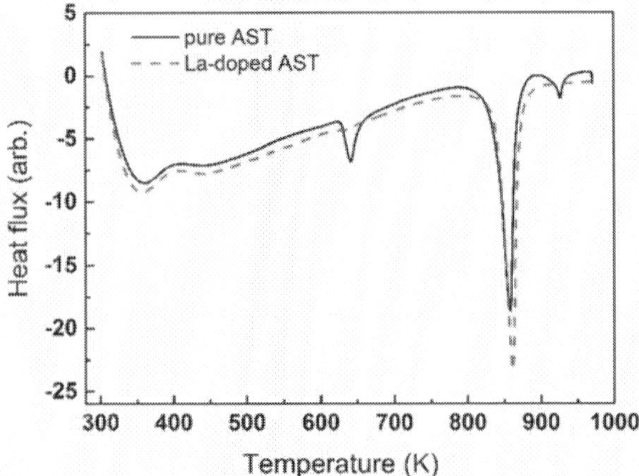

Figure 8. Differential scanning calorimetry (DSC) signals collected upon heating from La-free (continuous black curve) and La-doped (dashed red curve) samples.

Endothermic peak at around 630–650 K is observed for La-free material, which is associated with $Ag_2Te + Sb_2Te_3 \rightarrow \delta\text{-}AgSbTe_2$ phase transition at 360°C [101–103]. La-doped material, however, does not exhibit this transition. This corresponds well with thermal conductivity behavior reported by us earlier [58], in which temperature-dependent thermal conductivity of La-doped materials show up continuous trend, whereas La-free materials exhibit sharp drop of thermal conductivity around this temperature. This implies that La-additions help in stabilizing δ-phase against decomposition, which is expected to contribute to stability of TE device operation at service conditions. Sharp endothermic peak at ca. 860 K, which is common for both La-free and La-doped materials, is associated to melting.

5.3. Electrical property measurements

It was predicted from first-principles that La-doping reduces electrical conductivity and increases Seebeck coefficient, **Figure 3**. Measurements of electrical conductivity and Seebeck coefficients were carried out for both Series A and Series B materials. The samples of Series B are of our interest, since they dissolve La-atoms in δ-matrix; we will, therefore, introduce these results first. **Figure 9** displays experimentally collected electrical conductivity and Seebeck coefficient values of La-free and La-doped materials of Series B.

It is shown that electrical conductivity values decrease, e.g., from ca. 1400 down to 900 S cm^{-1} at room temperature, and Seebeck coefficient increase, e.g., from ca. 30 up to 70 μV K^{-1} at room temperature, due to La-doping. This behavior is, qualitatively, the same as that observed for calculated values shown in **Figure 3**. Moreover, temperature dependence, that is, electrical conductivity decreasing and Seebeck coefficient increasing with temperature for both La-free and La-doped materials, is identical to that indicated by calculated values shown in **Figure 3**. There are two major differences between experimental and calculated values appearing in **Figures 3** and **9**, respectively. First, the absolute values of measured Seebeck coefficient values are greater than calculated ones. Also, difference of electrical conductivity between La-doped and La-free materials is smaller for measured dataset than for calculated ones. This is probably due to difficulty to simulate low dopant concentrations in DFT [70]. Second, it is noteworthy that both values of electrical conductivity and Seebeck coefficients measured for La-free and La-doped materials converge at temperatures >650 K, **Figure 9**. Interestingly, these convergences occur due to sharp deviations of the values featured by La-free material, whereas the values of La-doped materials preserve their continuous trendline. This observation corresponds well with the behavior shown by DSC curves in **Figure 8**, where La-free compound decomposes at around 650 K, whereas La-doped compound seem to preserve its thermal stability. This also corresponds with converging thermal conductivities of the samples of Series B as discussed above [58]. Following our comparative discussion in Section 4.1.1.3, experimental values of electrical properties are found to be closer to experimental values reported in the literature than to calculated values [19, 25, 27, 98, 99].

To complement our understanding of the effects of La-doping on electronic properties, we measured temperature-dependent electrical conductivity and Seebeck coefficient values for the samples of Series A, as well. The results are plotted against temperature in **Figure 10**.

Comparison between the results attained for alloys of Series A and Series B is very instructive. As noted, the samples of Series A exhibited formation of LaTe$_2$-like precipitates, which "drain out" La atoms from δ-matrix, resulting in matrix compositions, that are nearly identical to each other for La-free and La-doped materials. For this reason, thermal conductivity values measured for La-free and La-doped materials seem to be practically identical in wide temperature range [58]. It is, therefore, not surprising to observe the same behavior for electrical properties, **Figure 10**.

It is indicated, that both electrical conductivity and Seebeck coefficient values measured for La-free and La-doped materials seem to be very close to each other in the entire temperature range, probably due to nearly identical matrix compositions for La-free and La-doped materials. Additionally, electrical conductivity and Seebeck coefficients featured by La-doped alloys exhibit relatively continuous temperature-dependent behavior, whereas values, measured for La-free alloys exhibit curled behavior. This, again, can be explained in terms of poor thermal stability of La-free materials, as discussed above.

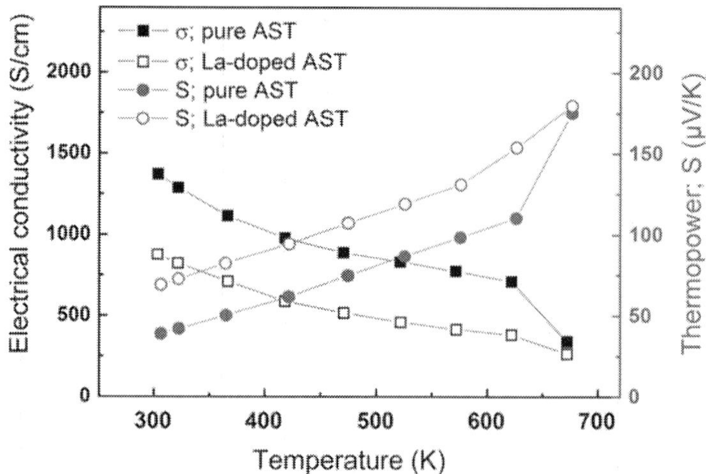

Figure 9. Electrical conductivity and Seebeck coefficient values measured for La-free (pure AST; filled black squares and blue circles, respectively) and La-doped (La-doped AST; empty black squares and blue circles, respectively) alloys of Series B in temperature range 300–673 K.

5.4. Implications for thermoelectric power conversion

It has been shown that La-doping has unequivocally positive effect on reducing lattice thermal conductivity, both computationally and experimentally. The effects on electrical properties, particularly electrical conductivity and Seebeck coefficient, are opposing each other. To assess the effects of La-doping on device's power capacity, TE PFs of La-free and La-doped materials of Series B are evaluated based on the data displayed in **Figure 9**. The results are shown in **Figure 11**.

It is clearly shown that La-doping affects positively PF for temperatures lower than 500 K, e.g., PF determined for room temperature increases from ca. 200 to 400 $\mu W\ m^{-1}\ K^{-2}$ due to La-doping. At higher temperatures, PFs of La-free and La-doped materials are practically identical. The maximum PF values observed are around 1000 $\mu W\ m^{-1}\ K^{-2}$. This trend is similar to that reported by Min et al. [27], that is, PF increasing from 300 up to 1500 $\mu W\ m^{-1}\ K^{-2}$ in respective temperature range from room temperature to 400°C for $AgSbTe_2$ alloy.

La-doping was tested by them for different compositions, where composition yielding the greatest PF values is $AgSb_{0.99}La_{0.01}Te_2$, with PF values around 1000–1200 $\mu W\ m^{-1}\ K^{-2}$ in the entire temperature range. Particularly, this La-doped material exhibits superior PF values up to ca. 325°C. This trend is similar to that reported by us in this study.

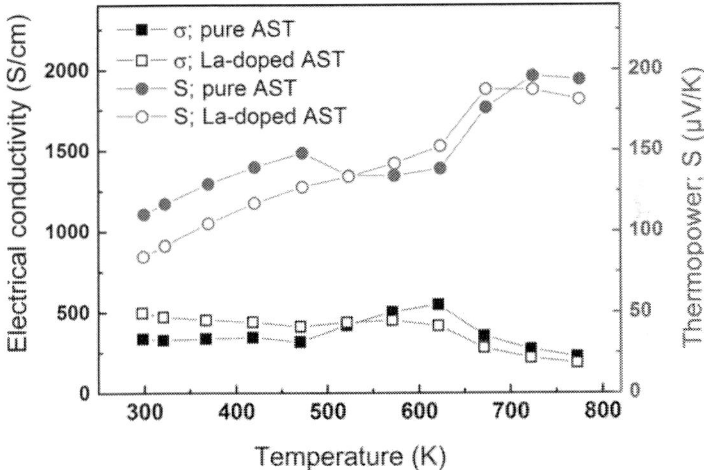

Figure 10. Electrical conductivity and Seebeck coefficient values measured for La-free (pure AST; filled black squares and blue circles, respectively) and La-doped (La-doped AST; empty black squares and blue circles, respectively) alloys of Series A in temperature range 300–773 K.

Finally, determination of TE figure-of-merit for both La-free and La-doped materials will provide us with the ultimate indication whether La-doping enhances TE power conversion efficiency. Based on thermal and electrical properties measured for Series A and Series B, temperature-dependent ZT values were determined and appear in **Figure 12**.

Most importantly, it is shown that La-doping increases ZT values markedly, **Figure 12b**, e.g., from ca. 0.3 to 0.45 at 473 K. This improvement is due to decrease in thermal conductivity in almost the entire temperature range and increase in PF at low-temperature regime due to La-doping. Above 600 K, again, both values of La-free and La-doped alloys converge due to poor thermal stability of La-free materials. ZT values of La-free and La-doped materials shown in **Figure 12b** correspond with those reported by Zhang et al. [101], where the effects of La-doping are comparable to those of stoichiometric variations about AgSbTe$_2$ composition. Similar values are reported by Mohanraman et al. [43] and Jovovic and Heremans [19] for Bi-doping, as well as, for Pb-doping [19]. Chen et al. obtain similar ZT values for Ge-doping [111] and for Sn-doping [112], depending on concentration. ZT values reported in the present study are, however, lower than those reported by Du et al. [25, 99], probably owing to different processing conditions yielding higher electrical conductivity values [113].

The picture, revealed for alloys of Series A, is, however, different; it is shown in **Figure 12a** that La-doping has little or no effect on ZT. This is, again, not surprising and follows the trends featured by Series A alloys for electrical conductivity and Seebeck coefficient, **Figure 9**, and thermal conductivity [58] associated to depletion of La-atoms from δ-matrix in La-alloyed materials hot-presses at 540°C.

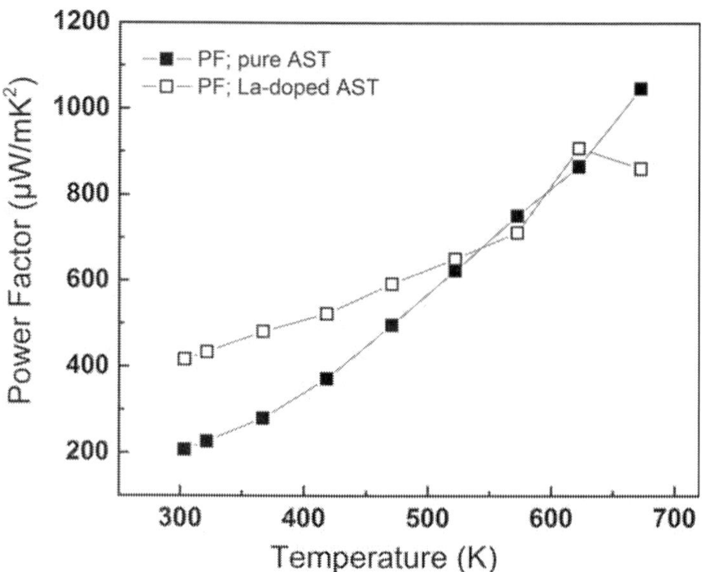

Figure 11. Thermoelectric power factor (PF) values evaluated for La-free (pure AST; filled black squares) and La-doped (La-doped AST; empty black squares) alloys of Series B in temperature range 300–673 K.

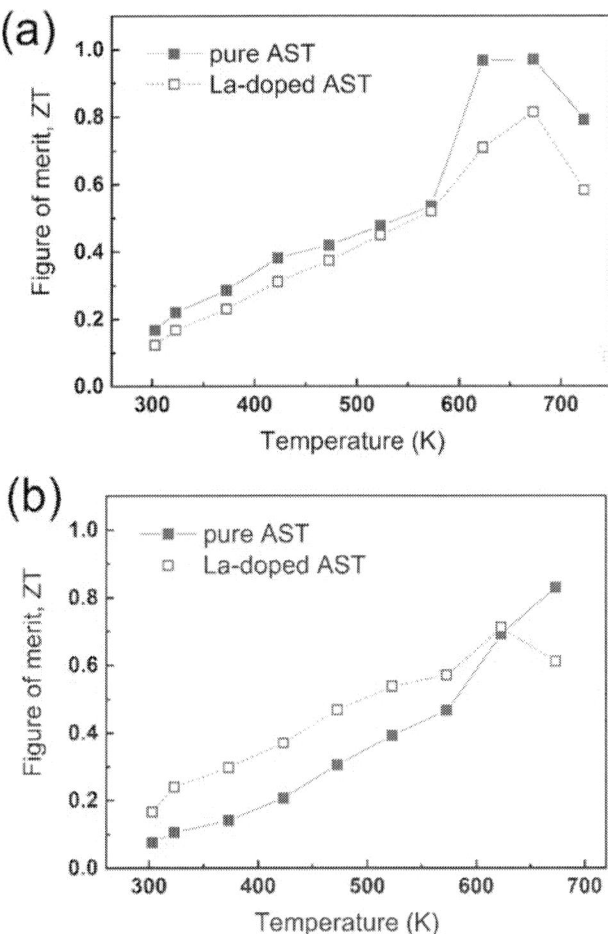

Figure 12. Thermoelectric figure of merit (ZT) values evaluated for La-free (pure AST; filled blue squares) and La-doped (La-doped AST; empty blue squares) alloys of (a) Series A and (b) Series B in temperature range 300–673 K.

6. Summary and concluding remarks

This chapter introduced the following findings. Computationally, total energy calculations for different polymorphs of AgSbTe$_2$ phase yield their Helmholtz free energies, implying, that P4/mmm space group symmetry is the most stable one at temperatures adequately higher than room temperatures. Predictions of the effects of doping on thermal conductivity are established on calculations of vibrational properties, such as phonon dispersion and density of states, from first-principles. Based on specific features in v-DOS curve, it is hypothesized that La-substitution for Ag-sites should result in reduced lattice thermal conductivity. These calculations predict reduction in average sound velocity from 1684 to 1563 m s^{-1} and of Debye temperature from 112 to 104 K due to La-doping. Applying Umklapp mechanism for phonon scattering with frequency-averaged inverse relaxation time, which is combined with boundary scattering, yields temperature-dependent functional forms for lattice thermal conductivity. Marked decrease due to La-doping, ranging between relative values of 11 and 19% depending on temperature, are observed. Then, calculations of electronic band structures of both La-free and La-doped lattices are performed, yielding TE transport coefficients applying Boltzmann transport theory. It is found that La-doping results in reduction in electrical conductivity (e.g., from ca. 1800 down to 250 S cm^{-1} at room temperature) at the same time with increase in Seebeck coefficient, e.g., from ca. 5 up to 40 μV K^{-1} at room temperature.

Attempts to infer conclusions with practical implications from DFT calculations must consider engineering aspects that extend further beyond single phase state having high symmetry unit cell, that maintains its physical properties with time. For example, considerations, such as long-term device operation under elevated service temperatures, should be taken into account. Such case requires original solution for simplified (or, sometimes, over-simplified) approach offered by DFT. Particularly, for the case of thermal stability, exposure of Sb-rich δ-phase to elevated temperatures results in precipitation of Sb$_p$Te$_q$-based phases at the same time with decrease in Sb/Ag ratio in δ-matrix. In this manner, thermal stability issues can be addressed by dividing the realistic conditions into a set of simplified problems, each can be handled by DFT. To this end, we first consider the case in which Sb$_2$Te$_3$ and Sb$_8$Te$_3$ phases precipitate inside AgSbTe$_2$-matrix. It is found that both electrical conductivity and Seebeck coefficient values of Sb$_2$Te$_3$-phase are larger, than those of Sb$_8$Te$_3$-phase in wide temperature range, e.g., ca. 2100 S cm^{-1} and 85 μV K^{-1} for Sb$_2$Te$_3$ compared to1380 S cm^{-1} and 29 μV K^{-1} for Sb$_8$Te$_3$ at 300 K, respectively. Moreover, it is estimated that precipitation of Sb$_2$Te$_3$-phase in AgSbTe$_2$-matrix is expected to improve the total values of both electrical conductivity and Seebeck coefficient. Concerning nucleation sequence of Sb$_2$Te$_3$ and Sb$_8$Te$_3$ phases in AgSbTe$_2$, their molar formation energies and interfacial free energies were calculated, suggesting that Sb$_8$Te$_3$ nucleates first as metastable phase, prior to the formation of equilibrium Sb$_2$Te$_3$ phase. Second, to address the influence of deviations from AgSbTe$_2$ stoichiometry on electron transport properties, off-stoichiometric model alloys Ag$_3$SbTe$_4$ and AgSb$_3$Te$_4$ were simulated. It is found that increase in Sb/Ag ratio results in decrease in electrical conductivity simultaneously with increase in Seebeck coefficient. Considering both effects of Sb$_2$Te$_3$ precipitation accompanied by simultaneous decrease in Sb/Ag ratio in δ-matrix taking place with aging time at temperatures below δ-solvus, it is expected that electrical conductivity of two-phase δ+Sb$_2$Te$_3$ alloy should increase with aging time, disregarding effects, such as electron boundary scattering. Interestingly, these two effects have opposite consequences regarding Seebeck coefficient, so that, it is difficult to assess resulting Seebeck coefficient.

Experimentally, model ternary (AgSbTe$_2$) and quaternary (3 at.% La-AgSbTe$_2$) alloys were synthesized by vacuum melting followed by quenching and hot-pressing. The appropriate conditions enabling formation of AgSbTe$_2$-matrix that dissolves La-atoms with no La-rich precipitates were established. DSC tests enable observation of Ag$_2$Te+Sb$_2$Te$_3 \rightarrow \delta$-AgSbTe$_2$ phase transition at 360°C for La-free alloys only, indicating improvement of alloy's thermal stability due to La-additions. Temperature-dependent thermal conductivity of both alloys indicate reduction in thermal conductivity as a result of La-alloying from 0.92 to 0.71 W m^{-1} K^{-1} at 573 K, which corresponds with the trend predicted from first-principles. Measurements of temperature-dependent electrical conductivity and Seebeck coefficients indicate that La-doping reduces electrical conductivity and increases Seebeck coefficients, as predicted from first-principles. Eventually, it is shown that La-doping has positive effects on TE figure-of-merit ZT, which is improved, e.g., from 0.35 up to 0.50 at 260°C.

We demonstrate how first-principles calculations serve as trustworthy tool for predicting TE performance of materials, screening the best candidates for application in TE devices. It is noteworthy that such DFT routines prove to be very efficient by prediction of TE properties in a way saving expensive and time-consuming experiments. The resulting materials that seem to possess improved performance are, eventually, processed in laboratory. We show how simple physical considerations can be implemented in DFT calculations and lead to improvement of power conversion efficiency. La-doping improves the alloys' thermal stability and reduces their thermal conductivity, as well as enhances TE power factor in certain temperature range. As a result, the total TE figure-of-merit improves significantly. We, finally, emphasize the universal aspects of this approach that can be applied for other TE materials, as well.

Acknowledgements

The author wishes to acknowledge generous support from the Israel Science Foundation (ISF), Grant no. 698/13, as well as, from the German-Israeli Foundation for Research and Development (GIF), Grant no. I-2333-1150.10/2012. Partial support from the Nancy and Stephen Grand Technion Energy Program (GTEP), the Russell Berrie Nanotechnology Institute (RBNI), Technion, and the Adelis Foundation for renewable energy research are greatly acknowledged, as well.

Author details

Yaron Amouyal

Address all correspondence to: amouyal@technion.ac.il

Department of Materials Science and Engineering, Technion—Israel Institute of Technology, Haifa, Israel

References

[1] Martin RM.: Electronic Structure – Basic Theory and Practical Methods. Cambridge, UK: Cambridge University Press; 2004.

[2] Argaman N, Makov G.: Density functional theory: An introduction. Am. J. Phys. 2000;**68**:69–79. doi:10.1119/1.19375

[3] Mishin Y, Asta M, Li J.: Atomistic modeling of interfaces and their impact on microstructure and properties. Acta. Mater. 2010;**58**:1117–51. doi:10.1016/j.actamat.2009.10.049

[4] Parr RG, Yang W.: Density-Functional Theory of Atoms and Molecules. New York: Oxford University Press; 1989.

[5] Hsu KF, Loo S, Guo F, Chen W, Dyck JS, Uher C, Hogan T, Polychroniadis EK, Kanatzidis MG.: Cubic $AgPb_mSbTe_{2+m}$: bulk thermoelectric materials with high figure of merit. Science. 2004;**303**:818–821. doi:10.1126/science.1092963

[6] Kanatzidis MG.: Nanostructured thermoelectrics: The new paradigm? Chem. Mater. 2010;**22**:648–659. doi:10.1021/cm902195j

[7] Dadda J, Muller E, Klobes B, Pereira PB, Hermann R.: Electronic properties as a function of Ag/Sb ratio in $Ag_{1-y}Pb_{18}Sb_{1+z}Te_{20}$ compounds. J. Electron. Mater. 2012;**41**:2065–72. doi:10.1007/s11664-012-2111-9

[8] Dadda J, Muller E, Perlt S, Hoche T, Pereira PB, Hermann RP.: Microstructures and nanostructures in long-term annealed $AgPb_{18}SbTe_{20}$ (LAST-18) compounds and their influence on the thermoelectric properties. J. Mater. Res. 2011;**26**:1800–1812. doi:10.1557/jmr.2011.142

[9] Perlt S, Hoche T, Dadda J, Muller E, Pereira PB, Hermann R, Sarahan M, Pippel E, Brydson R.: Microstructure analyses and thermoelectric properties of $Ag_{1-x}Pb_{18}Sb_{1+y}Te_{20}$. J. Solid State Chem. 2012;**193**:58–63. doi:10.1016/j.jssc.2012.03.064

[10] Cook BA, Kramer MJ, Harringa JL, Han MK, Chung DY, Kanatzidis MG.: Analysis of nanostructuring in high figure-of-merit $Ag_{1-x}Pb_mSbTe_{2+m}$ thermoelectric materials. Adv. Funct. Mater. 2009;**19**:1254–1259. doi:10.1002/adfm.200801284

[11] Sootsman JR, Chung DY, Kanatzidis MG.: New and old concepts in thermoelectric materials. Angew. Chem-Int. Edit. 2009;**48**:8616–8639. doi:10.1002/anie.200900598

[12] Steimer C, Welnic W, Kalb J, Wuttig M.: Towards an atomistic understanding of phase change materials. J. Optoelectron. Adv. Mater. 2006;**8**:2044–2050.

[13] Wang K, Steimer C, Detemple R, Wamwangi D, Wuttig M.: Assessment of Se based phase change alloy as a candidate for non-volatile electronic memory applications. Appl. Phys. A. 2005;**81**:1601–1605. doi:10.1007/s00339-005-3358-2

[14] Siegert KS, Lange FRL, Sittner ER, Volker H, Schlockermann C, Siegrist T, Wuttig M.: Impact of vacancy ordering on thermal transport in crystalline phase-change materials. Rep. Prog. Phys. 2015;**78**:013001. doi:10.1088/0034-4885/78/1/013001

[15] Wuttig M, Yamada N.: Phase-change materials for rewriteable data storage. Nat. Mater. 2007;**6**:1004. doi:10.1038/nmat2077

[16] Ma J, Delaire O, May AF, Carlton CE, McGuire MA, VanBebber LH, Abernathy DL, Ehlers G, Hong T, Huq A, Tian W, Keppens VM, Shao Horn Y, Sales BC.: Glass-like phonon scattering from a spontaneous nanostructure in AgSbTe$_2$. Nat. Nano. 2013;**8**:445–451. doi:10.1038/nnano.2013.95

[17] Specht ED, Ma J, Delaire O, Budai JD, May AF, Karapetrova EA.: Nanoscale structure in AgSbTe$_2$ determined by diffuse elastic neutron scattering. J. Electron. Mater. 2015;**44**:1536–9. doi:10.1007/s11664-014-3447-0

[18] Shinya H, Masago A, Fukushima T, Katayama-Yoshida H.: Inherent instability by antibonding coupling in AgSbTe$_2$. Jpn. J. Appl. Phys. 2016;**55**:6. doi:10.7567/jjap.55.041801

[19] Jovovic V, Heremans JP.: Doping effects on the thermoelectric properties of AgSbTe$_2$. J. Electron. Mater. 2009;**38**:1504–1509. doi:10.1007/s11664-009-0669-7

[20] Morelli DT, Jovovic V, Heremans JP.: Intrinsically minimal thermal conductivity in cubic I-V-VI$_2$ semiconductors. Phys. Rev. Lett. 2008;**101**:035901. doi:10.1103/PhysRevLett.101.035901

[21] Nielsen MD, Ozolins V, Heremans JP.: Lone pair electrons minimize lattice thermal conductivity. Energy Environ. Sci. 2013;**6**:570–578. doi:10.1039/C2EE23391F

[22] Ye L-H, Hoang K, Freeman AJ, Mahanti SD, He J, Tritt TM, Kanatzidis MG.: First-principles study of the electronic, optical, and lattice vibrational properties of AgSbTe$_2$. Phys. Rev. B. 2008;**77**:245203. doi:10.1103/PhysRevB.77.245203

[23] Lencer D, Salinga M, Grabowski B, Hickel T, Neugebauer J, Wuttig M.: A map for phase-change materials. Nat. Mater. 2008;**7**:972–977. http://www.nature.com/nmat/journal/v7/n12/suppinfo/nmat2330_S1.html

[24] Shportko K, Kremers S, Woda M, Lencer D, Robertson J, Wuttig M.: Resonant bonding in crystalline phase-change materials. Nat. Mater. 2008;**7**:653–658. doi:10.1038/nmat2226

[25] Du B, Li H, Xu J, Tang X, Uher C.: Enhanced figure-of-merit in Se-doped p-type AgSbTe$_2$ thermoelectric compound. Chem. Mater. 2010;**22**:5521–5527. doi:10.1021/cm101503y

[26] Ma H, Su T, Zhu P, Guo J, Jia X.: Preparation and transport properties of AgSbTe$_2$ by high-pressure and high-temperature. J. Alloys Compd. 2008;**454**:415–418. doi:10.1016/j.jallcom.2006.12.126

[27] Min BK, Kim BS, Kim IH, Lee JK, Kim MH, Oh MW, Park SD, Lee HW.: Electron transport properties of La-doped AgSbTe$_2$ thermoelectric compounds. Electron. Mater. Lett. 2011;**7**:255–260. doi:10.1007/s13391-011-0914-0

[28] Sharma PA, Sugar JD, Medlin DL.: Influence of nanostructuring and heterogeneous nucleation on the thermoelectric figure of merit in AgSbTe$_2$. J. Appl. Phys. 2010;**107**:113716. doi:10.1063/1.3446094

[29] Sugar JD, Medlin DL.: Precipitation of Ag_2Te in the thermoelectric material $AgSbTe_2$. J. Alloys Compd. 2009;**478**:75–82. doi:10.1016/j.jallcom.2008.11.054

[30] Pereira PB. Structure and Lattice Dynamics of Thermoelectric Complex Chalcogenides. Julich Forschungszentrum: Universite de Liege; 2012.

[31] Barabash SV, Ozolins V, Wolverton C.: First-principles theory of competing order types, phase separation, and phonon spectra in thermoelectric $AgPb_mSbTe_{m+2}$ alloys. Phys. Rev. Lett. 2008;**101**:155704. doi:10.1103/PhysRevLett.101.155704

[32] Hoang K, Mahanti SD, Salvador JR, Kanatzidis MG.: Atomic ordering and gap formation in Ag-Sb-based ternary chalcogenides. Phys. Rev. Lett. 2007;**99**:156403.doi:10.1103/PhysRevLett.99.156403

[33] Rezaei N, Hashemifar SJ, Akbarzadeh H.: Thermoelectric properties of $AgSbTe_2$ from first-principles calculations. J. Appl. Phys. 2014;**116**:103705. doi:10.1063/1.4895062

[34] Singh DJ, Terasaki I.: Thermoelectrics - nanostructuring and more. Nat. Mater. 2008;**7**:616–617. doi:10.1038/nmat2243

[35] Toberer ES, Zevalkink A, Snyder GJ.: Phonon engineering through crystal chemistry. J. Mater. Chem. 2011;**21**:15843–15852. doi:10.1039/c1jm11754h

[36] He JQ, Girard SN, Kanatzidis MG, Dravid VP.: Microstructure-lattice thermal conductivity correlation in nanostructured $PbTe_{0.7}S_{0.3}$ thermoelectric materials. Adv. Funct. Mater. 2010;**20**:764–772. doi:10.1002/adfm.200901905

[37] He JQ, Sootsman JR, Girard SN, Zheng JC, Wen JG, Zhu YM, Kanatzidis MG, Dravid VP.: On the origin of increased phonon scattering in nanostructured PbTe based thermoelectric materials. J. Am. Chem. Soc. 2010;**132**:8669–8675. doi:10.1021/ja1010948

[38] Kim W, Singer SL, Majumdar A, Zide JMO, Klenov D, Gossard AC, Stemmer S.: Reducing thermal conductivity of crystalline solids at high temperature using embedded nanostructures. Nano Lett. 2008;**8**:2097–2099. doi:10.1021/nl080189t

[39] Du B, Li H, Tang X.: Effect of Ce substitution for Sb on the thermoelectric properties of $AgSbTe_2$ compound. J. Electron. Mater. 2014;**43**:2384–2389. doi:10.1007/s11664-014-3076-7

[40] Du BL, Li H, Tang XF.: Enhanced thermoelectric performance in Na/Se doped p-type $AgSbTe_2$ compound. J. Inorg. Mater. 2011;**26**:680–684. doi:10.3724/sp.j.1077.2011.00680

[41] Du BL, Li H, Tang XF.: Enhanced thermoelectric performance in Na-doped p-type non-stoichiometric $AgSbTe_2$ compound. J. Alloys Compd. 2011;**509**:2039–2043. doi:10.1016/j.jallcom.2010.10.131

[42] Mohanraman R, Sankar R, Boopathi KM, Chou FC, Chu CW, Lee CH, Chen Y-Y.: Influence of In doping on the thermoelectric properties of an $AgSbTe_2$ compound with enhanced figure of merit. J. Mater. Chem A. 2014;**2**:2839–2844. doi: 10.1039/c3ta14547f

[43] Mohanraman R, Sankar R, Chou FC, Lee CH, Chen Y-Y.: Enhanced thermoelectric performance in Bi-doped p-type $AgSbTe_2$ compounds. J. Appl. Phys. 2013;**114**. doi:10.1063/1.4828478

[44] Wu H-j, Chen S-w, Ikeda T, Snyder GJ.: Reduced thermal conductivity in Pb-alloyed $AgSbTe_2$ thermoelectric materials. Acta Mater. 2012;**60**:6144–6151. doi:10.1016/j.actamat.2012.07.057

[45] Zhang H, Luo J, Zhu HT, Liu QL, Liang JK, Li JB, Liu GY.: Synthesis and thermoelectric properties of Mn-doped $AgSbTe_2$ compounds. Chin. Phys. B. 2012;**21**:6. doi:10.1088/1674-1056/21/10/106101

[46] Zhang H, Luo J, Zhu HT, Liu QL, Liang JK, Rao GH.: Phase stability, crystal structure and thermoelectric properties of Cu doped $AgSbTe_2$. Acta Phys. Sin. 2012;**61**:7.

[47] Zhang SN, Jiang GY, Zhu TJ, Zhao XB, Yang SH.: Doping effect on thermoelectric properties of nonstoichiometric $AgSbTe_2$ compounds. Int. J. Miner. Metall. Mater. 2011;**18**:352–356. doi:10.1007/s12613-011-0446-5

[48] Mohanraman R, Sankar R, Chou F-C, Lee C-H, Iizuka Y, Muthuselvam IP, Chen Y-Y: Influence of nanoscale Ag_2Te precipitates on the thermoelectric properties of the Sn doped P-type $AgSbTe_2$ compound. APL Mater. 2014;**2**:096114. doi:10.1063/1.4896435

[49] Du B, Xu J, Zhang W, Tang X.: Impact of in situ generated Ag_2Te nanoparticles on the microstructure and thermoelectric properties of $AgSbTe_2$ compounds. J. Electron. Mater. 2011;**40**:1249–53. doi:10.1007/s11664-011-1620-2

[50] Zhang SN, Zhu TJ, Yang SH, Yu C, Zhao XB.: Improved thermoelectric properties of $AgSbTe_2$ based compounds with nanoscale Ag_2Te in situ precipitates. J. Alloys Compd. 2010;**499**:215–220. doi:10.1016/j.jallcom.2010.03.170

[51] Du B, Yan Y, Tang X.: Variable-temperature in situ X-ray diffraction study of the thermodynamic evolution of $AgSbTe_2$ thermoelectric compound. J. Electron. Mater. 2015;**44**:2118–2123. doi:10.1007/s11664-015-3682-z

[52] Li HY, Jing HY, Han YD, Lu GQ, Xu LY, Liu T.: Interfacial evolution behavior of $AgSbTe_{2.01}$/nanosilver/Cu thermoelectric joints. Mater. Des. 2016;**89**:604–610. doi:10.1016/j.matdes.2015.09.163

[53] Wyzga PM, Wojciechowski KT.: Analysis of the influence of thermal treatment on the stability of $Ag_{1-x}Sb_{1+x}Te_{2+x}$ and Se-doped $AgSbTe_2$. J. Electron. Mater. 2016;**45**:1548–1554. doi:10.1007/s11664-015-4102-0

[54] Pei Y, Lensch-Falk J, Toberer ES, Medlin DL, Snyder GJ.: High thermoelectric performance in PbTe due to large nanoscale Ag_2Te precipitates and La doping. Adv. Funct. Mater. 2011;**21**:241–249. doi:10.1002/adfm.201000878

[55] Takagiwa Y, Pei Y, Pomrehn G, Snyder GJ.: Dopants effect on the band structure of PbTe thermoelectric material. Appl. Phys. Lett. 2012;**101**:092102–3. doi:10.1063/1.4748363

[56] Min B-K, Kim B-S, Oh M-W, Ryu B-K, Lee J-E, Joo S-J, Park S-D, Lee H-W, Lee H-s.: Effect of La-doping on AgSbTe$_2$ thermoelectric compounds. J. Korean Phys. Soc. 2016;**68**:164–169. doi:10.3938/jkps.68.164

[57] Amouyal Y.: On the role of lanthanum substitution defects in reducing lattice thermal conductivity of the AgSbTe$_2$ (P4/mmm) thermoelectric compound for energy conversion applications. Comput. Mater. Sci. 2013;**78**:98–103.doi:10.1016/j.commatsci.2013.05.027

[58] Amouyal Y.: Reducing lattice thermal conductivity of the thermoelectric compound AgSbTe$_2$ (P4/mmm) by lanthanum substitution: computational and experimental approaches. J. Electron. Mater. 2014;**43**:3772–3779. doi:10.1007/s11664-014-3145-y

[59] Quarez E, Hsu KF, Pcionek R, Frangis N, Polychroniadis EK, Kanatzidis MG.: Nanostructuring, compositional fluctuations, and atomic ordering in the thermoelectric materials AgPb$_m$SbTe$_{2+m}$. The myth of solid solutions. J. Am. Chem. Soc. 2005;**127**:9177–9190. doi:10.1021/ja051653o

[60] Kresse G, Furthmuller J.: Efficiency of ab-initio total energy calculations for metals and semiconductors using a plane-wave basis set. Comput. Mater. Sci. 1996;**6**:15–50. doi:10.1016/0927-0256(96)00008-0

[61] Kresse G, Furthmueller J.: Efficient iterative schemes for Ab Initio total-energy calculations using a plane-wave basis set. Phys. Rev. B: Condens. Matter Mater. Phys. 1996;**54**:11169–11186. doi:10.1103/PhysRevB.54.11169

[62] Kresse G, Hafner J.: Ab initio molecular-dynamics simulation of the liquid-metal-amorphous-semiconductor transition in Germanium. Phys. Rev. B: Condens. Matter Mater. Phys. 1994;**49**:14251–14269. doi:10.1103/PhysRevB.49.14251

[63] MedeA®: Materials Design, Inc. 2010;v. **2.74**. Angel Fire, NM, USA.

[64] Perdew JP, Ruzsinszky A, Csonka GI, Vydrov OA, Scuseria GE, Constantin LA, Zhou X, Burke K.: Restoring the density-gradient expansion for exchange in solids and surfaces. Phys. Rev. Lett. 2008;**100**:136406. doi:10.1103/PhysRevLett.100.136406

[65] Kresse G, Joubert D.: From ultrasoft pseudopotentials to the projector augmented-wave method. Phys. Rev. B. 1999;**59**:1758. doi:10.1103/PhysRevB.59.1758

[66] Blöchl PE, Jepsen O, Andersen OK.: Improved tetrahedron method for Brillouin-zone integrations. Phys. Rev. B. 1994;**49**:16223–16233. doi:10.1103/PhysRevB.49.16223

[67] Madsen GKH, Singh DJ.: BoltzTraP. A code for calculating band-structure dependent quantities. Comput. Phys. Commun. 2006;**175**:67–71. doi:10.1016/j.cpc.2006.03.007

[68] Scheidemantel TJ, Ambrosch-Draxl C, Thonhauser T, Badding JV, Sofo JO.: Transport coefficients from first-principles calculations. Phys. Rev. B. 2003;**68**:125210. doi:10.1103/PhysRevB.68.125210

[69] Ashcroft NW, Mermin ND. Solid State Physics: Holt, Rinehart and Winston; 1976.

[70] Joseph E, Amouyal Y.: Towards a predictive route for selection of doping elements for the thermoelectric compound PbTe from first-principles. J. Appl. Phys. 2015;**117**:175102. doi:10.1063/1.4919425

[71] Joseph E, Amouyal Y.: Enhancing thermoelectric performance of PbTe-based compounds by substituting elements: a first principles study. J. Electron. Mater. 2015;**44**:1460–1468. doi:10.1007/s11664-014-3416-7

[72] Poudeu PFP, Kanatzidis MG.: Design in solid state chemistry based on phase homologies. Sb_4Te_3 and Sb_8Te_9 as new members of the series $(Sb_2Te_3)_m \cdot (Sb_2)_n$. Chem. Commun. 2005:2672–2674. doi:10.1039/B500695C

[73] Kifune K, Fujita T, Kubota Y, Yamada N, Matsunaga T.: Crystallization of the chalcogenide compound Sb_8Te_3. Acta Crystallogr. Sect. B. 2011;**67**:381–385. doi:10.1107/S0108768111033738

[74] Amram D, Amouyal Y, Rabkin E.: Encapsulation by segregation – a multifaceted approach to gold segregation in iron particles on sapphire. Acta Mater. 2016;**102**:342–351. doi:10.1016/j.actamat.2015.08.081

[75] Ahmad S, Mahanti SD, Hoang K, Kanatzidis MG.: Ab initio studies of the electronic structure of defects in PbTe. Phys. Rev. B. 2006;**74**:155205. doi:10.1103/PhysRevB.74.155205

[76] Graff A, Amouyal Y.: Effects of lattice defects and niobium doping on thermoelectric properties of calcium manganate compounds for energy harvesting applications. J. Electron. Mater. 2016;**45**:1508–1516. doi:10.1007/s11664-015-4089-6

[77] de Boor J, Stiewe C, Ziolkowski P, Dasgupta T, Karpinski G, Lenz E, Edler F, Mueller E.: High-temperature measurement of seebeck coefficient and electrical conductivity. J. Electron. Mater. 2013;**42**:1711–1718. doi:10.1007/s11664-012-2404-z

[78] Edler F, Lenz E.: Metrology for energy harvesting. AIP Conference Proceedings. 2012;**1449**:369–372. doi:10.1063/1.4731573

[79] Cowan RD.: Pulse method of measuring thermal diffusivity at high temperatures. J. Appl. Phys. 1963;**34**:926–927. doi:10.1063/1.1729564

[80] Rowe DM.: Thermoelectrics Handbook: Macroto Nano. Boca Ravton: Taylor & Francis Group; 2006.

[81] Klemens PG.: The scattering of low-frequency lattice waves by static imperfections. Proc. Phys. Soc. Section A. 1955;**68**:1113.

[82] Klemens PG.: Thermal resistance due to point defects at high temperatures. Phys. Rev. 1960;**119**:507–509. doi:10.1103/PhysRev.119.507

[83] Abeles B.: Lattice thermal conductivity of disordered semiconductor alloys at high temperatures. Phys. Rev. 1963;**131**:1906–1911. doi:10.1103/PhysRev.131.1906

[84] Abeles B, Beers DS, Cody GD, Dismukes JP.: Thermal conductivity of Ge-Si alloys at high temperatures. Phys. Rev. 1962;**125**:44–46. doi:10.1103/PhysRev.125.44

[85] Hurd AJ, Kelly RL, Eggert RG, Lee M-H.: Energy-critical elements for sustainable development. MRS Bull. 2012;**37**:405–410. doi:10.1557/mrs.2012.54

[86] Callaway J.: Model for lattice thermal conductivity at low temperatures. Phys. Rev. 1959;**113**:1046. doi:10.1103/PhysRev.113.1046

[87] Callaway J, von Baeyer HC.: Effect of point imperfections on lattice thermal conductivity. Phys. Rev. 1960;**120**:1149. doi:10.1103/PhysRev.120.1149

[88] He J, Zhao L-D, Zheng J-C, Doak J, Wu H, Wang H-Q, Lee Y, Wolverton C, Kanatzidis M, Dravid V.: Role of sodium doping in lead chalcogenide thermoelectrics. J. Am. Chem. Soc. 2013;**135**:4624–4627. doi:10.1021/ja312562d

[89] He JQ, Girard SN, Zheng JC, Zhao LD, Kanatzidis MG, Dravid VP.: Strong phonon scattering by layer structured $PbSnS_2$ in PbTe based thermoelectric materials. Adv. Mater. 2012;**24**:4440–4444. doi:10.1002/adma.201201565

[90] Kim W, Zide J, Gossard A, Klenov D, Stemmer S, Shakouri A, Majumdar A.: Thermal conductivity reduction and thermoelectric figure of merit increase by embedding nanoparticles in crystalline semiconductors. Phys. Rev. Lett. 2006;**96**:045901. doi:10.1103/PhysRevLett.96.045901

[91] Wu LJ, Zheng JC, Zhou J, Li Q, Yang JH, Zhu YM.: Nanostructures and defects in thermoelectric $AgPb_{18}SbTe_{20}$ single crystal. J. Appl. Phys. 2009;**105**: 094317. doi:10.1063/1.3124364

[92] Lo S-H, He J, Biswas K, Kanatzidis MG, Dravid VP.: Phonon scattering and thermal conductivity in p-type nanostructured PbTe-BaTe bulk thermoelectric materials. Adv. Funct. Mater. 2012;**22**:5175–5184. doi:10.1002/adfm.201201221

[93] Graff A, Amouyal Y.: Reduced thermal conductivity in niobium-doped calcium-manganate compounds for thermoelectric applications. Appl. Phys. Lett. 2014;**105**:181906. doi:10.1063/1.4901269

[94] Goldsmid HJ.: Introduction to Thermoelectricity. Heidelberg Dordrecht London New York: Springer-Verlag; 2009.

[95] Tritt TM. Thermal Conductivity: Theory, Properties, and Applications. Clemson University, Clemson: Kluwer Academic/Plenum Publishers; 2004.

[96] Tritt TM.: Thermoelectric phenomena, materials, and applications. Annu. Rev. Mater. Res. 2011;**41**:433–448. doi:10.1146/annurev-matsci-062910-100453

[97] Morelli DT, Heremans JP, Slack GA.: Estimation of the isotope effect on the lattice thermal conductivity of group IV and group III-V semiconductors. Phys. Rev. B. 2002;**66**:195304. doi:10.1103/PhysRevB.66.195304

[98] Jovovic V, Heremans JP.: Measurements of the energy band gap and valence band structure of $AgSbTe_2$. Phys. Rev. B. 2008;**77**:245204. doi:10.1103/PhysRevB.77.245204

[99] Du BL, Li H, Xu JJ, Tang XF, Uher C.: Enhanced thermoelectric performance and novel nanopores in $AgSbTe_2$ prepared by melt spinning. J. Solid State Chem. 2011;**184**:109–114. doi:10.1016/j.jssc.2010.10.036

[100] Medlin DL, Sugar JD.: Interfacial defect structure at Sb_2Te_3 precipitates in the thermoelectric compound $AgSbTe_2$. Scripta Materialia. 2010;**62**:379–382. doi:10.1016/j.scriptamat.2009.11.028

[101] Zhang SN, Zhu TJ, Yang SH, Yu C, Zhao XB.: Phase compositions, nanoscale microstructures and thermoelectric properties in $Ag_{2-y}Sb_yTe_{1+y}$ alloys with precipitated Sb_2Te_3 plates. Acta Mater. 2010;**58**:4160–4169. doi:10.1016/j.actamat.2010.04.007

[102] Ayral-Marin RM, Brun G, Maurin M, Tedenac JC.: Contribution to the study of $AgSbTe_2$. Euro. J. Solid State Chem. 1990;**27**:747–757.

[103] Wu H-J, Chen S-W.: Phase equilibria of Ag–Sb–Te thermoelectric materials. Acta Mater. 2011;**59**:6463–6472. doi:10.1016/j.actamat.2011.07.010

[104] Sugar JD, Medlin DL.: Solid-state precipitation of stable and metastable layered compounds in thermoelectric $AgSbTe_2$. J. Mater. Sci. 2011;**46**:1668–1679. doi:10.1007/s10853-010-4984-4

[105] Kifune K, Fujita T, Tachizawa T, Kubota Y, Yamada N, Matsunaga T.: Crystal structures of X-phase in the Sb–Te binary alloy system. Cryst. Res. Technol. 2013;**48**:1011–1021. doi:10.1002/crat.201300252

[106] Kifune K, Kubota Y, Matsunaga T, Yamada N.: Extremely long period-stacking structure in the Sb-Te binary system. Acta Crystallographica. Section B. 2005;**61**:492–497. doi:10.1107/S0108768105017714

[107] Shelimova LE, Karpinskii OG, Kretova MA, Kosyakov VI, Shestakov VA, Zemskov VS, Kuznetsov FA.: Homologous series of layered tetradymite-like compounds in the Sb-Te and $GeTe-Sb_2Te_3$ systems. Inorg. Mater. 2000;**36**:768–775. doi:10.1007/bf02758595

[108] Sutton AP, Balluffi RW. Interfaces in Crystalline Materials, Oxford: Clarendon Press; 1995.

[109] Porter DA, Easterling KE. Phase Transformations in Metals and Alloys. 2ndEd. London: Chapman & Hall; 1992.

[110] Stowe K.: Crystal structure and electronic band structure of $LaTe_2$. J. Solid State Chem. 2000;**149**:155–166. doi:10.1006/jssc.1999.8514

[111] Chen Y, He B, Zhu TJ, Zhao XB.: Thermoelectric properties of non-stoichiometric AgSbTe$_2$ based alloys with a small amount of GeTe addition. J. Phys. D: Appl. Phys. 2012;**45**:115302. doi:10.1088/0022-3727/45/11/115302

[112] Chen Y, Nielsen MD, Gao YB, Zhu TJ, Zhao XB, Heremans JP.: SnTe-AgSbTe$_2$ thermoelectric alloys. Adv. Energy Mater. 2012;**2**:58–62. doi:10.1002/aenm.201100460

[113] Su T, Jia X, Ma H, Yu F, Tian Y, Zuo G, Zheng Y, Jiang Y, Dong D, Deng L, Qin B, Zheng S.: Enhanced thermoelectric performance of AgSbTe$_2$ synthesized by high pressure and high temperature. J. Appl. Phys. 2009;**105**: 3106102. doi:10.1063/1.3106102

Nanostructured State-of-the-Art Thermoelectric Materials Prepared by Straight-Forward Arc-Melting Method

Federico Serrano-Sánchez, Mouna Gharsallah,

Julián Bermúdez, Félix Carrascoso,

Norbert M. Nemes, Oscar J. Dura,

Marco A. López de la Torre, José L. Martínez,

María T. Fernández-Díaz and José A. Alonso

Additional information is available at the end of the chapter

http://dx.doi.org/10.5772/65115

Abstract

Thermoelectric materials constitute an alternative to harvest sustainable energy from waste heat. Among the most commonly utilized thermoelectric materials, we can mention Bi_2Te_3 (hole and electron conductivity type), PbTe and recently reported SnSe intermetallic alloys. We review recent results showing that all of them can be readily prepared in nanostructured form by arc-melting synthesis, yielding mechanically robust pellets of highly oriented polycrystals. These materials have been characterized by neutron powder diffraction (NPD), scanning electron microscopy (SEM) and electronic and thermal transport measurements. Analysis of NPD patterns demonstrates near-perfect stoichiometry of above-mentioned alloys and fair amount of anharmonicity of chemical bonds. SEM analysis shows stacking of nanosized sheets, each of them presumably single-crystalline, with large surfaces parallel to layered slabs. This nanostructuration affects notably thermoelectric properties, involving many surface boundaries (interfaces), which are responsible for large phonon scattering factors, yielding low thermal conductivity. Additionally, we describe homemade apparatus developed for the simultaneous measurement of Seebeck coefficient and electric conductivity at elevated temperatures.

Keywords: thermoelectrics, nanostructuration, lattice thermal conductivity, thermopower, neutron powder diffraction

1. Introduction

Thermoelectric materials possess the remarkable capability to transform temperature differences between two ends of a material sample directly and reversibly into a electrical potential difference. Waste heat recovery, which implies around 70% of primary energy production, exploited as a new source of power generation, could mean significant progress worldwide [1, 2]. Thermoelectric generators are able to perform this task, but currently they are not yet cost-effective. Several advantages featured by thermoelectric power generation devices, such as the absence of moving parts, reliability, endurance, quiet operation and no pollutant emission, make these devices valuable from an energy and environmental point of view and useful in a wide range of applications.

The dimensionless figure of merit $ZT = (S^2 \sigma)T/k$, where S stands for Seebeck coefficient, σ is the electrical conductivity and κ is the total thermal conductivity, evaluates thermoelectric performance of materials and serves as a reference value in thermoelectric materials research [3–8]. Maximization of ZT requires high Seebeck coefficient and low electrical resistivity and thermal conductivity values. This physical value is closely linked to the power generation efficiency of thermoelectric devices:

$$\varepsilon = \frac{T_H - T_C}{T_H} \frac{\sqrt{1 + ZT_M} - 1}{\sqrt{1 + ZT_M} + \frac{T_C}{T_H}},$$

(1)

where T_H, T_C and T_M are the temperature of the hot and cold ends and the average temperature. From this, we can abstract, that larger mean ZT along with larger temperature differences return better conversion efficiencies. Current commercial devices based on BiTe alloys reach efficiencies of ~6%, while new materials based on recent advances to improve ZT are expected to reach ~12–17%. These approaches are mainly focused on lowering lattice thermal conductivity by bulk nanostructuring and enhancing the power factor, $S^2\sigma$, by band engineering.

Experiments with nanostructured thermoelectric materials prove that highly efficient thermoelectric energy conversion could be forthcoming [9, 10]. Bulk samples, containing nanoscale constituents or inhomogeneity, exhibit enhanced thermoelectric phenomena, which are connected with the latest advances in optimizing thermoelectric figure of merit. Materials featuring these characteristics have been found among compounds, where nano-inclusions are inherently formed by using preparation methods to induce the nanostructured morphology. The main effect of nanostructuration is to affect the lattice thermal conductivity. Phonons are effectively scattered, when separation of defects or grain-sizes is similar to phonons' mean-free path. Consequently, bearing in mind the difference in electronic scattering length, structural unit-cells, comparable in size to heat carrying phonon wavelength, will improve the performance. On the other hand, quantum confinement effects could allow to treat S, σ and k quasi-independently and achieve higher power factors, defined as $S^2\sigma$ product [10–14].

Usually, thermoelectric nanocomposites are prepared initially and then assembled into bulk solids. Several methods for nanostructuring bulk materials have been developed; the most commonly used are spark plasma sintering (SPS), hot pressing, ball milling and wet chemical reactions [15]. All of them present different advantages and disadvantages, but they share drawbacks of long reaction and sample preparation times. For instance, it is expected, that the SPS method will be very beneficial for the reduction in lattice thermal conductivity due to retention of low-dimensional grains. On the other hand, it requires long annealing times and, as expected, results in more pronounced equiaxed morphology of powder particles with decrease in their size [16]. Chemical methods are convenient in terms of particle size, shape and crystallinity; nevertheless, removal of insulating organic capping ligands from nanocrystals is essential before consolidation into bulk pellets, and most of the chemically prepared materials present lower ZT values due to unsuitable charge carriers' concentrations and low intergranular connectivity achieved during compaction [17, 18].

Our present work deals with a straightforward and fast technique based on arc-melting synthesis. We have been able to prepare by this technique different families of thermoelectric materials including $Bi_{2-x}Sb_xTe_3$ and $Bi_2(Te_{1-x}Se_x)_3$ alloys, SnSe and related alloys, PbTe and GeSe compounds [19–21]. This method yields highly nanostructured samples prepared in really short times, which require no further processing and are directly implementable into devices. Highly oriented polycrystalline pellets are obtained with extremely low thermal conductivity, probably linked to the nanostructured nature of polycrystalline domains. This chapter describes synthesis by arc-melting and structural and thermoelectric characterization of these materials. Structural characterization has been carried out by X-ray diffraction (XRD) and neutron powder diffraction (NPD), which complements the study of thermoelectric properties and is used as the basis for density functional theory (DFT) calculations. Therefore, we review transport results of various bismuth telluride and tin selenide-related alloys, as well as, some other alloys. Bi_2Te_3 forms the basis for the most widely used thermoelectrics near room temperature. Arc-melting is a good technique affording rapid production with various doping and alloying. But how good is the resulting material? Thermoelectric property measurements by various procedures can reveal it.

2. Experimental section

2.1. Preparation by arc-melting

Intermetallic alloys of different above-mentioned families were prepared in an Edmund Buhler Compact Arc Melter MAM-1 (**Figure 1a**). Stoichiometric amounts of grinded mixture of reacting elements were pelletized in a glove box. Pellets were molten under Ar atmosphere in water-cooled Cu crucible (**Figure 1b**), leading to intermetallic ingots (**Figure 1c**), which can be ground to powder for structural characterization or cut with a diamond saw in bar-shaped samples for transport measurements. Complete characterization of these novel materials has included structural study by XRD and NPD and detailed examination of thermoelectric parameters.

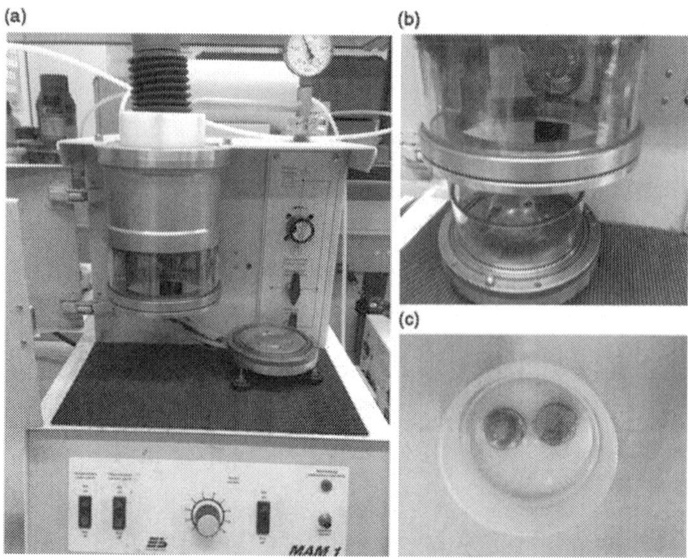

Figure 1. (a) Compact arc-melting furnace utilized for synthesis of nanostructured materials. (b) Water-cooled copper crucible, where sample can be quenched after melting process. (c) Typical aspect of as-grown ingots of intermetallic alloys.

2.2. Structural characterization

Initial characterization of products was carried out by laboratory XRD (Cu Kα, λ = 1.5406 Å). NPD diagrams were collected either at HRPT diffractometer of SINQ (The Swiss Spallation Neutron Source) spallation source at Paul Scherrer Institut or at D2B high-resolution diffractometer at Institut Laue-Langevin, Grenoble. Patterns were collected at room temperature with a wavelength of 1.494 Å (HRPT) or 1.594 Å (D2B). The high-flux mode was used ($\Delta d/d \approx 5 \times 10^{-4}$); typical collection time was 2 h. For some selected samples ($Sn_{0.8}Ge_{0.2}Se$), temperature-dependent NPD experiment was also carried out at D2B diffractometer. About 2 g of the sample was contained in a vanadium can and placed in the isothermal zone of the furnace with a vanadium resistor operating under vacuum ($P_{O2} \approx 10^{-6}$ torr). Measurements were carried out upon heating at 25, 200, 420 and 580°C, and NPD data were collected in diffractometer D2B. Diffraction data were analyzed by Rietveld method with FULLPROF program [22]. Line shape of diffraction peaks was generated by pseudo-Voigt function. The following parameters were refined: background points, zero shift, half-width, pseudo-Voigt, scale factor and unit-cell parameters. Positional and occupancy factors and anisotropic displacement factors were also refined for NPD data. Coherent scattering lengths for Bi, Te, Sn, Ge and Se were 8.532, 5.800, 6.225, 8.185 and 7.970 fm, respectively. A preferred orientation correction was applied, considering platelets perpendicular to the [001] for Bi_2Te_3-related alloys and the [100] direction for SnSe-related alloys.

2.3. Microstructural characterization

Surface texture of as-grown pellets is studied by field emission SEM (FE-SEM) in ZEISS 55 model. FE-SEM provides very focused energy electron beam, which improves greatly the spatial resolution and allows working at very low potentials, (from 0.02 to 5 kV); this helps to minimize effects of charge load on nonconductive samples and to avoid any damage to electron beam sensitive samples. It offers typical SEM image of surface topography of the sample with large depth of field. It is best suited for middle and low resolutions with high acceleration potential. It is mainly used to browse with low magnification looking for points of interest and to study samples with much topographical information. It also carries a secondary electron detector in lens: located inside the electron column, and it works with low-energy secondary electrons and offers higher resolution images. It is very sensitive to surface characteristics of the sample, so it is very suitable for surface characterization of any material.

2.4. Transport measurements

2.4.1. Physical properties measurement system (PPMS)

Three basic properties of thermoelectric materials (Seebeck coefficient, electrical resistivity and thermal conductivity) can be characterized simultaneously over a broad temperature range, between 2 and 400 K, by the thermal transport option (TTO) of the physical properties measurement system (PPMS) of Quantum Design Inc. This system allows for four electrical and thermal contacts with the sample. It uses two small thin-film temperature sensors (Cernox) mounted on small brass holders to measure both voltage and temperature drop across the sample. It uses another small brass piece with 2 kOhm resistive chip heater both to inject electrical current for resistivity measurement and to supply a known amount of heating power for Seebeck and thermal conductivity measurements. The fourth contact of the samples is the large brass baseplate of the sample holder, which thermally anchors it to the cryostat.

Measurements are carried out as follows: we use bar-shaped samples with $10 \times 3 \times 2$ mm^3 dimensions, prepared either by directly cutting from as-prepared ingots or by directly cold-pressing it after arc-melting in a properly shaped die. Four copper leads are wrapped around the bar and then fixed with silver epoxy. We use the following thermal protocol: cool the sample from 300 K to low temperature (either 2 K or 10 K), then warm it to 395 K and then cool it again to base temperature. We use slow sweep rate of 0.3 K/min (over 2 days for each sample) and gather data continuously. Electrical resistivity is measured by applying sinusoidal current with typical frequency of 17 Hz and amplitude between 10 µA and 10 mA. Seebeck coefficient is then measured by establishing a temperature gradient of typically 3% of the sample sink temperature. Here, the only important source of experimental error may arise from the fact, that Cernox temperature sensors are not touching the sample directly, but are thermally connected to it via a few millimeter-length of copper wires. Thermal conductivity is measured by dynamically modeling the temperature gradient on the sample between two Cernox sensors as known heater power (between 10 µW and 50 mW, adjusted to achieve 3% gradient) is supplied, and then removed, to one end of the sample. Above ~150 K, errors related to radiative heat losses become important. These are ameliorated to a certain extent by careful application of the correction software of TTO. Another, less important, source of heat losses is conduction through copper wires of voltage/temperature gradient leads. Finally, convective heat losses are minimized by the high vacuum option of PPMS, establishing pressure below 10^{-5} torr.

For Hall coefficient measurements, we use thin pellets (~1 mm thick, 10 mm diameter) in van der Pauw geometry, with four contacts placed along the perimeter, and pass a DC current with alternating sign to eliminate thermoelectric voltages along diagonal. Measured voltage is still dominated by ohmic resistance, which is then removed by comparing values obtained in large positive and negative magnetic fields.

2.4.2. High-temperature Seebeck coefficient measured in micro-miniature refrigerator (MMR) device

Seebeck coefficient is the ratio of voltage difference produced from applied temperature gradient. In principle, this concept is easy to understand, and one might think that this thermoelectric property is easy to measure. However, it can be difficult to evaluate [23, 24]. Basically, a temperature gradient is established between hot and cold ends of the sample and then the voltage drop, appearing between them, is measured. If the temperature gradient is kept constant during measurement, the method is called steady-state method.

Here we describe a method used by the commercial system of MMR Technologies Inc. (MMR device). This system allows measuring Seebeck coefficient of semiconductors and metals between 70 and 730 K. This method employs two pairs of thermocouples. One pair is formed by junctions of copper and reference material (constantan) with a well-known value of Seebeck coefficient. The other pair is formed of junctions of copper and sample, in which Seebeck coefficient is to be determined (**Figure 2a**). This method requires that both materials, reference and sample, have similar thermal conductance in order to ensure similar thermal transport through both of them. Thus, constantan wires with different diameters are used as a reference to allow evaluating materials with very different thermal conductivities.

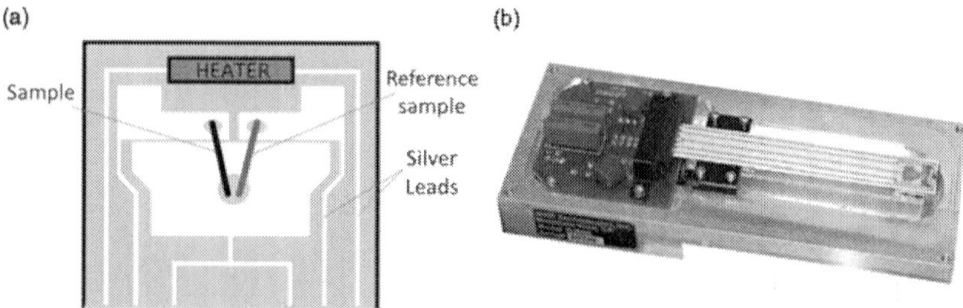

Figure 2. (a) Seebeck thermal stage supplied by MMR technologies. Black line indicates place of unknown sample and red line indicates that of reference sample. (b) Chamber with refrigerator, which provides temperature range between 70 and 730 K.

Another interesting characteristic of the method provided by MMR system is the double reference measurement technique, which gives more accurate and reproducible results. By using this feature, the equipment subtracts any instrumental offset voltage due to thermo-voltage effects from wires or connections. Different stages, made up of alumina or polyamide, allow covering a broad temperature range of measurement from 70 to 730 K (**Figure 2a** and **b**).

2.4.3. High-temperature transport device: high-temperature Seebeck measurements in homemade apparatus

As commented above, measurement of Seebeck coefficient seems like one of the less challenging tasks: create temperature gradient, measure voltage … What could be simpler? Yet, at elevated temperatures, this poses a challenge, mainly due to large, hard to control, temperature gradients. The largest systematic errors arise from the fact, that it is virtually impossible to detect temperatures at exactly the same spot where voltage difference is measured. Additional difficulty is strong chemical and metallurgic reactivity of typical thermoelectric materials at temperatures above a few hundred degree celsius, limiting the choice of building blocks for any instrument. Even such precious workhorse material as Pt is out of the question.

There are two main approaches to measure Seebeck effect in thermoelectric materials: integral and differential methods [24]. In the integral method, one end is maintained at a fixed temperature T_1, while another end is heated to induce, sometimes large, temperature gradient $\Delta T = T_2 - T_1$. According to the definition of Seebeck coefficient, we can write:

$$V_{TC} = V_T(T_1,T_2) - V_C(T_1,T_2) = \int_{T_1}^{T_2} (S_T(T) - S_C(T))dT, \tag{2}$$

where V_T and S_T are, respectively, Seebeck voltage and Seebeck coefficient of the thermoelectrics, while V_C and S_C are those of connecting lead. Normally, these conductors are metals with low Seebeck coefficient, such as Cu or Pt with known thermoelectric properties. Thus, it is possible to infer Seebeck voltage of interest, V_T, from voltage measured, V_{TC}, using the following expression:

$$V_{TC} + V_C(T_1,T_2) = V_T(T_1,T_2) = \int_{T_1}^{T_2} S_T(T)dT. \tag{3}$$

In the differential technique, small thermal gradient is used to estimate Seebeck coefficient at mean sample temperature. We can distinguish two approaches to this technique. In the DC method, constant thermal gradient is maintained, while mean temperature is varied, whereas in the AC method mean temperature is stabilized followed by the change in temperature gradient, usually in a sinusoidal form [25].

In the following, we describe a high-temperature homemade instrument to measure Seebeck coefficient with both integral and differential techniques. We based our design on the system described in Ref. [25], with some modifications, which are indicated below. Typical sample dimensions are pellets of around 10 mm diameter and 2 mm thickness, also ideal for laser flash measurements of thermal diffusivity.

Figure 3 shows the design scheme of measurement. This instrument is composed of two thermocouples, two blocks of niobium (Nb), two cartridge heaters with built-in thermocouples and a radiation furnace with yet another thermocouple. The whole assembly is placed in a vacuum chamber equipped with turbo-pump station and operates in a vacuum of around 7.5 × 10^{-7} torr and between 300 and 950 K. Baseplate can be water-cooled. We now describe important components of the homemade system.

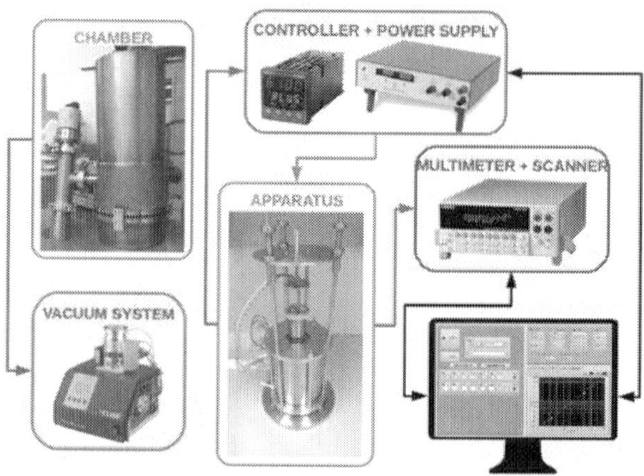

Figure 3. The top panel shows the main pieces of homemade instrument: vacuum chamber with turbo-pump station, furnace assembly and electronics for temperature control and voltage measurements. Screenshot of data acquisition software shows typical data resulting in differential AC measurement.

2.4.3.1. Thermocouples

Thermocouples are used to measure both temperature of the sample at either side and Seebeck voltage. These consist of two different wires, one of chromel and another of niobium, of 0.1 mm diameter, inserted in a four-bore ceramic tube (1.2 mm diameter with 0.2 mm bores) in such a way that where these wires cross Nb ones make contacts with the sample [25]. Thus, the difference between temperature and voltage measurements is minimized to the diameter of Nb wire. These four-bore ceramic tubes are mounted on spring-loaded mechanisms to press them dynamically against the sample. Springs are located outside the hot-zone to maintain their elasticity. We found, that this system works well enough without the sophisticated strain gauge mechanism of Ref. [25]. Niobium is a good choice for its low Seebeck coefficient—we use two Nb leads to detect Seebeck voltage of the sample.

2.4.3.2. Niobium blocks

Niobium blocks form the basis of apparatus. These cylinders, whose diameter and length are 20 mm, have two functions: they house cartridge heaters (and thus heat the sample differentially) and thermocouples, and press the sample during the measurement. After finding copper to be a poor choice due to its reactivity, we chose niobium for its chemical inertness against typical thermoelectric materials and for its good thermal conductivity. An advantage of using a typical metallic block instead of a ceramic one, as in Ref. [25], is that they can act as a pair of electrical contacts, to inject electrical current in 4-probe resistivity measurement, while ohmic voltage can be measured on contacts used for Seebeck voltage. Metallic blocks are also easier to machine. These blocks are electrically insulated from the rest of the equipment by ceramic (MACOR®) collars and are pressed against the sample by two sets of three springs on support bars.

2.4.3.3. Cartridge heaters

Cartridge heaters create controlled temperature gradient across the sample. They are inserted into niobium blocks off-center and are in good thermal contact with them. Cartridge heaters (from Watlow Ltd, 6.3 mm diameter, 25 mm length, 150 W) have individually incorporated J-type thermocouples. This enables us to smoothly control temperature gradient across the sample, even though cartridge heaters can deviate from the sample temperature by tens of degrees. We use programmable DC power supplies to control the temperature of cartridge heaters with high accuracy. Power supplies act as highly linear power amplifiers at the output of simple PID temperature controllers. Use of 0–5 V DC control logic helps us avoid introducing electronic noise into measurement, which would be rather troublesome in the typical solid state relay (SSR) scheme.

2.4.3.4. Furnace

The sample and Nb-blocks are surrounded by a small tubular furnace to establish the average temperature and crucially to reduce radiative heat losses. Homemade furnace is composed of an aluminum oxide tube of 34 mm internal diameter with rolled-up Kanthal wire. This heating element is surrounded by mineral fiber and stainless steel sheet with low emissivity. After placing the sample between the Nb-blocks, the whole furnace can slide up on three steel posts to position.

2.4.3.5. Breakout connector

The system has several thermocouples, and one of the crucial advantages of such an instrument, according to Ref. [25], is the possibility to easily swap and test different thermocouple wires for various samples. However, this poses technical challenge: these thermocouple wires must be run through the vacuum chamber. Although there exist commercial feedthroughs for thermocouples, these are costly and would eliminate flexibility to change the type of wire used. Therefore, we installed a break-out connector (DB-25) within the vacuum chamber and use copper wires from this post toward feedthrough (also DB-25, gold-plated pins) and outside toward electronics. As the break-out connector gets warm during operation (up to 80°C), we monitor its temperature with a resistive sensor (Pt100) and use this as "cold junction" in thermocouple measurements, in order to minimize spurious thermoelectric voltages.

2.4.3.6. Electronics

Seebeck voltage, thermocouples monitoring the sample temperature and any resistance measurements (such as Pt100 of breakout connector or the sample itself) are measured by Keithley-2700 scanning multimeter. Other thermocouples (cartridge heaters, furnace) are connected to three West-P6100 PID controllers. Their 0–5 V DC programming output acts on DeltaES150 DC power supplies. Smooth voltage output reduces electronic noise on sensitive Seebeck voltage and resistivity measurements from heating system.

2.4.3.7. Data acquisition and temperature set-point control

Data acquisition and temperature set-point control are handled by a LabVIEW program, which communicates via GPIB with multimeter and via RS485 serial protocol with PID controllers.

Salient features of this instrument are its operating range from slightly above room temperature up to 900 K and its flexibility to perform three different types of measurement schemes: quasi-integral, differential DC and differential AC. Each scheme has its own compromise between accuracy and overall required time for Seebeck coefficient measurements.

- **Quasi-integral** method is based on integral method and provides quick measurements, with low accuracy. We apply a step input to one heater, while the other is turned off. Although temperature of neither side of the sample is fixed, it is possible to extract the whole temperature-dependent Seebeck coefficient curve as polynomial function:

$$S(T) = S_0 + S_1 \cdot T + S_2 \cdot T^2 + S_3 \cdot T^3 + S_4 \cdot T^4 + S_5 \cdot T^5 + \cdots \tag{4}$$

According to Eq. (3) Seebeck voltage is:

$$V_T\left(T_1, T_2\right) = S_0 \cdot (T_2 - T_1) + \frac{S_1}{2} \cdot \left(T_2^2 - T_1^2\right) + \frac{S_2}{3} \cdot \left(T_2^3 - T_1^3\right) + \frac{S_3}{4} \cdot \left(T_2^4 - T_1^4\right) + \cdots \tag{5}$$

Therefore, from broad set of data, S_n polynomial coefficients can be numerically approximated. This technique does not generate particularly accurate results, but in as little as one hour, full $S(T)$ curve may be obtained up to 900 K, which is very helpful for screening purposes, when faced with large number of samples. The peculiarity of this technique is that resulting $S(T)$ curves are completely (and thus perplexingly) smooth, since they result from Eq. (4). Also, near the lowest and highest measured temperatures, $S(T)$ curves behave anomalously due to the way the numerical fit works. In a sense, integral method is closest to real-life operation with large temperature differences present.

- **Differential DC** method is the closest to definition of Seebeck coefficient:

$$S(T) = \frac{dV}{dT} \, . \tag{6}$$

This method provides more accurate results than quasi-integral one, but is somewhat slower. We maintain more-or-less constant temperature gradient of a few degrees using cartridge heaters, while raising overall temperature smoothly with the furnace. This method also works without the surrounding furnace, but temperature differences between cartridge heaters and respective sample sides are quite large (several tens of degrees). It is also difficult to gauge any systematic errors arising from voltage and temperature offsets. Differential DC method is the closest operation mode to MMR system.

- *Differential AC* is the slowest method, but also the most accurate, as described in Ref. [25]. It takes several hours (easily a full day) to obtain a few Seebeck coefficient data points, for example, every 50 K. This is because furnace and cartridge heater temperatures must be stabilized first and this takes a long time at high temperatures, and then cartridge heater temperatures are oscillated (with a period of tens of minutes) to obtain complete data sets (e.g., in **Figure 4**). The great advantage of this method is that any voltage offsets arising from stray thermoelectric emfs or temperature offsets, for example, due to miscalibration of thermocouples, are eliminated by linear regression to collected data. Differential AC method is reminiscent of the operation mode of PPMS.

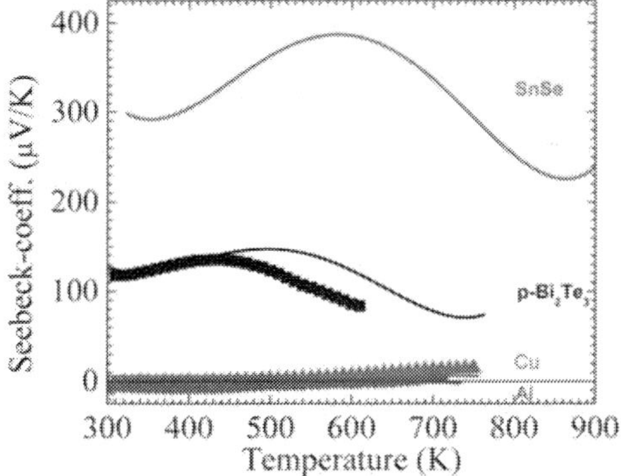

Figure 4. Typical temperature-dependent Seebeck coefficient data gathered with both differential DC method (full symbols) and integral method (lines) for low Seebeck metals, aluminum and copper, commercial p-type Bi_2Te_3 ingot and test pellet of SnSe produced by arc-melting.

We end this section by presenting a few temperature-dependent Seebeck coefficient curves for test materials. We used pieces of commercial aluminum and copper to check effects of voltage and temperature offsets. These metals have minimal Seebeck coefficient values at room temperature, around 3.5 µV K^{-1} for Al and 6.5 µV K^{-1} for Cu. These would vary by only a few µV K^{-1} in the studied temperature range. We found, that differential DC method gives Seebeck coefficient values differing by less than 10 µV K^{-1}, as absolute error. Relative error is less flattering, as for example, in the case of Al we measure negative values. Interestingly, integral method yields 5–6 µV K^{-1} for Cu, very close to tabulated value. Nevertheless, instrument is designed to study thermoelectric materials with large Seebeck coefficient values. We can take

± 10 μV K^{-1} as a rough estimate for systematic errors of the instrument. Results for commercial ingot of p-type Bi$_2$Te$_3$ above 100 μV K^{-1}, with characteristic maximum above 400 K, are similar to expected behavior. Finally, preliminary data from a pellet of SnSe produced by arc-melting and measured by quasi-integral method are also shown. These values are quite a bit lower than expected. The reason is still to be explored, but it is probably related to the fact, that data were recorded upon second heating run, and the sample may have suffered some chemical alteration. Whatever the case may be, it demonstrates the capabilities of the instrument.

2.4.4. Thermal conductivity: laser flash thermal diffusivity method

So-called laser flash thermal diffusivity technique is a useful method to determine thermal properties of bulk samples and also thin films. This method allows measuring thermal diffusivity (α) of a sample in very broad temperature range (80–2500 K) and diffusivity (0.001– 10 cm^2/s) range. For a given material, α is directly related to the speed at which the material can change its temperature. Thus, thermal conductivity of the sample is obtained from measurements of diffusivity (α), specific heat (C_p) and density (ϱ) of the sample by means of this relation:

$$\kappa = \alpha \cdot C_P \cdot \rho. \tag{7}$$

Commonly used systems, which measure α, usually allow also measuring C_p. In spite of this feature, it is recommended to measure specific heat by means of a separate technique, like differential scanning calorimetry (DSC), in order to obtain more accurate estimation of the final value. However, at high temperature, where the specific heat reaches a constant value, diffusivity provides the essential parameter to estimate thermal conductivity.

As shown in **Figure 5a**, use of this method to measure α implies illumination of one face of the sample by a laser pulse of length below 1 ms. An infrared (IR) detector placed behind rear face detects the signal, which is proportional to temperature rise. Thermal diffusivity value is obtained from IR signal rise against time. Example of IR signal profile vs. time is shown in **Figure 5b** corresponding to a graphite reference sample. Original Parker method [27] considers adiabatic conditions, and therefore, diffusivity value is obtained from thickness of the sample and time ($t_{1/2}$), where IR signal profile reaches half of maximum rise:

$$\alpha = 0.1388 \left(\frac{L^2}{t_{1/2}} \right), \tag{8}$$

where L is the thickness of the sample. This method is valid only for adiabatic conditions; however, different methods have been designed in order to consider effects of finite laser pulse time and radiative losses in nonadiabatic conditions [27–30]. Correction fits and evaluation models based on this method are also provided by manufacturers of state-of-the-art thermal diffusivity measuring systems.

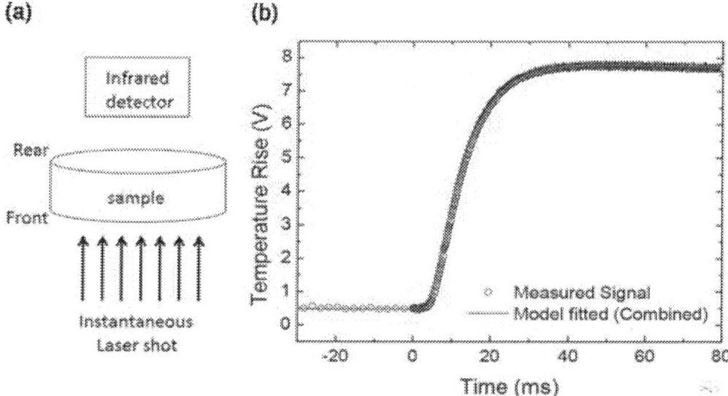

Figure 5. (a) Schematic illustration of laser flash method (adapted from *Thermoelectric Handbook: Macro to Nano*) [26], (b) signal profile and model fitted of graphite reference sample obtained in Laser flash system Linseis LFA 1000.

In addition, laser flash method imposes certain requirements on samples to be measured. It is mandatory to prepare plate samples with flat parallel planes to ensure correct acquisition of temperature profile at the rear face of the sample. Another important point about the sample preparation is to ensure the highest emissivity/absorption in the rear/front surface of the sample. For this purpose, thin coating of graphite well adhered over the sample's surface is commonly used (see **Figure 6a**).

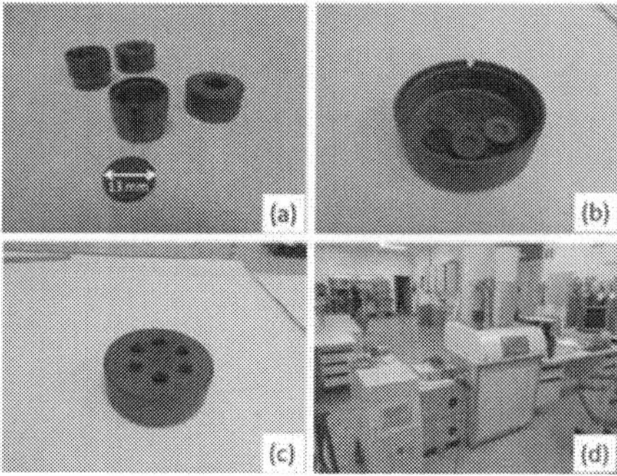

Figure 6. Different parts and sample holder (a, b), carrousel for six samples (c), and general view of laser flash equipment: Linseis LFA 1000 (d).

2.4.5. Thermal conductivity: 3ω method

Alternative procedure to determine thermal conductivity is the so-called 3ω (3 omega) method [31]. We sandwiched a thin gold wire (diameter $d = 25$ μm) between two identical, disk-shaped commercial Bi_2Te_3 samples. We used BN spray to electrically insulate the wire from the sample. This is a slight variation in the original method as heating/sensing wire is completely surrounded by the material. This method relies on applying AC current along the gold thread (heater) at frequency ω, which generates thermal oscillations finally resulting in a voltage at the third harmonic (3ω), which is closely related to thermal conductivity (κ):

$$V_{3\omega} = -\frac{V_{1\omega}^3 \cdot \beta}{8\pi \cdot l \cdot \kappa \cdot R_0} \cdot \left[\ln(2\omega) + \ln\left(\frac{d^2}{4\alpha}\right) - 2 \cdot \ln(2) \right] - i \cdot \frac{V_{1\omega}^2 \cdot \beta}{16 \cdot l \cdot \kappa \cdot R_0} , \tag{9}$$

where $V_{1\omega}$ is the amplitude of the ohmic voltage, l is the length of the heater, β is the temperature coefficient of gold, R_0 is the nominal wire electrical resistance, α is the thermal diffusivity of the sample. This formula is valid as long as thermal penetration depth, $\sqrt{\alpha/2\omega}$ is greater than five times the wire radius. Performing linear fit to above expression, of $V_{3\omega}$ vs $\ln(2\omega)$, thermal conductivity can be determined.

Here we present the results of this frequency-dependent third harmonic measurement for commercial p-type Bi_2Te_3 at 300 K [31]. Performing linear fit to the expression indicated in this section, of $V_{3\omega}$ vs $\ln(2\omega)$, as shown in **Figure 7**, thermal conductivity obtained is κ = 1.32 W m^{-1} K^{-1} at 300 K, matching reasonably well the results with TTO method of PPMS.

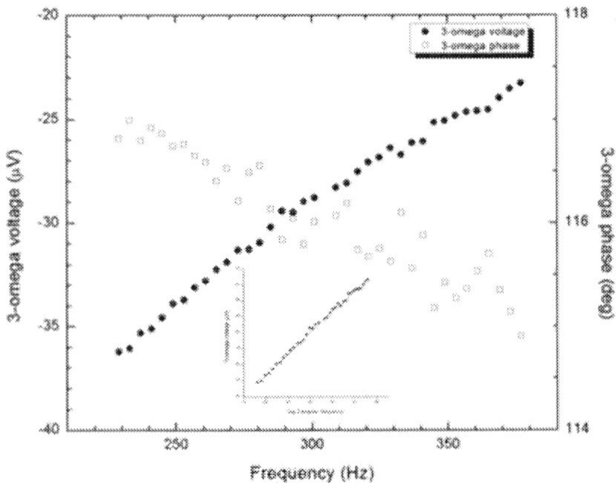

Figure 7. 3ω method for extracting thermal conductivity of commercial p-type Bi_2Te_3 at room temperature, with frequency-dependent third harmonic voltage, $V_{3\omega}$ (full symbols), and its phase shift (empty symbols) and with respect to logarithmic frequency (inset).

2.5. DFT calculations

We calculated the electronic density of states based on the experimentally determined unit cells in generalized gradient approximation (GGA) DFT scheme with Perdue-Burke-Emzerhof

(PBE) pseudopotentials in CASTEP using Materials Studio package [32, 33]. We considered 1×2×2 minimal supercells replacing between 0 and 4 Sn atoms with Ge or Sb out of 16. For Sb alloying, we also considered structures with one Sn atom missing. We checked two different configurations for each composition.

3. Results and discussion

3.1. Bi_2Te_3

Pristine Bi_2Te_3 was successfully prepared by arc-melting as reported in Ref. [20]. Bi_2Te_3 both n-type and p-type presents the best thermoelectric properties for applications near room temperature. Consequently, preparation of these compounds by arc-melting and verification of their thermoelectric properties was an interesting milestone. We found, that samples present high electrical conductivity, while they retain low thermal conductivity, which is strongly affected by morphology. On the other hand, Seebeck coefficient values are similar to previously reported results.

As shown in **Figure 8**, typical microstructure of samples can be described as piles of stacked sheets parallel to plane defined by **b** and **c** crystallographic axis, providing easy cleavage of materials. Thickness of individual sheets is well below 0.1 µm (typically 20–40 nm). Thermo-electric properties of these materials are strongly influenced by this micro- or nanostructura-tion, involving many surface boundaries (interfaces) that are responsible for scattering of both charge carriers and phonons.

Figure 8. Typical microstructure of Bi_2Te_3 prepared by arc-melting, showing nanostructuration in sheets perpendicular to crystallographic c axis. Magnification (a) 7000×, (b) 25,000×, (c) 50,000×, and (d) 80,000×.

Transport properties in pure Bi_2Te_3 were measured in PPMS device [20] and high-temperature device (described in Section 2.4.3) on cold-pressed pellets. In the temperature range between 300 and 540 K, absolute value of Seebeck coefficient increases continuously until a maximum of −93 µV K^{-1}, as shown in **Figure 9a**. These measurements were repeated in different samples.

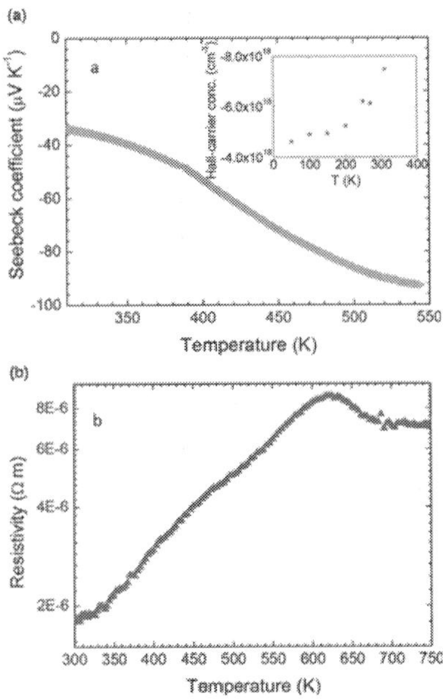

Figure 9. Temperature dependences of (a) Seebeck coefficient and (b) electrical resistivity of Bi_2Te_3 prepared by arc-melting.

Published data for undoped Bi_2Te_3 offer a wide range of Seebeck coefficient values from −50 to −260 µV K^{-1} [15, 34–36] depending on material's nature and charge carriers concentration. Compounds prepared by wet chemical methods demonstrate similar Seebeck coefficient values [18, 34], whereas for samples prepared by ball-milling and hot-pressing technique, this coefficient is as high as −190 µV K^{-1} [37]. In bismuth telluride samples, charge carriers concentration is strongly affected by antisite Bi_{Te} and Te_{Bi} defects, which are randomly created during synthesis process. This feature yields a wide range of Seebeck coefficient values measured in samples prepared by different methods [15, 34–39].

Electrical resistivity (**Figure 9b**) suggests semimetallic behavior, and, thus, we observe an increase in electrical resistivity with temperature. As temperature increases, charge carrier' scattering is augmented. We found resistivity values of 2 µOhm m at 320 K, whereas for samples prepared by mechanical alloying and SPS or hot pressing, they display higher values of 30 and 7 µOhm m, respectively [37, 38]. Samples synthesized by chemical processes show resistivity values between 13 and 5 µOhm m [34, 39]. These results indicate improvement compared to other preparation procedures. The number of thermally excited charge carriers increases with temperature, as shown by Hall concentration measurements. Charge carriers' density at 310 K is equal to 7.46×10^{18} cm^{-3}, similar to other literature values [35]. Hall mobility of charge carriers is calculated by $\mu_H = R_H \sigma$, resulting in 4514 cm^2 V^{-1} s^{-1}, which is an extremely high value due to low resistivity. Preferred orientation of sheets might be the main cause of this great electron mobility along measurement direction, resulting from the exceptionally anisotropic nature of Bi_2Te_3, with in-plane conductivity [40] over double that of out of plane.

Low thermal conductivity is critical for good thermoelectric performance. Thermal conductivity vs temperature is displayed in **Figure 10a** for Bi_2Te_3. At 365 K, thermal conductivity reaches its minimum value of 1.2 W m^{-1} K^{-1} after decreasing along the whole temperature range. This is an excellent value for Bi_2Te_3 bulk material, comparable to one of the lowest values presented in literature of 0.9 W m^{-1} K^{-1}, Ref. [15]. It implies good compromise with excellent electrical resistivity and its electronic contribution to thermal conductivity. Layered nanostructured morphology of pellets is probably playing important role in reducing thermal conductivity, where abundant grain boundaries along phonon path (between layers or block of layers) increase phonon scattering. This effect compensates for improved electrical resistivity. Other procedures leading to nanostructured samples, i.e., ball-milling and hot pressing, yield thermal conductivity value of 1.2 W m^{-1} K^{-1} at 330 K [37], while for chemical synthesis values of 0.8 W m^{-1} K^{-1} at 380 K are obtained [34], however, in alloys with higher electrical resistivity. Eventually, then, our arc-melting technique produces Bi_2Te_3 with ZT approaching 0.3 (**Figure 10b**).

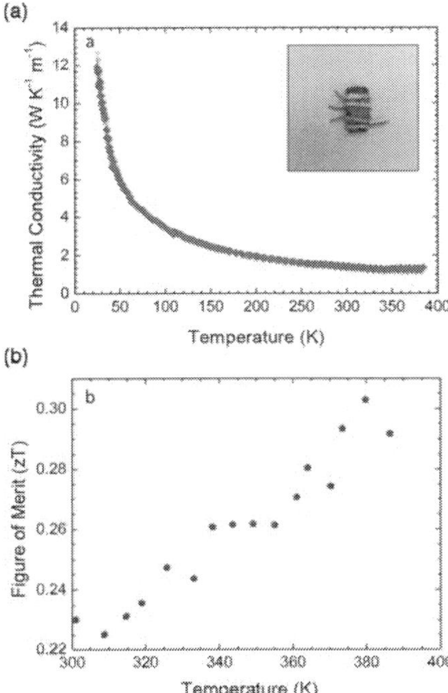

Figure 10. (a) Thermal conductivity measured in PPMS device; inset shows pellet used with TTO setup and (b) figure of merit of Bi_2Te_3 prepared by arc-melting.

3.2. Bi$_2$Te$_3$ based alloys: Bi$_2$(Te$_{1-x}$Se$_x$)$_3$

Preparation of Bi$_2$Te$_3$-Bi$_2$Se$_3$ solid solutions by alloying Bi$_2$Se$_3$ with Bi$_2$Te$_3$, where stronger Se-Bi interactions are created, enlarges band-gap energy and forms new donor levels. This may lead to enhanced electrical conductivity [41–43] and improve thermoelectric performance. Besides, decrease in lattice thermal conductivity is expected due to point defects induced by alloying. On the other hand, Bi$_2$(Te$_{1-x}$Se$_x$)$_3$ alloys are extremely susceptible to anisotropic effects, which have been notable disadvantages in preparation of bulk samples that have not been able to keep high electrical resistivity. Attempts for preparation of oriented grains in bulk samples have been made to overcome this issue [44].

We prepared Bi$_2$(Te$_{0.8}$Se$_{0.2}$)$_3$ pellets by arc-melting. NPD study was essential to investigate structural details of this doped sample. Neutrons are particularly suitable to study these intermetallic alloys having texturized nature of powder, which gives XRD patterns with large and untreatable preferred orientation effects. Neutrons provide bulk analysis, with good penetration; also the way of filling sample holders, (vanadium cylinders), helps to reduce unwanted preferred orientation, which is additionally minimized by rotation of sample holders. Moreover, lack of form factors for neutrons as diffraction probe enables accessing remote regions of reciprocal space, thus yielding accurate anisotropic displacement factors, which may give hints of the origin of phonon propagation across these materials, characterized by low thermal conductivity.

For Bi$_2$(Te$_{0.8}$Se$_{0.2}$)$_3$, NPD data were collected at RT at HRPT diffractometer of SINQ spallation source at PSI with λ = 1.494 Å. Crystal structure refinement was carried out in Bi$_2$Te$_3$-type model [45] in hexagonal setting of rhombohedral R-3m space group (no. 166), Z = 3, with Bi located at 6c (00z) Wyckoff site and Te/Se distributed at random over two different crystallo-graphic sites, (Te,Se)1 at 3a positions and (Te,Se)2 at 6c. There was excellent agreement between observed and calculated profiles, as shown in **Figure 11**; minor preferred orientation correction was effective in improving refinement for all reflections in the whole angular range, reaching low Bragg discrepancy factors of 5.15%. **Tables 1** and **2** include lattice and atomic parameters and anisotropic displacements factors, as well as discrepancy factors after refinement. Unit-cell parameters are **a** = 4.3315(1) Å and **c** = 30.208(7) Å. Unit-cell size is substantially smaller than that of parent Bi$_2$Te$_3$ compound (with unit-cell parameters: **a** = 4.385915 (6) Å, **c** = 30.495497 (1) Å, upon incorporation of smaller Se atoms.

Figure 12 shows two views of refined crystal structure of Bi$_2$(Te$_{0.8}$Se$_{0.2}$)$_3$, along c axis (left panel) and perpendicular to c axis (right panel). It consists of hexagonal close-packed sheets, each layer being composed of fivefold stacking sequence of covalently bonded (Te,Se)2-Bi-(Te,Se)1-Bi-(Te,Se)2 atoms, whereas interatomic forces between adjacent layers ((Te,Se)2-(Te,Se)2 interactions) are mainly van der Waals type. As a consequence, crystals of these alloys are easily cleaved perpendicular to **c**-direction. Bi atoms are coordinated to 3 (Te,Se)1 at distances of 3.129(7) Å and 3 (Te,Se)2 at 3.102(9) Å in distorted octahedral configuration. Distance between terminal (Te,Se)2 of adjacent layers is 3.634(9) Å. It is noteworthy that anisotropic displacement ellipsoids are strongly flattened with short axis perpendicular to bonding directions, i.e., along [110] directions as shown in the left panel in **Figure 12**.

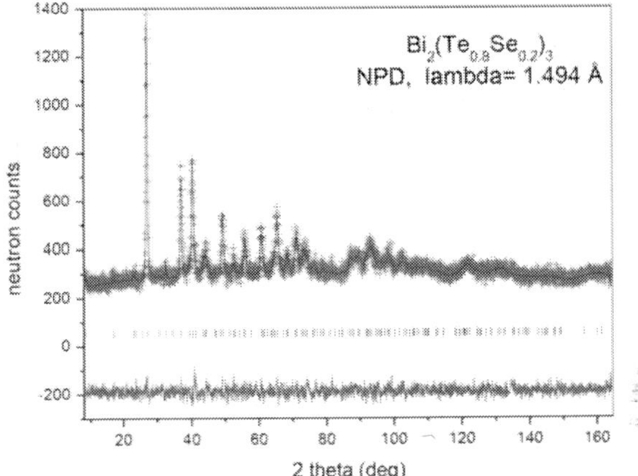

Figure 11. NPD profiles for Bi$_2$(Te$_{0.8}$Se$_{0.2}$)$_3$. Crosses are experimental points, solid line is calculated fit and difference is at the bottom. Vertical marks correspond to allowed Bragg reflections.

	x	y	z	U_{eq}	Occupation (<1)
Bi	0.00000	0.00000	0.3956 (4)	0.010 (4)	
Te1	0.00000	0.00000	0.00000	0.025 (11)	0.80000
Se1	0.00000	0.00000	0.00000	0.025 (11)	0.20000
Te2	0.00000	0.00000	0.7897 (3)	0.017 (7)	0.80000
Se2	0.00000	0.00000	0.7897 (3)	0.017 (7)	0.20000

Unit-cell parameters: a = 4.3315 (4) Å, c = 30.208 (5) Å, 490.83 (10) Å^3, Z = 3. U_{eq} and U^{ij} are, respectively, the equivalent and anisotropic atomic displacement parameters. Discrepancy factors after refinement are also included.

Table 1. Structural parameters for Bi$_2$(Te$_{0.8}$Se$_{0.2}$)$_3$ refined in R-3m space group (hexagonal setting) from NPD data collected at RT with λ = 1.494 Å.

	U^{11}	U^{22}	U^{33}	U^{12}	U^{13}	U^{23}
Bi	0.009 (2)	0.009 (2)	0.010 (8)	−0.005 (2)	0.00000	0.00000
Te1	0.020 (6)	0.020 (6)	0.04 (2)	−0.010 (6)	0.00000	0.00000
Se1	0.020 (6)	0.020 (6)	0.04 (2)	−0.010 (6)	0.00000	0.00000
Te2	0.017 (5)	0.017 (5)	0.018 (12)	−0.008 (5)	0.00000	0.00000
Se2	0.017 (5)	0.017 (5)	0.018 (12)	−0.008 (5)	0.00000	0.00000

Discrepancy factors: R_p = 4.688%, R_{wp} = 5.895%, R_{exp} = 5.359%, R_{Bragg} = 5.151%, χ^2 = 1.21.

Table 2. Anisotropic displacement parameters (Å^2).

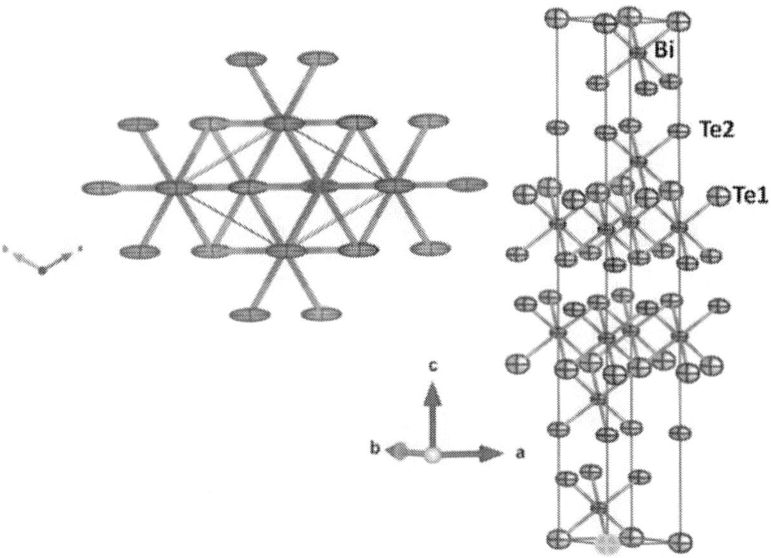

Figure 12. Two projections of crystal structure, along c axis (left panel) and perpendicular to c axis (right panel). Strongly anisotropic displacement ellipsoids (95% probability), with the short axis along [110] direction are illustrated.

Transport properties were evaluated in our homemade apparatus (Section 2.4.5). Seebeck coefficient vs temperature curve is plotted in **Figure 13a**.

Slow decrease in S is observed between 300 and 550 K, where average value is −75 μV K^{-1}. There is small improvement in thermopower regarding pure Bi_2Te_3 samples, but it is still low in contrast to other $Bi_2(Te_{1-x}Se_x)_3$ alloys, where values of −190 μV K^{-1} are reported [46] or even −259 μV K^{-1} at room temperature in samples with optimized composition [47]. Hall concentration of charge carriers is determined as 3.1×10^{19} cm^{-3} at 300 K (inset in **Figure 13a**), which is somewhat higher than in pure Bi_2Te_3 [20, 35], as a result of donor feature of $Bi_2(Te_{1-x}Se_x)_3$ alloys.

As expected, temperature dependence of electrical resistivity (**Figure 13b**) exhibits the same semimetallic behavior as pure compound, but abrupt reduction in resistivity is observed at 530 K, until minimum value of 55 μOhm m is reached. Compared with pure compound (Section 3.1), with extremely low electrical resistivity (2 μOhm m at 300 K) prepared by arc-melting, these results indicate deterioration of thermoelectric performance. The drop in electrical conduction is possibly a consequence of the enormous anisotropy of this alloy, meaning that orientations of layered structures are not aligned for improved electron mobility. Analogous relationship between charge carriers scattering, doping and electrical conductivity is reported by Ajay Soni et al. [47] for $Bi_2Te_{2.2}Se_{0.8}$ nanocomposite, which shows metallic and semiconductor behavior throughout their measurement range with values around 75 μOhm m at room temperature.

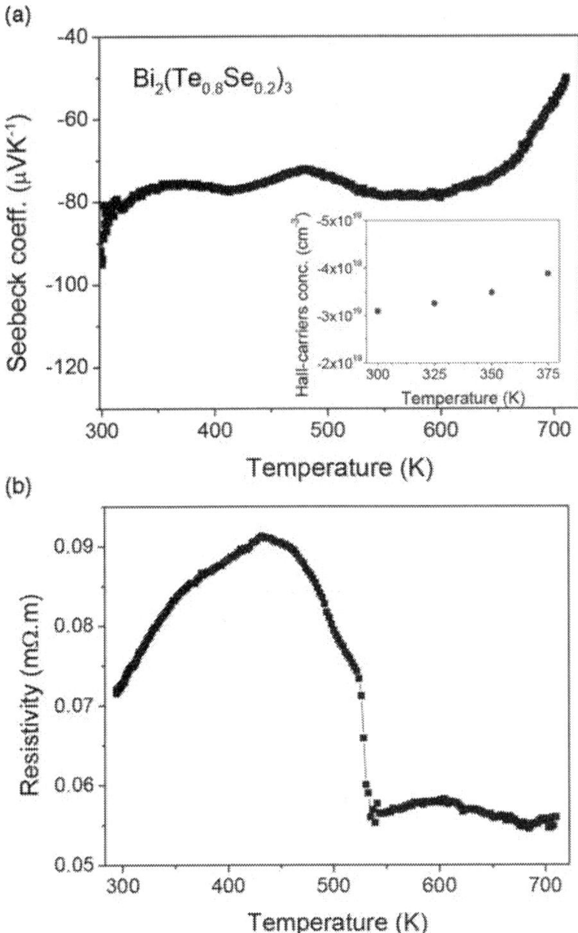

Figure 13. Temperature dependences of (a) Seebeck coefficient and (b) electrical resistivity for $Bi_2(Te_{0.8}Se_{0.2})_3$ measured in homemade apparatus.

Figure 14 shows dependence of total thermal conductivity on temperature determined in PPMS device.

Expected Umklapp maximum at low temperature appears with subsequent decrease over the whole measurement range until minimum value of 0.8 W m^{-1} K^{-1} is reached at 300 K. This is better (lower) thermal conductivity than reported for pristine nanostructured Bi_2Te_3 obtained by arc-melting, reaching 1.2 W m^{-1} K^{-1} at 365 K [20]. Excellent present values could be related to stronger anisotropy, higher electrical resistivity and point defects induced by alloying. In Bi_2Te_3-Bi_2Se_3 alloys produced by encapsulated melting and hot pressing, the best thermal conductivity value was equal to 1.04 W m^{-1} K^{-1} at 323 K [46], while for samples made by large-scale zone melting, values of 1.2 W m^{-1} K^{-1} at 323 K are observed [48].

In samples prepared by polyol method followed by SPS, which leads to nanocomposite materials, low values of thermal conductivity are found: 0.9 W m^{-1} K^{-1} at 300 K for $Bi_2Te_{2.2}Se_{0.8}$ and 0.7 W m^{-1} K^{-1} at 300 K for $Bi_2Te_{2.7}Se_{0.3}$ nanocomposites [47]. The literature survey shows, that our thermal conductivity values are among the best reported, which is probably related to the nanostructuration effects, that we describe for arc-melted samples, with advantage of the simplicity of our one-step straightforward method.

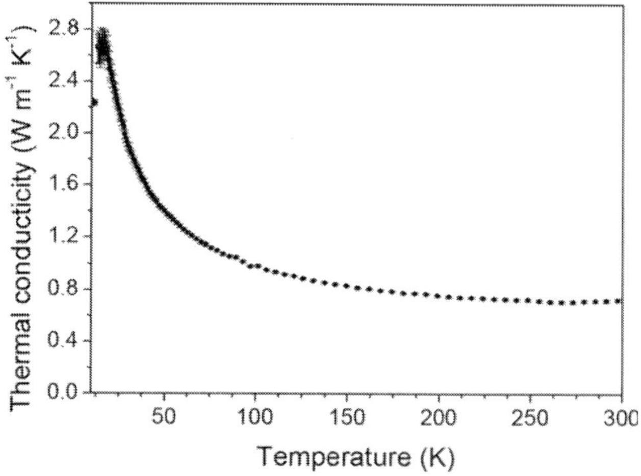

Figure 14. Temperature dependence of thermal conductivity of $Bi_2(Te_{0.8}Se_{0.2})_3$ measured in PPMS device. Extremely low thermal conductivity of 0.8 W m^{-1} K^{-1} is observed at 300 K.

3.3. SnSe and related alloys

Tin selenide and its various alloys show good thermoelectric performance at temperatures well above the maximum attainable in PPMS. Therefore, the main use of these results, measured with PPMS, is to guide us in search for candidate compositions. However, as we show, we found surprisingly high values of Seebeck coefficient, whereas thermal conductivity of tin selenide produced by arc-melting is considerably lower than that of single crystals. Both are highly promising results for thermoelectric materials.

3.3.1. Structural characterization

We have prepared by simple and straightforward arc-melting technique highly textured SnSe samples with record Seebeck coefficient values and extremely low thermal conductivity [19]. Test NPD study was essential to investigate structural details of SnSe, since this bulk study is by far much less sensitive to the preferred orientation effects. Not only pristine SnSe, but other several novel series of SnSe-based alloys, namely $Sn_{1-x}M_xSe$ (M = Sb, Ge) prepared by arc-melting, have been investigated by neutron diffraction. We will illustrate these studies with description of *in situ* structural evolution of $Sn_{0.8}Ge_{0.2}Se$ in the temperature range of maximum thermoelectric efficiency. NPD data were collected in diffractometer D2B. Measurements were taken at 25, 200, 420 and 580°C.

Figure 15 illustrates NPD patterns of $Sn_{0.8}Ge_{0.2}Se$ at 420 and 580°C. Crystal structure can be Rietveld-refined in orthorhombic *Pnma* space group below 420°C. At this temperature, an orthorhombic (*Pnma*) to orthorhombic (*Cmcm*) phase transition takes place. **Figure 16** shows phase diagram displaying temperature dependence of unit-cell parameters of both ortho-rhombic phases. A dramatic rearrangement of atoms is observed along with phase transitions, bearing a more ordered structure. **Figure 17** displays crystal structures at room temperature, 200, 420 and 580°C. It is noteworthy, that change in displacement ellipsoids directions with temperature presents its largest axis along **c** direction in *Pnma* space group, while at high temperature it is oriented along **b** axis in *Cmcm* space group. In *Pnma*, structure consists of trigonal pyramids $SnSn_3$ forming layers perpendicular to [100] direction, with thermal ellipsoids oriented within the layers, whereas across the transition to *Cmcm* the coordination environment changes to tetragonal pyramid, where large Sn ellipsoids in the basal square-plane adopt a configuration with the longest axis perpendicular to four closer chemical bonds, oriented along **b** axis of orthorhombic structure. Such high thermal displacements indicate strong rattling effect of Sn in a pentacoordinated cage, accounting for the observed decrease in thermal conductivity and good thermoelectric performance of this material.

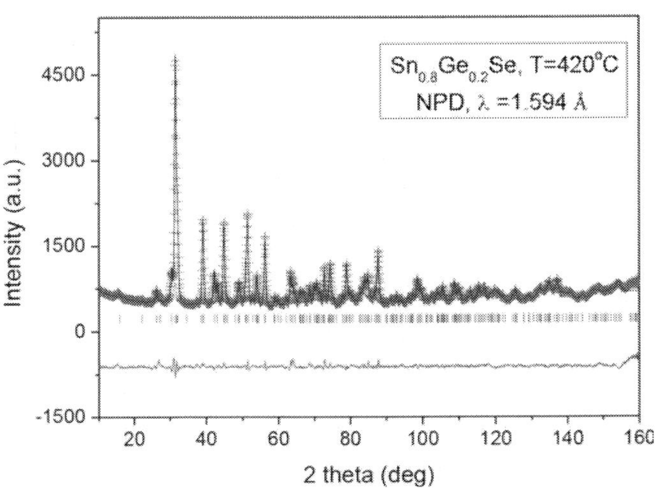

Figure 15. Observed (crosses), calculated (full line) and difference (at the bottom) NPD profiles for $Sn_{0.8}Ge_{0.2}Se$ at 420 °C, just below the phase transition. Vertical markers correspond to allowed Bragg reflections.

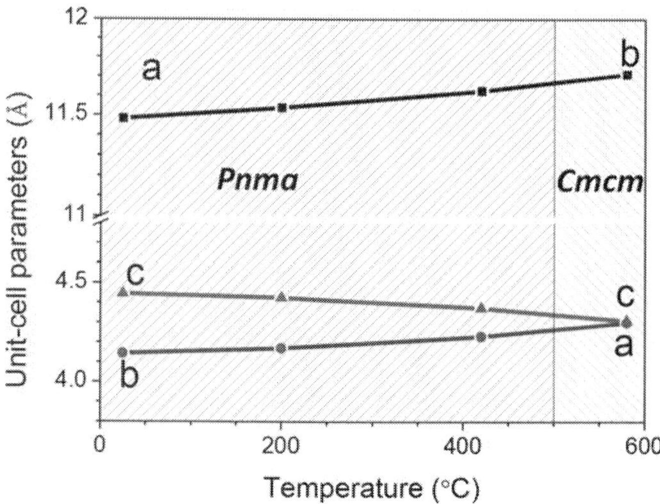

Figure 16. Phase diagram showing thermal evolution of unit-cell parameters.

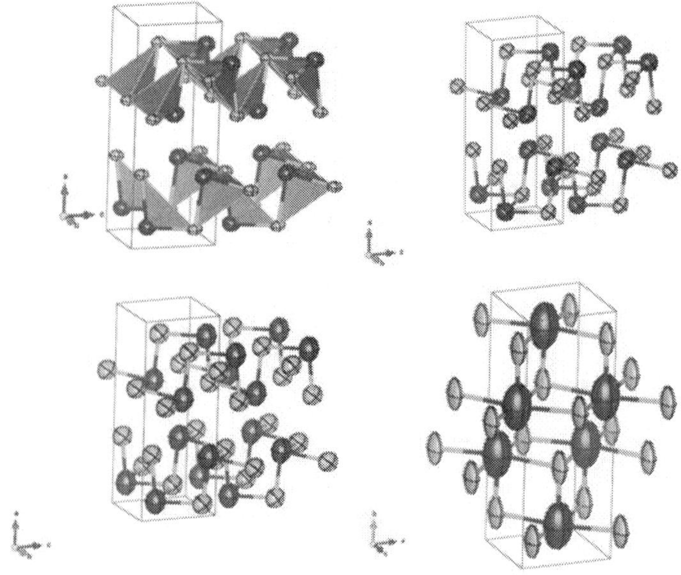

Figure 17. Crystal structures of orthorhombic phases at 25 (upper left), 200 (upper right), 420 (bottom left) and 580°C (bottom right).

3.3.2. Thermoelectric characterization

We first discuss our results on pure (i.e., unalloyed) SnSe, produced by arc-melting. The bar-shaped sample was cut directly from as-produced ingot as described above. **Figure 18** shows transport properties obtained in PPMS between 2 and 380 K. Seebeck coefficient (in middle panel) exhibits monotonic increase in positive (i.e., p-type) values reaching a maximum of 668 μV K^{-1} [19]. At the time, this was the highest Seebeck coefficient value reported in SnSe. We have observed this behavior in various samples on repeated measurements, consistently. While preliminary studies reported values only around 50 μV K^{-1} [49], the ground-breaking single crystal results of Zhao et al. [50] reach around 580 μV K^{-1}, quite independent of crystallographic direction [50]. Meanwhile, other reported values of polycrystalline samples are close to this value [51–53]. Uncontrolled differences in hole concentration may play a role in these variations. Zhao et al. [50] found, that Seebeck coefficient value reaches its maximum in SnSe around 525 K. When we extrapolate our results (limited below 400 K in PPMS), we could expect as much as 800 μV K^{-1}, and as we show below, indeed, we do find such high values with MMR device (described in Section 2.4.2).

Temperature dependence of electrical resistivity is shown in the bottom panel of **Figure 18**; Hall concentration of charge carriers at 300 K is 7.95 $\times$ 10^{15} cm^{-3} [19]. Resistivity decreases exponentially with temperature, as expected for a semiconductor. We consistently find, that resistivity of SnSe and its alloys produced by arc-melting is rather high. This is a persistent problem of the technique, which we must address in the future. Nevertheless, we must also bear in mind, that these results from PPMS are limited to a temperature range well below that, where SnSe functions as a good thermoelectric. For comparison, our arc-melting produced pellet has bulk resistivity at room temperature (295 K) of around 64 mOhm m, significantly higher than expected: the single crystals of Zhao et al. [50] have 1 mOhm m within bc plane and 5 mOhm m along a-direction, whereas Sassi et al. [52] report 11 mOhm m along the pressing direction in polycrystals and 5 mOhm m perpendicular to it. Hall effect measurements at 300 K yields p-type hole concentration of 7.95 $\times$ 10^{15} cm^{-3}. This is much lower than that of typical thermoelectrics and, indeed, of other SnSe reports (e.g., 4 $\times$ 10^{17} cm^{-3} by Zhao et al. [50]). Again, this is a persistent finding in our SnSe alloys: free charge carriers' concentration is much lower than expected, indicating the presence of strong traps for charge carriers. It also explains large electrical resistivity along with strong nanostructuration, producing abundant grain boundaries. Surprisingly, large Seebeck coefficient value is also related to low charge density, through Pisarenko relation [54].

Thermal conductivity of SnSe is shown in the top panel of **Figure 18**. It is overwhelmingly dominated by lattice contribution, due to low charge carriers' concentration.

Thermal conductivity peaks around 25 K due to UmKlapp scattering and then starts to decrease monotonically throughout the measurement range, reaching a value as low as 0.2 W m^{-1} K^{-1} at 395 K. This is a strikingly low value. Admittedly, direct heat-flow technique employed by PPMS is strongly affected by heat loss problems discussed above and these become acute above room temperature. Nevertheless, values are highly reliable below 100 K, and they are consistently very low there, too, in several samples. The intrinsic lattice thermal conductivity of SnSe is very low, probably an outcome of anharmonicity of chemical bonds. High Grüneisen parameters and strong phonon-phonon interactions are expected as a result of the presence of lone-electron pairs of both Se^{2-} and Sn^{2+} ions [49, 55]. In fact, lone pairs of p-block elements play an important role deforming lattice vibration, which results in strong anharmonicity; significantly, anisotropic vibrations of both Sn and Se atoms are determined by NPD, with the main ellipsoid axes directed along chemical bonds (**Figure 17**), which is also indicative of such anharmonicity, as vibrations are hindered out of bonding direction by voluminous electron pairs filling empty space in the crystal structure. Moreover, thermal conductivity is significantly lower than those reported in single crystals (1.8 W m^{-1} K^{-1}) [50, 56] and even lower compared to those recently reported for polycrystalline samples [52]. Extremely small values measured in the present material are most likely related to strong texture obtained during synthesis process, that leads to layered nanostructuration along **a** axis (**Figure 17**). This is particularly effective to boost phonon scattering at nanoscale, thus resulting in record low thermal conductivity for this polycrystalline material.

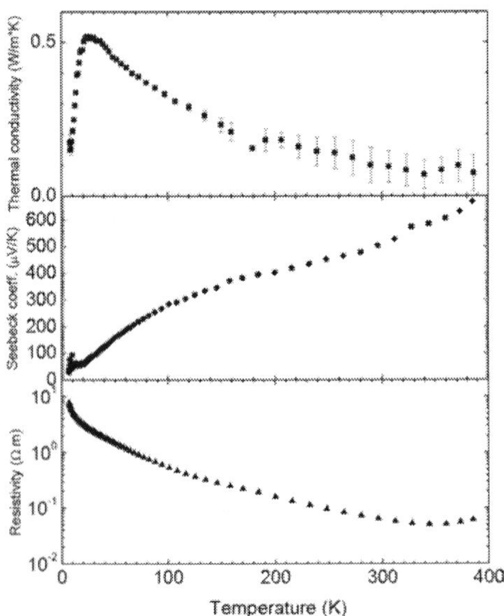

Figure 18. Temperature dependence of (top) thermal conductivity, (middle) Seebeck coefficient, (bottom) electrical resistivity of stoichiometric SnSe.

A second set of measurements of Seebeck coefficient at high temperature were carried out at MMR device, by comparing Seebeck effect of SnSe material with that of constantan wire, as described in Section 2.4.2. Seebeck coefficient was measured using $1 \times 1 \times 8$ mm^3 bar-shaped SnSe samples and reference constantan wire of 0.125 mm diameter. Reproducibility of measurements was confirmed by repeating them after making new contacts both on the sample and on reference constantan wire. This procedure warrants accuracy better than 5% over the whole temperature range.

Figure 19 shows Seebeck coefficient as function of temperature $S(T)$ measured using the method described above for SnSe. Initially, $S(T)$ increases with T from room temperature up to 400 K, where it reaches about 840 μV K^{-1}. Between 400 and 500 K, Seebeck coefficient is almost temperature independent, and above 500 K $S(T)$ decreases monotonically up to the maximum experimental temperature, which is lower than temperature corresponding to structural transition from *Pnma* to *Cmcm*, described above from NPD data. These values are slightly higher than those reported by Zhao et al. [50] on single-crystalline samples. This enhancement can be related to nanostructuration of the sample and presence of high density of boundaries.

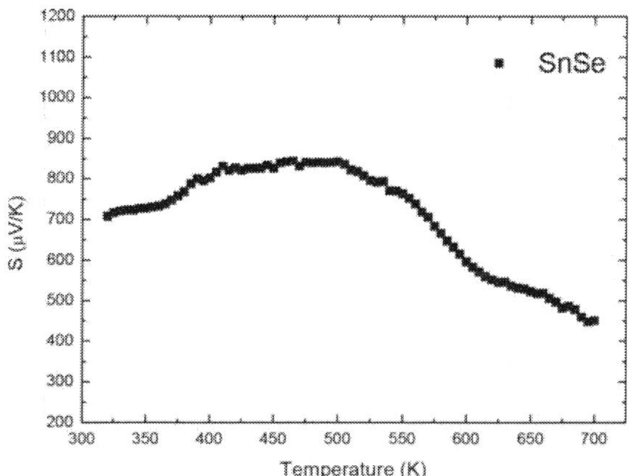

Figure 19. High-temperature Seebeck coefficient as function of temperature for SnSe measured in MMR device.

Morphology of the material produced by arc-melting is highly granular, nanostructured. This raises obvious concern about transport measurements. Do we measure intrinsic properties of the materials or are the data heavily distorted by grain-boundary effects? We can get an idea about this by comparing the three transport properties for the same material in two measurements. First, the sample is cut directly from the arc-molten ingot with a diamond saw and measured. Second, the material is directly cold-pressed after arc-melting and then measured. In both cases, the crystal structure, the platelet-like nanostructure and bulk bar-shaped form of samples are the same. What changes is the microstructure. The cold-pressed sample is denser with better grain-to-grain contacts. This is expected to raise both electrical and thermal conductivities. We use Sb-alloyed SnSe as an example. Thermal conductivity of SnSe and its alloys is not affected by the changed morphology. We also found that above room temperature, value of electrical resistivity has been improved by an order of magnitude, and Seebeck coefficient remains unchanged (not shown). This experiment demonstrates, that we are looking at the intrinsic thermal properties, whereas electrical connectivity of the material is in need of improving: measurements do not reflect the intrinsic electrical properties for tin selenide alloys.

Why would electrical and thermal conductivities respond differently to cold-pressing? Electrical conductivity is improved upon increasing grain-to-grain contacts and contact area. However, thermal conductivity is not affected. The reason is related to the nature of phonon scattering and the phonon mean-free path. We can estimate this by comparing low-temperature thermal conductivity (below UmKlapp peak) and specific heat (inset in the top panel of **Figure 20**), using phenomenological relation for phonon diffusion: $\kappa = \frac{1}{3}C_V \times v \times l$, that relates thermal conductivity to specific heat at constant volume C_V, sound velocity (v) and phonon mean-free path (l). By ignoring the rather complex phonon dispersion of SnSe and using phonon velocities as given by Zhao et al. [50], we can estimate phonon mean-free path to be between 2 and 10 nm; effectively at low temperature, it is limited by the nanostructured grain-size, but at higher temperatures intrinsic properties dominate.

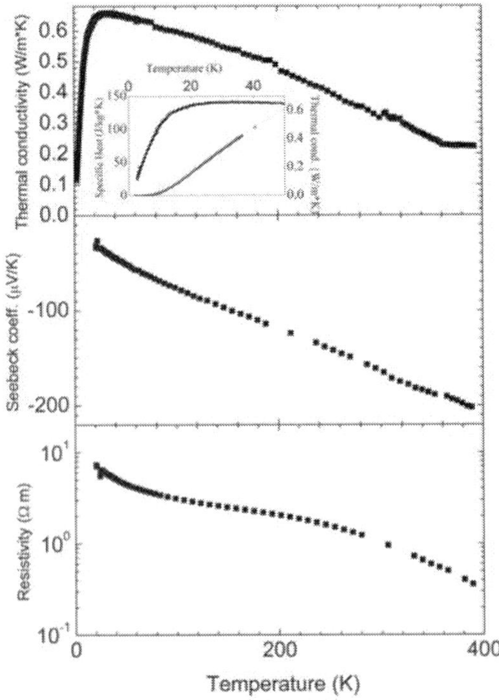

Figure 20. Temperature dependence of (top) thermal conductivity, (middle) Seebeck coefficient and (bottom) electrical resistivity and (inset) specific heat and thermal conductivity comparison at low temperature of $Sb_{0.2}Sn_{0.8}Se$ alloy.

The idea to study alloys of SnSe with SbSe is to control Fermi level and concentration of free charge carriers. To our dismay, we found, that SnSbSe alloys display very high resistivity, higher even, than nominally stoichiometric SnSe, with very low Hall concentration of charge carriers. We did manage to achieve n-type, negative, Seebeck coefficient. Indeed, absolute value of negative Seebeck coefficient reaches 100–200 μV K^{-1}, depending on composition and temperature. These results are summarized for $Sb_{0.2}Sn_{0.8}Se$ in **Figure 20**. Furthermore, for this composition, Hall effect measurements resulted in p-type, with hole concentration 3 × 10^{16} cm^{-3} at room temperature. Expected free electron concentration, from *ab initio* calculations of electronic density of states (**Figure 21**), is around 3 × 10^{21} cm^{-3} and obviously n-type for this level of Sb fraction in the compound. In order to resolve the apparent contradiction between measured and calculated free electron concentration and measured signs of Hall and Seebeck coefficients, we reconsidered the electronic structure calculation in view of strong Sn deficiency revealed by Rietveld refinement of NPD data. In our calculations, we use 1 × 2 × 2 minimal supercells with 16 Sn and 16 Se sites. Of these, we replace up to 3 Sn with Sb to approximate experimental alloying and remove one Sn to represent observed Sn site deficiency. Resulting band structure shows a striking narrow band in the gap above the valence band, appearing as a sharp peak in the density of states (DOS). This acts as shallow energy charge trap that localizes electrons transferred from Sb substitutes. Thus, Fermi level remains stuck in this band for a wide range of Sb concentration, invalidating our expectation of simple rigid-band charge transfer model. The complicated band structure is then responsible for different signs of Hall and Seebeck coefficients, too.

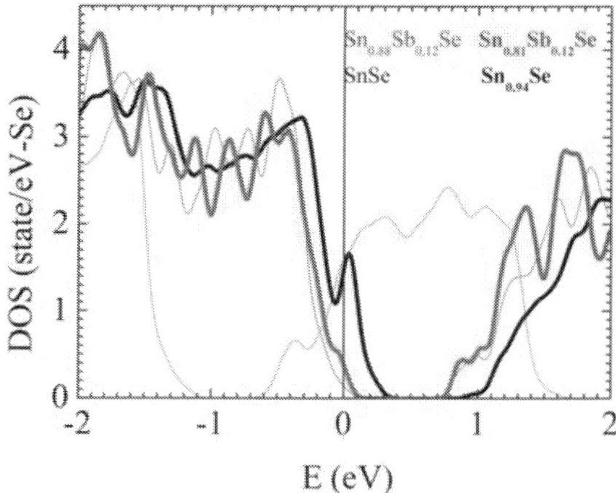

Figure 21. Electronic density of states of stoichiometric and Sn-deficient SnSe and Sb-SnSe with compositions indicated in labels. Fermi level is placed at E = 0, and thus band structures shift around with doping level.

The great advantage of arc-melting synthesis is that it affords a way to rapidly assay various compositions for thermoelectric properties. By introducing atoms of different radii, we can modify electronic structure, even without obvious charge transfer, this is so-called band engineering, for example, modification of bandwidth. Obvious candidates to do this are columnar neighbors of tin, lead and germanium. In the following we present results first on

SnPbSe alloys and then on SnGeSe alloys. In both series, we found p-type Seebeck coefficient values.

Studying three different SnPbSe alloy compositions with up to 30% Pb substitution on Sn site has shown no improvement on any of thermoelectric properties (**Figure 22**). Electrical resistivity increases by several orders of magnitude, as we have seen for SnSbSe alloys, too. Indeed, it is so high, that resistance of samples surpasses few MOhm limit of PPMS electronics at low temperature. The Seebeck coefficient value reaches up to 600 μV K^{-1} at 400 K, which is high, but no higher, than in stoichiometric SnSe produced by arc-melting. Finally, thermal conductivity shows the same UmKlapp peak at low temperature as SnSe and Sb$_{0.2}$Sn$_{0.8}$Se with no overall reduction.

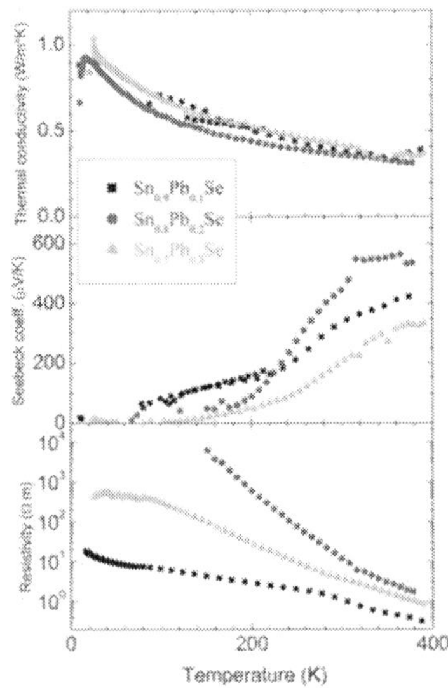

Figure 22. Temperature dependences of (top) thermal conductivity, (middle) Seebeck coefficient and (bottom) electrical resistivity of SnSe alloyed with Pb.

In contrast, by going to smaller ionic radius, SnGeSe alloy has several beneficial effects, although electrical resistivity is still too high. Most importantly, as shown in **Figure 23**, Seebeck coefficient value surpasses 1000 μV K^{-1} for low GeSe fraction. Curious, nonmonotonic change of Seebeck coefficient with increase in GeSe fraction is supported by electronic structure calculations, based on experimentally determined crystal structures. These reveal, that semiconducting gap also varies nonmonotonously with Ge substitution, first increasing with respect to the gap of SnSe and then decreasing with more Ge (**Figure 24**). Large Seebeck coefficient value is caused partly by low charge carriers concentration, as indicated by resistivity, that is 1–2 orders of magnitude above that of stoichiometric SnSe, following Pisarenko relation.

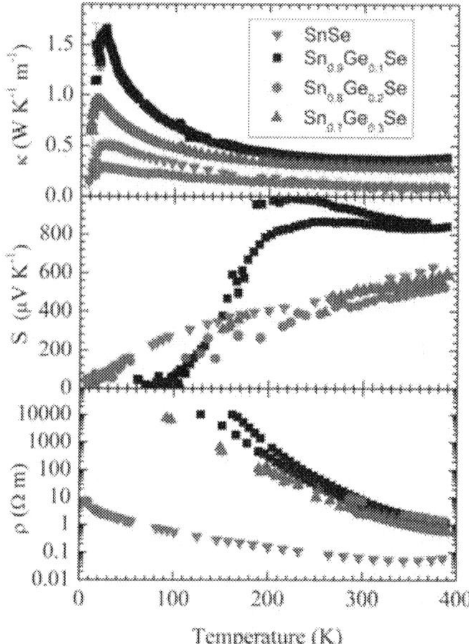

Figure 23. $Sn_{1-x}Ge_xSe$ (x = 0, 0.1, 0.2, 0.3) temperature dependences (top) thermal conductivity, (middle) Seebeck coefficient and (bottom) electrical resistivity—exhibiting characteristic semiconducting behavior. Two independent measurements are shown for x = 0.1 and 0.3. Below 150–200 K electrical resistance of Ge-doped SnSe samples increases beyond the limits (few MOhm) of the electronics of PPMS, and this influences Seebeck voltage, too (from Ref. [21]).

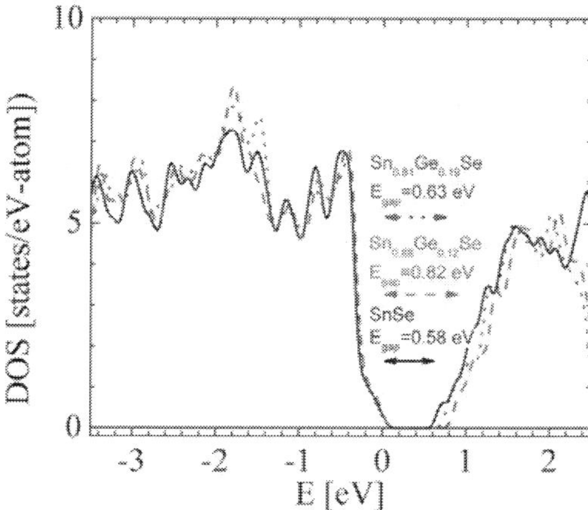

Figure 24. Calculated electronic density of states (DOS) of SnSe, black solid line, $Sn_{0.88}Ge_{0.12}Se$, red dashed line, and $Sn_{0.81}Ge_{0.19}Se$, blue dotted line (from Ref. [21]).

3.3.3. Thermal conductivity results from laser flash thermal diffusivity method

Figure 25 shows total thermal conductivity (κ) obtained by laser flash diffusivity method for different thermoelectric compounds: SnSe, $Sn_{0.8}Ge_{0.2}Se$ and $Sn_{0.8}Sb_{0.2}Se$; PbTe is used as a reference.

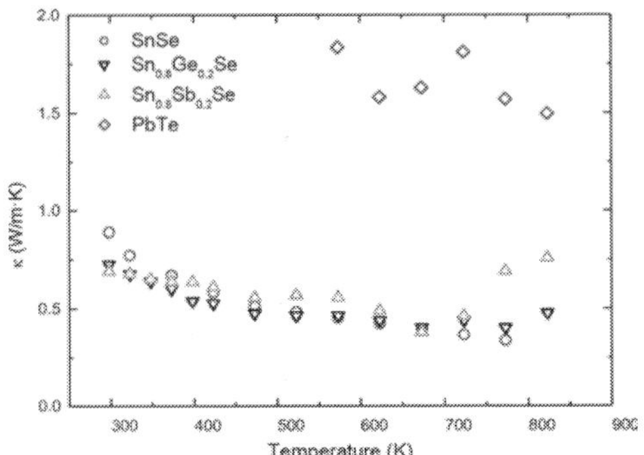

Figure 25. Thermal conductivity values of SnSe, $Sn_{0.8}Ge_{0.2}Se$, $Sn_{0.8}Sb_{0.2}Se$ and PbTe obtained by laser flash diffusivity method.

Lead telluride, PbTe, is a well-known thermoelectric material, that, due to its band gap of about 0.3 eV, is useful in intermediate temperature range of operation. Its thermal conductivity falls from room temperature with a 1/T dependence, which is a fingerprint of the enhancement of phonon-phonon interactions with increasing temperature. This behavior, together with its band-gap value, makes PbTe one of the most competitive thermoelectric materials for generators above 500 K. However, several efforts are being made to enhance its efficiency. The main ways to drive this goal are enhancement of thermoelectric properties by nanostructuring and modifications in the density of states to create resonant states in the conduction band [57].

The pure SnSe and related alloys ($Sn_{0.8}Ge_{0.2}Se$, $Sn_{0.8}Sb_{0.2}Se$) display significantly lower thermal conductivities in the high-temperature region. At room temperature, values of total thermal conductivity are 0.89 and 0.7 W m^{-1} K^{-1} for SnSe and $Sn_{0.8}Sb_{0.2}Se$ compounds, respectively. These values are further reduced with increasing temperature, reaching 0.4 W m^{-1} K^{-1} at 675 K. It is noteworthy, that above-described thermal conductivities determined in PPMS are considerably lower than values obtained by an indirect procedure, from thermal diffusivity.

Total thermal conductivity has two contributions, lattice thermal conductivity (κ_{lat}), due to phonon transport, and charge carriers' thermal conductivity (κ_{ch}), due to thermal transport of charge carriers (electrons and/or holes). As a first approximation, Wiedemann-Franz law [26] allows reasonable estimation of charge carriers thermal conductivity as a function of temperature, $\kappa_{ch} = (L_0T)/\rho$, where L_0 is the Lorentz number and ρ is the electrical resistivity. For the case of SnSe family, total thermal conductivity is almost fully dominated by lattice thermal conductivity. In fact, ratio of thermal conductivity due to charge carriers to total thermal conductivity is approximately 10^{-4}. The different anisotropic direction of both measurements might be related to the differences in the magnitude of determined thermal conductivity with respect to the direct measure provided by PPMS.

4. Conclusions

We have described a fast one-step procedure to prepare nanostructured intermetallic alloys belonging to the families of well-known Bi_2Te_3 and recently described SnSe, all of them showing similar nanostructure consisting of stacks of nanosheets, that perturb propagation of phonons and provide extremely low thermal conductivity. Crystal structure studies from NPD data reveal anisotropic displacement parameters, probably due to the presence of lone-electron pairs of the p-block elements (Bi, Te, Sn, Se…) also contributing to low lattice thermal conductivity. Seebeck coefficient values are enhanced in SnSe system, reaching extraordinary high values close to 1000 μV K^{-1} in SnGeSe alloys. As a drawback of nanostructuration, electrical resistivity values are much higher in this system than those described in single crystalline samples, probably arising from many grain boundaries, which perturb charge carriers path. We describe also a simple apparatus for the measurement of high-temperature transport properties, ideally conceived to determine S and σ in disk-shaped pellets directly obtained from the intermetallic ingots. The use of Nb pistons, chemically inert to reactive p elements like Bi, Te or Se, is particularly suitable given the weak S factor for Nb, yielding reproducible results for known materials like Cu or Al.

Abbreviations

BN	Boron nitride
DFT	Density functional theory
GGA	Generalized gradient approximation
HRPT	High-resolution powder diffractometer for thermal neutrons
MMR	Micro-miniature refrigerator
NPD	Neutron powder diffraction
PBE	Perdue-Burke-Emzerhof
PSI	Paul Scherrer Institut
RT	Room temperature
SEM	Scanning electron microscope
SSR	Sum of squares regression
XRD	X-ray diffraction
ZT	Figure of Merit
CASTEP	Cambridge Serial Total Energy Package
PID	proportional-integral-derivative

Acknowledgements

We thank the financial support of the Spanish Ministry of Science and Innovation to the project MAT2013-41099-R and by JCCM through Project PPII-2014-019-P. We thank the Institut Laue-Langevin (ILL) and Paul Scherrer Institut (PSI) for providing the neutron beam time.

Author details

Federico Serrano-Sánchez[1], Mouna Gharsallah[1,2], Julián Bermúdez[1], Félix Carrascoso[1], Norbert M. Nemes[1], Oscar J. Dura[3], Marco A. López de la Torre[3], José L. Martínez[1], María T. Fernández-Díaz[4] and José A. Alonso[1*]

*Address all correspondence to: ja.alonso@icmm.csic.es

1 Institute of Materials Science of Madrid, Madrid, Spain

2 National School of Engineers, Sfax University, Tunisia

3 Department of Applied Physics and INEI, University of Castilla La Mancha, Ciudad Real, Spain

4 Institut Laue Langevin, Grenoble, France

References

[1] Funahashi R, Barbier T, Combe E.: Thermoelectric materials for middle and high temperature ranges. J. Mater. Res. 2015;30:2544–2557. doi:10.1557/jmr.2015.145

[2] Zhang X, Zhao L-D.: Thermoelectric materials: energy conversion between heat and electricity. J. Materiomics. 2015;1(2):92–105. doi:10.1016/j.jmat.2015.01.001

[3] Goldsmid HJ.: Thermoelectric Refrigeration. Springer, London; 1964. doi: 10.1007/978-1-4899-5723-8

[4] Rowe D.M., editor. Thermoelectrics and Its Energy Harvesting. CRC Press, Boca Raton, Florida; 2012.

[5] Bell LE.: Cooling, heating, generating power, and recovering waste heat with thermoelectric systems. Science. 2008;321(5895):1457–1461. doi:10.1126/science.1158899

[6] Zhao L-D, Dravid VP, Kanatzidis MG.: The panoscopic approach to high performance thermoelectrics. Energy Environ. Sci. 2014;7(1):251. doi:10.1039/C3EE43099E

[7]   Snyder GJ, Toberer ES.: Complex thermoelectric materials. Nat. Mater. 2008;7(2):105–114. doi:10.1038/nmat2090

[8]   Amatya R, Mayer PM, Ram RJ.: High temperature Z-meter setup for characterizing thermoelectric material under large temperature gradient. Rev. Sci. Instrum. 2012;83(7): 075117. doi:10.1063/1.4731650

[9]   Chen G, Dresselhaus MS, Dresselhaus G, Fleurial J-P, Caillat T.: Recent developments in thermoelectric materials. Int. Mater. Rev. 2003;48:45–66. doi: 10.1179/095066003225010182

[10]   Dresselhaus MS, Chen G, Tang MY, Yang R, Lee H, Wang D, et al.: New directions for low-dimensional thermoelectric materials. Adv. Mater. 2007;19(8):1043–1053. doi: 10.1002/adma.200600527

[11]   Bilc DI, Hautier G, Waroquiers D, Rignanese G-M, Ghosez P.: Low-dimensional transport and large thermoelectric power factors in bulk semiconductors by band engineering of highly directional electronic states. Phys. Rev. Lett. 2015;114(13):1–5. doi:10.1103/PhysRevLett.114.136601

[12]   Wang Y, Huang H, Ruan X.: Decomposition of coherent and incoherent phonon conduction in superlattices and random multilayers. Phys. Rev. B. 2014;90(16):48–50. doi:10.1103/PhysRevB.90.165406

[13]   Mu X, Zhang T, Go DB.: Coherent and incoherent phonon thermal transport in isotopically modified graphene superlattices. Carbon. 2014;83:208–216. doi:10.1016/j.carbon.2014.11.028

[14]   Wang Y, Gu C, Ruan X.: Optimization of the random multilayer structure to break the random-alloy limit of thermal conductivity. Appl. Phys. Lett. 2015;106(7):073104. doi: 10.1063/1.4913319

[15]   Lan Y, Minnich AJ, Chen G, Ren Z.: Enhancement of thermoelectric figure-of-merit by a bulk nanostructuring approach. Adv. Funct. Mater. 2010;20(3):357–376. doi:10.1002/adfm.200901512

[16]   Bomshtein N, Spiridonov G, Dashevsky Z.: Thermoelectric, structural, and mechanical properties of spark-plasma-sintered submicro- and microstructured p-type $Bi_{0.5}Sb_{1.5}Te_3$. J. Electron. Mater. 2012;41(6):1546–1553. doi:10.1007/s11664-012-1950-8

[17]   Cao YQ, Zhao XB, Zhu TJ, Zhang XB, Tu JP.: Syntheses and thermoelectric properties of Bi2 Te3 Sb2 Te3 bulk nanocomposites with laminated nanostructure. Appl. Phys. Lett. 2008;92(14):143106. doi:10.1063/1.2900960

[18]   Zhao Y, Dyck JS, Hernandez BM, Burda C.: Enhancing thermoelectric performance of ternary nanocrystals through adjusting carrier concentration. J. Am. Chem. Soc. 2010;132(14):4983–4983. doi:10.1021/ja100020m

[19] Serrano-Sánchez F, Gharsallah M, Nemes NM, Mompean FJ, Martínez JL, Alonso JA.: Record Seebeck coefficient and extremely low thermal conductivity in nanostructured SnSe. Appl. Phys. Lett. 2015;106(8):083902. doi:10.1063/1.4913260

[20] Gharsallah M, Serrano-Sánchez F, Bermúdez J, Nemes NM, Martínez JL, Elhalouani F, et al.: Nanostructured Bi_2Te_3 prepared by a straightforward arc-melting method. Nanoscale Res. Lett. 2016;11:142. doi:10.1186/s11671-016-1345-5

[21] Gharsallah M, Serrano-Sánchez F, Nemes NM, Mompeán FJ, Martínez JL, Fernández-Díaz MT, et al.: Giant Seebeck effect in Ge-doped SnSe. Sci. Rep. 2016;6:26774. doi: 10.1038/srep26774

[22] Rodríguez-Carvajal J.: Recent advances in magnetic structure determination by neutron powder diffraction. Phys. B. 1993;192(1–2):55–69. doi:10.1016/0921-4526(93)90108-I

[23] Borup KA, de Boor J, Wang H, Drymiotis F, Gascoin F, Shi X, et al.: Measuring thermoelectric transport properties of materials. Energy Environ. Sci. 2015;8:423–435. doi: 10.1039/c4ee01320d

[24] Martin J.: Protocols for the high temperature measurement of the Seebeck coefficient in thermoelectric materials. Meas. Sci. Technol. 2013;24:085601. doi: 10.1088/0957-0233/24/8/085601

[25] Iwanaga S, Toberer ES, Lalonde A, Snyder GJ.: A high temperature apparatus for measurement of the Seebeck coefficient. Rev. Sci. Instrum. 2011;83(6):063905. doi: 10.1063/1.3601358

[26] Rowe DM.: Thermoelectrics Handbook: Macro to Nano. Taylor & Francis ed. Boca Raton, FL: CRC Press; 2006. doi:10.1201/9781420038903

[27] Parker WJ, Jenkins RJ, Butler CP, Abbott GL.: Flash method of determining thermal diffusivity, heat capacity, and thermal conductivity. J. Appl. Phys. 1961;32(9):1679–1684. doi:10.1063/1.1728417

[28] Dusza L.: Combined solution of the simultaneous heat loss and finite pulse corrections with the laser flash method. High Temp. High Press. 1995;27–28(5):467–473.

[29] Cowan RD.: Pulse method of measuring thermal diffusivity at high temperatures. J. Appl. Phys. 1963;34(4):926–927. doi:10.1063/1.1729564

[30] Clark LM, Taylor RE.: Radiation loss in the flash method for thermal diffusivity. J. Appl. Phys. 1975;46(2):714–719. doi:10.1063/1.321635

[31] Cahill DG.: Thermal conductivity measurement from 30 to 750 K: The 3Ω method. Rev. Sci. Instrum. 1990;61(2):802–808. doi:10.1063/1.1141498

[32] Clark SJ, Segall MD, Pickard CJ, Hasnip PJ, Probert MJ, Refson K, et al.: First principles methods using CASTEP. Zeitschrift für Krist. 2005;220(5–6):567–570. doi:10.1524/zkri. 220.5.567.65075

[33] Perdew JP, Burke K, Ernzerhof M.: Generalized gradient approximation made simple. Phys. Rev. Lett. 1996;77(18):3865–3868. doi:10.1103/PhysRevLett.77.3865

[34] Saleemi M, Toprak MS, Li S, Johnsson M, Muhammed M.: Synthesis, processing, and thermoelectric properties of bulk nanostructured bismuth telluride (Bi_2Te_3). J. Mater. Chem. 2012;22(2):725–730. doi:10.1039/C1JM13880D

[35] Feng S-K, Li S-M, Fu H-Z.: Probing the thermoelectric transport properties of n-type Bi_2Te_3 close to the limit of constitutional undercooling. Chinese Phys. B. 2014;23(11): 117202. doi:10.1088/1674-1056/23/11/117202

[36] Peranio N, Eibl O, Bäßler S, Nielsch K, Klobes B, Hermann RP, et al.: From thermoelectric bulk to nanomaterials: current progress for Bi_2Te_3 and $CoSb_3$. Phys. Status Solidi A. 2015;11:1–11. doi:10.1002/pssa.201532614

[37] Yang JY, Fan XA, Chen RG, Zhu W, Bao SQ, Duan XK.: Consolidation and thermoelectric properties of n-type bismuth telluride based materials by mechanical alloying and hot pressing. J. Alloys Compd. 2006;416(1–2):270–273. doi:10.1016/j.jallcom.2005.08.054

[38] Zhao LD, Zhang B-P, Liu WS, Zhang HL, Li J-F.: Effects of annealing on electrical properties of n-type Bi_2Te_3 fabricated by mechanical alloying and spark plasma sintering. J. Alloys Compd. 2009;467(1–2):91–97. doi:10.1016/j.jallcom.2007.12.063

[39] Scheele M, Oeschler N, Meier K, Kornowski A, Klinke C, Weller H.: Synthesis and thermoelectric characterization of Bi_2Te_3 nanoparticles. Adv. Funct. Mater. 2010;19(21): 3476–3483. doi:10.1002/adfm.200901261

[40] Ohsugi IJ, Kojima T, Sakata M, Yamanashi M, Nishida IA.: Evaluation of anisotropic thermoelectricity of sintered Bi_2Te_3 on the basis of the orientation distribution of crystallites. J. Appl. Phys. 1994;76(4):2235–2239. doi:10.1063/1.357641

[41] Scheele M, Oeschler N, Veremchuk I, Reinsberg KG, Kreuziger AM, Kornowski A, et al.: ZT enhancement in solution-grown $Sb_{(2-x)}Bi_xTe_3$ nanoplatelets. ACS Nano. 2010;4(7): 4283–4291. doi:10.1021/nn1008963

[42] Sharp J, Goldsmid HJ, Nolas GS.: Thermoelectrics: Basic Principles and New Materials Developments. New York, NY: Springer; 2001. doi:10.1007/978-3-662-04569-5

[43] Pei Y, Shi X, LaLonde A, Wang H, Chen L, Snyder GJ.: Convergence of electronic bands for high performance bulk thermoelectrics. Nature. 2011;473(7345):66–69. doi: 10.1038/nature09996

[44] Choi S-M, Lee KH, Lim YS, Seo W-S, Lee S.: Effects of doping on the positional uniformity of the thermoelectric properties of n-type $Bi_2Te_{2.7}Se_{0.3}$ polycrystalline bulks. J. Korean Phys. Soc. 2016;68(1):17–21. doi:10.3938/jkps.68.17

[45] Adam A.: Rietveld refinement of the semiconducting system $Bi_{2-x}Fe_xTe_3$ from X-ray powder diffraction. Mater. Res. Bull. 2007;42(12):1986–1994. doi:10.1016/j.materresbull.2007.02.027

[46] Lee G-E, Eum A-Y, Song K-M, Kim I-H, Lim YS, Seo W-S, et al.: Preparation and thermoelectric properties of n-type $Bi_2Te_{2.7}Se_{0.3}$. Dm. J. Electron. Mater. 2015;44(6):1579–1584. doi:10.1007/s11664-014-3485-7

[47] Ajay S, Zhao Y, Yu L, Michael Khor KA, Dresselhaus MS, Qihua X.: Enhanced thermoelectric properties of solution grown $Bi_2Te_{3-x}Se_x$ nanoplatelet composites. Nano Lett. 2012;12:1203–1209. doi:10.1021/nl2034859

[48] Wang S, Li H, Lu R, Zheng G, Tang X.: Metal nanoparticle decorated n-type Bi_2Te_3-based materials with enhanced thermoelectric performances. Nanotechnology. 2013;24(28): 285702. doi:10.1088/0957-4484/24/28/285702

[49] Patel TH, Vaidya R, Patel SG.: Effect of pressure on electrical properties of SnS_xSe_{1-x} single crystals. High Press. Res. 2003;23(4):417–423. doi:10.1080/0895795031000114368

[50] Zhao L-D, Lo S-H, Zhang Y, Sun H, Tan G, Uher C, et al.: Ultralow thermal conductivity and high thermoelectric figure of merit in SnSe crystals. Nature. 2014;508(7496):373–377. doi:10.1038/nature13184

[51] Chen C-L, Wang H, Chen Y-Y, Day T, Snyder J. Thermoelectric properties of p-type polycrystalline SnSe doped with Ag. J. Mater. Chem. A. 2014;2(29):11171–11176. doi:10.1039/c4ta01643b

[52] Sassi S, Candolfi C, Vaney JB, Ohorodniichuk V, Masschelein P, Dauscher A, et al.: Assessment of the thermoelectric performance of polycrystalline p-type SnSe. Appl. Phys. Lett. 2014;104(21):212105. doi:10.1063/1.4880817

[53] Fu Y, Xu J, Liu G-Q, Yang J, Tan X, Liu Z, et al.: Enhanced thermoelectric performance in p-type polycrystalline SnSe benefiting from texture modulation. J. Mater. Chem. C. 2016;4:1201–1207. doi:10.1039/C5TC03652F

[54] Ioffe AF.: Physics of Semiconductors. New York, NY: Academic Press; 1960.

[55] Katsuyama S, Maezawa F, Tanaka T.: Synthesis and thermoelectric properties of sintered skutterudite $CoSb_3$ with a bimodal distribution of crystal grains. J. Phys. Conf. Ser. 2012;379(1):12004. doi:10.1088/1742-6596/379/1/012004

[56] Stitzer DP.: Lattice thermal conductivity of semiconductors: a chemical bond approach. J. Phys. Chem. Solids. 1970;31:19–40. doi:10.1016/0038-1098(69)90198-7

[57] Sootsman JR, Chung DY, Kanatzidis MG.: New and old concepts in thermoelectric materials. Angew. Chem. Int. Ed. 2009;48(46):8616–8639. doi:10.1002/anie.200900598

Nanometer Structured Epitaxial Films and Foliated Layers Based on Bismuth and Antimony Chalcogenides with Topological Surface States

Lidia N. Lukyanova, Yuri A. Boikov, Oleg A. Usov, Mikhail P. Volkov and Viacheslav A. Danilov

Additional information is available at the end of the chapter

http://dx.doi.org/10.5772/65750

Abstract

The thermoelectric and galvanomagnetic properties of nanometer structured epitaxial films and foliated layers based on bismuth and antimony chalcogenides were investigated, and an increase in the figure of merit Z up to $3.85 \times 10^{-3}\,\mathrm{K^{-1}}$ was observed in the $Bi_{0.5}Sb_{1.5}Te_3$ films over the temperature range of 180–200 K. It is shown that an increase in the Seebeck coefficient and the change in the slope on temperature, associated with changes in the effective scattering parameter of charge carriers and strong anisotropy of scattering in the films, lead to enhance power factor due to the growth of the effective mass of the density of states. These features are consistent with the results of research of oscillation effects in strong magnetic fields at low temperatures and research of Raman scattering at normal and high pressures in the foliated layers of solid solutions $(Bi, Sb)_2(Te, Se)_3$, in which the topological Dirac surface states were observed. The unique properties of topological surface states in the investigated films and layers make topological insulators promising material for innovation nanostructured thermoelectrics.

Keywords: thermoelectric films, topological surface states, power factor, scattering on interphase, block boundaries

1. Introduction

Thermoelectrics based on bismuth and antimony chalcogenides are well known and have been extensively studied for their excellent thermoelectric properties [1–3]. Recently, the nanostruc-

tured $(Bi,Sb)_2(Te,Se)_3$ epitaxial films were shown to possess an enhanced thermoelectric figure of merit Z compared to corresponding bulk crystals due to mechanical stresses and intensive phonon scattering at the grain boundaries. Lattice thermal conductivity of heteroepitaxial nanostructured films may be substantially diminished, as compared to corresponding bulk crystals, due to acoustic phonon scattering by grain and interface boundaries. Boundaries and strain may be easily induced in thin Bi_2Te_3-based films grown on mismatched substrates. In contrast to point defects, which suppress heat transfer by short wave phonons, grain and interface boundaries are efficient in scattering of long wave ones. That is why, investigation of thin films with different structure and level of mechanical stresses looks quite important for development of thermoelectric materials with enhanced performance.

At the current stage, a new property of these materials, known as topological insulator, became one of the important subjects of the investigations [4–6]. These novel quantum states are the result of the electronic band inversion due to strong spin-orbit interaction, so the bulk becomes insulating and the surface displays an unusual metallic electronic surface state of Dirac fermions with linear dispersion and spin texture that ensures high mobility of charge carriers due to the lack of backscattering on defects. The nanostructured films of topological materials [7, 8] theoretically proved the possibility of formation of topological excitons that can condense in a wide temperature range with the appearance of superfluidity and thus a significant increase in the mobility of the charge carriers. Currently, there is only a preliminary report on the pilot study of the heterostructure Sb_2Te_3/Bi_2Te_3, which assumes the realization of exciton condensation [9]. The theoretical model [10] for topological thermoelectrics based on the Landauer transport theory shows that the value of the Seebeck coefficient and thermoelectric efficiency is determined by the ratio of the mean free path of Dirac fermions to the mean free path of bulk electrons. According to the proposed model, an increase in the contribution of the surface states and the energy dependence of the lifetime of the electronic states provide the maximum amount of the Seebeck coefficient and the increase in the thermoelectric efficiency. Experimental research of transport of thin films $(Bi_{1-x}Sb_x)_2Te_3$ [11] carried out in a wide range of compositions, and temperatures have confirmed validity of this model. It follows that the study of surface states of Dirac fermions is promising for thermoelectricity to improve the energy conversion efficiency of nanostructured thermoelectrics. This chapter includes the thermoelectric properties under normal conditions and under high pressure, galvanomagnetic and optical properties of nanostructured films based on Bi_2Te_3, obtained by different methods, in order to determine the possible effect of the topological surface states of Dirac fermions depending on the composition, the Seebeck coefficient and temperature.

2. Features of formation and structure of the grown $(Bi,Sb)_2Te_3$ films

Because of incongruent evaporation/sublimation of $(Bi,Sb)_2Te_3$ solid solutions and pronounced volatility of tellurium at temperatures higher than 400°C, formation of stoichiometric epitaxial films of bismuth and antimony chalcogenides is a nontrivial task. Hot wall technique [12, 13] was used to grow stoichiometric $Bi_{0.5}Sb_{1.5}Te_3$ layers with a thickness of 30–500 nm.

Structure of the grown thermoelectric films was investigated by X-ray (Philips X'pert MRD, CuK$_{\alpha1}$, ω/2Θ- and ϕ-scans). Surface morphology of the grown (Bi,Sb)$_2$Te$_3$ films was studied by atomic force microscopy (Nanoscope IIIa, tipping mode). Nanostructured epitaxial thermoelectric films of (Bi,Sb)$_2$Te$_3$ were grown by hot wall technique on mica (muscovite) substrates. Usage of a mica substrate promotes small in-plane misorientation of the blocks in the BST film, but relatively high deposition temperature is in a favor of low density of defects in their volume. Thickness of the grown thermoelectric layers was in the range of 30–500 nm. Substrate temperature during thermoelectric film formation was roughly 70°C less then temperature of sublimating stoichiometric Bi$_{0.5}$Sb$_{1.5}$Te$_3$ burden. X-ray ω/2Θ scans were traced for the grown Bi$_{0.5}$Sb$_{1.5}$Te$_3$ films when plane including incident and reflected Roentgen beams was in plane normal to (000.1) of mica or (10$\bar{1}$.5) of the thermoelectric layer (see **Figures 1** and **2**). From obtained scans follow that c-axis in the (Bi,Sb)$_2$Te$_3$ films grown on mica by hot wall technique was normal to substrate plane. Driving force for preferential orientation of c-axis in the grown films along normal to a substrate plane was substantial anisotropy of a surface free energy of the bismuth and antimony chalcogenides. From X-ray, ϕ-scan traced for a (10$\bar{1}$.5) Bi$_{0.5}$Sb$_{1.5}$Te$_3$ reflex, see insert in **Figure 1**, follows that thermoelectric films grown on mica were well in-plane preferentially oriented as well. In-plane disorientation of blocks in the films was ~0.3°. (The estimation obtained from full width at half of a maximum of a peak on the ϕ-scan. Roughly, equidistant system of growth steps was clear detectable at the film surface (see **Figure 2**), at AFM image of free surface of the (Bi,Sb)$_2$Te$_3$ film grow on mica. Height of the growth steps was about 1 nm.

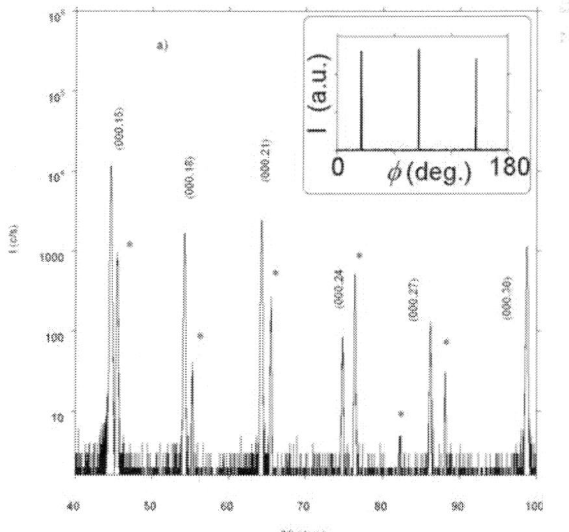

Figure 1. X-ray ω/2Θ scan traced for the grown Bi$_{0.5}$Sb$_{1.5}$Te$_3$ film when plane including incident and reflected Roentgen beams was in plane normal to (000.1) plane of substrate. Insert plots ϕ-scan of a (10$\bar{1}$.5) reflex from the same film.

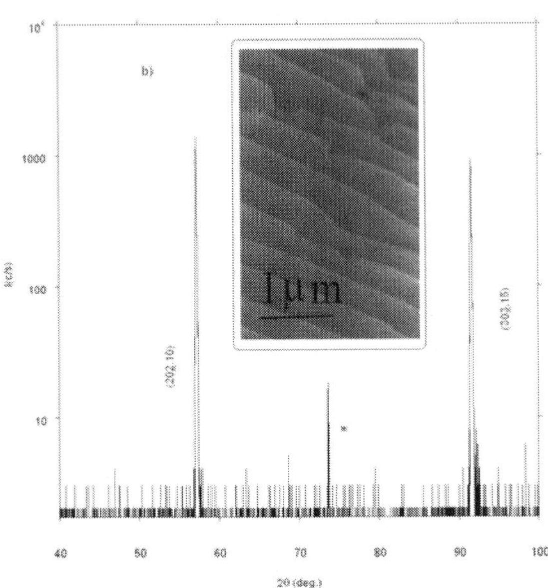

Figure 2. X-ray ω/2Θ scan traced for the grown $Bi_{0.5}Sb_{1.5}Te_3$ film when plane including incident and reflected Roentgen beams was normal to (101.5) plane of the thermoelectric layer. AFM image of free surface of the grown thermoelectric layer is shown on insert.

3. Thermoelectric properties

Efficiency of thermoelectric energy conversion is dependent on figure of merit (Z) of the used materials ($Z = S^2 \sigma/\kappa$, where S—Seebeck coefficient, σ—electrical conductivity, and κ—thermal conductivity) with electron and hole conductance. Thermoelectric properties of bismuth telluride and related solid solution $Bi_{0.5}Sb_{1.5}Te_3$ heteroepitaxial nanostructured films were investigated below room temperature. The temperature dependences of the Seebeck coefficient S and the electroconductivity σ of the Bi_2Te_3 and $Bi_{0.5}Sb_{1.5}Te_3$ films are shown in **Figure 3**. The electrical conductivity of bulk samples grows more sharply with temperature decrease for both Bi_2Te_3 and $Bi_{0.5}Sb_{1.5}Te_3$ solid solution than for films (**Figure 3**, curves 6, 8 and 5, 7). The observed decrease in electrical conductivity in the films is related to the influence of scattering on interphase and intercrystallite grain boundaries.

The temperature dependences of the Seebeck coefficient (**Figure 3**), unlike such electrical conductivity dependences, are located higher for films than for bulk Bi_2Te_3 (curves 1, 2) and $Bi_{0.5}Sb_{1.5}Te_3$ (curves 3, 4). The highest power factor values were obtained in submicrometer $Bi_{0.5}Sb_{1.5}Te_3$ film at S = 242 μV K⁻¹ over the temperature range of 80–300 K and in the $Bi_{0.5}Sb_{1.5}Te_3$ film at S = 234 μV K⁻¹ over the range of 130–260 K (**Figure 4**, curves 5, 3). An enhancement of the Seebeck coefficient and change in its temperature dependence slope both indicate the changes of charge carrier scattering mechanisms in grown films [14, 15] compared those to the bulk thermoelectric materials (**Figure 4**, curves 1–5 and curves 6–9).

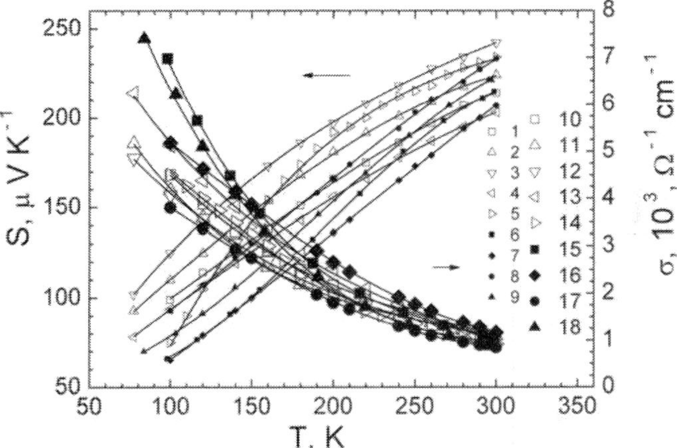

Figure 3. Temperature dependences of the Seebeck coefficient S (1–9), electroconductivity σ (10–18) in heteroepitaxial films (1–5, 10–14), and bulk samples (6–9 and 15–18) of $Bi_{0.5}Sb_{1.5}Te_3$ (1–3, 6–9, 10–12, 16–18) and Bi_2Te_3 (4, 5, 13, 14).

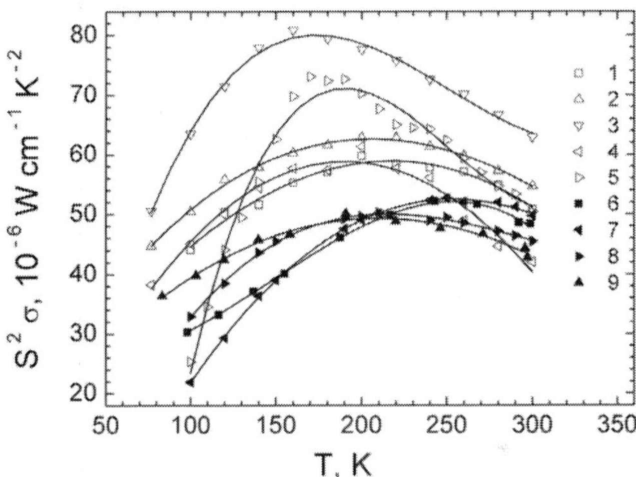

Figure 4. Temperature dependences of power factor $S^2\sigma$ for heteroepitaxial films (1–5) and bulk samples (6–9) of $Bi_{0.5}Sb_{1.5}Te_3$ (1–3, 6–9) and Bi_2Te_3 (4, 5) for S (μV K^{-1}) at room temperatures: $1-214$, $2-224$, $3-242$, $4-203$, $5-234$, $6-214$, $7-207$, $8-233$, $9-221$.

The results of the study of galvanomagnetic and thermoelectric properties in epitaxial films have been used to determine the average effective mass of the density of states m/m_0 and mobility μ_0 of charge carriers taking into account the effective scattering parameter in the model with an isotropic relaxation time similar to the bulk thermoelectrics [14, 16, 17]. Calculations of the effective mass m/m_0 and mobility m_0 have showed that the effective mass of the films is higher than in the bulk samples (**Figure 5**, curves 1–5 and 6, 7) with slight reduction of the mobility in the films (**Figure 5**, curves 13, 14). From the study of the galvano-magnetic properties, the behavior of the effective mass and mobility in the films was found to depend on the scattering mechanism of charge carriers and the parameters of the ellipsoidal constant energy surfaces [13, 18].

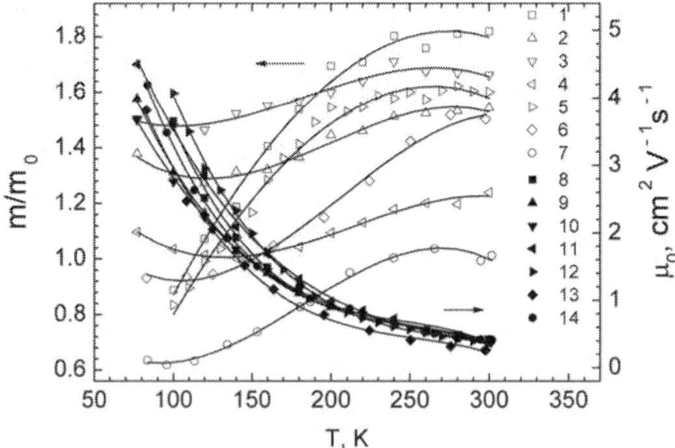

Figure 5. Temperature dependences of the density-of-states effective mass m/m_0 (1–7) and charge carrier mobilitym_0 (8–14) for films (1–5, 8–12) and bulk samples (6, 7, 13, 14) of $Bi_{0.5}Sb_{1.5}Te_3$ (1–3, 6–7, 8–10, 13–14) and Bi_2Te_3 (4, 5, 11, 12).

The product $(m/m_0)^{3/2}\mu_0$, proportional to the figure of merit Z, is higher for films than bulk thermoelectrics due to growth of the effective mass of the density states, which determines an increase in power factor of the films (**Figure 6**). At temperatures below 200 K, an increase in the $(m/m_0)^{3/2}\mu_0$ was observed in the solid solution $Bi_{0.5}Sb_{1.5}Te_3$ at a value of the Seebeck coefficient $S = 242 \ \mu V \ K^{-1}$ at room temperature (**Figure 6**, curve 3).

An estimated value of the figure of merit Z in heteroepitaxial $Bi_{0.5}Sb_{1.5}Te_3$ film increases to $3.85 \times 10^{-3} \ K^{-1}$ over the temperature range of 180–200 K. Such increase in Z is approximately by 60% compared to conventional bulk materials and by 20% compared with multicomponent bulk thermoelectrics optimized for temperatures below 200 K [19, 20]. Reduction in thermal conductivity in the $Bi_{0.5}Sb_{1.5}Te_3$ films can reach 20–30% due to additional scattering of charge carriers in the intercrystallite and interphase boundaries [3, 21] that give an additional rise in the figure of merit.

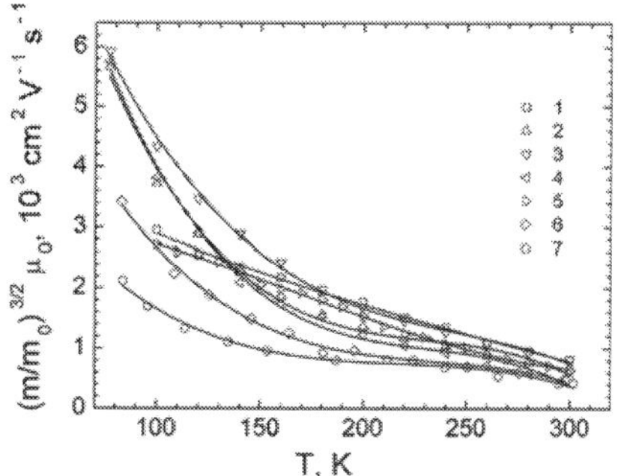

Figure 6. Temperature dependence of the $(m/m_0)^{3/2}$ μ_0 parameter for films (1–5) and bulk samples (6, 7) of $Bi_{0.5}Sb_{1.5}Te_3$ (1–3, 5–7) and Bi_2Te_3 (4).

4. Mechanisms of charge carriers scattering

The charge carrier scattering mechanisms of Bi_2Te_3 and solid solution $Bi_{0.5}Sb_{1.5}Te_3$ films were investigated from analysis of the galvanomagnetic coefficients including transverse and longitudinal components of magnetoresistivity tensor r_{ijkl}, electroresistivity r_{ij}, and Hall coefficient r_{ijk} within many-valley model of energy spectrum for isotropic scattering mechanism [14, 18, 22]. A relaxation time for isotropic carrier scattering depends on energy E by power law: $\tau = \tau_0\,E^r$, where τ_0 is an energy independent factor and r is the scattering parameter. The least square analysis of experimental galvanomagnetic coefficients for isotropic scattering mechanism permits to determine the degeneracy parameter β_d [13, 15]. The dependence of β_d on temperature obtained in magnetic field at $B = 10\,T$ is in agreement with that at $B = 14\,T$ (**Figure 7**, curves 1, 2).

The parameter β_d depends on temperature more sharply in the film than in the bulk $Bi_{2-x}Sb_xTe_3$ solid solutions (**Figure 7**, points 3, 4). The temperature dependence of the β_d of the films is supposed to be explained by an additional charge carrier scattering on interphase and intercrystallite boundaries of monocrystalline grains of the films. As shown in **Figure 8**, the dependence of the degeneracy parameter β_d on the reduced Fermi level η shows that the β_d values in films are less than in bulk materials. Therefore, the degeneracy of films is smaller than bulk thermoelectrics [13].

The effective scattering parameter r_{eff} and the reduced Fermi level η were calculated by Nelder-Mead least square method from the temperature dependences of the degeneracy parameter β_d and the Seebeck coefficient S [14, 24]. As compared to bulk materials, the values of the parameter r_{eff} are considerably different from the value $r = -0.5$, specific for an acoustic phonon scattering mechanism due to sharper energy dependence of electron relaxation time in the films, that is, explained by an additional charge carriers scattering on interphase and intercrystallite boundaries of epitaxial films (**Figure 8**).

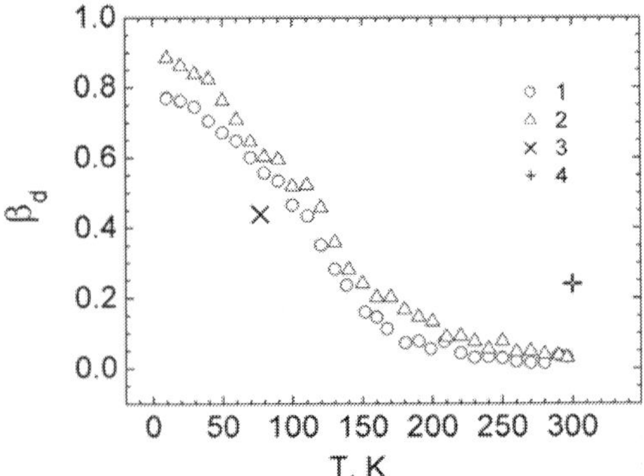

Figure 7. Temperature dependence of the degeneracy parameter β_d (1–2) in the $Bi_{0.5}Sb_{1.5}Te_3$ film. β_d is (1) 10 T and (2) 14 T. Points for bulk solid solutions: 3 [23], 4 [14].

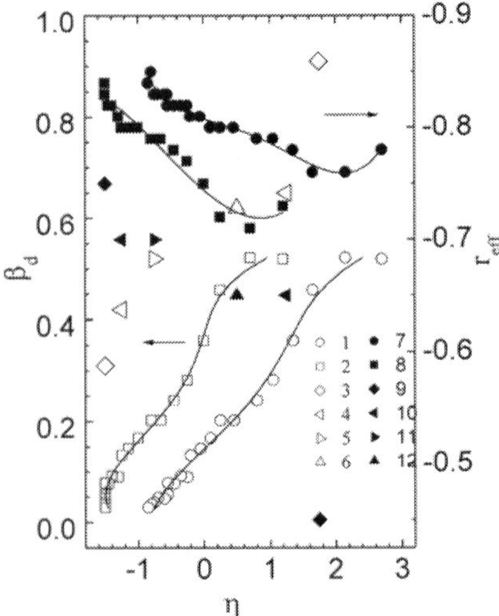

Figure 8. The degeneracy parameter β_d (1–6) and the effective scattering parameter r_{eff} (7–12) on reduced Fermi level η in Bi_2Te_3 (1, 7), $Bi_{0.5}Sb_{1.5}Te_3$ (2, 8) films, and $Bi_{2-x}Sb_xTe_{3-y}Se_y$ (x = 1.2, y = 0.09), (3, 4, 9, 10); $Bi_{2-x}Sb_xTe_{3-y}Se_y$ (x = 1.3, y = 0.07), (5, 11); $Bi_{2-x}Sb_xTe_3$ (x = 1.6), (6, 12) bulk solid solutions.

The materials under study exhibit both anisotropy of the transport properties and anisotropy of charge carrier scattering. In a six-valley model of the energy spectrum with anisotropic scattering of charge carriers, the components of the relaxation time tensor $\overset{\leftrightarrow}{\tau}(\varepsilon)$ can be presented as $\tau_{ij} = \varphi(\varepsilon)\tilde{\tau}_{ij}$ where $\varphi(\varepsilon)$ is an isotropic function depending on μ and τ_{ij} is an anisotropic multiplier that is independent on energy. The ratios of the $\overset{\leftrightarrow}{\tau}(\varepsilon)$ tensor components were determined in the temperature interval from 10 to 300 K [15, 17, 25, 26]. The relation between the $\overleftrightarrow{\tau}(\varepsilon)$ components was found as follows: $\tau_{22} > \tau_{11} > \tau_{33}$, and charge carrier scattering along bisector directions was dominant in the film as in the bulk thermoelectrics at low temperatures [17, 25, 27]. The ratio τ_{22}/τ_{11} in the $Bi_{2-x}Sb_xTe_3$ film is increased along bisector axes, but the ratio τ_{33}/τ_{11} is diminished along the trigonal direction in contrast to corresponding bulk thermoelectrics at low [17, 25] and room [27] temperatures. The value τ_{23} is near the same as τ_{11} for the films, while for bulk materials τ_{23} is less than τ_{11}. These specific features of charge carriers scattering lead to increase in the slope of the Seebeck coefficient dependence on temperature (**Figure 3**) and to enhance the thermoelectric power factor for films. Optimization of structure and charge state of the grains and/or interface boundaries might be in favor of a large thermoelectric power factor and figure of merit.

5. Thermoelectric properties under high pressure

The thermoelectric properties of n-$Bi_2Te_{3-x-y}Se_xS_y$ solid solutions with atomic substitutions in the tellurium sublattice at ($x = 0.27$, 0.3, $y = 0$, and $x = y = 0.09$) and p-Bi_2Te_3 were studied under pressure of 8 GPa on submicron layer samples at room temperature using the technique described in Ref. [28–30]. It was found that the Seebeck coefficient decreases and the electro-conductivity increases with increasein pressure, but the power factor $S^2\sigma$ increases for all compositions, and becomes maximum at pressures of 3–4 GPa (**Figure 9**). The effective mass m/m_0 and mobility μ_0 in the n-$Bi_2Te_{3-x-y}Se_xS_y$ and p-Bi_2Te_3 films were obtained taking into account of the change in the scattering mechanism depending on the solid solution composition and carrier density [14, 31]. With increasing pressure P, the effective mass m/m_0 [29] in the n- and p-type compositions decreases (**Figure 10**). For p-Bi_2Te_3 and for composition at x = y = 0.09, the dependence of m/m_0 and μ_0 on P has an inflection at pressures about 3–4 GPa [29, 30]. These inflections in the dependences of m/m_0 and μ_0 on P, observed at nearly the same pressure as for the maximum value of power factor $S^2\sigma$, were explained by the influence of topological phase transition at room temperature (**Figure 10**, curves 3, 7).

The existence of the topological transition in Bi_2Te_3 is confirmed by precise diffraction studies of the pressure dependence of lattice parameters [32], abrupt change in the elasticity modulus and its derivative [33], and the change in the Fermi surface section from study of de Haas-van Alphen [32, 34]. This topological transition in Bi_2Te_3 was also confirmed by study of Raman spectroscopy under high pressure [35]. The maximum of the product $(m/m_0)^{3/2}\mu_0$, proportional to the figure of merit, was observed at about the same pressure range as for the topological transition (**Figure 11**). The estimations of the thermal conductivity κ in the n- and p-type materials show that increase in κ in the pressure range $\sim$3–4 GPa is not higher than 50% [36]. But the power factor of n-$Bi_2Te_{3-x-y}Se_xS_y$ solid solutions and the p-type compositions $Bi_{2-x}Sb_xTe_3$ [29] increases under pressure more significantly, and thus, enhancement of the figure of merit values can reach 50–70% taking into account the influence of topological transition at room temperature.

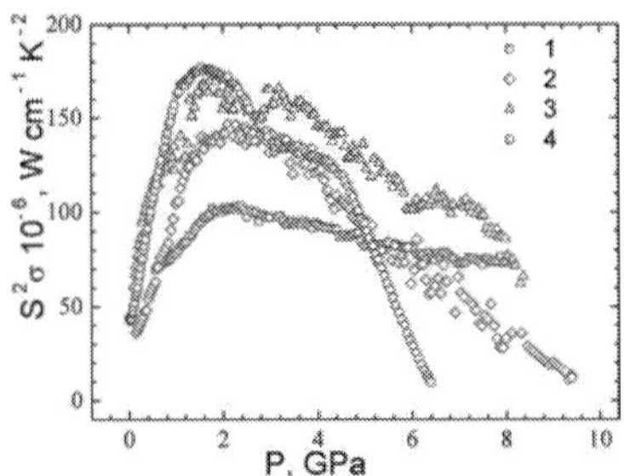

Figure 9. Pressure dependences of power factor of the n-Bi$_2$Te$_{3-x-y}$Se$_x$S$_y$ solid solution layers; x, y are (1) 0.27, 0; (2) 0.3, 0; (3) 0.09, 0.09 and p-Bi$_2$Te$_3$ (4).

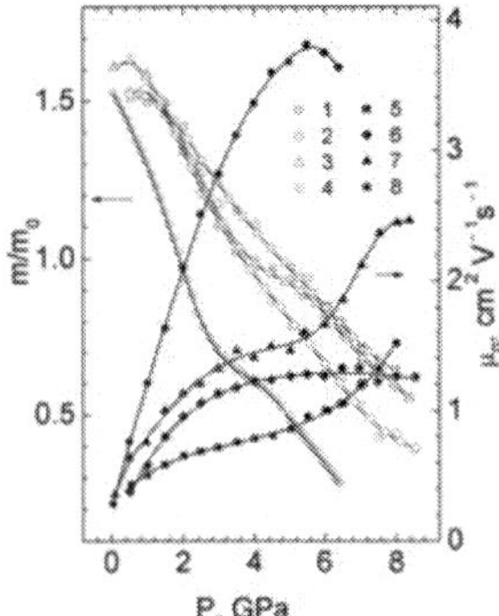

Figure 10. Pressure dependences of the effective mass m/m$_0$ and the mobility μ_0 of the n-Bi$_2$Te$_{3-x-y}$Se$_x$S$_y$ solid solutions; x, y are (1, 5) 0.27, 0; (2, 6) 0.3, 0; (3, 7) 0.09, 0.09 and p-Bi$_2$Te$_3$ (4, 8).

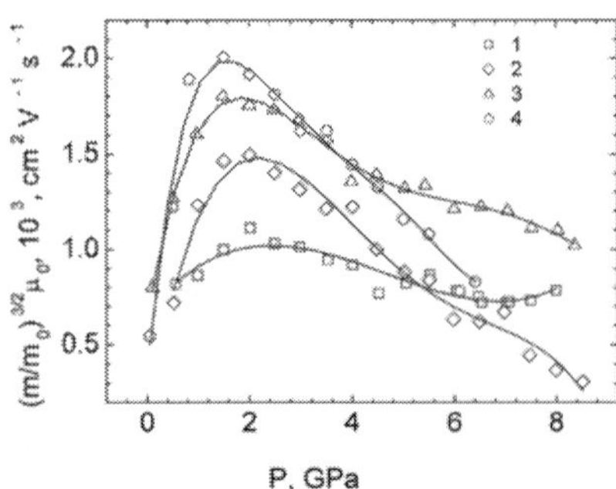

Figure 11. Pressure dependences of the product $(m/m_0)^{3/2}\mu_0$ of the n-Bi$_2$Te$_{3-x-y}$Se$_x$S$_y$ solid solutions; x, y are (1) 0.27, 0; (2) 0.3, 0; (3) 0.09, 0.09 and p-Bi$_2$Te$_3$ (4).

6. Quantum oscillations of magnetoresistance

Quantum oscillations of the magnetoresistance associated with surface electronic states in three-dimensional topological insulators have been studied in p-type Bi_2Te_3 films [37] in strong magnetic fields from 6 to 14 T at low temperatures (**Figure 12**). The main parameters of surface charge carriers in the films (**Tables 1** and **2**) were determined by analyzing the temperature dependences of normalized amplitude of magnetoresistance oscillations [38]. The phase shift of oscillation period, evaluated by extrapolation of dependence of the Landau level indexes (n = 2, 3, 4, 5) (**Figure 12**) on inverse magnetic field in the limit of 1/B = 0, was found to be consistent with the value of π Berry phase, which is characteristic of surface states of Dirac fermions with linear dispersion [37]. The obtained parameters of the surface states of charge carriers in the nanostructured materials under consideration are important for the development of new high-performance thermoelectrics. The parameters of the surface states of Dirac fermions, such as the mean free path and the energy dependence of lifetime of the charge carriers at the surface, the Fermi energy and respective position of the Fermi level, have a specific influence on the Seebeck coefficient and power factor.

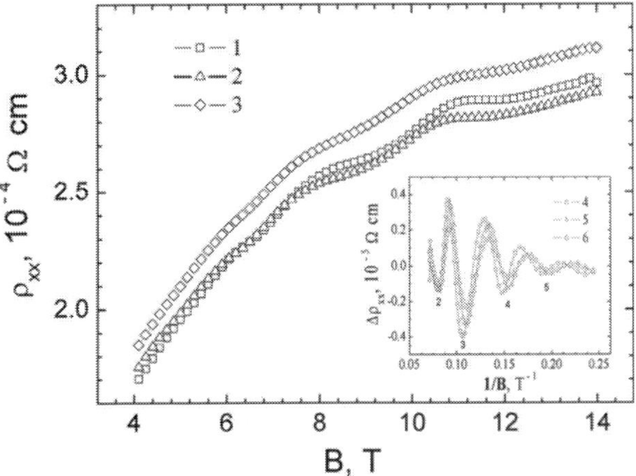

Figure 12. Magnetoresistance dependence ϱ_{xx} (1–3) on magnetic field B and quantum oscillations $\Delta\varrho_{xx}$ (4–6) dependence on inverse magnetic field 1/B; ϱ_{xx} at the temperatures: 1, 4–1.6 K, 2, 5–4.2 K, 3, 6–10 K and the Landau level index ($n = 2$, 3, 4, 5) corresponding to the minimum amplitudes of the oscillations.

v, THz	v_F, 10^5, m/s	E_F, meV	μ, m²/Vs	l_F, nm	τ, 10^{-13}, s	T_D, K
2.0	1.06	19.3	0.33	59	5.5	2.2

Table 1. Cyclotron resonance frequency v, Fermi velocity v_F, Fermi energy E_F, charge carrier mobility μ, mean free path of charge carriers l_F, the relaxation time τ, and Dingle temperature T_D of the Bi_2Te_3 films.

F, T	$S(k_F)$, nm^{-2}	k_F, nm	$n_{F_s} \times 10^{12}$, cm^{-2}
24	0.23	0.27	0.58
30 [39]	0.29	0.30	0.72
41.7 [40]	0.40	0.36	1.0
50 [39]	0.48	0.39	1.21

Table 2. Frequency of quantum oscillations of the magnetoresistance F, cross section of the Fermi surface $S(k_F)$, Fermi wave vector k_F, and surface concentration of charge carriers n_{F_s} in the Bi$_2$Te$_3$ films.

7. Raman spectra

The resonance Raman scattering and morphology of an interlayer van der Waals surface (0001) in thin layer films of chalcogenides based on bismuth and antimony were studied in dependence on the composition, the Seebeck coefficient, and the thickness of the samples. Raman spectra of optical phonons E_g^2, A_{1u}^2, and A_{1g}^2 in thin layers of binary compound n-Bi$_2$Te$_3$ and alloys Bi$_2$Te$_{3-y}$Se$_y$, Bi$_{2-x}$Sb$_x$Te$_{3-y}$Se$_y$, in heteroepitaxial films of p-Bi$_2$Te$_3$ and chemically foliated solid solutions Bi$_{2-x}$Sb$_x$Te$_{3-y}$Se$_y$ are shown in **Figures 13** and **14**. The morphology of the interlayer surfaces (0001) of these samples was studied by atomic force microscopy.

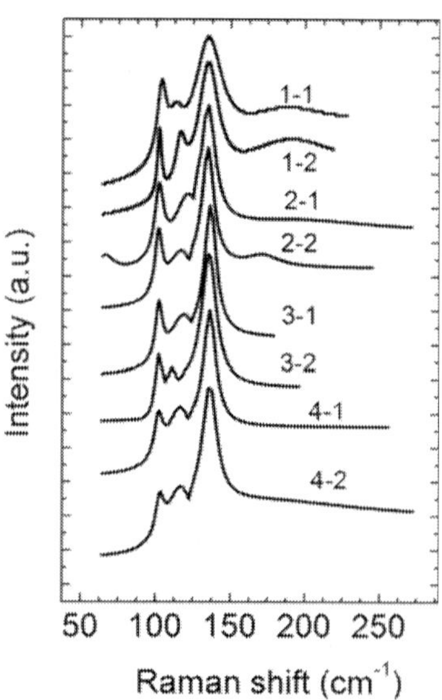

Figure 13. Raman spectra of mechanically split thin layers of n-Bi$_2$Te$_3$ (1-1, 1-2), n-Bi$_2$Te$_{2.88}$Se$_{0.12}$ (2-1, 2-2) and n-Bi$_2$Te$_{2.7}$Se$_{0.3}$ (3-1, 3-2, 4-1, 4-2). (1-1, 1-2): S = -270 µV K^{-1}, R$_q$ = 0.45 nm, H$_a$ = 1.75 nm. (2-1, 2-2): S = -285 µV K^{-1}, R$_q$ = 3.8 nm, H$_a$ = 8 nm. (3-1, 3-2): S = -305 µV K^{-1}, R$_q$ = 4.6 nm, H$_a$ = 15 nm. (4-1, 4-2): S = -315 µV K^{-1}, R$_q$ = 2.9 nm, H$_a$ = 2.3 nm.

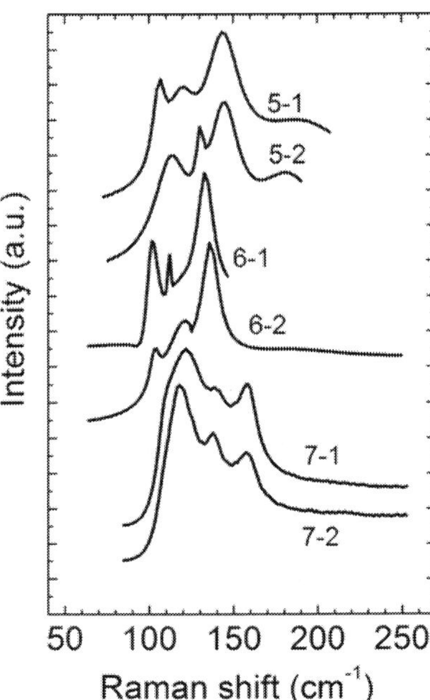

Figure 14. Raman spectra of the thin films of n-$Bi_{1.6}Sb_{0.4}Te_{2.91}Se_{0.09}$ (5-1, 5-2), p-Bi_2Te_3 (6-1, 6-2) and foliated $Bi_{0.9}Sb_{1.1}Te_{2.94}Se_{0.06}$ (7-1, 7-2) solid solution. (5-1, 5-2): S = -280 μV K^{-1}, R_q = 0.36 nm, H_a = 1.55 nm. (6-1, 6-2): R_q = 0.56 nm, H_a = 1.8 nm. (7-1, 7-2): S = -280 μV K^{-1}. (7-1) dissolution time t = 150 h, R_q = 36 nm, H_a = 180 nm. (7-2) t = 200 h, R_q = 26 nm, H_a = 80 nm.

The roughness R_q and H_a corresponding to the maximum of the distribution function of nanofragment heights on the surface of samples [41] are indicated in the captions of **Figures 13** and **14**.

The appearance of the inactive phonons A_{1u}^2 in the Raman spectra, caused by a violation of the inversion symmetry of the crystal, was revealed at decreasing sample thickness and also at high pressure for which topological phase transition in Bi_2Te_3 [35] was observed. Therefore, the occurrence of the A_{1u}^2 was explained by the behavior of surface electronic states of Dirac fermions [42, 43]. The relative intensities $I(A_{1u}^2)/I(E_g^2)$ have maximal values in the most thin layers of solid solutions of n-$Bi_2Te_{2.7}Se_{0.3}$, n-$Bi_{1.6}Sb_{0.4}Te_{2.91}Se_{0.09}$, and epitaxial film of the p-B_2Te_3 with high Seebeck coefficients (**Figure 15**) with high quality of the interlayer (0001) surface with small roughness R_q and H_a values (**Figures 13** and **14**). In the samples n-$Bi_2Te_{2.88}Se_{0.12}$, prepared by the Czochralski technique and by the chemical foliated p-$Bi_{0.9}Sb_{1.1}Te_{2.94}S_{0.06}$ the ratio $I(A_{1u}^2)/I(E_{g2})$, is quite lower (**Figure 15**, curves 2, 7). So the ratio $I(A_{1u}^2)/I(E_{g2})$ is strongly affected by the used technology, composition and thickness of the samples. The analysis of the Raman spectra of bismuth telluride and its solid solutions allows to optimize the Seebeck coefficients, composition, sample thickness, and morphology of the surface, which provide the significant role of surface states of Dirac fermions at room temperature.

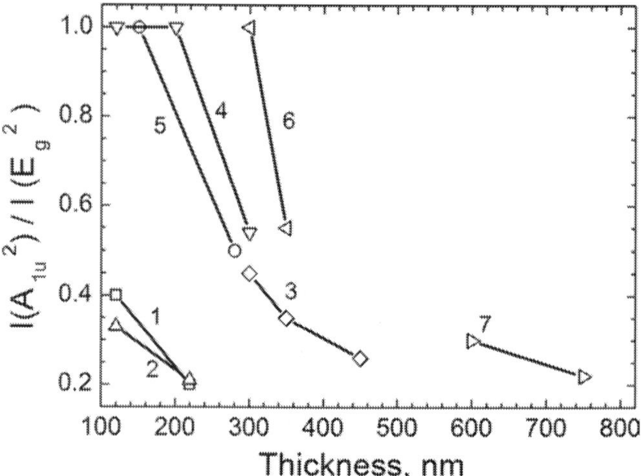

Figure 15. The dependence of the relative intensities $I(A_{1u}^2)/I(E_g^2)$ on the thickness of the layers of n-Bi_2Te_3 (1), n-$Bi_2Te_{2.88}Se_{0.12}$ (2), n-$Bi_2Te_{2.7}Se_{0.3}$ (3, 4), n-$Bi_{1.6}Sb_{0.4}Te_{2.91}Se_{0.09}$ (5), p-Bi_2Te_3 (6), and p-$Bi_{0.9}Sb_{1.1}Te_{2.94}Se_{0.06}$ (7).

8. Conclusion

The thermoelectric and galvanomagnetic properties of heteroepitaxial films based on bismuth telluride, grown by the hot wall epitaxy method, were investigated. The highest power factor was obtained in the $Bi_{0.5}Sb_{1.5}Te_3$ films. An enhancement of the Seebeck coefficient and change in its temperature dependence slope both indicate the variation in the charge carrier scattering mechanisms compared to the bulk thermoelectric materials. The increase in the power factor and $(m/m_0)^{3/2}\mu_0$, associated with the increase in the effective mass of density of states m/m_0, and the reduction in thermal conductivity lead to an increase in the figure of merit Z in the $Bi_{0.5}Sb_{1.5}Te_3$ films up to 3.85×10^{-3} K^{-1} over the temperature range of 180–200 K. Such increase in Z is approximately more by 60% compared to similar bulk materials. The charge carrier scattering mechanisms of the $Bi_{0.5}Sb_{1.5}Te_3$ films were investigated from the data on galvanomagnetic properties for isotropic and anisotropic scattering within many-valley model of energy spectrum. For isotropic scattering, the degeneracy parameter β_d, the effective scattering parameter r_{eff}, and the reduced Fermi level η were calculated. As compared to bulk materials, the values of the parameter r_{eff} are considerably different from r = -0.5, specific for an acoustic phonon scattering in the epitaxial films. The difference of the r_{eff} values is related to an additional charge carriers scattering on interphase and intercrystallite boundaries in the films.

The account of anisotropy of the carrier scattering mechanism has shown that scattering along bisector crystallographic axes is main as compared with corresponding bulk thermoelectrics. Revealed charge carrier scattering peculiarities affect transport properties of the films and might be in favor of a large thermoelectric power factor and the figure of merit.

The studies of the thermoelectric properties of the n-$Bi_2Te_{3-x-y}Se_xS_y$ solid solutions under pressure have shown the increase in power factor and product $(m/m_0)^{3/2}\mu_0$, which is determined by the growth of the effective mass m/m_0. In the composition n-$Bi_2Te_{3-x-y}Se_xS_y$ ($x = y = 0.09$) and p-Bi_2Te_3, the change in the slopes of the pressure dependence of the effective mass and the mobility in the range of 3–4 GPa are coincident with maximum of power factor and explained by an influence of the topological transition. An increase in the thermoelectric figure of merit, as compared to normal conditions, was estimated as 50–70% under pressure due to effect of the topological transitions.

Quantum oscillations of the magnetoresistance were revealed at low temperatures T below 10 K in the range of magnetic field from 6 to 14 T in nanostructured submicron Bi_2Te_3 films grown by hot wall technique. From the analysis of the magnetoresistance oscillations, the cyclotron resonance frequency, cross-sectional Fermi surface, Landau level indexes and π Berry phase, effective cyclotron mass, Fermi wave vector, velocity and Fermi energy, surface charge carrier concentration, lifetime, and mobility of charge carriers were evaluated. The estimated parameters of topological electronic surface states of nanostructured chalcogenides of bismuth and antimony are of special interest for development of new high-performance thermoelectrics, because the Seebeck coefficient and hence power factor are significantly determined by the basic parameters such as charge carriers lifetime, the mean free path of charge carriers, and Fermi energy level position.

The resonance Raman scattering and morphology of the interlayer surface (0001) of bismuth and antimony chalcogenides were studied at room temperature depending on the composition, the Seebeck coefficient, and the thickness of the layers. Raman shifts and the relative intensities of phonon modes were studied in mechanical and chemical foliated thin layers and epitaxial films of bismuth telluride and its solid solutions. The increase in relative intensity ratio of Raman inactive phonons $I(A_{1u}^2)/I(E_g^2)$, sensitive to the topological surface states, was observed for thin layers of n-$Bi_2Te_{2.7}Se_{0.3}$ and n-$Bi_{1.6}Sb_{0.4}Te_{2.91}Se_{0.09}$ solid solutions at low carrier density with Seebeck coefficient values more than -280 μV K^{-1} and in epitaxial p-B_2Te_3 film grown by the hot wall method on mica with perfect interlayer surface. Thus, the resonance Raman spectra analysis allows to optimize the composition, thickness, Seebeck coefficient values, and morphology of the layers and films with enhanced contribution of the topological surface states at room temperature, which increases the prospects of application of these thermoelectrics.

Author details

Lidia N. Lukyanova*, Yuri A. Boikov, Oleg A. Usov, Mikhail P. Volkov and Viacheslav A. Danilov

*Address all correspondence to: lidia.lukyanova@mail.ioffe.ru

Ioffe Institute, Saint Petersburg, Russia

References

[1] Rowe DM, editor. Thermoelectrics Handbook: Macro to Nano. CRC Press, Boca Raton; 2006. 954 p. Print ISBN: 978-0-8493-2264-8. eBook ISBN: 978-1-4200-3890-3

[2] Nolas GS, Sharp J, Goldsmid HJ. Thermoelectrics: Basic Principles and New Materials Developments. Springer, New York; 2001. 295 p. ISBN: 978-3-662-04569-5

[3] Rowe DM, editor. Modules, Systems, and Applications in Thermoelectrics. CRC Press, Boca Raton; 2012. 581 p. ISBN: 978-1-4398-7472-1

[4] Hasan MZ, Kane CL. Colloquium: topological insulators. Rev. Mod. Phys. 2010; 82: 3045–67. doi:10.1103/RevModPhys.82.3045

[5] Qu DX, Hor YS, Xiong J, Cava RJ, Ong NP. Quantum oscillations and hall anomaly of surface states in the topological insulator Bi_2Te_3. Science. 2010; 325: 821–4. doi:10.1126/science.1189792

[6] Chen YL, Analytis JG, Chu JH, Liu ZK, Mo SK, Qi XL, Zhang HJ, Lu H, Dai X, Fang Z, Zhang SC, Fisher IR, Hussain Z, Shen ZX. Experimental realization of a three-dimensional topological insulator Bi_2Te_3. Science. 2009; 325: 178–81. doi:10.1126/science.1173034

[7] Seradjeh B, Moore JE, Franz M. Exciton condensation and charge fractionalization in a topological insulator film. Phys. Rev. Lett. 2009; 103: 066402/1–4. doi:10.1103/PhysRevLett.103.066402

[8] Chen Y. Topological Insulator Based Energy Efficient Devices. Proc. SPIE. 2012; 8373: 83730B/1–5. doi:10.1117/12.920513

[9] Eschbach M, Młyńczak E, Keller J, Kemmerer J, Lanais M, Neumann E, Erich C, Gellman M, Gospodarič P, Döring S, Mussler G, Demarina N, Luysberg M, Bihlmayer G, Schäpers T, Plucinski L, Blügel S, Morgenstern M, Schneider CM, Grützmacher D. Realization of a vertical topological p-n junction in epitaxial Sb_2Te_3/Bi_2Te_3 heterostructures. Nat. Commun. 2015; 6: 8816/1–7. doi:10.1038/ncomms9816

[10] Xu Y, Gan Z, Zhang SC. Enhanced thermoelectric performance and anomalous Seebeck effects in topological insulators. Phys. Rev. Lett. 2014; 112: 226801/1–5. doi:10.1103/PhysRevLett.112.226801

[11] Zhang J, Feng X, Xu Y, Guo M, Zhang Z, Ou Y, Feng Y, Li K, Zhang H, Wang L, Chen X, Gan Z, Zhang SC, He K, Ma X, Xue QK, Wang Y. Disentangling the magnetoelectric and thermoelectric transport in topological insulator thin films. Phys. Rev. B. 2015; 91: 075431/1–11. doi:10.1103/PhysRevB.91.075431

[12] Lopes-Otero A. Hot wall epitaxy. Thin Sol. Films. 1978; 49: 3–57. doi:10.1016/0040-6090(78)90309-7

[13] Boikov Yu A, Lukyanova LN, Danilov VA, Volkov MP, Goltsman BM, Kutasov VA. Features of growth and galvanomagnetic properties of the Bi_2Te_3-based epitaxial films. AIP Conf. Proc. 2012; 1449: 107–10. doi:10.1063/1.4731508

[14] Lukyanova LN, Kutasov VA, Popov VV, Konstantinov PP. Galvanomagnetic and thermoelectric properties of p-$Bi_{2-x}Sb_xTe_{3-y}Se_y$ solid solutions at low temperatures (<220 K). Phys. Solid State. 2004; 46: 1404–9. doi:10.1134/1.1788770

[15] Lukyanova LN, Boikov Yu A, Danilov VA, Volkov MP, Kutasov VA. Parameters of the constant-energy surface and features of charge carrier scattering of Bi_2Te_3-based epitaxial films. J. Electron. Mater. 2013; 42: 1796–800. doi:10.1007/s11664-012-2432-8

[16] Lukyanova LN, Kutasov VA, Konstantinov PP, Popov VV. Optimization of Solid Solutions Based on Bismuth and Antimony Chalcogenides above Room Temperature. In: Rowe DM, editor. Modules, Systems, and Applications in Thermoelectrics. CRC Press, Boca Raton; 2012. pp. 7.1–18. ISBN: 978-1-4398-7472-1

[17] Lukyanova LN, Kutasov VA, Popov VV, Konstantinov PP, Fedorov MI. Anisotropic Scattering in the $(Bi, Sb)_2(Te, Se, S)_3$ Solid Solutions. In: Proceedings of the XXV International Conference on Thermoelectric. IEEE, Austria, Vienna; 2006. pp. 496–9. doi:10.1109/ICT.2006.331342

[18] Lukyanova LN, Boikov Yu A, Danilov VA, Usov OA, Volkov MP, Kutasov VA. Thermoelectric and galvanomagnetic properties of bismuth chalcogenide nanostructured heteroepitaxial films. Semicond. Sci. Technol. 2015; 30: 015011–6. doi:10.1088/0268-1242/30/1/015011

[19] Kutasov VA, Lukyanova LN, Vedernikov MV. Shifting the Maximum Figure-of-Merit of $(Bi, Sb)_2(Te, Se)_3$ Thermoelectrics to Lower Temperatures. In: Rowe DM, editor. Thermoelectrics Handbook: Macro to Nano. CRC Press, Boca Raton; 2006. pp. 37.1–18. ISBN: 978-0-8493-2264-8

[20] Lukyanova LN, Kutasov VA, Popov VV, Konstantinov PP. Galvanomagnetic and thermoelectric properties of multicomponent n-type solid solutions based on Bi and Sb chalcogenides. Phys. Solid State. 2006; 48: 647–53. doi:10.1134/S1063783406040068

[21] Boikov Yu A, Gol'tsman BM, Kutasov VA. Structure effect on thermoconductivity of Bi_2Te_3 and $Bi_{0.5}Sb_{1.5}Te_3$ films. Sov. Phys. Solid State. 1978; 20: 757–60.

[22] Lukyanova LN, Kutasov VA, Popov VV, Konstantinov PP. Galvanomagnetic properties of multicomponent solid solutions based on Bi and Sb chalcogenides. In: Proceedings of the IEEE XXIV International Conference on Thermoelectrics, ICT, Clemson University, SC, USA; 2005. pp. 426–9. doi:10.1109/ICT.2005.1519978

[23] Lukyanova LN, Kutasov VA, Konstantinov PP, Popov VV. Features of the behavior of the figure of merit for p-type solid solutions based on bismuth and antimony chalcogenides. J. Electron. Mater. 2010; 39: 2070–3. doi:10.1007/s11664-009-1007-9

[24] Lagarias JC, Reads JA, Wright MN, Wright PE. Convergence properties of the Nelder-Mead simplex method in low dimensions. SIAM J. Optim. 1998; 9: 112–47.

[25] Ashworth HA, Rayne JA, Ure RW. Transport properties of Bi_2Te_3. Phys Rev. B. 1971; 3: 2646–61. doi:10.1103/PhysRevB.3.2646

[26] Lukyanova LN, Kutasov VA, Konstantinov PP, Popov VV. Effect of charge scattering anisotropy on the thermoelectric properties of the $(Bi,Sb)_2(Te,Se,S)_3$ solid solutions. Phys. Solid State. 2008; 50: 597–602. doi:10.1134/S106378340804001X

[27] Testardi LR, Burstein E. Low- and high-field galvanomagnetic properties of p-type Bi_2Te_3. Phys. Rev. B. 1972; 6: 460–9. doi:10.1103/PhysRevB.6.460

[28] Ovsyannikov SV, Shchennikov VV. Pressure-tuned colossal improvement of thermo-electric efficiency of PbTe. Appl. Phys. Lett. 2007; 90: 122103/1–3. doi:10.1063/1.2715123

[29] Ovsyannikov SV, Grigoreva YA, Vorontsov GV, Lukyanova LN, Kutasov VA, Shchen-nikov VV. Thermoelectric properties of p-$Bi_{2-x}Sb_xTe_3$ solid solutions under pressure. Phys. Solid State. 2012; 54: 2/261–6. doi:10.1134/S1063783412020254

[30] Korobeinikov IV, Luk'yanova LN, Vorontsov GV, Shchennikov VV, Kutasov VA. Thermoelectric properties of n-$Bi_2Te_{3-x-y}Se_xS_y$ solid solutions under high pressure. Phys. Solid State. 2014; 56: 263–9. doi:10.1134/S1063783414020152

[31] Lukyanova LN, Kutasov VA, Konstantinov PP. Multicomponent n-$(Bi,Sb)_2(Te,Se,S)_3$ solid solutions with different atomic substitutions in the Bi and Te sublattices. Phys. Solid State. 2008; 50: 2237–44. doi:10.1134/S1063783408120020

[32] Polvani DA, Meng JF, Chandra Shekar NV, Sharp J, Badding JV. Large improvement in thermoelectric properties in pressure-tuned p-type $Sb_{1.5}Bi_{0.5}Te_3$. Chem. Mater. 2001; 13: 2068–71. doi:10.1021/cm000888q

[33] Sologub VV, Shubnikov ML, Itskevich ES, Kashirskaya LM, Parfen'ev RV, Goletskaya AD. Change of Bi,Te, band structure under hydrostatic compression. Sov. Phys. JETP. 1980; 52: 1203–6.

[34] Itskevich E. S., Kashirskaya L. M., Kraidenov V. F. Anomalies in the low-temperature thermoelectric power of p-Bi_2Te_3 and Te associated with topological electronic transi-tions under pressure. Semiconductors. 1997; 31: 276–278. doi: 10.1134/1.1187126

[35] Ovsyannikov SV, Morozova NV, Korobeinikov IV, Lukyanova LN, Manakov AY, Likhacheva AY, Ancharov AI, Vokhmyanin AP, Berger IF, Usov OA, Kutasov VA, Kulbachinskii VA, Okada T, Shchennikov VV. Enhanced power factor and high-pressure effects in $(Bi,Sb)_2(Te,Se)_3$ thermoelectrics. Appl. Phys. Lett. 2015; 106 14 143901/1–5. doi: 10.1063/1.4916947

[36] Jacobsen MK, Sinogeikin SV, Kumar RS. Cornelius AL High pressure transport characteristics of Bi_2Te_3, Sb_2Te_3, and $BiSbTe_3$ J. Phys. Chem. Solids 2012; 73: 1154–1158. doi: 10.1016/j.jpcs.2012.05.001

[37] Lukyanova LN., Boikov YuA, Danilov VA, Usov OA, Volkov MP, Kutasov VA. Surface States of Charge Carriers in Epitaxial Films of the Topological Insulator Bi_2Te_3. Phys. Solid State. 2014; 56 5 941–947. doi: 10.1134/S1063783414050163

[38] Shoenberg D. Magnetic Oscillations in Metals. Series: Monographs on Physics. Cambridge University Press, Cambridge; 2009 596 p. ISBN: 9780521118781

[39] Taskin A., Ren Z., Sasaki S., Segawa K., Ando Y. Observation of Dirac Holes and Electrons in a Topological Insulator. Phys. Rev. Lett. 2011; 107: 016801/1–4. doi: 10.1103/PhysRevLett.107.016801

[40] Dong-Xia Qu, Hor YS, Xiong J, Cava RJ., Ong NP. Quantum Oscillations and Hall Anomaly of Surface States in the Topological Insulator Bi_2Te_3. Science. 2010; 39: 821-824. doi: 10.1126/science.1189792

[41] Lukyanova LN, Bibik AYu, Aseev VA, Usov OA, Makarenko IV, Petrov VN, Nikonorov NV, Kutasov VA. Surface Morphology and Raman Spectroscopy of Thin Layers of Antimony and Bismuth Chalcogenides Phys. Solid State. 2016; 58 7 1440–7. doi: 10.1134/S1063783416070258

[42] Shahil K M F, Hossain M Z, Teweldebrhan D and Balandin A A. Effect of injection current density on electroluminescence in silicon quantum dot light-emitting diodes. Appl. Phys. Let. 2010; 95: 153103/1–3. doi: 10.1063/1.3248025

[43] Glinka Yu D, Babakiray S, Johnson T A, Lederman D. Thickness tunable quantum interference between surface phonon and Dirac plasmon states in thin films of the topological insulator Bi_2Se_3. J. Phys.: Condens. Matter. 2015; 27: 052203/1–7 doi: 10.1088/0953-8984/27/5/052203

Chapter 10

Thermoelectric Power Generation by Clathrates

Andrei V. Shevelkov

Additional information is available at the end of the chapter

http://dx.doi.org/10.5772/65600

Abstract

Clathrate compounds combine aesthetic beauty of their crystal structures with promising thermoelectric properties that have made them one of the most explored family of compounds deemed as base for thermoelectric generators for mid- and high-temperature application. This chapter surveys crystal and electronic structure and structure-related transport properties of selected types of clathrates and discusses their thermoelectric performance and prospects of their future applications.

Keywords: thermoelectric materials, thermoelectric power generation, clathrates, phonon glass-electronic crystal, charge carrier transport, heat transport

1. Introduction

No compound is able to outplay properly doped bismuth telluride as material for thermoelectric cooling. Since the pioneer works of A.F. Ioffe in the 1950, this material solely holds the position in the industry [1]. Situation is different, when it comes to thermoelectric power generation, where traditional materials based on Bi_2Te_3 are giving way to new state-of-the-art materials. Among the latter, there are clathrates; these compounds combine low, glass-like thermal conductivity with high electrical conductivity and Seebeck coefficient and are demanded as perspective thermoelectric materials that convert temperature gradient into electric power [2–4].

Clathrates are different from many other prospective materials for thermoelectric power generators, because they feature the spatial separation of two substructures known as host clathrate framework and rattling guests [4–6]. The framework is based on strong covalent bonds, four for each atom, that ensure effective transport of charge carriers leading to high values of both electrical conductivity and thermopower, whereas the rattling of guests inside oversized cages of the framework causes low thermal conductivity owing to either scattering

---

of heat-carrying phonons or reducing phonon group velocity due to avoided crossing of rattling modes and branches of acoustic phonons. The spatial host-guest separation provides the base for practical utilization of "phonon glass-electron crystal" (PGEC) concept introduced by Slack, according to which decoupling of heat and charge carriers transport enables their independent optimization [7]. But despite the spatial separation, thermal and charge carriers transport properties are not truly independent, which makes optimization of thermoelectric efficiency a very intricate and delicate task. Recent years have witnessed appreciable progress in enhancing thermoelectric efficiency of clathrates at mid- and high-temperature regions. The phonon engineering approaches, including introduction of rare earth guests and formation of complex superstructures, have led to extremely low thermal conductivity for narrow-gap clathrate semiconductors. New synthetic approaches have enabled accurate tuning of charge carriers' concentration by extremely precise doping, as well as providing very high densities of properly consolidated ceramic materials. Finally, new compositions of clathrates have emerged, that allow combination of reasonably high thermoelectric efficiency and utmost chemical and thermal stability. Already now, there are examples of clathrate compounds displaying high values of figure-of-merit, even surpassing unity at T > 470 K, and further progress is highly expected.

This chapter surveys recent progress in developing thermoelectric materials for power generation on the base of inorganic clathrate compounds. We consider crystal and electronic structures of these compounds, the underlying physics of their thermoelectric properties, synthetic methods of their preparation, and, as a central issue, their thermoelectric performance. Current achievements and future prospects are discussed.

2. Clathrates as inclusion compounds

2.1. Crystal structures

Clathrates belong to plentiful class of inclusion compounds. Their discovery is traced back to the beginning of the nineteenth century, when Sir Davy observed formation of solid chlorine hydrate upon passing gaseous chlorine through water cooled to +5°C. Other hydrates came soon after, and by the middle of the twentieth century, quite a number of hydrates of various gases and liquids were discovered, and their crystal structures were solved. Despite clear differences in their chemical composition and crystal structures, these compounds shared a single common feature, which is complete sequestering of a guest moiety inside cages of framework. Another distinct feature of those compounds is the absence of strong host-guest bonds. In 1965 [8], Kasper, Hagenmuller, and Pouchard reported two new sodium silicides, whose crystal structures were identical to hydrates of various gases, proving that host-guest size matching had the primary role in their formation and stability, rather than the details of chemical bonding. Since then, almost three hundred compounds belonging to ten structure types were documented [3–5, 9]. They involve almost 50 chemical elements constituting more than 50% of all stable chemical elements (**Figure 1**).

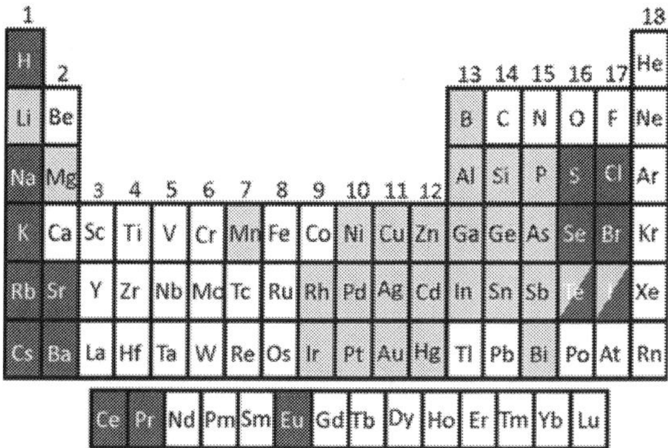

Figure 1. "Clathrate Periodic Table."

All clathrate crystal structures are based on closed polyhedra having from 20 to 28 vertices (**Figure 2**). Various combinations of these polyhedra lead to complete filling of the space, which is distinctive feature of clathrates. In this review, we will focus on four clathrates types, known as type-I, type-II, type-III, and type-VIII clathrates, as many of them demonstrate high thermoelectric figure-of-merit, in part stemming from the details of their crystal structure.

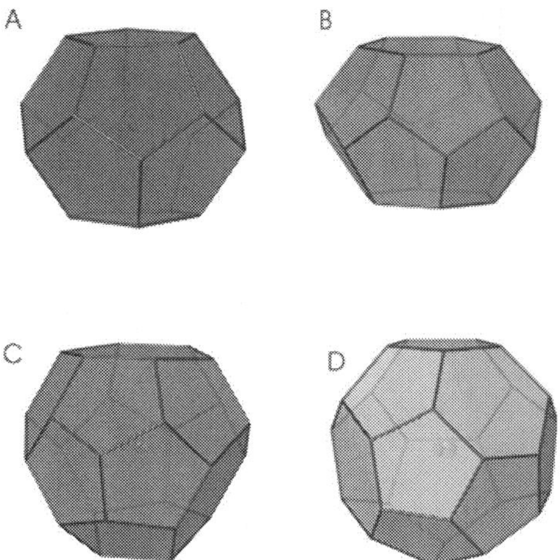

Figure 2. Clathrate-forming polyhedra: (a) 20-vertex pentagonal dodecahedron; (b) 24-vertex tetrakaidecahedron; (c) 26-vertex pentakaidecahedron; and (d) 28-vertex hexakaidecahedron.

Clathrates of type-I are the most numerous; recent review lists over 100 representatives of this structure type [10]. Their crystal structure consists of two types of polyhedra, 20-vertex dodecahedron and 24-vertex tetrakaidecahedron. The latter polyhedra form three-dimensional cubic array in such a way, that centers of adjacent 6-member rings form close rod packing, as in the crystal structure of Cr_3Si, whereas the smaller polyhedra fill the remaining empty space (**Figure 3**). Guest atoms fill the centers of the polyhedra, forming only long, non-covalent contacts with atoms forming polyhedral framework. The resulting crystal structure belongs to cubic space group Pm3n and has general chemical formula $E_{46}G_8$, which emphasizes that there are 46 framework atoms and eight guest atoms per unit cell. Whereas the guest atoms occupy two positions, sixfold inside the larger polyhedral cage and twofold inside the smaller one, and have very large coordination numbers of 24 and 20, respectively, all atoms of the framework (24-fold, 16-fold, and sixfold) have tetrahedron environment.

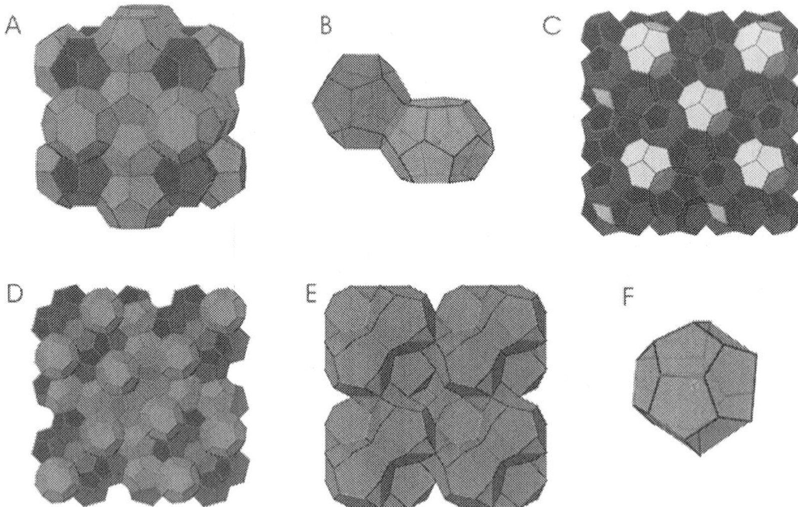

Figure 3. Crystal structure of clathrates: (A) type-I clathrate; (B) two adjacent polyhedra in type-I clathrate; (C) type-II clathrate; (D) type-III clathrate; (E) type-VIII clathrate; and (F) asymmetric cage in type-VIII clathrate.

The nature of chemical elements that form type-I clathrates is quite diverse. In general, type-I clathrates are classified into two groups depending on charge of the framework. The most numerous are anionic clathrates, in which the framework bears negative charge compensated by guest cations. A reverse of the host-guest polarity leads to cationic (also known as inversed) clathrates. As a rule, atoms that form framework come from *p*-block of Mendeleev periodic table; however, inclusion of *d*-metals is also possible. Guest atoms are different depending on charge of the framework. In anionic clathrates, guests are cations of large alkali or alkali earth metals; only a few examples of clathrates hosting rare earth metals are documented [11–13]. In the case of cationic clathrates, halogens and chalcogens of 3–5 periods of Mendeleev table serve as anionic guests.

Type-I clathrates frequently feature deviations from ideal crystal structure described above. This includes mixed occupancy of positions by atoms of different chemical nature, partially vacant positions, splitting of positions into two or three closely lying partial occupied sites, and various types of atom and vacancy ordering that lead to formation of superstructures and reduction in symmetry [10]. In most cases, these crystallographic details affect the electronic structure of clathrates and invoke properties that enhance thermoelectric efficiency.

Other clathrate types are less numerous. Their crystal structures are also built of different high-coordination polyhedra. For instance, type-II clathrate is made of combination of 20-vertex dodecahedra with 28-vertex hexakaidecahedra in such a fashion that cubic face-centered structure is formed (**Figure 3**). Crystal structure of type-III clathrates is the only clathrate structure that contains three types of polyhedra at the time; they are 20-vertex dodecahedra, 24-vertex tetrakaidecahedra, and 26-vertex pentakaidecahedra. They share faces to form a tetragonal crystal structure displayed in **Figure 3**. Type-VIII clathrates are slightly different as they have only one type of polyhedra, which is substantially distorted. It can be viewed as dodecahedron, in which three E–E bonds are broken, and three extra E atoms are inserted instead. The resulting polyhedron has rather low symmetry, but its packing within cubic unit cell brings about clathrate type of the crystal structure. As long as distorted polyhedra cannot fill the entire space, additional 8-vertex polyhedra are left unfilled in this crystal structure (**Figure 3**).

2.2. Application of Zintl scheme and electronic structures

Chemical composition of clathrates frequently looks unusual in terms of the stoichiometry of phases. For instance, the following compounds displaying promising thermoelectric proper-ties are formulated as $Sr_8Ga_{16}Ge_{30}$, $Ba_8Ga_{16}Sn_{30}$, $K_8In_8Sn_{38}$, and $Si_{30}P_{16}Te_8$. These formulas can be rationalized on the basis of Zintl electron-counting scheme, which, in fact, shows that these compounds should behave as semiconductors [14].

Application of Zintl scheme rests on the tetrahedral coordination of all atoms of the framework. They all form four two-center, two-electron (2c–2e) bonds, thus forming electronic octet. Let us consider clathrate compound with formula $Ba_8Ga_{16}Sn_{38}$. Its framework comprises tetrahe-drally bonded Ga and Sn atoms, whereas Ba guests compensate for the charge of the frame-work. Each Sn atom forms four 2c–2e bonds, for which it uses four own electrons and four electrons shared with four neighbors. Therefore, it does not require loss or gain of further electrons, which means that under Zintl scheme its formal oxidation state is zero. Similarly, Ga atom, having three valence electrons, is one electron short of forming four 2c–2e bonds. It must gain one electron to achieve an octet, thus acquiring formal oxidation state of −1. There are 16 Ga atoms per formula, which requires compensating for 16 negative charges. Ba atoms with coordination numbers of 20 and 24 clearly exist as Ba^{2+} cations. There are eight +2 cations that compensate for the charge of the framework and ensure the overall electroneutrality of $Ba_8Ga_{16}Sn_{30}$.

The overall electroneutrality of clathrate compound along with formation of electron octets makes these compounds semiconductors unless limitations of Zintl scheme are overcome. This may happen under various circumstances, including an introduction of *d*-metal into frame-work, combination of elements, that would lead to overlap of valence and conduction bands, and energy gain of accepting or expelling an electron in favor of formation of chemical bond of high bond energy, as in the case of Si–Si bond [15]. In such cases, metal-to-insulator transition (MIT) may occur, leading to temperature-dependent properties, with prospective thermoelectric parameters at the verge of MIT.

Electronic structure of clathrates, albeit having little in common with Zintl counting scheme, still shows the propensity of this compounds to behave as semiconductors [14]. Electronic structure of various clathrates has been assessed in numerous reports and discussed in several reviews [3–6]. However, the majority of the studies are dealing with calculations at different levels. For type-I clathrates, it was shown that fulfillment of Zintl rule shows up in the following way: All bonding states and, if necessary, nonbonding states (lone pairs neighboring vacancies) are filled and lie below Fermi level, whereas all antibonding states are empty and compose conduction band. Common feature of band structure is that the states in vicinity of Fermi level are composed predominantly by individual contributions that are the most sensitive to various substitutions within clathrate framework. For instance, in type-I clathrate, $Sn_{24-x}In_xP_{22}I_8$ indium orbitals have the largest contribution to the states just below Fermi level [16]. In $Sn_{24}P_{19.3}I_8$, another type-I clathrate, but with vacancies in the positions of phosphorus, lone pairs on tin atoms, that surround vacancies, cluster together to form sharp states at the top of valence band [17] (**Figure 4**). Therefore, minor changes in concentration of vacancies or doping element can substantially alter transport properties of clathrates.

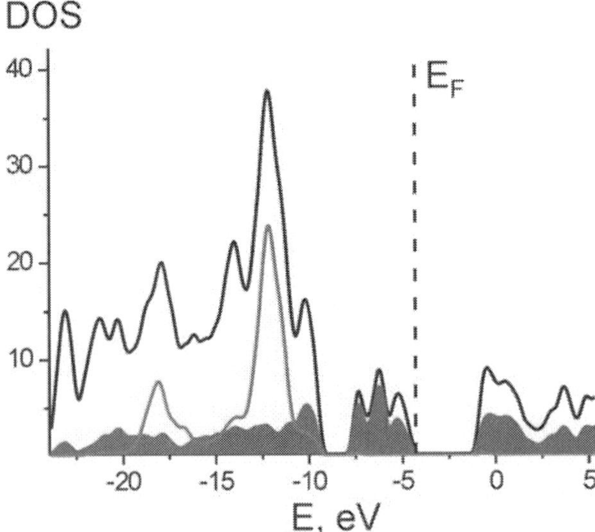

Figure 4. Scheme of the band structure of $Sn_{24}P_{19.3}I_8$ presented as density of states (DOS) versus energy. Black, total DOS; green, contribution of 4-bonded Sn; red, contribution of 3 + 3-bonded Sn.

Experimental studies of the electronic structure of clathrates are very rare. This is explained by necessity to have rather large single crystals and clean surface to investigate electronic structure by means of X-ray photoelectron spectroscopy (XPS). Recently, these obstacles were overcome, and comprehensive picture of electronic band structure of type-I clathrate $Sn_{24-x}In_xAs_{22}I_8$ was obtained [18]. This study proves that chemical bonding has different nature; within the framework, strong covalent bonds are present, whereas the host-guest interactions have pronounced electrostatic nature with clear transfer of electrons from the framework atoms toward guest iodine species. Further, it is shown that top of valence band is composed of shallow I $5p$, As $4p$, In $5p$, and Sn $5p$ orbitals that are largely mixed (**Figure 5**).

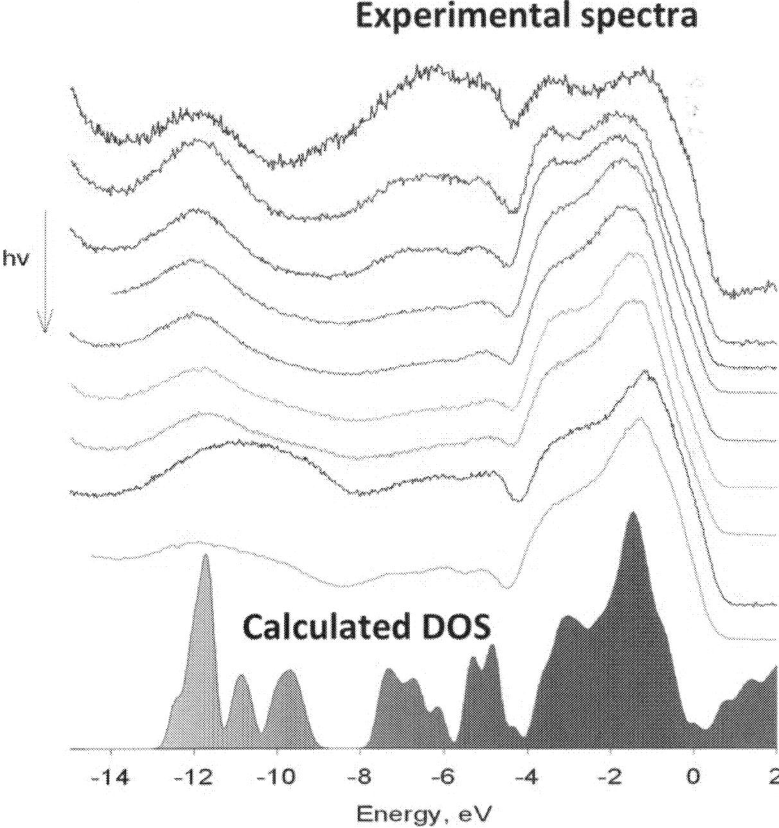

Figure 5. Experimental and calculated electronic band structure of $Sn_{22-x-\delta}In_xAs_{22-y}I_8$ for $x = 12$. Reprinted with permission from Inorg. Chem. 2015, 54, 11542–11549. Copyright 2015 American Chemical Society.

In contrast to majority of clathrate types, where semiconducting properties are hardly violated, in the case of type-II clathrates, metallic behavior is more norm than exception. Most of type-II clathrates feature frameworks made of single kind of atoms, $Na_{24-x}Si_{136}$ being a typical example [19]. In these clathrates, the strength of Si-Si bond (226 kJ/mol) outplays energy loss associated with filling the bottom of conduction band by electrons upon occupation of the guest sites by sodium. Depending on concentration of the guest atoms, MIT is expected, which may lead to various interesting properties, including high thermoelectric performance [20].

3. Sample preparation

3.1. Synthesis and crystal growth

Synthetic routes to clathrates are different. They largely depend on the nature of elements constituting a particular compound. High-temperature ampoule synthesis is the most common method for preparing clathrate compounds, the exact temperature depending on the chemical system. The highest temperatures are explored in the case of silicon-based clathrates owing to very low reactivity of silicon. Heating up to 1500 K might be necessary to enroll this element into reaction; for instance, type-III clathrate $Si_{130}P_{42}Te_{21}$ was synthesized by heating the stoichiometric mixture of elemental components at 1425 K for 18 days [21]. Further prolonged annealing with intermediate regrinding is always required to achieve homogeneous product. Lower temperatures, between 800 and 1250 K, are required by less inert germanium. For comparison with the previous example, we note that to synthesize isostructural type-III clathrate $Ge_{130}P_{42}Te_{21}$, temperature of 953 K was sufficient. Completely different scenario is realized in the case of tin. The latter element has low melting point of 505 K, and preparation of tin-based clathrates is associated with formation of melts rich in tin. This frequently becomes an obstacle, because surface of melted tin becomes covered with poorly reactive compounds, such as Sn_4P_3 or SnAs, leading to incomplete reaction of precursors [18]. This obstacle can be overcome by introducing vapor transport agents. For instance, elemental iodine or SnI_4 tend to facilitate reactions owing to formation of volatile intermediates [22, 23].

Other synthetic methods include flux synthesis, precursor decomposition, high-pressure synthesis, and oxidation in ionic liquids [14]. They are used in selected cases depending on the properties of desired clathrates. Of those methods, flux synthesis is rather intensively used both for synthesis and for crystal growth when such low-melting metals as gallium or tin or even aluminum are included in chemical composition of clathrates. Metals themselves produce flux and at the same time are used as reactants. In some cases, large crystals with mass up to 60 mg were prepared by pulling from the melt [24]. A peculiar variation of this method was used for growing crystals of thermoelectric clathrate $Ba_8Ga_{16}Sn_{30}$, where two p-metals, gallium and tin, were used as common flux, and properties of the resulting crystals strongly depended on which metal was taken in excess [25].

3.2. Sample densification

As clathrates are deemed as prospective thermoelectric materials, the problem of sample densification is put forward. Only for a limited number of clathrates, cold pressing produces samples with the density up to 85% of theoretical. These cases are limited to tin-based compounds that exhibit less rigid clathrate frameworks [26].

In recent years, major success in preparing dense samples of various clathrates has been achieved by using of spark plasma sintering (SPS). This method is based on a simultaneous application of temperature, pressure, and DC pulses to sample under inert atmosphere or vacuum (**Figure 6**). High-energy DC pulses are believed to excite plasma nearest to intergrain contacts, leading to high local overheating and consequent bridging of grains with formation of larger uniform particles. Although the exact mechanism is not known and the very formation of plasma is sometimes questioned, this method has been successfully used for preparation of many types of materials [27]. In particular, SPS allows synthesis of clathrates at lower temperatures and lower pressures compared to standard high-pressure method, which is very advantageous as long as clathrates cannot withstand too high pressure because of readily collapse of their tracery framework [28]. For instance, compact and dense pellets of $Ge_{30}P_{16}Se_8$ could be prepared at temperature of 773 K and pressure of 60 MPa that already provided sample density of 96% relative to theoretical one [29]. Similarly, silicon-based clathrates were densified at significantly harsher condition of 1100 K and 110 MPa to achieve sample density of 95% [21, 30]. In both cases, no degradation of the initial sample was observed, proving that densification does not change composition and structure of clathrates and that concomitant thermoelectric measurements are performed on the samples of desired nature.

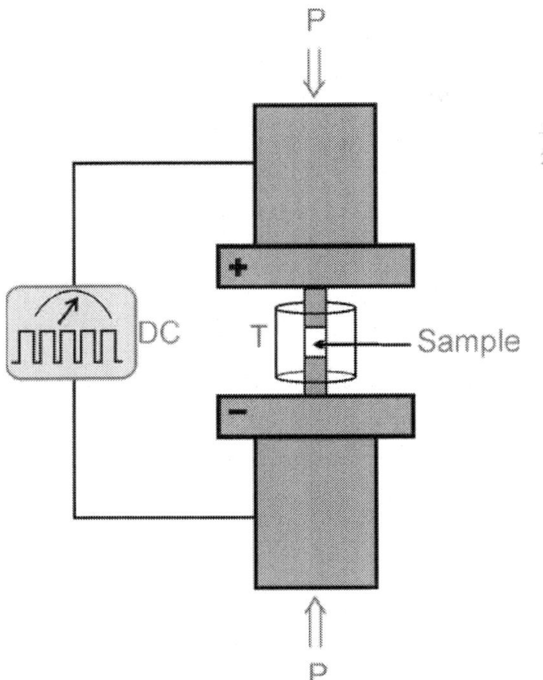

Figure 6. Scheme of the SPS method.

4. Transport properties

4.1. Charge carriers transport

As long as clathrates belong to the family of Zintl compounds, they frequently display activation type of conductivity typical for proper semiconductors. They possess rather high values of electrical conductivity, σ, and Seebeck coefficient, S, giving rise to moderately high values of power factor, $S^2\sigma$. The latter describes transport of charge carries and depends largely on details of the band structure of given compounds.

The advantageous property of clathrates is that their crystal structure, in particular, the spatial separation of host and guest substructures, provides opportunities for tuning charge carriers transport almost independently of phonon transport.

Electrical conductivity of type-I clathrates ranges from several S m^{-1} for ideally balanced compounds to nearly 10^5 S m^{-1} for properly doped semiconductors. For instance, $Sn_{20.5}As_{22}I_8$ has room-temperature electrical conductivity just below 1 S m^{-1}, whereas introduction of In as doping element pushes electrical conductivity to 135–461 S m^{-1} depending on concentration of indium and corresponding vacancies in clathrate framework [16, 31]. Similarly, stoichiometric $Si_{30}P_{16}Te_8$ is not good electrical conductor with room-temperature value of 63.3 S m^{-1} [32]. However, its band structure can be altered upon creating vacancies in guest positions with concomitant change in the Si:P ratio. As a result, band gap was decreased from 1.24 eV to minimum of 0.12 eV and electrical conductivity was increased up to $(1–4) \times 10^4$ S m^{-1} depending on actual composition of clathrate [30]. Importantly, electronic structure and, hence, conducting properties are only weakly sensitive to isovalent substitution, provided that the substituting atoms reside on similar crystallographic sites. For instance, $K_8M_8Sn_{38}$ (M = Al, Ga, and In) exhibits almost the same room-temperature conductivity of $(6.5–12.5) \times 10^4$ S m^{-1} [33]. In these compounds, small change in electrical conductivity can be attributed to shrinkage of clathrate framework upon going from In to Ga and to Al. Another example of sensitivity of transport properties to the framework structural modification is provided by $Sn_{20}Zn_4P_{21.2}X_8$ (X = Br, I). When Br is a guest, the shrinkage of the framework leads to relaxation of atoms residing next to vacancies causing a significant shortage of Sn–P and Zn–P bonds compared to I-based compound. Accordingly, the framework becomes more conductive as band gap decreases from 0.25 to 0.11 eV. As a result, $Sn_{20}Zn_4P_{21.2}Br_8$ displays much greater room-temperature conductivity of 250 S m^{-1} compared to 0.4 S m^{-1} for I-based analog prepared under the same conditions [23].

Basically, electrical conductivity is product of charge, charge carriers' concentration, and mobility. The former is constant, but two other parameters vary with both temperature and chemical nature of clathrate. However, charge carriers' concentration is intrinsic property of a given composition, whereas their mobility is sensitive to grain boundaries. Therefore, observed conductivity of compound with a given composition may depend upon preparation and compacting methods. Type-I clathrate $Sn_{24}P_{19.3}I_8$ provides example of drastic change in electrical conductivity in response to different preparation routes. As-prepared and cold-pressed samples display room-temperature conductivity of 335 S m^{-1}, whereas SPS-treated sample shows much higher conductivity of 6.5×10^3 S m^{-1} [17, 26]. Temperature-dependent impedance spectroscopy measurements showed that for SPS-compacted sample of high density (92% of theoretical), intergrain contacts start to contribute significantly to total impedance only below 75 K, while above this temperature only activation part could be detected [34].

Importantly, electrical conductivity can be suppressed significantly by significant disorder of crystal structure, which is exemplified by very low value of $\sigma \approx 1$ S m^{-1} at 300 K for $Sn_{20.5}As_{22}I_8$, which is four orders of magnitude smaller than for phosphorus analog. The only reason for such difference is reported to be tremendous disorder in crystal structure of As-based compound, leading to significant scattering of charge carriers on flaws of crystal structure [31].

At high temperatures, many clathrates demonstrate very high electrical conductivity, showing that no other mechanism than activation has any noticeable contribution. There are rare cases of pure metallic properties, where electrical conductivity decreases with temperature as for $Na_{22}Si_{136}$ [19]; considerably more numerous are examples of clathrates lying at the border of metallic and semiconducting regimes and showing slight increase in electrical conductivity with temperature. For instance, type-III clathrate $Si_{132}P_{42}Te_{21}$ displays only threefold increase in electrical conductivity upon heating from 300 to 1100 K [35]. At low temperatures, majority of clathrates display very high electrical resistivity. Noticeably, several Si-based clathrates possess transition into superconducting states below 10 K. For instance, type-I clathrate Ba_8Si_{46} has T_C of 8 K [36], and type-IX clathrate Ba_6Ge_{25} turns on superconducting below 3.8 K [37].

Type-II clathrates are different from those of other types in displaying metallic type of electrical conductivity, and many of them behave as normal metals. In particular, $Cs_8Na_{16}Si_{136}$ and $Cs_8Na_{16}Ge_{136}$ combine high electrical conductivity manifested by smooth increase in electrical resistivity with temperature-independent Pauli paramagnetism; such combination is typical for good metals [38].

Clathrates demonstrate different types of majority carriers, and therefore, Seebeck coefficient can be positive (holes) or negative (electrons). Absolute values of Seebeck coefficients vary from one clathrate to another and depend on multifold factors. They include band gap width, concentration of charge carriers, degree of the framework disorder, and many others. In most cases, as generally observed for proper semiconductors, the higher the electrical conductivity is, the lower the Seebeck coefficient is, which stems from the opposite trend of their dependence upon charge carriers' concentration [1, 3]. This is exemplified by several clathrates of different structure types. Whereas type-I $Sn_{24}P_{19.3}I_8$ demonstrates at 300 K high electrical conductivity of 6.5×10^3 S $\times$ m^{-1}, but also exhibits rather low Seebeck coefficient of only +80 μV $\times$ K^{-1}, formally isostructural compound $Ge_{38}Sb_8I_8$ displays very high Seebeck coefficient of about +800 μV $\times$ K^{-1}, and its electrical conductivity does not exceed 10^{-1} S $\times$ m^{-1} at the same temperature [39]. Some kind of compromise between values of electrical conductivity and Seebeck coefficient is achieved for charge carriers' concentration of 10^{19} cm^{-3}. For instance, type-VIII clathrate $Ba_8Ga_{16}Sn_{30}$ doped with small amounts of Cu demonstrates $S = 350$ μV $\times$ K^{-1} coexisting with $\sigma = 3 \times 10^4$ S $\times$ m^{-1} at 300 K [40].

Important value describing the entire charge carriers' transport is so-called power factor, *PF*, which is related to other properties as $PF = S^2 \sigma$ [2, 3]. Therefore, for more effective transport of charge carriers, both electrical conductivity and Seebeck coefficient should be maximized, which is impossible for intrinsic semiconductors. Consequently, attempts have been made to optimize charge carriers' concentration by multiple doping and/or vacancy formation. This may lead to altering the band structure by introducing donor and/or acceptor levels, which may be broad enough to cause their overlap with both conduction and valence bands, giving rise to properties of "bad metal" and, provided the optimal tuning is achieved, to metal-to-semiconductor transition. As a result of this strategy, combination of $S = 170\ \mu V \times K^{-1}$ with $\sigma = 4.75 \times 10^4\ S \times m^{-1}$ at 300 K was achieved for $Si_{46-x}P_xTe_{8-y}$, leading to $PF = 0.14 \times 10^{-3}\ W \times m^{-1} \times K^{-2}$, which is almost four orders of magnitude greater than that for ideally stoichiometric compound $Si_{30}P_{16}Te_8$ [30, 32].

As temperature increases, both electrical conductivity and Seebeck coefficient tend to grow (**Figure 7**), and therefore, power factor also increases; for instance, *PF* for $Ge_{31}P_{15}Se_8$ is three orders of magnitude higher at 650 K compared to 300 K [29].

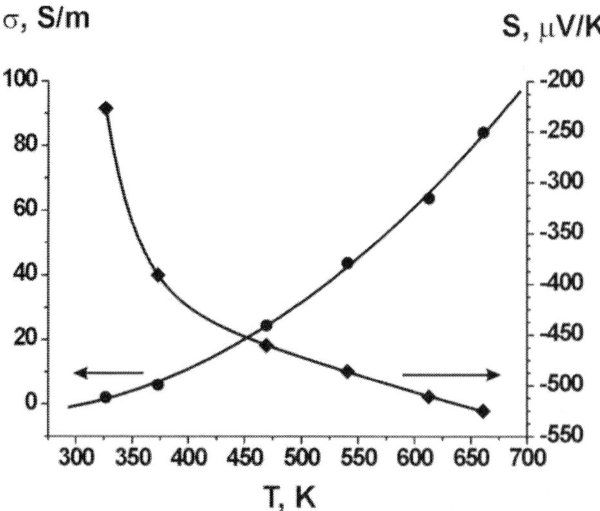

Figure 7. Electrical conductivity σ and Seebeck coefficient S of clathrate $Ge_{31}P_{15}Se_8$ as function of temperature.

4.2. Guest dynamics and heat transport

Clathrates are famous for their low, glass-like thermal conductivity, which originates from the details of their crystal structure, namely from the motion of guest atoms inside oversized cages of the framework (see **Figure 3b**). Such a motion is known as rattling; it provides pseudo-localized vibrations that are alien to concerted (Debye) vibrations of atoms composing the framework.

Analysis of atomic displacement parameters (ADPs) shows that in all types of clathrate compounds guest atoms have the highest values of ADPs and that absolute values depend on the nature of guest atom and degree of host-guest mismatch. As a rule, temperature dependence of ADPs is linear, which provides an opportunity to estimate characteristic Debye and Einstein temperatures, θ_D and θ_E, that are proportional to the slope of $<U^2>(T)$ function, where $<U^2>$ is the mean square atomic displacement either taken for any particular guest atom or averaged over all framework atoms. These characteristic temperatures describe dynamics of clathrate compounds. In particular, θ_D characterizes the framework; the higher the Debye temperature, the more rigid the framework. Value of θ_D depends primarily on the nature of atoms composing the framework. Si-based clathrates are known to be the most rigid, and their θ_D values may exceed 500 K [20]. Frameworks based on tin or germanium are less rigid, and θ_D value falls in the range of 150–320 K largely depending on the nature and concentration of doping element.

Einstein characteristic temperature provides information on pseudo-localized vibrations of guest atoms inside the framework. In general, its characteristics depend on type of clathrate crystals structure, on atomic mass and size of guest atom, and on host-guest mismatch for given clathrate compound.

Further analysis shows that in all clathrates, ADPs for guest atoms are always greater than for the framework ones. For instance, **Figure 8** displays temperature dependence of ADPs for crystal structure of cationic clathrate, in which framework is composed of silicon and phosphorus atoms in approximate ratio 2:1, whereas tellurium and selenium atoms jointly occupy guest positions.

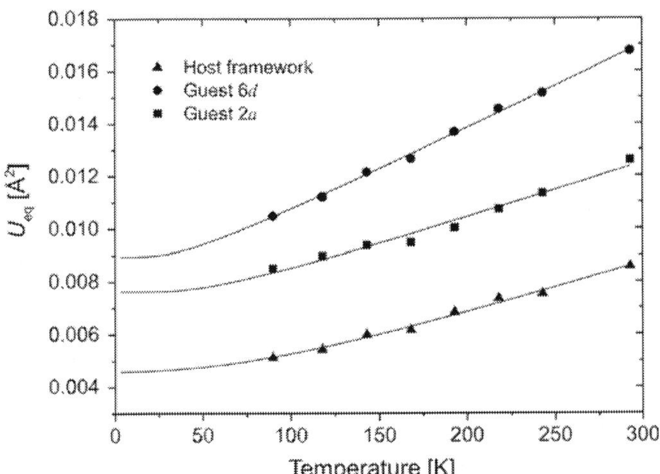

Figure 8. Temperature dependence of ADPs in crystal structure of type-I clathrate [Si,P]$_{46}$Te$_{6.78}$Se$_{1.22}$. Reprinted with permission from Inorg. Chem. 2012, 51, 11396–11405. Copyright 2012 American Chemical Society.

Clearly, ADPs averaged over the framework atoms are the lowest in the structure, ADP for guests in *2a* position comes next, and that for guest in *6d* position is the highest. The difference between two guest positions is related to structural features. Effective volume of 20-vertex cage centered by *2a* site is lower than that of 24-vertex cage centered at *6d*. Moreover, 20-vertex cage is perfectly isotropic, whereas in 24-vertex cage (*cf.* **Figure 2**), motion in the direction of two hexagonal faces and that in perpendicular direction should occur at different frequencies. Such an anisotropy was clearly demonstrated for type-I clathrate $Sn_{24}P_{19.3}I_8$ [41]. **Figure 9** shows that, firstly, ADP of I2 atom residing in the center of 24-vertex cage is the largest in the system. Secondly, whereas motion of I1 atoms is described by single Einstein temperature of 76 K, displacement of I2 is characterized by two Einstein modes because of anisotropy of vibrations. In particular, axial movement in direction to hexagonal faces of tetrakaidecahedron occurs at lower frequency than that in perpendicular direction; respective values of θ_E are 79 and 63 K.

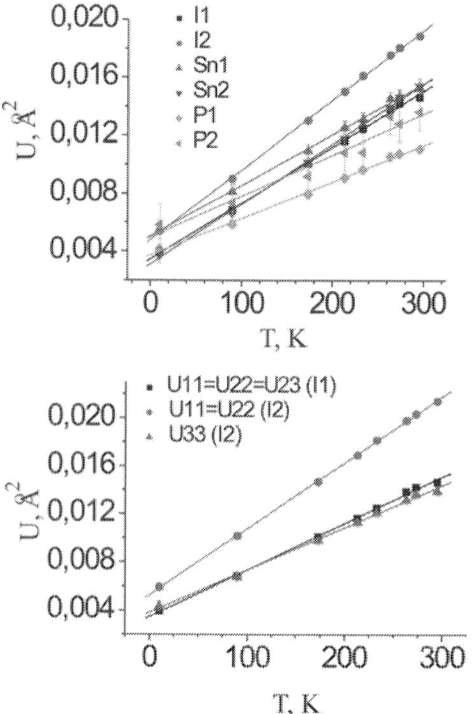

Figure 9. Temperature dependence of ADPs for $Sn_{24}P_{19.3}I_8$. (top) Equivalent ADPs for all atoms. (bottom) Guest atom ADPs in an anisotropic mode. Reprinted with permission from J. Alloys Compd. 2012, 520, 174–179. Copyright 2012 Elsevier.

Guest dynamics can be probed by various methods, ADPs analysis being just a most typical example. Other methods include direct or indirect observation of guest vibration frequencies by means of Raman spectroscopy, inelastic neutron scattering, resonance ultrasound spectro-scopy, heat capacity data, and other tools. Of them, low temperature examination of heat capacity data is frequently used to analyze jointly Debye and Einstein modes. Such analysis was performed for quite a number of clathrates. It was shown that no anomaly is observed below room temperature pointing at the absence of phase transitions, which is corroborated by linearity of $U(T)$ dependencies. At low temperatures, heat capacity of clathrates does not obey Debye law of cubes owing to significant contributions of Einstein modes. For type-I clathrate $Sn_{24}P_{19.3}I_8$ [41] described above, low-T part of $C_P(T)$ dependence could be circum-scribed only by taking into account three different contributions, one Debye and two Ein-stein, that account for concerted vibrations of the entire framework and for two localized modes (**Figure 10**). Extracted values of θ_D (265 K) and θ_E (60 and 78 K) match to values ob-tained from ADPs [42].

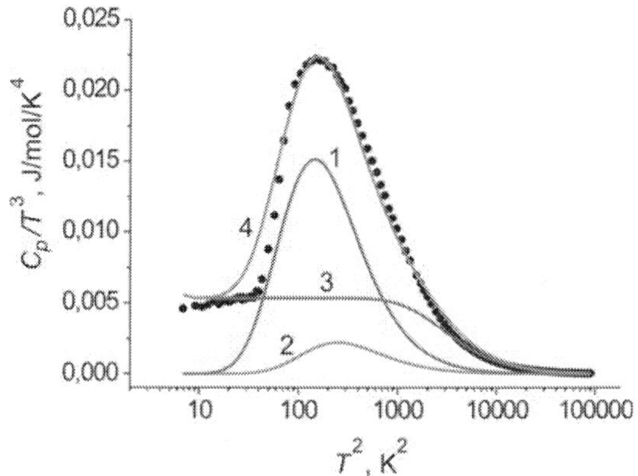

Figure 10. Plot of C_P/T^3 versus T^2 in semi-logarithmic coordinates. Two Einstein (1, 2) and one Debye (3) contributions to total C_P/T^3 values (4) are given in comparison with experimental data (filled circles). Reprinted with permission from J. Alloys Compd. 2012, 520, 174–179. Copyright 2012 Elsevier.

Lattice dynamics defines the principal contribution to thermal conductivity of clathrates. Although the majority of clathrates are low-gap semiconductors, they display very low values of thermal conductivity, which ranges at room temperature from 0.4 to 2.0 W × m^{-1} × K^{-1}. Rattling of guest atoms is the primary reason of reducing thermal conductivity of clathrates due to either lowering of phonon group velocity because of avoided crossing of acoustic modes or resonant scattering of phonons by rattling modes. However, other features of particular clathrate compounds can be added to the mechanism of reducing thermal conductivity. First, vacancy formation within the clathrate framework makes it less rigid leading to reducing Debye temperature, which, in turn, is proportional to velocity of sound, v_s, that is related to thermal conductivity as $\kappa_L = 1/3(v_s C_P \lambda)$, where κ_L is lattice part of thermal conductivity, C_P is heat capacity, and λ is phonon mean free path. Second, formation of superstructures gives rise to high unit volumes causing less concerted vibrations of framework atoms, thus reducing thermal conductivity. Third, mass alternation within guest substructure alters phonon mean free path without affecting individual rattling modes, thus also reducing thermal conductivity. Finally, in real systems, any combination of these scenarios is possible.

Mass alternation leads to low thermal conductivity of mixed-guest clathrates $Sn_{24}P_{19.3}I_{8-x}Br_x$ (x = 2–4) [26]. For any composition x, thermal conductivity is lower than for single-guest compounds, although the latter phases already exhibit low thermal conductivity due to both guest rattling and vacancies within the framework. The lowest value of $0.5\ W \times m^{-1} \times K^{-1}$ is observed at 300 K for composition with I:Br ratio of 1:1 (**Figure 11**), proving that mass alternation is the driving force for reducing thermal conductivity. Mass alternation brings about another peculiar effect as thermal conductivity of such clathrates is glass-like. Whereas for typical crystalline semiconductors thermal conductivity increases until temperature of about 30–50 K and then decreases as $\kappa = f(T^{-1})$, glass-like clathrates show smooth increase in thermal conductivity and then temperature-independent regime in the range of about 50–300 K (**Figure 11**).

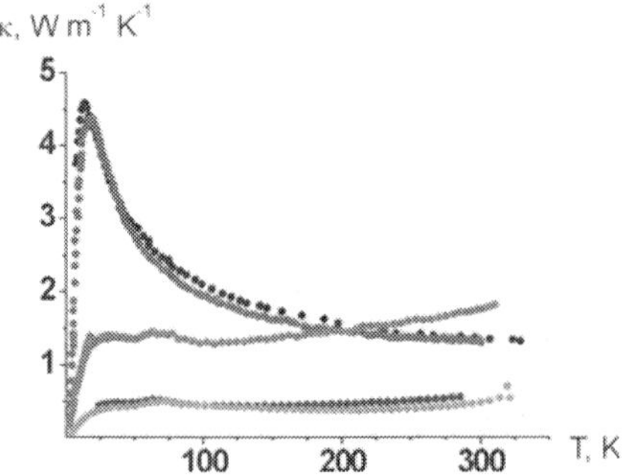

Figure 11. Temperature dependence of thermal conductivity for several clathrates: black, Cs_8Sn_{44}; red, $Ba_8Ga_{16}Ge_{30}$; green, $Sn_{24}P_{19.3}I_8$; blue, $Sn_{24}P_{19.3}I_4Br_4$; cyan, $Sn_{20.5}As_{22}I_8$.

Recently, it was shown that off-center displacement of guest atoms adds significantly to glass-like character of thermal conductivity; in particular, thermal conductivity of $Sr_8Ga_{16}Ge_{30}$ turns from crystalline-like to glass-like upon increasing off-center displacement of guest atoms sitting on 6d site [43].

Clathrate $Sn_{20.5}As_{22}I_8$ displays combination of eightfold cubic superstructure of type-I clathrates with vacancies and mixed occupancies of sites within the framework [31]. In response to structural features, this compound exhibits very low thermal conductivity with room-temperature value slightly over $0.4\ W \times m^{-1} \times K^{-1}$. Increasing complexity of the crystal structure by partial substitution of indium for tin results in further diminishing of thermal conductivity down to $0.36\ W \times m^{-1} \times K^{-1}$ [16] (**Figure 11**).

$Ba_8Au_{16}P_{30}$ provides an example of peculiar orthorhombic superstructure of type-I structure with fivefold increase in the unit volume. In the region of 40–400 K, this compound demonstrates low, almost temperature-independent, thermal conductivity of $0.6\ W \times m^{-1} \times K^{-1}$ [44]. However, this compound is not Zintl phase. It demonstrates metallic-like electrical conductance with resistivity slightly increasing with increased temperature. Therefore, another mechanism of thermal conductivity has substantial contribution to total thermal conductivity, which is electronic thermal conductivity. The latter is proportional to electrical conductivity, σ, according to Wiedemann-Franz equation $\kappa_e = L_0\sigma T$, where $L_0 = 2.45 \times 10^{-8}$ W $\times$ Ohm $\times$ K^{-2} is ideal temperature-independent Lorentz number and T is absolute temperature. It was shown that electronic part of thermal conductivity in $Ba_8Au_{16}P_{30}$ increases from 0.2 W $\times$ m^{-1} $\times$ K^{-1} at 100 K to slightly over 0.5 W $\times$ m^{-1} $\times$ K^{-1} at 400 K, meaning that at the same time lattice part of thermal conductivity decreases in the same interval from about 0.4 to even below 0.2 W $\times$ m^{-1} $\times$ K^{-1} at 400 K, which is the lowest documented value of lattice thermal conductivity for clathrates.

In rare cases, electronic part of thermal conductivity may play dominating role provided clathrate shows properties of good metallic conductor. Type-II clathrate $Na_{24}Si_{136}$ is example, showing dominating contribution of electronic thermal conductivity amounting at 24 W $\times$ m^{-1} $\times$ K^{-1} at room temperature [20].

4.3. Thermoelectric figure-of-merit

Analysis of transport properties of clathrates leads to conclusion that they possess high electrical conductivity up to 6.5×10^4 S/m, high absolute values of Seebeck coefficient up to ± 800 μV/K, and low thermal conductivity down to 0.4 W $\times$ m^{-1} $\times$ K^{-1}. Were these values pertinent to single compound, its thermoelectric figure-of-merit would reach unbelievable values largely exceeding $ZT = 1$ at room temperature, which is benchmark of current state-of-the-art thermoelectric materials. However, due to the significant unavoidable coupling of charge carriers and heat transport, ZT values for clathrate compounds are quite low at room temperature, scarcely surpassing $ZT = 0.1$. Interestingly, the highest room-temperature ZT values are achieved for type-VIII clathrates. For instance, Sb-doped p-type $Ba_8Ga_{16}Sn_{30}$ demonstrates ZT = 0.6 and 300 K, whereas n-type $Ba_8Ga_{16}Sn_{30}$ displays $ZT = 0.5$ at the same temperature [45].

At higher temperature, as both electrical conductivity and Seebeck coefficient tend to grow, whereas thermal conductivity remains essentially constant (combination that is true for the majority of semiconducting clathrates), ZT increases with increasing temperature.

Type-I and type-II clathrates are the most studied species. Their thermoelectric properties have been reported in numerous papers, and, in general, it was shown that type-II clathrates rarely show promising thermoelectric properties due to the metallic properties that evoke low Seebeck coefficients of these compounds [6]. On the contrary, type-I clathrates demonstrate higher ZT with increasing temperature, with $Ba_8Ga_{16}Ge_{30}$ being the record holder displaying $ZT = 1.35$ at 900 K for Czochralski-pulled crystals [46].

Up to date, type-VIII clathrates demonstrate the highest values of ZT at elevated temperatures. These compounds are far less numerous than type-I and type-II counterparts, but, nevertheless, provide good examples of well-studied thermoelectric materials. In mid-temperature region, properly doped $Ba_8Ga_{16}Sn_{30}$ holds the record of the highest ZT. For n-type crystals grown from Ga flux and p-type crystals grown from Sn-flux display the highest thermoelectric efficiency. When properly doped, these compounds exhibit appreciable high values of ZT

reaching 1.45 at 500 K for Cu-doped n-type material and 1.0 at 480 K for Sb-doped n-type material [40, 47]. In general, prominent figures-of-merit can be reached only in the case of doped materials, even if doping is homovalent, but affords appropriate change in electronegativity and host-guest mismatch due to the adjustment of atomic radii. For instance, type-VIII clathrate $Sr_8Ga_{18}Ge_{30}$ does not display intriguing thermoelectric properties; however, partial substitution of Al for Ga affords $ZT = 0.56$ at 800 K [48]. Interestingly, replacement of guest Sr atoms by Eu ones leads to much poorer thermoelectric efficiency despite clearly similar atomic radii of these M^{2+} cations. The reason of this effect is not clear; probably, it is associated with magnetic structure of Eu-based analog. Moreover, this compound was reported to undergo second-order phase transition upon cooling to below 13 K followed by antiferromagnetic ordering that triggers giant magnetocaloric effect with magnetic entropy of $11.3 J \times kg^{-1} \times K^{-1}$ [49]. Another example of increasing ZT upon introduction of magnetic cation is provided by $Ba_{6.9}Ce_{1.1}Au_6Si_{40}$, for which realization of Kondo interactions is believed to enhance the figure-of-merit by factor of 2 [11].

Because type-VIII clathrates demonstrate relatively poor thermal stability, their possible applications are limited by about 800 K, and they cannot be regarded as candidates for high-temperature thermoelectric power generation. Instead, silicon-based type-I and type-III clathrates are being investigated at high temperatures because of their utmost stability against oxidation in air [35]. In particular, cationic clathrates $Si_{31.9}P_{7.1}Te_{7.0}$ (type-I) and $Si_{132}P_{40}Te_{21.5}$ (type-III) are chemically and thermally stable up to 1200 K owing to several nanometers thin layers of phosphorus-doped silicon dioxide, which protects bulk samples from penetrating oxygen, that would lead to oxidation. Reported values of ZT for these Si-based clathrates do not exceed 0.4 (**Figure 12**); however, no attempts to increase the figure-of-merit have been performed so far.

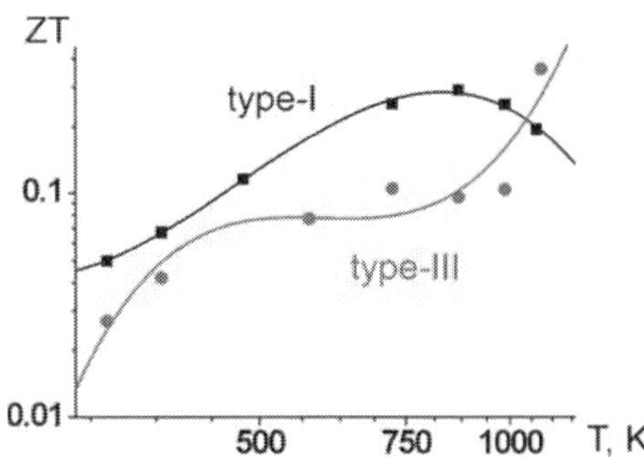

Figure 12. Figure-of-merit as function of temperature in double-logarithmic coordinates for type-I and type-III clathrates in Si-P-Te system.

Summarizing this section, it is worth noting that thermoelectric figure-of-merit for several clathrates of different structure types reaches 1.4–1.45 in the region of 500–800 K. The main tool for achieving such high values lies in the subtle doping of various low-gap clathrates that causes simultaneous increase in electrical conductivity and Seebeck coefficient caused by proper doping accompanied by minor decrease in thermal conductivity caused by slight mass alteration.

5. Conclusion and outlook

Clathrates have been an attractive family of compounds primarily because of their fascinating structures. Within decades, it has become clear that clathrates are unique compounds combining spatial separation of host and guest substructures with very narrow (if any) band gaps, which allows almost independent optimization of charge carriers and thermal transport by tuning charge carriers' concentration and host-guest mismatch. Many chemical elements are known to take part in building clathrates frameworks of several types and serving as guests, making the property tuning plentiful and multifarious. With many instruments in hand, this tuning has already led to discovery of many clathrate compounds with carefully and wisely altered properties. Thermoelectric property optimization has been the central topic of clathrate research and resulted in various intriguing and promising achievements. They include, importantly, thermoelectric figure-of-merit almost reaching $ZT = 1.5$ in mid-T range and discovery of clathrates that demonstrate utmost stability in moist air at higher temperatures.

Nowadays, clathrates, albeit showing promising thermoelectric performance, are still far from commercial production and applications. Waiting for their explorations are elaboration of fabrication methods leading to n- and p-type legs of thermoelectric device, investigation of their compatibility at working temperature (from 500 to 1100 K), and engineering of contact and isolation layers. However, emerging sphere of automotive thermoelectric power generation requires new and more efficient thermoelectric materials capable of working at mid-T range being environmentally benign, whereas new trends in solar energy harvesting call for new thermoelectric materials exhibiting combination of high efficiency with outstanding chemical and thermal stability.

Nevertheless, clathrate research is an ongoing exploration. More than 300 papers are being published per annum in this decade on the topics ranging from the property optimization to uncovering of the underlying physics to elaboration of synthetic pathways and to discovery of new clathrates and related materials. Whereas the former topic works for near-future applications, the latter one is still of basic research. However, it shows that many new clathrates, including those of rare or even new types, are awaiting their discovery and property investigation.

Acknowledgements

This work is supported in part by the Russian Science Foundation under Grant # 16-12-00004.

Author details

Andrei V. Shevelkov

Address all correspondence to: shev@inorg.chem.msu.ru

Department of Chemistry, Lomonosov Moscow State University, Moscow, Russia

References

[1] Ioffe AF. Semiconductor Thermoelements and Thermoelectric Cooling. London: Infosearch Ltd.; 1957.

[2] Nolas GS, Cohn JL, Slack GA, Schjuman SB. Semiconducting Ge clathrates: promising candidates for thermoelectric applications. Appl. Phys. Lett. 1998; 73:178–180.

[3] Snyder GJ, Toberer ES. Complex thermoelectric materials. Nat. Mater. 2008; 7:105–114.

[4] Christensen M, Lock N, Overgaard J, Iversen BB. Crystal structures of thermoelectric n- and p-type $Ba_8Ga_{16}Ge_{30}$ studied by single crystal, multitemperature, neutron diffraction, conventional X-ray diffraction and resonant synchrotron X-ray diffraction. J. Amer. Chem. Soc. 2006; 128:15657–15665.

[5] Shevelkov AV, Kovnir KA, Zaikina JV. Chemistry and Physics of Inverse (Cationic) Clathrates and Tin Anionic Clathrates. In: The Physics and Chemistry of Inorganic Clathrates. Nolas GS, editor. Dorchester: Springer; 2014. pp. 125–167.

[6] Beekman M, Nolas GS. Inorganic clathrate-II materials of group 14: synthetic routes and physical properties. J. Mater. Chem. 2008; 18:842–851.

[7] Slack GA. New Materials and Performance Limits for Thermoelectric Cooling In: CRC Handbook of Thermoelectrics. Rowe DM, editor. Boca Raton, FL: CRC Press; 1995.

[8] Kasper JS, Hagenmuller P, Pouchard M, Cros C. Clathrate structure of silicon Na_8Si_{46} and Na_xSi_{136} (x < 11). Science. 1965; 150:1713–1714.

[9] Kirsanova MA, Olenev AV, Abakumov AM, Bykov MA, Shevelkov AV. Extension of the clathrate family: type-X clatrhate $Ge_{79}P_{29}S_{18}Te_6$. Angew. Chem. Int. Ed. 2011; 50:2371–2374.

[10] Kirsanova MA, Shevelkov AV. Clathrates and semiclathrates of Type-I: crystal structure and superstructures. Z. Kristallogr. 2013; 228:215–227.

[11] Prokofiev A, Sidorenko A, Hradil K, Ikeda M, Svagera R, Waas M, Winkler H, Neumaier K, Paschen S. Thermopower enhancement by encapsulating cerium in clathrate cages. Nat. Mater. 2013; 12:1096–1101.

[12] Kovnir K, Stockert U, Budnyk S, Prots Y, Baitinger M, Paschen S, Shevelkov AV, Grin Y. Introducing a magnetic guest to a tetrel-free clathrate: synthesis, structure, and properties of $Eu_xBa_{8-x}Cu_{16}P_{30}$ ($0 \leq x \leq 1.5$). Inorg. Chem. 2011; 50:10387–10396.

[13] Bentien A, Pacheco V, Paschen S, Grin Y, Steglich F. Transport properties of composition tuned α- and β–$Eu_8Ga_{16-x}Ge_{30+x}$. Phys. Rev. B. 2005; 71:165206.

[14] Shevelkov AV, Kovnir K. Zintl Clathrates. In: Structure and Bonding, Vol. 139 (Zintl Phases). Fässler TF, editor. Berlin, Heidelberg: Springer-Verlag; 2011. pp. 97–142.

[15] Mingos DM. Essential Trends in Inorganic Chemistry. Oxford: Oxford University Press; 1998.

[16] Kelm EA, Olenev AV, Bykov MA, Sobolev AV, Presniakov IA, Kulbachinskii VA, Kytin VG, Shevelkov AV. Synthesis, crystal structure, and thermoelectric properties of clathrates in the Sn-In-As-I system. Z. Anorg. Allg. Chem. 2011; 637:2059–2067.

[17] Shatruk MM, Kovnir KA, Shevelkov AV, Presniakov IA, Popovkin BA. First tin pnictidehalides $Sn_{24}P_{19.3}I_8$ and $Sn_{24}As_{19.3}I_8$: synthesis and the clathrate-I type of the crystal structure. Inorg. Chem. 1999; 38:3455–3457.

[18] Yashina LV, Volykhov AA, Neudachina VS, Aleksandrova NV, Reshetova LN, Tamm ME, Pérez-Dieste V, Escudero C, Vyalikh DV, Shevelkov AV. Experimental and computational insight into the chemical bonding and electronic structure of clathrate compounds in the Sn–In–As–I system. Inorg. Chem. 2015; 54:11542–11549.

[19] Beekman M, Sebastian CP, Grin Yu, Nolas GS. Synthesis, crystal structure, and transport properties of $Na_{22}Si_{136}$. J. Electron. Mater. 2009; 38:1136–1141.

[20] Beekman M, Schnelle W, Borrmann H, Baitinger M, Grin Y, Nolas GS. Intrinsic electrical and thermal properties from single crystals of $Na_{24}Si_{136}$. Phys. Rev. Lett. 2010; 104:018301.

[21] Zaikina JV, Kovnir KA, Haarmann F, Schnelle W, Burkhardt U, Borrmann H, Schwarz U, Grin Y, Shevelkov AV. The first silicon-based cationic clathrate III with high thermal stability: $Si_{172-x}P_xTe_y$ ($x=2y$, $y>20$). Chem. Eur. J. 2008; 14:5414–5422.

[22] Kovnir KA, Sobolev AV, Presniakov IA, Lebedev OI, Van Tendeloo G, Schnelle W, Grin Y, Shevelkov AV. $Sn_{19.3}Cu_{4.7}As_{22}I_8$: a new clathrate-I compound with transition-metal atoms in the cationic framework. Inorg. Chem. 2005; 44:8786–8793.

[23] Kovnir KA, Shatruk MM, Reshetova LN, Presniakov IA, Dikarev EV, Baitinger M, Haarmann F, Schnelle W, Baenitz M, Grin Y, Shevelkov AV. Novel compounds $Sn_{20}Zn_4P_{22-v}I_8$ ($v=1.2$), $Sn_{17}Zn_7P_{22}I_8$, and $Sn_{17}Zn_7P_{22}Br_8$: synthesis, properties, and special features of their clathrate-like crystal structures. Solid State Sci. 2005; 7:957–968.

[24] Condron CL, Martin J, Nolas GS, Piccoli PMB, Schultz AJ, Kauzlarich SM. Structure and thermoelectric characterization of $Ba_8Al_{14}Si_{31}$. Inorg. Chem. 2006; 45:9381–9386.

[25] Avila MA, Suekuni K, Umeo K, Fukuoka H, Yamanaka S, Takabatake T. Glasslike versus crystalline thermal conductivity in carrier-tuned $Ba_8Ga_{16}X_{30}$ clathrates (X=Ge,Sn). Phys. Rev. B. 2006; 74:125109.

[26] Zaikina JV, Schnelle W, Kovnir KA, Olenev AV, Grin Y, Shevelkov AV. Crystal structure, thermoelectric and magnetic properties of the type-I clathrate solid solutions $Sn_{24}P_{19.3(2)}Br_xI_{8-x}$ ($0 \leq x \leq 8$) and $Sn_{24}P_{19.3(2)}Cl_yI_{8-y}$ ($y \leq 0.8$). Solid State Sci. 2007; 9:664–671.

[27] Dumont-Botto E, Bourbon C, Patoux S, Rozier P, Dolle M. Synthesis by spark plasma sintering: a new way to obtain electrode materials for lithium ion batteries. J. Power Sources. 2011; 196:2274–2278.

[28] Jaussaud N, Toulemonde P, Pouchard M, San Miguel A, Gravereau P, Pechev S, Goglio G, Cros C. High pressure synthesis and crystal structure of two forms of a new tellurium-silicon clathrate related to the classical type I. Solid State Sci. 2004; 6:401–404.

[29] Kirsanova MA, Mori T, Maruyama S, Matveeva M, Batuk D, Abakumov AM, Gerasimenko AV, Olenev AV, Grin Y, Shevelkov AV. Synthesis, structure, and transport properties of type-I derived clathrate $Ge_{46-x}P_xSe_{8-y}$ (x = 15.4(1); y = 0–2.65) with diverse host-guest bonding. Inorg. Chem. 2013; 52:577–588.

[30] Zaikina JV, Kovnir KA, Burkhardt U, Schnelle W, Haarmann F, Schwarz U, Grin Y, Shevelkov AV. Cationic clathrate I $Si_{46-x}P_xTe_y$ ($6.6(1) \leq y \leq 7.5(1)$, $x \leq 2y$): crystal structure, homogeneity range, and physical properties. Inorg. Chem. 2009; 48:3720–3730.

[31] Zaikina JV, Kovnir KA, Sobolev AV, Presniakov IA, Prots Y, Baitinger M, Schnelle W, Olenev AV, Lebedev OI, Van Tendeloo G, Grin Y, Shevelkov AV. $Sn_{20.53.5}As_{22}I_8$: a largely disordered cationic clathrate with a new type of superstructure and abnormally low thermal conductivity. Chem. Eur. J. 2007; 13:5090–5099.

[32] Kishimoto K, Koyanagi T, Akai K, Matsuura M. Synthesis and thermoelectric properties of type-I clathrate compounds $Si_{46-x}P_xTe_8$. Jap. J. Appl. Phys. 2007; 46:L746–L748.

[33] Hayashi M, Kishimoto K, Akai K, Asada H, Kishio KT, Koyanagi K. Preparation and thermoelectric properties of sintered n-type $K_8M_8Sn_{38}$ (M = Al, Ga and In) with the type-I clathrate structure. J. Phys. D Appl. Phys. 2012; 45:455308.

[34] Yakimchuk AV, Zaikina JV, Reshetova LN, Ryabova LI, Khokhlov DR, Shevelkov AV. Impedance of $Sn_{24}P_{19.3}Br_xI_{8-x}$ semiconducting clathrates. Low Temp. Phys. 2007; 33:276–279.

[35] Zaikina JV, Mori T, Kovnir K, Teschner D, Senyshyn A, Schwarz U, Grin Y, Shevelkov AV. Bulk and surface structure and high-temperature thermoelectric properties of inverse clathrate-III in the Si-P-Te system. Chem. Eur. J. 2010; 16:12582–12589.

[36] Fukuoka H, Kiyoto J, Yamanaka S. Superconductivity of metal deficient silicon clathrate compounds, Ba8-xSi46 ($0 < x < 1.4$). Inorg. Chem. 2004; 42:2933–2937.

[37] Yuan HQ, Grosche FM, Carrillo-Cabrera W, Paschen S, Sparn G, Baenitz M, Grin Y, Steglich F. High-pressure studies on a new superconducting clathrate: Ba_6Ge_{25}. J. Phys. Condens. Matter. 2002; 14:11249.

[38] Bobev S, Sevov SC. Synthesis and characterization of stable stoichiometric clathrates of silicon and germanium: $Cs_8Na_{16}Si_{136}$ and $Cs_8Na_{16}Ge_{136}$. J. Amer. Chem. Soc. 1999; 121:3795–3796.

[39] Kishimoto K, Arimura S, Koyanagi T. Preparation and thermoelectric properties of sintered iodine-containing clathrate compounds $Ge_{38}Sb_8I_8$ and $Sn_{38}Sb_8I_8$. Appl. Phys. Lett. 2006; 88:222115.

[40] Saiga Y, Dua B, Deng SK, Kajisa K, Takabatake T. Thermoelectric properties of type-VIII clathrate $Ba_8Ga_{16}Sn_{30}$ doped with Cu. J. Alloys Compd. 2012; 537:303–307.

[41] Novikov VV, Matovnikov AV, Avdashchenko DV, Mitroshenkov NV, Dikarev EV, Takamizawa S, Kirsanova MA, Shevelkov AV. Low-temperature structure and lattice dynamics of the thermoelectric clathrate $Sn_{24}P_{19.3}I_8$. J. Alloys Compd. 2012; 520:174–179.

[42] Kovnir KA, Zaikina JV, Reshetova LN, Olenev AV, Dikarev EV, Shevelkov AV. Unusually high chemical compressibility of normally rigid type-I clathrate framework: synthesis and structural study of $Sn_{24}P_{19.3}Br_xI_{8-x}$ solid solution, the prospective thermoelectric material. Inorg. Chem. 2004; 43:3230–3236.

[43] Christensen S, Schmøkel MS, Borup KA, Madsen GKH, McIntyre GJ, Capelli SC, Christensen M, Iversen BB. "Glass-like" thermal conductivity gradually induced in thermoelectric Sr8Ga16Ge30 clathrate by off-centered guest atoms. J. Appl. Phys. 2016; 119:185102.

[44] Fulmer J, Lebedev OI, Roddatis VV, Kaseman DC, Sen S, Dolyniuk JA, Lee K, Olenev AV, Kovnir K. Clathrate $Ba_8Au_{16}P_{30}$: the "Gold Standard" for lattice thermal conductivity. J. Am. Chem. Soc. 2013; 135:12313–12323.

[45] Saiga Y, Suekuni K, Deng SK, Yamamoto T, Kono Y, Ohya N, Takabatake T. Optimization of thermoelectric properties of type-VIII clathrate $Ba_8Ga_{16}Sn_{30}$ by carrier tuning. J. Alloys Compd. 2010; 507:1–5.

[46] Saramat A, Svensson G, Palmqvist AEC, Stiewe C, Mueller E, Platzek D, Williams SGK, Rowe DM, Bryan JD, Stucky GD. Large thermoelectric figure of merit at high temperature in Czochralski-grown clathrate Ba8Ga16Ge30J. Appl. Phys. 2006; 99:023708.

[47] Kono Y, Ohya N, Saiga Y, Suekuni K, Takabatake T, Akai K, Yamamoto S. Carrier doping in the type VIII clathrate $Ba_8Ga_{16}Sn_{30}$. J. Electr. Mat. 2011; 40:845–850.

[48] Sasaki Y, Kishimoto K, Koyanagi T, Asada H, Akai K. Synthesis and thermoelectric properties of type-VIII germanium clathrates $Sr_8Al_xGa_yGe_{46-x-y}$. J. Appl. Phys. 2009; 105:073702.

[49] Phan MH, Woods GT, Chaturvedi A, Stefanoski S, Nolas GS, Srikanth H. Long-range ferromagnetism and giant magnetocaloric effect in type VIII $Eu_8Ga_{16}Ge_{30}$ clathrates. Appl. Phys. Lett. 2008; 93:252505.

Chapter 11

Efficient Thermoelectric Materials Based on Solid Solutions of Mg$_2$X Compounds (X = Si, Ge, Sn)

Vladimir K. Zaitsev, Grigoriy N. Isachenko and
Alexander T. Burkov

Additional information is available at the end of the chapter

http://dx.doi.org/10.5772/65864

Abstract

The silicides have obvious attractive characteristics that make them promising materials as thermoelectric energy converters. The constituting elements are abundant and have low price, many of compounds have good high temperature stability. Therefore, considerable efforts have been made, especially in the past 10 years, in order to develop efficient silicide-based thermoelectric materials. These efforts have culminated in creation of Mg$_2$(Si-Sn) n-type thermoelectric alloys with proven maximum thermoelectric figure of merit ZT of 1.3. This success is based on combination of two approaches to maximize the thermoelectric performance: the band structure engineering and the alloying. In this chapter, we review data on crystal and electronic structure as well as on the thermoelectric properties of Mg$_2$X compounds and their solid solutions.

Keywords: silicides, magnesium silicide, thermoelectricity, figure of merit

1. Introduction

Among the large family of silicon-based compounds, semiconducting silicides have received particular interest as thermoelectric materials because they are potentially cheap and mostly stable materials. Comparatively, low charge carriers' mobility in these semiconductors is compensated by high electron state density, i.e. high effective mass of charge carriers. Therefore, silicides were the main focus of thermoelectric research community since the 1950s [1]. Investigations of these materials were especially active during the past 10 years. The most important results have been achieved for Mg$_2$X (X = Si, Sn, Ge)-based alloys. Based on the Zaitsev et al. [2] work, n-type Mg$_2$(Si-Sn) solid solutions with thermoelectric figure of merit

$ZT = \frac{S^2\sigma}{\kappa}T$ (S is thermopower or Seebeck coefficient; σ is electrical conductivity; κ is thermal conductivity, and T is absolute temperature) up to 1.3 were obtained by several research groups [3–7]. Many researchers believe that there is possibility for further improvement. Now considerable efforts are directed to the development of a matching p-type material.

Already in the 1960s, it was shown that Mg_2X compounds (X = Si, Ge, Sn) and their solid solutions are promising compounds for thermoelectric energy conversion [8, 9]. Very high values of ZT are reported in Refs. [10, 11]. However, later the interest to these compounds has been almost vanished until the last decade. A new wave of research activity on Mg_2X compounds was initiated by information about high figure of merit achieved in Mg_2Si-Mg_2Sn solid solutions and growing interest to environment-friendly materials for thermoelectric energy conversion.

The maximum conversion efficiency of thermoelectric generator η is determined by dimensionless figure of merit ZT [12]:

$$\eta = \frac{T_H - T_C}{T_H} \frac{\sqrt{\overline{ZT} + 1} - 1}{\sqrt{\overline{ZT} + 1} + \dfrac{T_C}{T_H}}, \tag{1}$$

where T_H and T_C are temperatures at hot and at cold junctions of thermoelectric generator thermopile. $\overline{ZT}$ is the dimensionless figure of merit, averaged over working temperature range $\Delta T = T_H - T_C$. The semiconductor physics theory gives the following estimate for parameter Z [13]:

$$Z_{\max} = \frac{(m^*)^{\frac{3}{2}}\mu}{\kappa_{\text{lat}}}, \tag{2}$$

where m^* is the effective mass of electron state density (DOS), μ is the free charge carriers' mobility, and κ_{lat} is the lattice thermal conductivity. One can see that a good thermoelectric material will have heavy effective mass, high charge carriers' mobility, and low lattice thermal conductivity. However, in fact coefficients determining Z are strongly interdependent. Thermoelectric materials with high DOS typically have low mobility. Introducing a disorder to suppress the thermal conductivity usually leads to decrease of charge carriers' mobility. This is the reason of slow progress in the development of efficient thermoelectrics.

The unique characteristics of an electronic band structure of Mg_2X compounds make possible to explore the combination of two approaches to optimize the thermoelectric performance of such materials: the band structure engineering and the alloying [2, 5]. The combination allows to simultaneously maximize electronic parameters, characterized by power factor $S^2\sigma$, and to minimize lattice thermal conductivity, yielding high values of parameter Z.

In this chapter, we summarize the present state of the knowledge on the crystal and electronic structure of Mg$_2$X compounds and their alloys, and review experimental data on thermoelectric properties of compounds.

2. Properties of Mg$_2$X compounds

2.1. Physical properties and crystal structure of Mg$_2$X

The basic properties of Mg$_2$X compounds are shown in **Table 1**. Melting temperature and energy gap, E_g, are typical for so-called middle temperature range thermoelectrics (600 < T < 1200 K). The materials, especially Mg$_2$Si, have very low density, d. Therefore, the ratio $\frac{ZT}{d}$ for Mg$_2$Si is the highest among commercial thermoelectrics. This is advantage for applications, where weight is a significant factor. High ZT in Mg$_2$Si can be related to high electron (μ_n) and low hole (μ_p) mobility. Their values at room temperature are shown in **Table 1**. Mg$_2$Ge has the highest electron mobility, but the electron to hole mobility ratio is lower in comparison to that for Mg$_2$Si. Mg$_2$Sn has the highest effective mass of DOS. It should be noted that among Mg$_2$X compounds, Mg$_2$Sn has the highest hole mobility with small difference between election and hole mobility. This suggests that p-type thermoelectrics based on Mg$_2$X alloys should contain a large fraction of Mg$_2$Sn.

Compo-und	Melting temperature, T_m	Lattice constant, a	Density, d	Bandgap E_g (0 K)	Mobility (300 K)		Lattice thermal conductivity κ_{lat} (300K)
					μ_n	μ_p	
	(K)	(Å)	(g cm^{-3})	(eV)	(cm^2 V$^{-1}\cdot$s^{-1})		(W m^{-1} K^{-1})
Mg$_2$Si	1375 [14]	6.338 [15]	1.88 [18]	0.77 [18]	405	65	7.9 [19]
Mg$_2$Ge	1388 [14]	6.3849 [16]	3.09 [18]	0.74 [18]	530	110	6.6 [19]
Mg$_2$Sn	1051 [14]	6.765 [17]	3.59 [18]	0.35 [18]	320	260	5.9 [19]

Table 1. Some parameters of Mg$_2$X compounds.

Phase diagrams for the systems of magnesium and carbon groups of elements are well known [14]. Each phase diagram contains only one chemical compound of Mg$_2$X-type and two eutectic points. Mg$_2$X compounds crystallize with cubic, CaF$_2$-type, structure (space group Fm3m) [16, 20]. In Mg$_2$X structure, the fluorine atom is replaced by the magnesium atom and the calcium atom is replaced by X atom (**Figure 1**). Each atom of the X group is surrounded by eight magnesium atoms in a regular cube. The bond in all these compounds is covalent [18]. Lattice parameters of compounds are presented in **Table 1**.

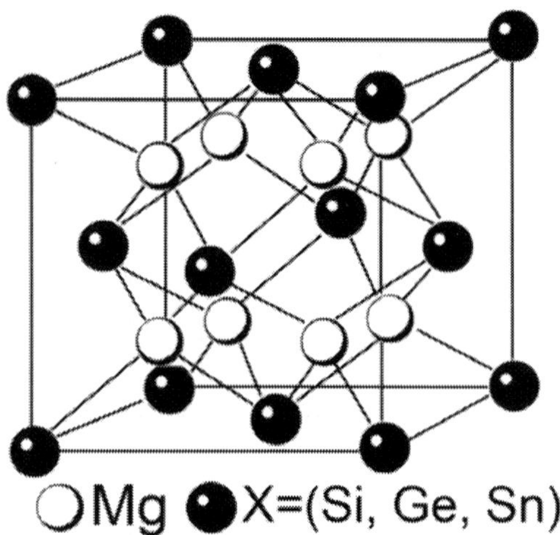

Figure 1. Mg$_2$X crystal structure.

2.2. Energy spectra of current carriers in Mg$_2$X

Fundamental parameters of the electronic structure of the Mg$_2$X compound can be obtained from optical and electronic transport property measurements on high quality single crystals. Comprehensive review of transport properties and electronic energy structure for Mg$_2$X compounds is given in Ref. [21].

Based on the analysis of optical and electronic transport data, supplemented by results of band structure calculations, the band structure of Mg$_2$X compounds was proposed [16, 22–27]. **Figure 2** shows schematically the most important characteristic of this band structure near to Fermi energy. The valence band of the compounds is similar to the valence band of Si and Ge. It consists of two degenerate bands (V_1, V_2) with different effective masses ($m^*_{V_1}$ and $m^*_{V_2}$) and a third band (V_3) split below the two other bands by gap E_2 due to spin-orbital interaction. The maximums of valence bands are located at Γ-point of a Brillouin zone. The conduction band consists of two subbands C_L and C_H of light ($m^*_{C_L}$) and heavy ($m^*_{C_H}$) electrons with their minimums located at X-point of a Brillouin zone. These subbands are separated by energy gap $E_1 = E_{C_H}(\text{X}) - E_{C_L}(\text{X})$. There is a third conduction band C with minimum at Γ-point, separated by gap E_0 from the top of valence bands. However, E_0 is considerably larger than indirect band gap E_g; therefore, C band has no direct effect on thermoelectric properties of Mg$_2$X compounds. Theoretical calculations confirmed this structure except for the fact that these calculations did not take into account spin-orbital interaction [16, 25–27].

Location of conduction band minimum at the X-point is favorable for thermoelectric performance of a material. In this case, the effective mass of DOS is six times heavier than inertial mass.

Because of that n-type Mg$_2$Si has high electrical conductivity and high thermopower. On the other hand, the valence band structure does not have such favorable thermoelectric features. The maximum of the valance band is at Γ-point; thus, the inertial mass and effective mass of DOS are not different. The valence band has three subbands, one of which split due to spin-orbital interaction [28]. This splitting extends with the increasing atom mass.

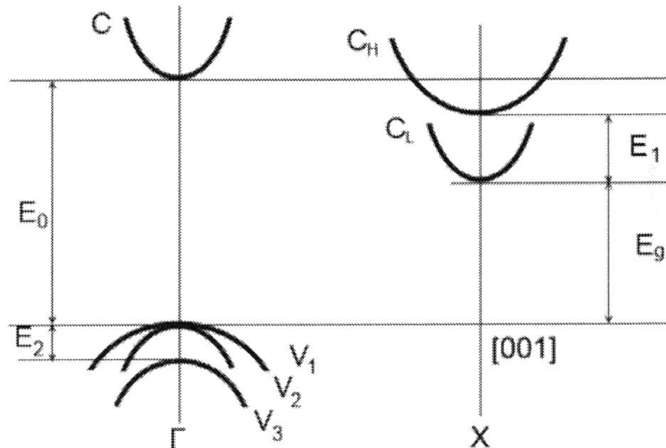

Figure 2. Schematic band structure of Mg$_2$X. For Mg$_2$Si and Mg$_2$Ge light electron band (C$_L$) lies below heavy electron band (C$_H$), as shown in the picture. In the case of Mg$_2$Sn the heavy electron band C$_H$ is below the light electron band C$_L$.

Parameters of band structure for Mg$_2$Sn, Mg$_2$Ge, and Mg$_2$Si are presented in **Table 2**. The values of indirect band gap E_g determined from electrical conductivity temperature dependence (E_g^T) and from optical data (E_g^0) are in good agreement. E_1 and E_2 are gaps between the conduction and valence subbands, respectively. E_0 is a direct band gap value. According to definition of E_1, it is positive for Mg$_2$Ge and Mg$_2$Si, where the low-lying conduction band has smaller effective mass. The opposite situation is in Mg$_2$Sn, where E_1 is negative. The effective mass of conduction band (m_C^*) is shown for a low-lying subband, i.e. $m_{C_L}^*$ for Mg$_2$Ge and Mg$_2$Si, while

$m_{C_H}^*$ —in the case of Mg$_2$Sn. The temperature coefficient of a band gap is shown in the last column.

Compound	E_g^T, eV	E_g^0, eV	E_1, eV	E_0, eV	E_2, eV	$\dfrac{m_C^*}{m_0}$	$\dfrac{m_{V_1V_2}^*}{m_0}$	$\dfrac{dE_g}{dT} \times 10^4$, (eV K^{-1})
Mg$_2$Sn	0.36 [18]	0.35 [23]	−0.16 [23]		0.48 [22]	1.2 [23]	1.3 [23]	−3.2 [23]
Mg$_2$Ge	0.74 [18]	0.57 [24]	0.58 [24]	1.80 [29]	0.20 [22]	0.18 [30]	0.31 [30]	−1.8 [24]
Mg$_2$Si	0.77 [18]	0.78 [22]	0.4 [22]		0.03 [22]	0.45 [31]	0.9 [31]	−6 [18]

Table 2. Parameters of Mg$_2$X band structure (presented in **Figure 3**).

2.3. Thermal conductivity of Mg_2X compounds

Figure 3 shows temperature dependencies of reciprocal thermal conductivity of pure Mg_2X compounds. One can see that reciprocal thermal conductivity can be described satisfactory by a linear law and residual reciprocal thermal conductivity is zero within experimental uncertainty. The most probable reason for observed difference in data of different authors is dependence of reciprocal thermal conductivity on deviation from stoichiometry.

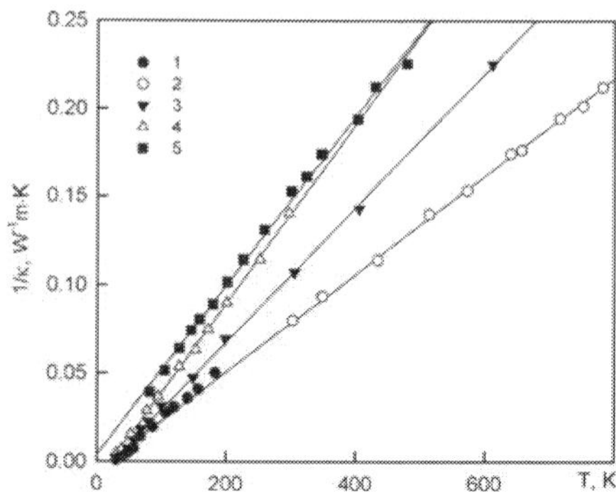

Figure 3. Temperature dependence of reciprocal thermal conductivity of pure Mg_2X compounds: 1, 2—Mg_2Si [19, 21]; 3 —Mg_2Ge[19]; 4, 5—Mg_2Sn[19, 32].

3. Solid solutions of Mg_2X compounds

3.1. Mg_2X-based solid solutions

As one can see from **Table 1**, Mg_2X compounds have relatively high thermal conductivity, which should be decreased to make these compounds practically useful thermoelectrics. However, decreasing in thermal conductivity should not lead to a considerable decrease of charge carriers' mobility. Thermal conductivity can be reduced by selective scattering of phonons and electrons by point defects through forming solid solutions (alloys) between these isostructural compounds.

There is a continuous series of solid solutions in the system Mg_2Si-Mg_2Ge [9]. Phase diagrams of Mg_2Si-Mg_2Sn and Mg_2Ge-Mg_2Sn have wide peritectic region in the middle composition range [15, 33]. Until recently, it was commonly accepted that solid solutions exist only at compositions $x < 0.4$ and $x > 0.6$ for the $Mg_2Si_{1-x}Sn_x$ system, and at $x < 0.3$, $x > 0.5$ for the $Mg_2Ge_{1-x}Sn_x$ system. However, it has been demonstrated that solid solutions of any composition can be produced avoiding liquid stage by mechanical alloying.

Figure 4 shows dependences of lattice parameter (a) on alloys composition (x). In Ref. [33], it was shown that $a(x)$ dependence follows to Vegard's law for the whole composition range of the Mg_2Si-Mg_2Sn system.

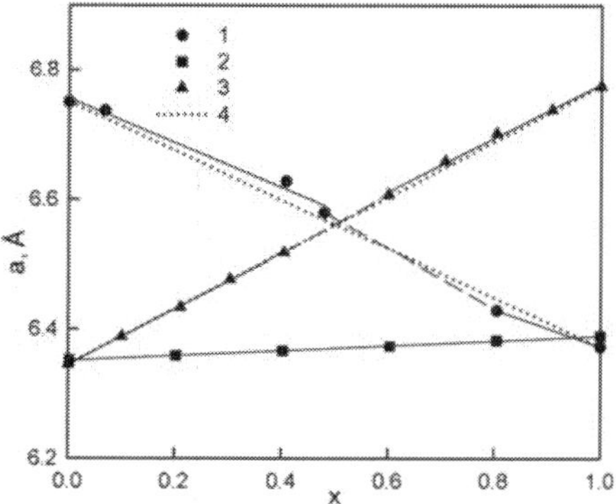

Figure 4. Lattice parameter a vs. solid solution composition x dependences: $1-$Mg$_2$Sn$_{1-x}$Ge$_x$ [15]; $2-$Mg$_2$Si$_{1-x}$Ge$_x$ [9]; $3-$Mg$_2$Si$_{1-x}$Sn$_x$ [33]; $4-$Vegard's law.

3.2. Thermal conductivity of Mg$_2$X solid solutions

Figure 5 shows the experimental values of lattice thermal conductivity of Mg$_2$Si$_{1-x}$Sn$_x$ [34], Mg$_2$Ge$_{1-x}$Sn$_x$[34], Mg$_2$Si$_{1-x}$Ge$_x$ [35] alloys, and the results of calculations according to procedure, described in Refs. [34, 36]. In alloys, thermal conductivity sharply decreases with the addition of a small amount of second compound, while it has a weak dependence on the composition in the middle composition range $0.2 < x < 0.8$. One can see that the lowest thermal conductivity can be achieved in the system Mg$_2$Si$_{1-x}$Sn$_x$ due to the maximum mass difference between the compounds. Consequently, this system is the most favorable from the point of view of thermoelectric energy conversion.

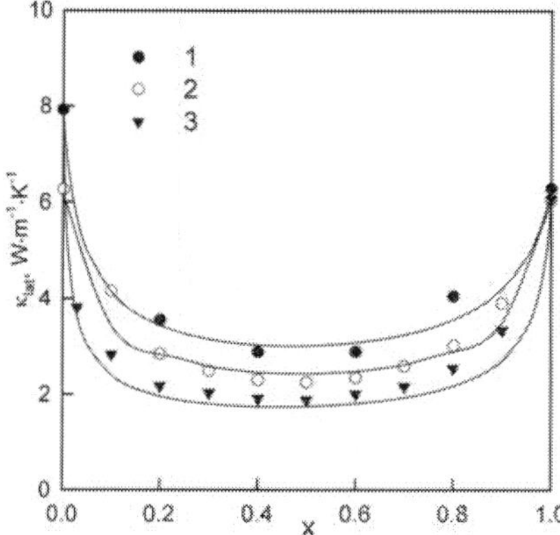

Figure 5. Lattice thermal conductivity of alloys at room temperature: 1–Mg$_2$Si$_{1-x}$Ge$_x$, 2–Mg$_2$Ge$_{1-x}$Sn$_x$, 3–Mg$_2$Si$_{1-x}$Sn$_x$. Symbols: experiment 1–[35]; 2, 3–[34]; lines–calculation [35].

3.3. Dependency of energy gap on solid solution composition

Besides lower thermal conductivity, the solid solutions of Mg_2X provide opportunity to further enhancement of thermoelectric properties by electronic band structure engineering. **Figure 6** shows dependences of energy gap of Mg_2X alloys vs. composition [9, 15, 37–40]. From the study of the Mg_2Si-Mg_2Ge system [9]—one can conclude that energy gap is practically independent of alloy composition.

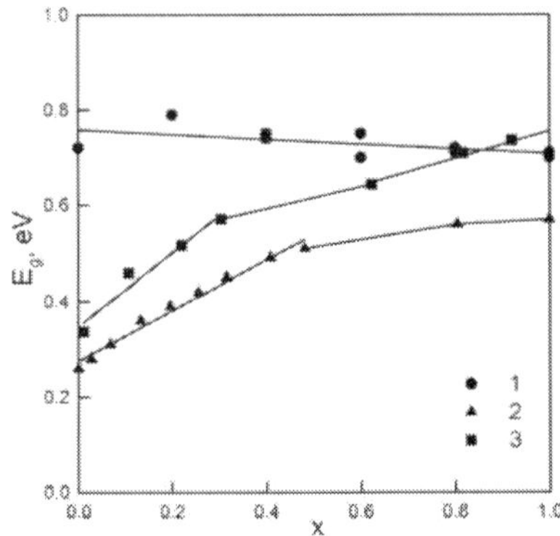

Figure 6. Energy gap E_g of alloys as a function of composition x. $1-Mg_2Si_{1-x}Ge_x$ [9]; $2-Mg_2Sn_{1-x}Ge_x$ [15]; $3-Mg_2Si_{1-x}Sn_x$ [37].

The situation is very different in other two alloy systems. The Mg_2Ge-Mg_2Sn system was studied by Busch et al. [15]. Notwithstanding the very narrow band gaps of Mg_2Sn and Mg_2Ge, it can be concluded that band gap dependence on solid solution composition is nonlinear. Our results of the band gap study of $Mg_2Ge_{1-x}Sn_x$ solid solutions confirm this behavior. The situation is the same in the Mg_2Si-Mg_2Sn system [37–40]. Zaitsev et al. [37] proposed that there is band inversion in $Mg_2Si_{1-x}Sn_x$ solid solutions. It means that in Mg_2Si and Mg_2Sn conduction bands C_L and C_H change their positions. Band inversion hypothesis was confirmed theoretically. Fedorov et al. [38] showed that the lowest conduction bands of Mg_2Si and Mg_2Ge are formed by Si or Ge states, whereas that of Mg_2Sn was formed by Mg states. **Figure 7** shows schematically dependence of relative positions of heavy electrons and light electrons conduction bands, as well as, the top of the valence band on the $Mg_2Si_{1-x}Sn_x$ alloy composition. The lower panel explains the occurrence of kink on dependence of the band gap on composition at the inversion point. According to calculation in Ref. [5], composition dependence on light electron subband position in $Mg_2Si_{1-x}Sn_x$ is nonlinear, while corresponding dependence of position of heavy electron subband is linear. Therefore, actual dependence of band gap on composition is more complex. Nevertheless, the scheme shown in **Figure 7** illustrates correctly the essential physics. At the composition value, corresponding to band inversion point, the minima of heavy and light electrons subbands have equal energy. From the point of view of thermoelectricity, such situation is favorable, because DOS increases without decreasing in electron mobility. Such degeneration of subbands occurs at certain composition and certain temperature, so this favorable situation is very limited.

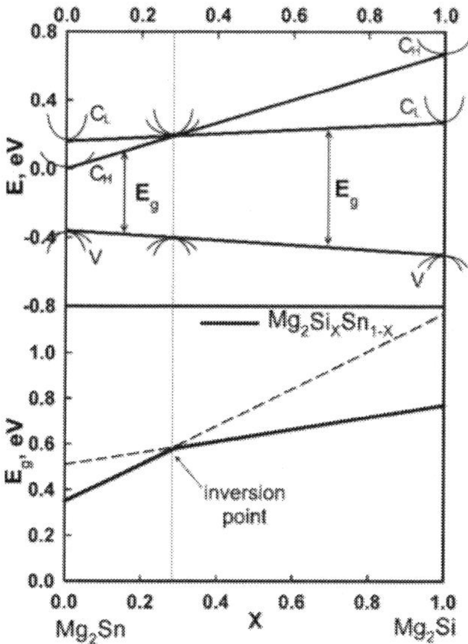

Figure 7. Schematic dependence of relative positions of heavy and light electron conduction subbands as well as the top of the valence band on the composition of the Mg$_2$Si$_{1-x}$Sn$_x$ alloy (the upper panel). Explanation of origin of kink on band gap dependence (the lower panel).

Calculations show that the most favorable situation realizes when heavy electrons subband lays higher [41]. Another advantage of this situation is the absence of interband scattering [40].

3.4. Synthesis technology and doping

There are several methods to produce Mg$_2$X compounds. One of them is direct co-melting [8, 10]. This method has some limitation due to the large difference in melting temperature of components and high magnesium vapor pressure. It is necessary to pay a special attention to magnesium losses due to evaporation and segregation of the components (especially for Mg$_2$Sn).

Another way to produce these compounds is through a solid-state reaction. Mg$_2$X compounds have negative heat of formation, i.e. the formation reaction is exothermic [42–45]. However, oxide films on Mg particles prevent the reaction. Therefore, it is necessary to pay attention to the purity of components and avoid oxidization during mixing. Alternative manufacturing route has been developed for magnesium silicide derivatives [46]. Elemental powders were mixed in stoichiometric proportions, cold pressed into cylindrical preforms and heated in an oxygen-free environment to initiate the exothermic reaction. Reaction products were additionally heat treated for homogenization. Dense sinters can be produced by hot uniaxial pressing of the obtained powders under moderate temperature and pressure conditions.

Several advantages were identified in the proposed technology: relatively short time of synthesis, possibility of *in-situ* or *ex-situ* doping and grain size control.

Single crystals of Mg_2X compounds can be easily produced by any methods of directed crystallization.

It is hard to produce homogeneous solid solutions via a liquid phase through co-melting of the components. One of the problems is related to large difference in masses of magnesium, silicon, germanium, and tin atoms. Without stirring, segregation by specific weight occurs. The other problem relates to phase diagrams of solid solutions, which have large difference in liquidus and solidus curves in a wide range of compositions [33]. Therefore, compositional segregation occurs during crystallization as well. In order to homogenize alloys, a long-term annealing is necessary. The necessary homogenization annealing time is determined by diffusion processes, which depend on temperature and crystallite size. Temperature cannot be high due to magnesium evaporation. In order to shorten the annealing time, hot pressing can be utilized. Ingots of alloys are crushed into powder and then powder is pressed in a vacuum. The finer grain size the less time for homogenization is needed [47]. Annealing is not required for the samples produced from nanosize particles.

Recently, mechanical alloying in the ball mill followed by spark plasma sintering (SPS) has become the most popular preparation technique for this solid solution.

As mentioned above, the figure of merit Z is function of free charge carriers' concentration. Optimal concentration yielding maximum ZT value is equal to about 10^{19} to 10^{20} cm^{-3}. Theoretical and experimental investigations of a doping impurity effect in Mg_2Si for a wide range of impurity elements (B, Al, N, P, Sb, Bi, Cu, Ag, Au) were made by Tani and Kido [48, 49]. As, P, Sb, Bi, Al, and N were suggested as n-type dopants whereas Ga is suggested as p-type dopant. For In, Ag, Cu, and Au, the doping effect, i.e. a resulting conduction type, depends on the site in lattice where a doping atom will occupy. Actually, Ag-doped samples show p-type of conductivity. In Mg_2Sn-rich solid solutions, impurities Na, Li, Ga, Ag and at low concentration Al and In act as p-type dopants [50–53].

4. Thermoelectric properties of Mg_2X and its alloys

4.1. Mg_2X composites

As it was already mentioned that the new wave of research activity on Mg_2X-based thermoelectrics was initiated by work of Zaitsev et al., who demonstrated stable Mg_2(Si-Sn) alloys with a maximum ZT value of about 1.1 at 800 K [2]. A systematic study of Mg_2Si thermoelectric properties was performed by Tani et.al. They found $(ZT)_{max}$ for Mg_2Si, doped with 2 at% of bismuth of about 0.86 at 820 K with samples, fabricated by SPS [54]. However, such large $(ZT)_{max}$ has been not supported by independent researchers [55]. Sb and P-doped Mg_2Si was investigated from 300 up to 900 K with the Sb content of up to 2% [56] and the P content up to 3% [57]. The samples were prepared from high-purity powder components by SPS. The maximum $ZT = 0.56$ was obtained at 860 K for sample with 2% of Sb due to the lowest thermal conductivity.

Samples of Mg₂Si, undoped and doped with Bi and Ag, were grown by a vertical Bridgman method [58, 59]. The n-type Bi-doped samples have a maximum ZT of 0.65 at 840 K, while Ag-doped samples are of p-type (below 650 K) and show a maximum ZT of 0.1 at 570 K.

A comprehensive study of a doping mechanism, i.e. location of dopants in Mg₂Si was undertaken by Farahi et al. [60]. Samples of Sb- and Bi-doped Mg₂Si were prepared via two-stage annealing of powder mixtures of individual components at 823 K for 3.5 days and at 873 K for 5 days, followed by hot pressing. It was shown that part of dopants replaces Si, while the rest forms, Mg₃Sb₂ and Mg₃Bi₂, found between the grains of doped Mg₂Si particles. As doping of Sb and Bi only partly led to Si substitution, experimentally determined charge carriers' concentration was lower than originally expected.

Using a technique of incremental milling, phase pure Mg₂Si was produced within a few hours with negligible oxygen contamination [61]. In this technique, to prevent agglomeration of ductile Mg during ball milling, Mg is added to Si + doped mixture by small portions followed a comparatively short milling period until the stoichiometric amount of Mg is attained. More effective Bi doping is achieved with higher mobility values at lower concentrations of dopant compared to previous work. A peak ZT value of about 0.7 is achieved at 775 K using an optimum doping level with only 0.15% of Bi, which is an order of magnitude lower than that mentioned in Ref. [54].

Temperature dependences of the figure of merit ZT of Mg₂Si samples doped with different dopants are shown in **Figure 8**.

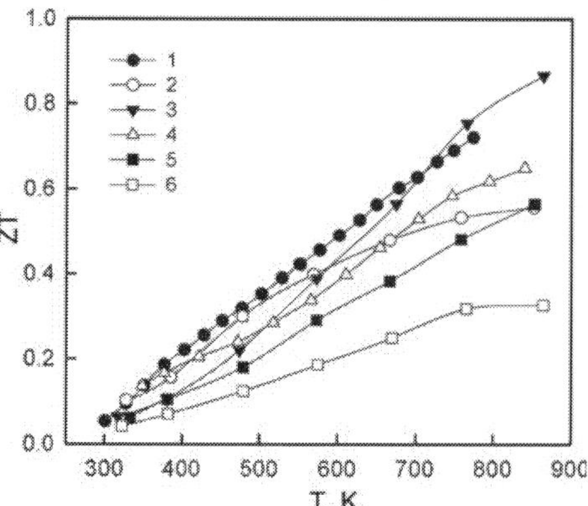

Figure 8. Figure of merit ZT temperature dependences of Mg₂Si. 1—Mg₂Si+0.15% Bi [61]; 2—Mg₂Si+0.5% Sb [58]; 3—Mg₂Si+2% Sb [54]; 4—Mg₂Si+1% Bi [59]; 5—Mg₂Si+2% Sb [56]; 6—Mg₂Si+3% P [57].

4.2. Figure of merit of n-type Mg$_2$X-based solid solutions

Analysis of transport properties and band structure features has shown that the Mg$_2$Si-Mg$_2$Sn system is the most promising for development of efficient n-type thermoelectrics. **Figure 9** shows the effect of high band degeneracy on ZT. Temperature dependences of ZT are shown for n-type Mg$_2$Si$_{0.4}$Sn$_{0.6}$ (left) and Mg$_2$Si$_{0.6}$Sn$_{0.4}$ (right) alloys. C_L and C_H subbands of the conduction band are close to each other in the Mg$_2$Si$_{0.4}$Sn$_{0.6}$ alloy, as the result it has higher ZT at lower temperatures. In Mg$_2$Si$_{0.6}$Sn$_{0.4}$, subbands C_L and C_H are separated by a narrow gap; therefore, at low temperatures C_H subband gives no contribution to electronic transport. Therefore, ZT of Mg$_2$Si$_{0.6}$Sn$_{0.4}$ at low temperatures is smaller in comparison with ZT of Mg$_2$Si$_{0.6}$Sn$_{0.4}$. However, at higher temperatures, the C_H subband in this alloy gives increasing contribution to electrical conductivity and thermopower, which gives rise to enhanced ZT values. Although both solid solutions have high maximum ZT values close to 1.2, the average value $(ZT)_{av}$ of Mg$_2$Si$_{0.4}$Sn$_{0.6}$, in the temperature range of 400–850 K, is higher (about 0.83 and 0.78, respectively). This study revealed the best compositions of n-type solid solutions and allowed reproducible synthesis of thermoelectrics with $ZT_{max} \approx 1.2$ and higher [2, 5]. Comparison of obtained results with the data for the state-of-the-art thermoelectrics revealed that these materials are among the best thermoelectrics of n-type in the temperature range of 600–870 K.

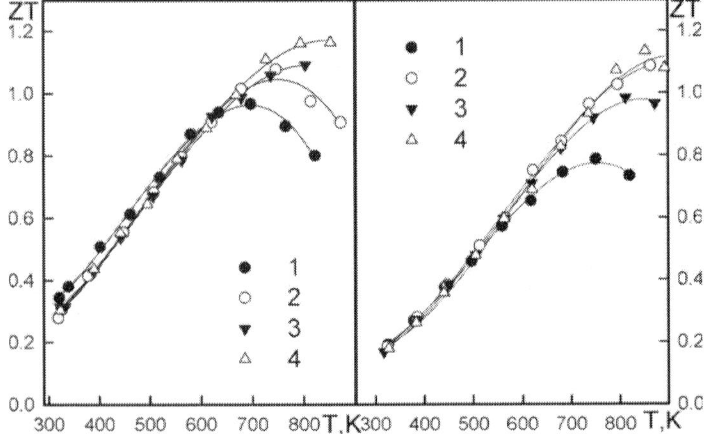

Figure 9. Figure of merit ZT temperature dependences of alloys Mg$_2$Si$_{0.6}$Sn$_{0.4}$ (right) and Mg$_2$Si$_{0.4}$Sn$_{0.6}$ (left). n, 10^{20} cm^{-3}: (right) 1−3.17; 2−3.30; 3−3.83; 4−4.54. n, 10^{20} cm^{-3}: (left) 1−2.31; 2−2.52; 3−2.99; 4−3.10.

Several approaches have been used in order to maximize the figure of merit, including optimization of alloy composition and doping level, various types of nanostructuring. The nanostructuring is currently considered as the most promising and universal approach to enhance the thermoelectric performance. There are a number of technological approaches for producing different kinds of nanostructured materials. Most important among them are nanocrystalline materials, materials with nanoprecipitates of second phase and materials with nanoinclusions of foreign substance. All these approaches were applied with different degree of success to Mg$_2$X compounds and related alloys.

Effects of nanostructuring on Mg$_2$Si were theoretically modeled and systematically analyzed in Ref. [62]. It was shown that nanostructuring limits the energy-dependent phonon mean free path in Mg$_2$Si, which results in significant reduction (50%) in lattice thermal conductivity. However, it was also concluded that nanostructuring in both p-type and n-type Mg$_2$Si increases significantly charge carrier scattering and leads to unfavorable reduction in electrical conductivity. A decrease in charge carriers' mobility of nanostructured Mg$_2$Si strongly affects the power factor, resulting in only minor enhancement in the overall figure of merit. In the case of nanostructured n-type Mg$_2$Si, an optimal doping concentration of 8.1×10^{19} cm^{-3} was estimated for achieving ZT of 0.83 at 850 K, which is less than 10% improvement in comparison with the maximum ZT of bulk Mg$_2$Si. On the other hand, in the case of nanostructured p-type Mg$_2$Si, a maximum ZT of 0.90 at 850 K was predicted, which is nearly 37% improvement over the maximum ZT of bulk Mg$_2$Si. The predicted optimum dopant concentration for p-type Mg$_2$Si was equal to 4.3×10^{20} cm^3. In practice, inherent challenge for p-type Mg$_2$Si is a charge carriers' compensation effect that limits the maximum charge carriers' concentration by value 10^{18} cm^{-3}.

A higher effect of nanostructuring on the efficiency of n-Mg$_2$Si and n-Mg$_2$Si$_{0.8}$Sn$_{0.2}$ alloys was predicted in another theoretical work [63]. It was shown that relatively higher depression of lattice thermal conductivity compared to decrease in electrical conductivity due to grain boundary scattering can lead to 10 and 15% increase of ZT at 850 K in nanostructured materials based on Mg$_2$Si and Mg$_2$Si$_{0.8}$Sn$_{0.2}$, respectively. A nanostructured alloy is more favorable for increase in the figure of merit than bulk Mg$_2$Si.

The presence of nanoinclusions is considered as an alternative approach to achieve nanoscale effects. Theoretical estimate of additional scattering on nanoinclusions of Mg$_2$Si and Mg$_2$Ge in the n-Mg$_2$Si$_{0.4}$Sn$_{0.6}$ matrix predicted a considerable increase in the figure of merit [64]. A small concentration of nanoparticles (about 3.4%) can lead to 60% reduction of thermal conductivity at 300 K and to 40% at 800 K with the optimal particle size of a few nanometers. The best value of ZT 1.9 at 800 K is predicted for Mg$_2$Si or Mg$_2$Ge nanoparticles in Mg$_2$Si$_{0.4}$Sn$_{0.6}$, which is considerably higher than the best experimental value for these alloys.

Various material synthesis technologies and alloy compositions were used in experiments in order to increase the figure of merit. Combination of induction melting, melt spinning (MS), and spark plasma sintering (SPS) methods were used to produce n-type Mg$_2$Si$_{0.4}$Sn$_{0.6}$ alloys doped with Bi [6]. Multiple localized nanostructures within the matrix containing nanoscale precipitates and mesoscale grains were formed, resulting in a remarkable decrease of lattice thermal conductivity, particularly for the samples with nanoscale precipitates having a size of 10–20 nm. Meanwhile, electrical resistivity was reduced and the Seebeck coefficient was increased by Bi-doping, causing improved electrical performance. Figure of merit ZT was significantly improved and the maximum value reaches 1.20 at 573 K for the Mg$_2$Si$_{0.4}$Sn$_{0.6}$+3% Bi sample, which is higher than that of nondoped samples. In comparison to samples of a similar composition, prepared by a conventional procedure, these samples have very low thermal conductivity, larger thermopower, and lower electrical conductivity.

Another way to increase the ZT value is use of quasi-quaternary alloys Mg_2Si-Mg_2Sn-Mg_2Ge. Although theoretical calculation did not predict noticeable influence of Ge on the lattice thermal conductivity of Mg_2Si-Mg_2Sn [65], it was demonstrated experimentally that ZT can be increased up to 1.4 in Bi-doped $Mg_2Si_{1-x-y}Sn_xGe_y$ ($x = 0.4$ and $y = 0.05$) alloys [3, 66]. Alloys were prepared by solid-state synthesis and sintering via hot pressing. Transmission electron microscopy (TEM) confirmed the coexistence of phases with different stoichiometry and yielded nanofeatures of the $Mg_2Si_{1-x-y}Sn_xGe_y$ phase. Thermoelectric properties of these materials were affected by different stoichiometry and the Sn-rich phase is believed to play a crucial role. High figure of merit could be attributed to a relatively high power factor that is related to contribution of the Sn-rich phase as well as low thermal conductivity that originates from nanostructuring.

Homogeneous alloys $Mg_2Si_{0.3}Sn_{0.7}$ were successfully prepared by nonequilibrium synthesis (melt spinning) followed by hot pressing and a plasma-assisted sintering (MS-PAS) technique [7]. Microstructure homogenization promotes charge carrier transport and effectively enhances the power factor. As a result, the MS-PAS sample achieved the highest figure of merit ZT of 1.30 at 750 K. However, the $Mg_2Si_{0.3}Sn_{0.7}$ alloy is intrinsically unstable at higher temperatures and tends to decompose into various Si-rich and Sn-rich phases even following the modest annealing at 773 K for 2 h.

The influence of grain size on thermoelectric properties of $Mg_2Si_{0.8}Sn_{0.2}$ doped with Sb was investigated using samples prepared by hot-pressing synthesized powders with grain sizes in the range from 100 to below 70 nm [67]. Contrary to expectation, no significant reduction of thermal conductivity in nanograined samples was found. ZT showed very weak dependence on grain's size with maximum values of about 0.8–0.9 at 900 K.

The best ZT results for n-type of Mg_2Si-based thermoelectrics are summarized in **Figure 10**.

4.3. Figure of merit of p-type Mg_2X solid solutions

To realize high performance of n-type Mg_2X-based alloys in practical applications, one needs to have a matching p-type material, preferably of the same base material. Therefore, considerable efforts have been made to the development of p-type Mg_2X-based alloys. However, progress with this development has been not so impressive as with n-type materials. At present, the maximum ZT of p-type Mg_2X-based alloys is about 0.5. There are several reasons for this. The high ZT of the n-type $Mg_2(Si$-$Sn)$ alloys is connected in part with high valley degeneracy of the conduction band that increases in alloys due to the band inversion effect. The valley degeneracy effect is absent for the valence band, since the top of this band is located at the Γ-point of the Brillouin zone. Furthermore, hole mobility is lower than the electron mobility in all Mg_2X compounds; hence, the onset of intrinsic conduction gives a negative impact on thermoelectric performance at lower temperatures in comparison with n-type alloys. The difference between hole and electron mobility is smallest in Mg_2Sn, where the electron-to-hole mobility ratio is about 1.5. Therefore, one can expect that the most efficient p-type alloy will contain a large fraction of Mg_2Sn.

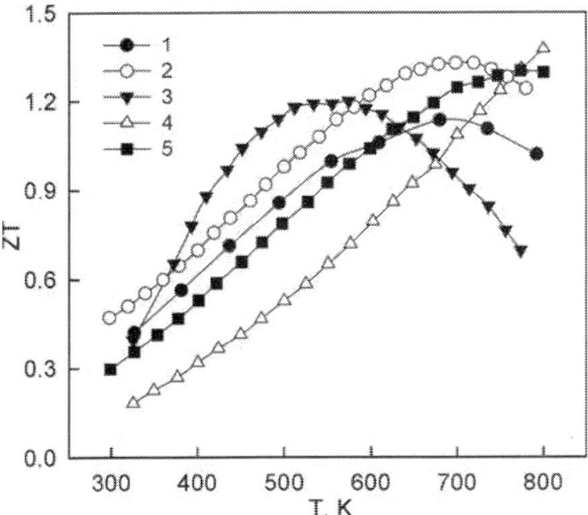

Figure 10. The best figure of merit of the Mg$_2$Si-Mg$_2$Sn alloy. 1—Mg$_2$Si$_{0.4}$Sn$_{0.6}$+Sb [2]; 2—Mg$_2$Si$_{0.3}$Sn$_{0.7}$+0.6% Sb [5]; 3—Mg$_2$Si$_{0.4}$Sn$_{0.6}$+1% Bi [6]; 4—Mg$_2$Si$_{0.53}$Sn$_{0.4}$Ge$_{0.05}$Bi$_{0.02}$ [3]; 5—Mg$_2$(Si$_{0.3}$Sn$_{0.7}$)$_{0.98}$Sb$_{0.02}$ [7].

There are several potential p-type dopants for Mg$_2$X compounds. The most effective impurities for Mg$_2$Sn-rich alloys are Ga and Li. Both of these dopants provide hole concentrations higher than 10^{20} cm^{-3}. Our study shows that these impurities yield one hole per dopant atom up to 2.5% Ga and 1.5% Li. Alloy Mg$_2$Si$_{0.3}$Sn$_{0.7}$ doped with these impurities has a maximum ZT of up to 0.45 at 650 K [4, 68].

Experimental and theoretical studies of effects related to Ga doping of the Mg$_2$Si compound and the Mg$_2$Si$_{0.6}$Ge$_{0.4}$ alloy by measurements of electrical resistivity, thermopower, Hall coefficient, and thermal conductivity, supplemented by electronic band structure calculations, have shown that p-type materials with the maximum ZT value of 0.36 at 625 K can be obtained for Mg$_2$Si$_{0.6}$Ge$_{0.4}$:Ga (0.8%) [69].

Another p-type dopant is silver. The maximum figure of merit ZT of 0.38 was achieved at 675 K for Mg$_{1.98}$Ag$_{0.02}$Si$_{0.4}$Sn$_{0.6}$ [70]. It was found that the solubility of Ag in Mg$_2$Si$_{0.4}$Sn$_{0.6}$ is about 2%. Oversaturated Ag doping in Mg$_2$Si$_{0.4}$Sn$_{0.6}$ is unfavorable for the improvement of thermoelectric properties.

Investigation on the effect of Li doping on electrical and thermal transport properties of Mg$_2$Si$_{0.3}$Sn$_{0.7}$ alloys indicated that Li is an efficient dopant occupying Mg sites. Theoretical calculations as well as experiments indicate that Li doping preserves high hole mobility. While overall thermal conductivity increases with an increase in the Li content (due to enhanced electrical conductivity) at low to mid-range temperatures, the beneficial effect of Li doping is shifting the onset of bipolar conductivity to higher temperatures and thus extending the regime, where thermal conductivity benefits from Umklapp phonon scattering. As a consequence, thermoelectric performance is significantly improved with the figure of merit ZT

reaching a value of 0.50 at around 750 K at the Li doping level of 0.07 [71]. **Figure 11** summarizes ZT temperature dependences for the best p-type Mg_2X-based alloys.

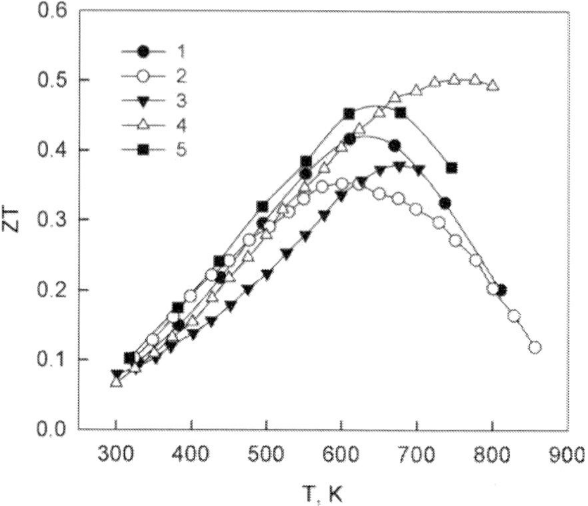

Figure 11. The figure of merit for the state-of-the-art p-type Mg_2Si-Mg_2Sn alloys. $1-Mg_2(Si_{0.3}Sn_{0.7})_{0.985}Ga_{0.015}$ [4]; $2-Mg_2Si_{0.6}Ge_{0.4}+0.8\%$ Ga [69]; $3-Mg_{1.98}Ge_{0.4}Sn_{0.6}Ag_{0.02}$ [70]; $4-Mg_{1.86}Si_{0.3}Sn_{0.7}Li_{0.14}$ [71]; $5-Mg_{1.99}Si_{0.3}Sn_{0.7}Li_{0.01}$ [68].

Known attempts to use nanostructuring have not yielded positive results for p-doped Mg_2X-based alloys. Another practically important problem with p-type alloys containing a large fraction of Mg_2Sb is their intrinsic instability.

5. Conclusion

The last decade comprehensive study of Mg_2X and Mg_2X-based alloys has yielded rather impressive results. Mg_2X-based n-type alloys are sufficiently stable at a temperature up to about 800 K and have maximum figure of merit close to 1.5. The combination of high-thermoelectric performance with low cost of raw elemental materials places these materials among the best thermoelectrics for temperature range from 300 to 800 K. However, there are still many problems to be solved in order to bring these alloys to the application stage. The most important problem is the failure to develop a matching p-type thermoelectric material. The best ZT value for p-type Mg_2X-based alloys does not exceed 0.5. Moreover, the p-type alloys are not sufficiently stable. Another problem is the absence of technology for making stable, high quality electrical contacts with the alloys. However, this problem certainly can be solved with adequate efforts and resources.

Author details

Vladimir K. Zaitsev[1], Grigoriy N. Isachenko[1,2*] and Alexander T. Burkov[1,2]

*Address all correspondence to: isachenko@inbox.ru

1 Ioffe Institute, Politeknicheskaya ul., Saint Petersburg, Russia

2 ITMO University, Kronverkskiy pr., Saint Petersburg, Russia

References

[1] Nikitin E.: Investigation on temperature dependencies of electrical conductivity and thermopower of silicides. Zhurnal Tekhnicheskoj Fiziki. 1958;28:23–25.

[2] Zaitsev VK, Fedorov MI, Gurieva EA, Eremin IS, Konstantinov PP, Samunin AY, Vedernikov MM.: Highly effective Mg$_2$Si$_{1-x}$Sn$_x$ thermoelectrics. Physical Review B. 2006;74(4):45207. DOI: 10.1103/PhysRevB.74.045207

[3] Khan A, Vlachos N, Kyratsi T.: High thermoelectric figure of merit of Mg$_2$Si$_{0.55}$Sn$_{0.4}$Ge$_{0.05}$ materials doped with Bi and Sb. Scripta Materialia. 2013;69(8):606–609. DOI: 10.1016/j.scriptamat.2013.07.008

[4] Fedorov MI, Zaitsev VK, Isachenko GN.: High effective thermoelectrics based on the Mg$_2$Si-Mg$_2$Sn solid solution. Solid State Phenomena. 2011;170:286–292. DOI: 10.4028/www.scientific.net/SSP.170.286

[5] Liu W, Tan X, Yin K, Liu H, Tang X, Shi J, Zhang Q, Uher C.: Convergence of conduction bands as a means of enhancing thermoelectric performance of n-type Mg$_2$Si$_{1-x}$Sn$_x$ solid solutions. Physical Review Letters. 2012;108(16):166601. DOI: 10.1103/PhysRevLett.108.166601

[6] Zhang X, Liu H, Lu Q, Zhang J, Zhang F.: Enhanced thermoelectric performance of Mg$_2$Si$_{0.4}$Sn$_{0.6}$ solid solutions by in nanostructures and minute Bi-doping. Applied Physics Letters. 2013;103(6):063901. DOI: 10.1063/1.4816971

[7] Zhang Q, Zheng Y, Su X, Yin K, Tang X, Uher C.: Enhanced power factor of Mg$_2$Si$_{0.3}$Sn$_{0.7}$ synthesized by a non-equilibrium rapid solidification method. Scripta Materialia. 2015;96:1–4. DOI: 10.1016/j.scriptamat.2014.09.009

[8] Nikitin EN, Bazanov VG, Tarasov VI.: The thermoelectric properties of solid solution Mg$_2$Si-Mg$_2$Sn. Soviet Physics of Solid State. 1961;3(12):2648–2651.

[9] Labotz RJ, Mason DR, O'Kane DF.: The thermoelectric properties of mixed crystals Mg$_2$Ge$_x$Si$_{1-x}$. Journal of the Electrochemical Society. 1963;110(2):127–134.

[10] Nicolau MC. Material for direct thermoelectric energy conversion with a high figure of merit. In: Proceedings of International Conference on Thermoelectric Energy Conversion; Arlington, Texas. 1976. p. 59.

[11] Noda Y, Kon H, Furukawa Y, Nishida IA, Masumoto K.: Temperature dependence of thermoelectric properties of $Mg_2Si_{0.6}Ge_{0.4}$. Materials Transactions, JIM. 1992;33(9):851–855.

[12] Ioffe AF. Semiconductor Thermoelements and Thermoelectric Cooling. London: Infosearch; 1957. 184 p.

[13] Chasmar RP, Stratton R.: The thermoelectric figure of merit and its relation to thermoelectric generators. Journal of Electronics and Control. 1959;7(1):52–72. DOI: 10.1080/00207215908937186

[14] Baker H, editor. ASM Handbook, Volume 03 – Alloy Phase Diagrams. USA: ASM International; 1992. 512 p.

[15] Busch G, Winkler U.: Electrical conductivity of intermetallic solid solutions. Helvetica Physica Acta. 1953;26(5):578–583.

[16] Grosch GH, Range KJ.: Studies on AB2-type intermetallic compounds. I. Mg_2Ge and Mg_2Sn: single-crystal structure refinement and ab inito calculations. Journal of Alloys and Compounds. 1996;235(2):250255.

[17] Zintl E, Kaiser H.: On the ability of elements to bind negative ions. Zeitschrift für anorganische und allgemeine Chemie. 1933;211(1/2):113–131.

[18] Winkler U.: Electrical properties of intermetallic compounds Mg_2Si, Mg_2Ge, Mg_2Sn and Mg_2P. Helvetica Physica Acta. 1955;28(7):633–666.

[19] Martin JJ.: Thermal conductivity of Mg_2Si, Mg_2Ge and Mg_2Sn. Journal of Physics and Chemistry of Solids. 1972;33(5):1139–1148.

[20] Sacklowski A.: X-ray investigations of some alloys. Annalen der Physik. 1925;382(11):241–272.

[21] Zaitsev VK, Fedorov MI, Eremin IS, Gurieva EA.: Thermoelectrics on the base of solid solutions of Mg_2B^{IV} compounds (B^{IV}=Si, Ge, Sn). In: D.M. Rowe, editor. CRC Handbook of Thermoelectrics: Macro to Nano. New York: CRC Press; 2006. p. 29-1-29-11.

[22] Koenig P, Lynch DW, Danielson GC.: Infrared absorption in magnesium silicide and magnesium germanide. Journal of Physics and Chemistry of Solids. 1961;20(1/21961):122–126.

[23] Lipson HG, Kahan A.: Infrared absorption of Mg_2Sn. Physical Review. 1964;133A:800–810.

[24] Lott LA, Lynch DW.: Infrared absorption in Mg_2Ge. Physical Review. 1966;141(2):681–686.

[25] Au-Yang MY, Cohen ML.: Electronic structure and optical properties of Mg$_2$Si, Mg$_2$Ge and Mg$_2$Sn. Physical Review. 1969;178(3):1358–1364.

[26] Baranek P, Schamps J, Noiret I.: Ab initio studies of electronic structure, phonon modes, and elastic properties of Mg$_2$Si. Journal of Physical Chemistry B. 1997;101(45):9147–9152.

[27] Arnaud B, Alouani M.: Electron-hole excitations in Mg$_2$Si and Mg$_2$Ge compounds. Physical Review B. 2001;64(3):033202.

[28] Kutorasinski K, Wiendlocha B, Tobola J, Kaprzyk S. Importance of relativistic effects in electronic structure and thermopower calculations for Mg$_2$Si, Mg$_2$Ge, and Mg$_2$Sn. Physical Review B. 2014;89(11):115205. DOI: 10.1103/PhysRevB.89.115205

[29] Mead CA.: Photothresholds in Mg$_2$Ge. Journal of Applied Physics. 1964;35(8):2460–2462. DOI: 10.1063/1.1702881

[30] Redin RD, Morris RG, Danielson GC.: Semiconducting properties of Mg$_2$Ge single crystals. Physical Review. 1958;109(6):1916–1920. DOI: 10.1103/PhysRev.109.1916

[31] Morris RG, Redin RD, Danielson GC.: Semiconducting properties of Mg$_2$Si single crystals. Physical Review. 1958;109(6):1909–1915. DOI: 10.1103/PhysRev.109.1909

[32] Zaitsev VK, Nikitin EN.: Electrical properties, thermal conductivity and forbidden-band width of Mg$_2$Sn at high temperatures. Soviet Physics of Solid State. 1970;12:289–292.

[33] Nikitin EN, Tkalenko EN, Zaitsev VK, Zaslavskii AI, Kuznetsov AK.: Investigation of phase diagram and of some properties of Mg$_2$Si-Mg$_2$Sn solid solutions. Neorg. Mater. 1968;4(11):1902–1906.

[34] Zaitsev VK, Tkalenko EN, Nikitin EN.: Lattice thermal conductivity of Mg$_2$Si-Mg$_2$Sn, Mg$_2$Ge-Mg$_2$Sn and Mg$_2$Si-Mg$_2$Ge solid solutions. Soviet Physics of Solid State. 1969;11:221–224.

[35] Labotz RJ, Mason DR.: The thermal conductivities of Mg$_2$Si and Mg$_2$Ge. Journal of the Electrochemical Society. 1963;110(2):120–126.

[36] Fedorov MI, Zaitsev VK.: Optimization of thermoelectric parameters in some silicide based materials. In: Proceedings of XIX International Conference on Thermoelectrics (ICT 2000); Cardiff, UK, 2000. p. 17.

[37] Zaitsev VK, Nikitin EN, Tkalenko EN.: Width of forbidden band in solid solutions Mg$_2$Si-Mg$_2$Sn. Soviet Physics of Solid State. 1969;11:3000.

[38] Fedorov M.I., Gurieva E.A., Eremin I.S., Konstantinov P.P., Samunin A.Yu, Zaitsev V.K., Sano S., Rauscher L. Kinetic properties of solid solutions Mg$_2$Si$_{1-x-y}$Sn$_x$Ge$_y$, Proceedings of VIII European Workshop on thermoelectrics, Poland, Krakow, September 15–17, 2004, p.75.

[39] Fedorov MI, Zaitsev VK, Eremin IS, Gurieva EA, Burkov AT, Konstantinov PP, Veder-nikov MV, Samunin AYu, Isachenko GN.: Kinetic properties of p-type $Mg_2Si_{0.4}Sn_{0.6}$ solid solutions. In: Proceedings of Twenty-Second International Conference on Thermoelectrics, ICT'03; IEEE; 2003. p. 134.

[40] Fedorov MI, Pshenay-Severin DA, Zaitsev VK, Sano S, Vedernikov MV.: Features of conduction mechanism in n-type $Mg_2Si_{1-x}Sn_x$ solid solutions. In: Proceedings of Twenty-Second International Conference on Thermoelectrics, ICT'03; IEEE; 2003, p. 142.

[41] Pshenay-Severin DA and Fedorov MI.: Effect of the band structure on the thermoelec-tric properties of a semiconductor. Physics of the Solid State. 2007;49(9):1633–1637. DOI: 10.1134/S1063783407090053

[42] Clark CR, Wright C, Suryanarayana C, Baburaj EG, Froes FH.: Synthesis of Mg_2X (X = Si, Ge, or Sn) intermetallics by mechanical alloying. Materials Letters. 1997;33(1-2):71–75.

[43] Xiaoping Niu, Li Lu.: Formation of magnesium silicide by mechanical alloying. Advanced Performance Materials. 1997;4(3):275–283.

[44] Riffel M, Schilz J.: Influence of production parameters on the thermoelectric proper-ties of Mg_2Si. In: Proceedings of 16th International Conference on Thermoelectrics, ICT'97; 1997. p. 283.

[45] Schilz J, Muller E, Kaysser WA, Langer G, Lugscheider E, Schiller G, Henue R.: Graded thermoelectric materials by plasma spray forming. In: Shiota I, Miyamoto Y, editors. Functionally Graded Materials 1996. Netherlands: Elsevier Science B.V.; 1997. p. 563–568.

[46] Godlewska E, Mars K, Zawadzka K.: Alternative route for the preparation of $CoSb_3$ and Mg_2Si derivatives. Journal of Solid State Chemistry. 2012;193:109–113. DOI: 10.1016/j.jssc.2012.03.070

[47] Samunin AYu, Zaitsev VK, Konstantinov PP, Fedorov MI, Isachenko GN, Burkov AT, Novikov SV, Gurieva EA.: Thermoelectric properties of hot-pressed materials based on $Mg_2Si_nSn_{1-n}$. Journal of Electronic Materials. 2013;42(7):1676–1679. DOI: 10.1007/s11664-012-2372-3

[48] Tani J-I, Kido H.: First-principles and experimental studies of impurity doping into Mg_2Si. Intermetallics. 2008;16(3):418–423. DOI: 10.1016/j.intermet.2007.12.001

[49] Tani J-I, Kido H.: Thermoelectric properties of Al-doped $Mg_2Si_{1-x}Sn_x$ ($x \leqq 0.1$). Journal of Alloys and Compounds. 2008;466(1–2):335–340. DOI: 10.1016/j.jallcom.2007.11.029

[50] Tani J-I, Kido H.: Impurity doping into Mg_2Sn: A first-principles study. Physica B. 2012;407(17):3493–3498. DOI: 10.1016/j.physb.2012.05.008

[51] Isoda Y, Tada S, Nagai T, Fujiu H, Shinohara Y.: Thermoelectric properties of p-type Mg$_{2.00}$Si$_{0.25}$Sn$_{0.75}$ with Li and Ag double doping. Journal of Electronic Materials. 2010;39(9):1531–1535. DOI: 10.1007/s11664-010-1280-7

[52] Tada S, Isoda Y, Udono H, Fujiu H, Kumagai S, Shinohara Y.: Thermoelectric properties of p-type Mg$_2$Si$_{0.25}$Sn$_{0.75}$ doped with sodium acetate and metallic sodium. Journal of Electronic Materials. 2014;43(6):1580–1584. DOI: 10.1007/s11664-013-2797-3

[53] Fedorov M I, Zaitsev VK, Eremin IS, Gurieva EA, Burkov AT, Konstantinov PP, Vedernikov MV, Samunin AYu, Isachenko GN, Shabaldin AA.: Transport properties of Mg$_2$X$_{0.4}$Sn$_{0.6}$ solid solutions (X = Si, Ge) with p-type conductivity. Physics of the Solid State. 2006;48(8): 1486–1490. DOI: 10.1134/S1063783406080117

[54] Tani J-I, Kido H.: Thermoelectric properties of Bi-doped Mg$_2$Si semiconductors. Physica B: Condensed Matter. 2005;364(1-4):218–224. DOI: http://dx.doi.org/10.1016/j.physb.2005.04.017

[55] Choi SM, Kim KH, Kim IH, Kim SU, Seo WS.: Thermoelectric properties of the Bi-doped Mg$_2$Si system. Current Applied Physics. 2011;11(3):S388–S391. DOI: 10.1016/j.cap.2011.01.031

[56] Tani J-I, Kido H.: Thermoelectric properties of Sb-doped Mg$_2$Si semiconductors. Intermetallics. 2007;15(9):1202–1207. DOI: 10.1016/j.intermet.2007.02.009

[57] Tani J-I, Kido H.: Thermoelectric properties of P-doped Mg$_2$Si semiconductors. Japanese Journal of Applied Physics. 2007;46(6A):3309–3314. DOI: 10.1143/JJAP.46.3309

[58] Akasaka M, Iida T, Nemoto T, Soga J, Sato J, Makino K, Fukano M, Takanashi Y.: Non-wetting crystal growth of Mg$_2$Si by vertical Bridgman method and thermoelectric characteristics. Journal of Crystal Growth. 2007;304(1):196–201. DOI: 10.1016/j.jcrysgro.2006.10.270

[59] Akasaka M, Iida T, Matsumoto A, Yamanaka K, Takanashi Y, Imai T, Hamada N.: The thermoelectric properties of bulk crystalline n- and p-type Mg$_2$Si prepared by the vertical Bridgman method. Journal of Applied Physics. 2008;104(1):013703. DOI: 10.1063/1.2946722

[60] Farahi N, VanZant M, Zhao J, Tse JS, Prabhudev S, Botton GA, Salvador JR, Borondics F, Liu Z, Kleinke H.: Sb- and Bi-doped Mg$_2$Si: location of the dopants, micro- and nanostructures, electronic structures and thermoelectric properties. Dalton Transactions. 2014;43(40):14983–14991. DOI: 10.1039/C4DT01177E

[61] Bux SK, Yeung MT, Toberer ES, Snyder GJ, Kaner RB, Fleurial JP.: Mechanochemical synthesis and thermoelectric properties of high quality magnesium silicide. Journal of Materials Chemistry. 2011;21(33):12259–12266. DOI: 10.1039/C1JM10827A

[62] Satyala N, Vashaee D.: Detrimental influence of nanostructuring on the thermoelectric properties of magnesium silicide. Journal of Applied Physics. 2012;112(9):093716. DOI: 10.1063/1.4764872

[63] Pshenai-Severin DA, Fedorov MI, Samunin AYu. The influence of grain boundary scattering on thermoelectric properties of Mg_2Si and $Mg_2Si_{0.8}Sn_{0.2}$. Journal of Electronic Materials. 2013;42(7):1707–1710. DOI: 10.1007/s11664-012-2403-0

[64] Wang S, Mingo N.: Improved thermoelectric properties of $Mg_2Si_xGe_ySn_{1-x-y}$ nanoparticle-in-alloy materials. Applied Physics Letters. 2009;94(20):203109. DOI: 10.1063/1.3139785

[65] Bochkov LV, Fedorov MI, Isachenko GN, Zaitsev VK, Eremin IS.: Influence of germanium on lattice thermalconductivity of the magnesium silicide solid solutions. Vestnik MAH. 2014;3(52):26.

[66] Khan A, Vlachos N, Hatzikraniotis E, Polymeris G, Lioutas C, Stefanaki E, Paraskevopoulos K, Giapintzakis I, Kyratsi T.: Thermoelectric properties of highly efficient Bi-doped $Mg_2Si_{1-x-y}Sn_xGe_y$ materials. Acta Materialia. 2014;77:43–53. DOI: 10.1016/j.actamat.2014.04.060

[67] Samunin AYu, Zaytsev VK, Pshenay-Severin DA, Konstantinov PP, Isachenko GN, Fedorov MI, Novikov SV.: Thermoelectric properties of n-type Mg_2Si-Mg_2Sn solid solutions with different grain sizes. Physics of the Solid State. 2016;58(8):1528–1531. DOI: 10.1134/S1063783416080242

[68] Isachenko GN, Samunin AYu, Gurieva EA, Fedorov MI, Pshenay-Severin DA, Konstantinov PP, Kamolova MD.: Thermoelectric properties of nanostructured p-$Mg_2Si_xSn_{1-x}$ (x=0.2 to 0.4) solid solutions. Journal of Electronic Materials. 2016;45(3): 1982–1986. DOI: 10.1007/s11664-016-4345-4

[69] Ihou-Mouko H, Mercier C, Tobola J, Pont G, Scherrer H.: Thermoelectric properties and electronic structure of p-type Mg_2Si and $Mg_2Si_{0.6}Ge_{0.4}$ compounds doped with Ga. Journal of Alloys and Compounds. 2011;509(23):6503–6508. DOI: 10.1016/j.jallcom. 2011.03.081

[70] Jiang G, Chen L, He J, Gao H, Du Z, Zhao X, Tritt TM, Zhu T.: Improving p-type thermoelectric performance of $Mg_2(Ge,Sn)$ compounds via solid solution and Ag doping. Intermetallics. 2013;32:312–317. DOI: 10.1016/j.intermet.2012.08.002

[71] Zhang Q, Cheng L, Liu W, Zheng Y, Su X, Chi H, Liu H, Yan Y, Tang X, Uher C.: Low effective mass and carrier concentration optimization for high performance p-type $Mg_{2(1-x)}Li_{2x}Si_{0.3}Sn_{0.7}$ solid solutions. Physical Chemistry Chemical Physics. 2014;16(43): 23576–23583. DOI: 10.1039/C4CP03468F

Simulation of Phenomena Related to Thermoelectricity

Simulation of Morphological Effects on Thermoelectric Power, Thermal and Electrical Conductivity in Multi-Phase Thermoelectric Materials

Yaniv Gelbstein

Additional information is available at the end of the chapter

http://dx.doi.org/10.5772/65099

Abstract

Multi-phase thermoelectric materials are mainly investigated these days due to their potential of lattice thermal conductivity reduction by scattering of phonons at interfaces of the involved phases, leading to the enhancement of expected thermoelectric efficiency. On the other hand, electronic effects of the involved phases on thermoelectric performance are not always being considered, while developing new multi-phase thermoelectric materials. In this chapter, electronic effects resulting from controlling the phase distribution and morphology alignment in multi-phase composite materials is carefully described using the general effective media (GEM) method and analytic approaches. It is shown that taking into account the specific thermoelectric properties of the involved phases might be utilized for estimating expected effective thermoelectric properties of such composite materials for any distribution and relative amount of the phases. An implementation of GEM method for the IV–VI (including SnTe and GeTe), bismuth telluride (Bi_2Te_3), higher manganese silicides (HMS) and half-Heusler classes of thermoelectric materials is described in details.

Keywords: thermoelectric, GEM, multi-phase

1. Thermoelectrics

Climate changes, due to fossil fuels combustion and greenhouse gases emission, cause deep concern about environmental conservation. Another pressing issue is sustainable energy

production that is coupled with depletion of conventional energy resources. This concern might be tackled by converting the waste heat generated in internal-combustion vehicles, factories, computers, etc. into electrical energy. Converting this waste heat into electricity will reduce fossil fuel consumption and emission of pollutants. This can be achieved by direct thermoelectric (TE) converters, as was successfully demonstrated by development of various highly efficient TE material classes, including Bi_2Te_3 [1–3] for temperatures, T, of up to ~300°C, SnTe [4, 5], PbTe [6, 7] and GeTe [8–11], for temperatures range $300 \leq T \leq 500$°C, and higher manganese silicides (HMS) [12–14], half-Heuslers [15–20], which are capable to operate at higher temperatures. Such materials require unique combination of electronic (i.e. Seebeck coefficient, α, electrical resistivity, ρ, and electronic thermal conductivity, κ_e) and lattice (i.e. lattice thermal conductivity, κ_l) properties, enabling the highest possible TE figure of merit, $ZT = \alpha^2 T/[\rho(\kappa_e + \kappa_l)]$, values, for achieving significant heat to electricity conversion efficiencies. Due to the fact, that electronic TE properties are strongly correlated, and follow opposite trends upon modifying charge carriers' concentration, many of recently developed TE materials, were focused on nano-structuring methods, capable of κ_l reduction due to lattice modifications and correspondingly increasing ZT. Such methods included alloying (for PbTe, as an example, alloying with SrTe [21, 22], MgTe [23] and CdTe [24], resulted in strained endotaxial nano-structures), applying layered structures with increased interfaces population (e.g. SnSe [25]), and thermodynamically driven phase separation reactions, generating nano-scale modulations (e.g. $Ge_xPb_{1-x}Te$ [26–28] and $Ge_x(Sn_yPb_{1-y})_{1-x}Te$ [29, 30]). All of these approaches resulted in significant increase of ZT up to ~2.5 [25] due to effective scattering of phonons by associated generated nano-features. Nevertheless, although significant enhancement of TE properties was reported due to phonon scattering by nano-structured phases in such multi-phase TE materials, most of these researches did not investigate individual electronic contributions of each of the involved phases on effective TE transport properties.

2. Multi-phase thermoelectric materials

In the last few decades, major trend is to move from pristine single-crystal TE compositions towards polycrystalline multi-phase materials. One of the reasons for that is improved shear mechanical strength of polycrystalline materials compared to single crystals, exhibiting high compression, but very low transverse strengths, required to withstand high thermal and mechanical gradients applied in practical applications. Another reason is the possibility of phonon scattering by the involved interfaces as mentioned above. Most of the TE materials investigated these days are being synthesized by powder metallurgy approach under high uniaxial mechanical pressures, deforming involved grains and phases into anisotropic geometrical morphologies, which affect the electronic transport properties. Besides, a certain amount of porosity (as a second phase) is in many cases unavoidable, adversely affecting TE transport properties. Furthermore, many of currently employed TE materials (e.g. Bi_2Te_3 and HMS) are crystallographic anisotropic with optimal TE transport properties along preferred orientations. Some researches of such materials for TE applications do not consider crystallographic anisotropy, while assuming, that randomly oriented grains of different crystallographic planes cancel each other in polycrystalline samples. Yet, some anisotropy can exist also in such materials in case of highly anisotropic specific properties (e.g. mechanical properties), leading to textured polycrystal. For example, texture development of non-cubic polycrystalline alloys was attributed to multiple deformation modes applied in each grain, twinning resulting in grain reorientation and strong directional grain interactions [12]. Specifically, in Bi_2Te_3, for example, exhibiting highly anisotropic layered crystal structure consists of 15 parallel layers stacked along crystallographic c axis, the presence of van der Waals gap in the crystal lattice, divides crystal into blocks of five mono-atomic sheets [1]. In this case, retaining the crystallographic anisotropy is highly desired. This is due to the fact, that in transverse to crystallographic c axis, TE power factor (numerator in ZT expression) is considerably higher, than in parallel to this direction, mainly due to higher electrical conductivity values. For powder metallurgy synthesized Bi_2Te_3-based materials, it was shown that moderate powder grinding pressures, might retain some of the crystallographic anisotropy, due to the weaker van der Waals bonding of atoms located in adjacent layers along c-axis, compared to ionic/covalent bonding between atoms located in each of the layers [31]. In this example, higher ZT values in transverse to powder pressing direction are expected as in single crystals. This example highlights the significance of controlling phases' morphology for optimizing TE transport properties.

Besides of metallurgical phases, individual transport properties of two species (e.g. light and heavy holes in p-type PbTe [32]), in materials with complicated electronic band structures might contribute dramatically to effective TE transport properties.

In this chapter, effective TE properties (Seebeck coefficient, α electrical resistivity, ρ or conductivity, $\sigma = \rho^{-1}$ and thermal conductivity, κ) of general complex structure, consisting of at least two independent phases with any respective relative amount and geometrical alignment are derived by using the GEM method [4] and individual TE properties of each of the involved phases. This approach can be utilized for maximizing TE figure of merit of multi-phase composite materials, for example, by intentional alignment of the involved phases along the optimal TE direction.

We consider in this chapter a simple formulation for modelling of multi-phase TE materials, originated from materials science aspects, such as inter-diffusion, alloying, dissolution, phase transitions, phase separation, phase segregation, precipitation, recrystallization and other phenomena, that can take place in operation conditions of TE modules, especially TE power generation modules exposed to high thermo-mechanical stresses.

3. TE GEM effective equations for two-phase materials

Effective TE properties of two-phase composites can be accurately predicted by GEM method, Eqs. (1)–(3) [4, 33–35]:

$$\frac{\alpha_{\text{eff}} - \alpha_2}{\alpha_1 - \alpha_2} = \frac{\dfrac{\kappa_{\text{eff}}/\kappa_2}{\sigma_{\text{eff}}/\sigma_2} - 1}{\dfrac{\kappa_1/\kappa_2}{\sigma_1/\sigma_2} - 1}, \tag{1}$$

$$x_1 \frac{(\sigma_1)^{1/t} - (\sigma_{\text{eff}})^{1/t}}{(\sigma_1)^{1/t} + A(\sigma_{\text{eff}})^{1/t}} = (1 - x_1) \frac{(\sigma_{\text{eff}})^{1/t} - (\sigma_2)^{1/t}}{(\sigma_2)^{1/t} + A(\sigma_{\text{eff}})^{1/t}}, \tag{2}$$

$$x_1 \frac{(\kappa_1)^{1/t} - (\kappa_{\text{eff}})^{1/t}}{(\kappa_1)^{1/t} + A(\kappa_{\text{eff}})^{1/t}} = (1 - x_1) \frac{(\kappa_{\text{eff}})^{1/t} - (\kappa_2)^{1/t}}{(\kappa_2)^{1/t} + A(\kappa_{\text{eff}})^{1/t}}. \tag{3}$$

These three GEM equations, Eqs. (1)–(3), are usually employed for calculating effective Seebeck coefficient (α_{eff}) and effective electrical and thermal conductivities (σ_{eff} and κ_{eff}, respectively) for two-phase materials using individual electrical (σ_1 and σ_2) and thermal (κ_1 and κ_2) conductivity, as well as, individual Seebeck coefficient (α_1 and α_2) values of involved phases. Morphological parameters A, t can be derived by modelling of experimental results or from percolation equation [33, 34]. Parameter x_1 is volume fraction of one of the phases. Values of A and t are strongly affected by phase distribution and morphology. It was shown, that for homogeneously distributed second phase in continuous matrix, t value is equal to 1 [4] and the entire morphological alignment possibilities of the second phase related to the matrix phase are bounded by the so-called 'parallel' and 'series' alignment of the phases (relative to electrical potential or temperature gradients). Parameter A varies from 8 for parallel to 0 for series alignments. It can be seen, that for substituting $t = 1$ and $A = 8$ in Eqs. (2) and (3), as in the case of phases distribution in parallel to electrical current direction, reduces equations to Eq. (4), while substituting of Eq. (4) in Eq. (1) leads to Eq. (5):

$$\left(\sigma_{\text{eff}}, \kappa_{\text{eff}}\right) = \left(\sigma_1, \kappa_1\right)x_1 + \left(\sigma_2, \kappa_2\right)\left(1 - x_1\right), \tag{4}$$

$$\alpha_{\text{eff}} = \frac{\alpha_1 \sigma_1 x_1 + \alpha_2 \sigma_2 \left(1 - x_1\right)}{\sigma_1 x_1 + \sigma_2 \left(1 - x_1\right)}. \tag{5}$$

Similarly, substituting $t = 1$ and $A = 0$ in Eqs. (2) and (3), as in the case of series alignment as explained above, reduces them into Eq. (6):

$$\left(\sigma_{\text{eff}},\kappa_{\text{eff}}\right)=\frac{\left(\sigma_{1},\kappa_{1}\right)\left(\sigma_{2},\kappa_{2}\right)}{\left(\sigma_{1},\kappa_{1}\right)\left(1-x_{1}\right)+\left(\sigma_{2},\kappa_{2}\right)x_{1}}. \tag{6}$$

Please note that although for the case of parallel alignment, effective electrical and thermal conductivity, Eq. (4), follow a simple rule of mixture, a more complicated dependency is apparent for series alignment, Eq. (6). Yet, as shown in Eq. (7), for this latter case, effective electrical resistivity, $\rho_{\text{eff}}=\sigma_{\text{eff}}^{-1}$, follows the rule of mixture:

$$\rho_{\text{eff}}=\rho_{1}x_{1}+\rho_{2}\left(1-x_{1}\right). \tag{7}$$

Substituting of Eq. (6) in Eq. (1), leads in this case to Eq. (8):

$$\alpha_{\text{eff}}=\frac{\alpha_{1}\kappa_{2}x_{1}+\alpha_{2}\kappa_{1}\left(1-x_{1}\right)}{\kappa_{1}\left(1-x_{1}\right)+\kappa_{2}x_{1}}. \tag{8}$$

While investigating Eqs. (5) and (8), for the cases of parallel and series alignment, respectively, it can be easily seen, that for both cases, effective Seebeck coefficient depends not only on individual Seebeck coefficients of the two phases, but also on other electronic transport properties, electrical conductivity of the involved phases for the case of parallel alignment, Eq. (5), and thermal conductivity of the involved phases for the case of series alignment, Eq. (8). An explanation for this observation is given in the next section.

4. Analytical effective equations for multi-phase materials

In order to extend GEM, Eqs. (1)–(3) listed above for two-phase composite materials, into higher-ordered composites with three or more coexisting phases, a simple analytical model for calculating effective TE properties of several conductors, subjected to external electrical and thermal gradients, can be applied. For this purpose, two boundary conditions explained above, can be examined; one for conductors connected in parallel to both thermal and electrical applied gradients and the other for conductors connected in series.

4.1. Thermoelectric phases in parallel

In the case of three distributed conductors oriented in parallel to external temperature, $\Delta T = T_{\text{h}}-T_{\text{c}}$, and electrical potentials, V, gradients, shown schematically as 1, 2 and 3 in **Figure 1(a)**, each of them might be considered as a single phase with sample's length and perspective cross-section area according to its relative amount (**Figure 1b**). For this case, electrical analogue, shown in **Figure 1(c)**, includes three parallel branches, with power source reflecting the individual open circuit voltage developed according to Seebeck effect ($V_{1,2,3}=\alpha_{1,2,3}\Delta T$, where $\alpha_{1,2,3}$ – Seebeck coefficients of the involved phases) under applied temperature difference, connected serially to resistor $R_{1,2,3}$, reflecting internal total electrical resistance, of each of the phases. In this case, electrical currents $I_{1,2,3}$, flowing through connectors are given by Eq. (9):

$$I_{1,2,3} = \frac{V - \int_{T_c}^{T_h} \alpha_{1,2,3} dT}{R_{1,2,3}}. \tag{9}$$

Total electrical current I in three-phase system is given by Eq. (10):

$$I = I_1 + I_2 + I_3 = V\left(\frac{1}{R_1} + \frac{1}{R_2} + \frac{1}{R_3}\right) - \int_{T_c}^{T_h}\left(\frac{\alpha_1}{R_1} + \frac{\alpha_2}{R_2} + \frac{\alpha_3}{R_3}\right) dT. \tag{10}$$

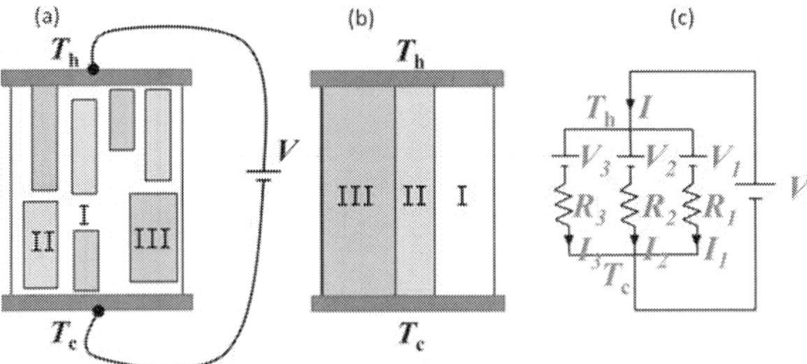

Figure 1. Schematical description of three phases, I–III, oriented in parallel to external temperature and electrical gradients, as distributed in the sample (a) and as combined entities with sample's length and perspective cross-section area according to their relative amount (b). The electrical analogue of this three-phase material is given in (c).

Considering definition of Seebeck coefficient as derivative of applied voltage with respect to temperature for non-current flowing condition, Eq. (11), a simple manipulation of Eq. (10) gives Eq. (12), which describes effective Seebeck coefficient, α_{eff}, of parallel connected three-phase structure:

$$\alpha_{eff} \stackrel{\text{def}}{=} \left.\frac{dV}{dT}\right|_{I=0}, \tag{11}$$

$$\alpha_{\text{eff}} = \frac{\dfrac{\alpha_1}{R_1} + \dfrac{\alpha_2}{R_2} + \dfrac{\alpha_3}{R_3}}{\dfrac{1}{R_1} + \dfrac{1}{R_2} + \dfrac{1}{R_3}} = \frac{\alpha_1 R_2 R_3 + \alpha_2 R_1 R_3 + \alpha_3 R_1 R_2}{R_2 R_3 + R_1 R_3 + R_1 R_2}. \tag{12}$$

Using specific parameters (resistivity $\rho_{1,2,3}$ and conductivity $\sigma_{1,2,3} = (\rho_{1,2,3})^{-1}$) instead of resistances $R_{1,2,3}$, as described in Eq. (13), expression for α_{eff} for parallel connected three-phase structures can be derived, Eq. (14):

$$R_{1,2,3} = \frac{\rho_{1,2,3} l_{\text{samp}}}{\tilde{A}_{1,2,3}},\tag{13}$$

$$\alpha_{eff} = \frac{\left(\frac{\alpha_1 \rho_2 \rho_3}{\tilde{A}_2 \tilde{A}_3} + \frac{\alpha_2 \rho_1 \rho_3}{\tilde{A}_1 \tilde{A}_3} + \frac{\alpha_3 \rho_1 \rho_2}{\tilde{A}_1 \tilde{A}_2}\right)}{\left(\frac{\rho_2 \rho_3}{\tilde{A}_2 \tilde{A}_3} + \frac{\rho_1 \rho_3}{\tilde{A}_1 \tilde{A}_3} + \frac{\rho_1 \rho_2}{\tilde{A}_1 \tilde{A}_2}\right)} = \frac{\alpha_1 \tilde{A}_1 \rho_2 \rho_3 + \alpha_2 \tilde{A}_2 \rho_1 \rho_3 + \alpha_3 \tilde{A}_3 \rho_1 \rho_2}{\tilde{A}_1 \rho_2 \rho_3 + \tilde{A}_2 \rho_1 \rho_3 + \tilde{A}_3 \rho_1 \rho_2} =$$

$$\frac{\alpha_1 \sigma_1 \tilde{A}_1 + \alpha_2 \sigma_2 \tilde{A}_2 + \alpha_3 \sigma_3 \tilde{A}_3}{\sigma_1 \tilde{A}_1 + \sigma_2 \tilde{A}_2 + \sigma_3 \tilde{A}_3},\tag{14}$$

where, $l_{\text{samp}} = l_1 = l_2 = l_3$ is the sample's length, $\tilde{A}_{1,2,3}$ is the cross-section area transverse to electrical current flow.

While considering, volume fractions, $x_{1,2,3}$ ($= \tilde{A}_{1,2,3} \cdot l_{\text{samp}}/V_{\text{samp}}$, where V_{samp} is sample's volume) of the respective phase, Eq. (15) can be easily derived:

$$\left(\alpha_{\text{eff}}\right)_{\text{parallel}} = \frac{\alpha_1 \sigma_1 x_1 + \alpha_2 \sigma_2 x_2 + \alpha_3 \sigma_3 x_3}{\sigma_1 x_1 + \sigma_2 x_2 + \sigma_3 x_3} = \frac{\sum \alpha_i \sigma_i x_i}{\sum \sigma_i x_i}.\tag{15}$$

From electrical analogue shown in **Figure 1(c)**, effective electrical and thermal conductivities can also be easily derived, as expressed in Eqs. (16) and (17), respectively:

$$\left(\sigma_{\text{eff}}\right)_{\text{parallel}} = \sigma_1 x_1 + \sigma_2 x_2 + \sigma_3 x_3 = \sum \sigma_i x_i,\tag{16}$$

$$\left(\kappa_{\text{eff}}\right)_{\text{parallel}} = \kappa_1 x_1 + \kappa_2 x_2 + \kappa_3 x_3 = \sum \kappa_i x_i.\tag{17}$$

It is noteworthy that applying the same approach for higher i-ordered multi-phase materials will follow the general-ordered right-hand side expressions of Eqs. (15)–(17). Furthermore, it can be easily seen that Eqs. (15)–(17) for the case of two-phase materials are reduced to Eqs. (5) and (4), respectively, derived from the GEM method.

4.2. Thermoelectric phases in series

Equivalent description for the case of three distributed conductors oriented in series to external temperature and electrical potentials gradients is shown in **Figure 2(a)**.

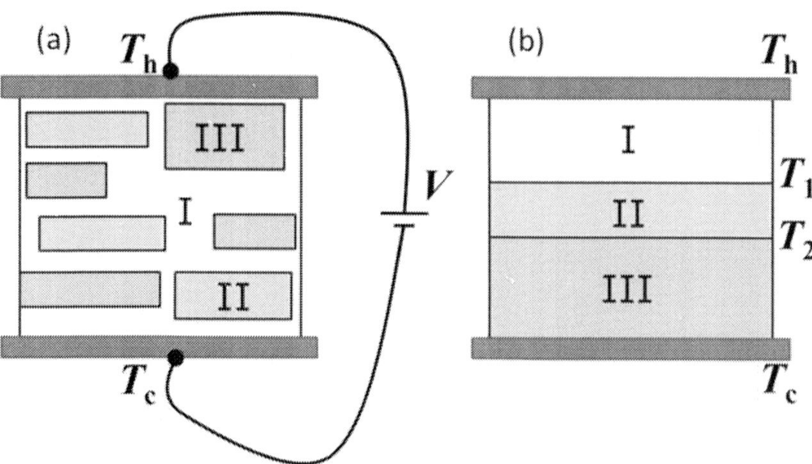

Figure 2. Schematical description of three phases, I–III, oriented in series to external temperature and electrical gradients, as distributed in the sample (a) and as combined entities with sample's diameter and perspective lengths according to their relative amount (b).

For this case, a similar analysis is presented, taking into account individual thermal gradients applied on each of the phases. Taking into account that the first, second and third phases are subjected to temperature differences of $(T_h - T_1)$, $(T_1 - T_2)$ and $(T_2 - T_c)$, respectively, as shown in **Figure 2(b)**, where $T_{1,2}$ are intermediate temperatures $(T_h > T_1 > T_2 > T_c)$, effective Seebeck coefficient of such serially aligned three-phase samples can be described in terms of Eq. (18):

$$\alpha_{\text{eff}} = \frac{\alpha_1(T_h - T_1) + \alpha_2(T_1 - T_2) + \alpha_3(T_2 - T_c)}{T_h - T_c}. \tag{18}$$

Under adiabatic heat conduction conditions, where no lateral heat losses are apparent, the heat flow, Q, through the entire sample and the individual phases can be described in terms of unidirectional Fourier heat conduction equation, Eq. (19):

$$Q = \frac{\kappa_1 \tilde{A}}{l_1}(T_h - T_1) = \frac{\kappa_2 \tilde{A}}{l_2}(T_1 - T_2) = \frac{\kappa_3 \tilde{A}}{l_3}(T_2 - T_c) = \frac{\kappa_{\text{eff}} \tilde{A}}{l_{\text{samp}}}(T_h - T_c), \tag{19}$$

where κ_{eff} is effective thermal conductivity of the three-phase material, $\tilde{A}$ is cross-section area transverse to heat flow and $\kappa_{1,2,3}$ and $l_{1,2,3}$ are thermal conductivity and effective length of each of the involved phases, respectively.

Using expression (19), the numerator terms of Eq. (18) can be easily described in terms of expressions (20):

$$\alpha_1\left(T_h - T_1\right) = \frac{Ql_1\alpha_1}{\kappa_1\tilde{A}}, \alpha_2\left(T_1 - T_2\right) = \frac{Ql_2\alpha_2}{\kappa_2\tilde{A}}, \alpha_3\left(T_2 - T_c\right) = \frac{Ql_3\alpha_3}{\kappa_3\tilde{A}}. \tag{20}$$

In the rightmost equation of expression (19), $\kappa_{\text{eff}}\tilde{A}/l_{\text{samp}}$ represents overall thermal conductance, K_{eff} of the three-phase sample, which is described in Eq. (21), in terms of serially connected thermal resistances, $R_{\text{th},1,2,3}$, specified in Eq. (22):

$$K_{\text{eff}} = \frac{\kappa_{\text{eff}}\tilde{A}}{l_{\text{samp}}} = \frac{1}{\left(R_{\text{th}}\right)_1 + \left(R_{\text{th}}\right)_2 + \left(R_{\text{th}}\right)_3}, \tag{21}$$

$$\left(R_{\text{th}}\right)_{1,2,3} = \frac{1}{\left(\dfrac{\kappa_{1,2,3}\tilde{A}}{l_{1,2,3}}\right)}. \tag{22}$$

Combining Eqs. (21) and (22) leads to Eq. (23):

$$K_{\text{eff}} = \frac{1}{\dfrac{l_1}{\kappa_1\tilde{A}} + \dfrac{l_2}{\kappa_2\tilde{A}} + \dfrac{l_3}{\kappa_3\tilde{A}}}. \tag{23}$$

Substitution of the expression of K_{eff}, Eq. (23) in the rightmost term of expression (19) results in the expression of $T_h - T_c$, presented in Eq. (24):

$$Q = \frac{1}{\dfrac{l_1}{\kappa_1\tilde{A}} + \dfrac{l_2}{\kappa_2\tilde{A}} + \dfrac{l_3}{\kappa_3\tilde{A}}}\left(T_h - T_c\right) or \left(T_h - T_c\right) = Q\left(\frac{l_1}{\kappa_1\tilde{A}} + \frac{l_2}{\kappa_2\tilde{A}} + \frac{l_3}{\kappa_3\tilde{A}}\right). \tag{24}$$

Substitution of temperature differences derived in Eqs. (20) and (24) into Eq. (18) results in the expression of α_{eff} for serially connected three-phase structures, Eq. (25):

$$\left(\alpha_{eff}\right)_{series} = \frac{\left(\dfrac{Ql_1\alpha_1}{\kappa_1\tilde{A}} + \dfrac{Ql_2\alpha_2}{\kappa_2\tilde{A}} + \dfrac{Ql_3\alpha_3}{\kappa_3\tilde{A}}\right)}{Q\left(\dfrac{l_1}{\kappa_1\tilde{A}} + \dfrac{l_2}{\kappa_2\tilde{A}} + \dfrac{l_3}{\kappa_3\tilde{A}}\right)} = \frac{\left(\dfrac{\alpha_1 x_1}{\kappa_1} + \dfrac{\alpha_2 x_2}{\kappa_2} + \dfrac{\alpha_3 x_3}{\kappa_3}\right)}{\left(\dfrac{x_1}{\kappa_1} + \dfrac{x_2}{\kappa_2} + \dfrac{x_3}{\kappa_3}\right)} = \frac{\sum \dfrac{\alpha_i x_i}{\kappa_i}}{\sum \dfrac{x_i}{\kappa_i}}. \tag{25}$$

Applying the same considerations described above, effective electrical and thermal conductivities can also be derived, as expressed in Eqs. (26) and (27), respectively:

$$\left(\sigma_{\mathrm{eff}}\right)_{\mathrm{series}} = \frac{1}{\dfrac{x_1}{\sigma_1} + \dfrac{x_2}{\sigma_2} + \dfrac{x_3}{\sigma_3}} = \frac{1}{\left(\sum \dfrac{x_i}{\sigma_i}\right)},$$

(26)

$$\left(\kappa_{\mathrm{eff}}\right)_{\mathrm{series}} = \frac{1}{\dfrac{x_1}{\kappa_1} + \dfrac{x_2}{\kappa_2} + \dfrac{x_3}{\kappa_3}} = \frac{1}{\left(\sum \dfrac{x_i}{\kappa_i}\right)}.$$

(27)

Similarly to the previous case of parallel-connected phases, i-ordered multi-phase materials will follow general-ordered right-hand side expressions of Eqs. (25)–(27). Furthermore, it can be easily seen, that Eq. (25) and Eqs. (26) and (27) for the case of two-phase materials are reduced to Eqs. (8) and (6), respectively, derived from the GEM method, highlighting validity of the analytic approach described here.

5. Practical examples and applications

Prior to describing the full potential of the GEM concept on optimizing performance of multi-phase TE materials, two general examples highlighting the potential of the method for monitoring the microstructure and phase morphology are described.

While analysing measured electrical and thermal conductivities of Cu following different spark plasma sintering (SPS) conditions, resulting in porosity levels in the range of 0–30%, a good agreement to GEM equations, Eqs. (2) and (3), was observed while assuming homogeneous dispersion ($t = 1$) and nearly spherical morphology ($A = 2$), as were observed by electronic microscopy, as well as σ_1, κ_1 values of pure Cu (the matrix phase), and σ_2, κ_2 equal to zero (the pores phase) [36]. This approach not just validated experimentally the GEM equations described above, but also paved a route for monitoring porosity amount during SPS consolidation process, which is widely applied in the synthesis of TE materials, as pointed out above, just by measuring electrical resistivity of the samples. For the SnTe system in the two-phase compositional range between pure Sn and SnTe compound, a parallel morphological alignment of the phases was identified both by electronic microscopy and by measuring Seebeck coefficient values of the samples [4]. The latter was validated by comparing measured α_{eff} to values, calculated by GEM equation, Eq. (1), with various A values. The best agreement was obtained for $A = 8$, indicating a parallel alignment of the phases. This approach validated the possibility to identify geometrical alignment of the phases just by measuring Seebeck coefficient values without any requirement of advanced electron microscopy.

Specifically, for TE materials, it was recently shown that upon introduction of $MoSe_2$ phase into layered n-type $Bi_2Te_{2.4}Se_{0.6}$ alloy for optimizing its TE performance, the best performance was obtained for oriented samples with $A = 0.3$, in Eqs. (1)–(3), as shown, for example, for ρ_{eff} in **Figure 3(a)** [3]. In this figure, the agreement of red experimental points with $A = 0.3$ curve can be clearly seen.

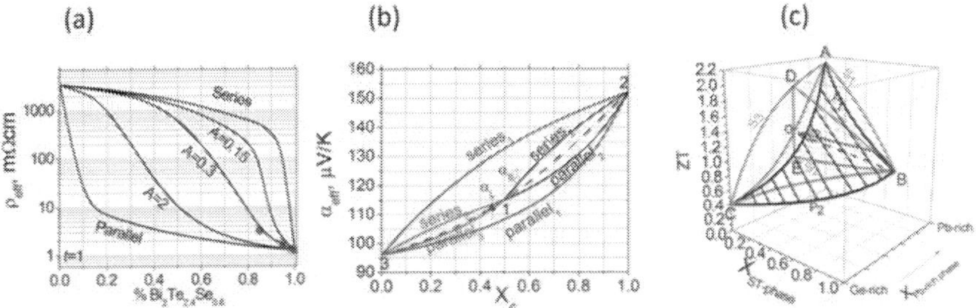

Figure 3. (a) Variations of effective electrical resistivity values upon introduction of $MoSe_2$ in $Bi_2Te_{2.4}Se_{0.6}$-$MoSe_2$ two-phase system [3]. (b) Room temperature GEM analysis of effective Seebeck coefficient upon homogeneous mixing ($t = 1$) of c-axis and a-axis oriented grains of HMS for different geometrical alignment ($0,series<A<\infty,parallel$) conditions [12]. (c) Interaction of ZT surfaces and volumes between three phases, solution treated (ST) matrix (B), Pb-rich (A) and Ge-rich (C) phases of $Pb_{0.25}Sn_{0.25}Ge_{0.5}Te$. The entire interaction volumes are bounded by ABC points, where each volume is bounded by two surfaces of series (S_1-S_2-S_3) and parallel (P_1-P_2-P_3) alignments [37].

A similar approach was recently applied for investigation of the morphological effects on TE properties of $Ti_{0.3}Zr_{0.35}Hf_{0.35}Ni_{1+\delta}Sn$ alloys following phase separation into half-Heulser $Ti_{0.3}Zr_{0.35}Hf_{0.35}NiSn$ and Heusler $Ti_{0.3}Zr_{0.35}Hf_{0.35}Ni_2Sn$ phases [15]. In this research, it was found that although phases' orientation was aligned in intermediate level ($A = 0.8$) between parallel ($A = 8$) and spherical ($A = 2$) alignments, enhanced TE performance is expected in a series alignment while substituting $A=0$ in Eqs. (1)–(3).

Another very interesting implementation of GEM approach was recently applied to estimate effective room temperature Seebeck coefficient and electrical resistivity values of a randomly morphological oriented homogeneous mixture of (001) and (hk0) grains in anisotropic polycrystalline HMS TE samples [12]. Applying GEM analysis to homogeneous distribution of (001) and (hk0) oriented grains ($t = 1$), for different alignment (A) conditions, resulted in the blue curves shown in **Figure 3(b)**. In this figure, the upper and lower blue curves represent series and parallel alignments of two configurations, respectively, points 2 and 3 represent c- and a-axis-oriented crystals, respectively, and intermediate dashed blue curve indicates a spherical distribution of two directions. Point 1 indicates 50% mixture of the directions for a spherical alignment, representing mixture of two directions, as in the case of non-textured polycrystalline HMS powder. The black and red curves of **Figure 3(b)** indicate interaction between c- and a-axis-oriented grains with randomly distributed polycrystalline powder (point 1 in **Figure 3b**), as was calculated by GEM approach. In that case, a partial c-axis preferred orientated powder, embedded in a homogeneous surrounding of macroscopic non-preferred-orientated powder is expected to exhibit α_{eff} values that are bounded in between the series$_2$ and parallel$_2$ black curves of the figures. Similarly, α_{eff} values for partial a-axis preferred orientated powder are expected to be bounded between series$_3$ and parallel$_3$ red curves of the figure. The experimentally measured $\alpha_{\parallel}$, $\alpha_{\perp}$ values while considering 10% preferred orientation, as was identified by XRD, are also shown in the figure. It can be seen that experimental points lie in the interaction zone between c- and a-axis-orientated powder and a randomly distributed powder, bounded by the black and red curves, respectively. This indicates the validity of proposed calculation route to estimate electronic transport properties of textured polycrystalline materials. It can be also seen that for HMS, ~10% preferred orientation of both of investigated directions is almost independent of the orientation of the grains, and, therefore, controlling the alignment of the grains morphology is not expected to affect the effective Seebeck coefficient.

Implementation of GEM concept in three-phase TE materials, based on Eqs. (15)–(17) and (25)–(27), was recently shown for quasi-ternary GeTe-PbTe-SnTe system [37, 38]. Specifically, it was shown that phase separation of solution-treated (ST) $Pb_{0.25}Sn_{0.25}Ge_{0.5}Te$ composition (phase B in **Figure 3c**) into Pb-rich, $Pb_{0.33}Sn_{0.3}Ge_{0.37}Te$ (phase A), and Ge-rich, $Pb_{0.1}Sn_{0.17}Ge_{0.73}Te$ (phase C) phases is apparent in the system. In this system, prolonged thermal treatments at each temperature resulted, at the first stages, in three phases, parent B phase and two decomposed A and C phases. This stage is terminated by full decomposition into A and C, where only these phases are apparent. Furthermore, a lamellar alignment of the phases was observed at the first 24 h of thermal treatment, while prolonged treatments were resulted in spheroidization, due to reduced surface area free energy at this configuration. It was also observed that ZT values were increased during the first 24 h while reduced at more prolonged durations. For explaining these experimental evidences, GEM approach was applied, as shown in **Figure 3(c)**. In this figure, triangle BDE indicates the specific interaction surface for separation of the phase B into the phases A and C, where BD side of triangle represents series ('lamellar') alignment morphology and BE represents parallel alignment of the phases. The dashed BO line represents spherical alignment. It can be easily shown that measured ZT values, indicated by the blue line, indeed follow the series alignment (BD line) at the first decomposition stages, but from this point on approach the dashed BO line until a full spheroidization is occurred (at point o). From this analysis, it was concluded that any theoretical possibility for retaining the lamellar morphology in this system would result in even higher ZT values of up to ~1.8 after a complete decomposition of the matrix into the two involved separation phases.

6. Concluding remarks

In this chapter, the potential of GEM approach to optimize electronic properties of multi-phase thermoelectric materials in terms of compositional or morphological considerations is shown in details. This approach already proved itself in monitoring of the densification rate of powder metallurgy processed materials, as well as in the determination of compositional modifications in binary systems just by measuring one of the transport properties. It is just beginning to approach the true potential to optimize thermoelectric transport properties of multi-phase materials, such as those containing embedded nano-features for reduction of the lattice thermal conductivity, where electronic contribution of the involved phase is usually neglected. It was shown that method does not just explain unexpected electronic trends in such materials, but might be employed for prediction of synthesis routes for optimizing thermoelectric figure of merit based on different compositions or alignment morphologies.

Based on the pointed above examples, it is obvious that for TE power generators operating at low (<300°C), intermediate (300–500°C) and high (>500°C) temperature ranges, Bi_2Te_3, PbTe/

GeTe and HMS/half-Heusler-based compositions might be employed. In such systems, identifying compositions enabling phase separation or precipitation into multi-phases, according to specific phase diagram, has a potential to reduce lattice thermal conductivity. Yet, for maximizing TE potential, optimal geometrical alignment of the phases should be identified. Using the proposed approach, based on individual TE transport properties of the involved phases, optimal geometrical alignment direction might be identified, leading to enhanced TE performance, enabling a real contribution to the society by reducing our dependence on fossil fuels and by minimizing emission of greenhouse gases.

Acknowledgements

The work was supported by the Ministry of National Infrastructures, Energy and Water Resources grant (3/15), No. 215-11-050.

Author details

Yaniv Gelbstein

Address all correspondence to: yanivge@bgu.ac.il

Department of Materials Engineering, Ben-Gurion University of the Negev, Beer-Sheva, Israel

References

[1] Meroz O, Ben-Ayoun D, Beeri O, Gelbstein Y. Development of $Bi_2Te_{2.4}Se_{0.6}$ Alloy for thermoelectric power generation applications. Journal of Alloys and Compounds. 2016; 679:196–201.

[2] Vizel R, Bargig T, Beeri O, Gelbstein Y. Bonding of Bi_2Te_3 based thermoelectric legs to metallic contacts using $Bi_{0.82}Sb_{0.18}$ alloy. Journal of Electronic Materials. 2016; 45(3):1296–1300.

[3] Shalev T, Meroz O, Beeri O, Gelbstein Y. Investigation of the influence of $MoSe_2$ on the thermoelectric properties of n-type $Bi_2Te_{2.4}Se_{0.6}$. Journal of Electronic Materials. 2015; 44(6):1402.

[4] Gelbstein Y. Thermoelectric power and structural properties in two phase Sn/SnTe Alloys. Journal of Applied Physics. 2009; 105:023713.

[5] Guttmann GM, Dadon D, Gelbstein Y. Electronic tuning of the transport properties of off-stoichiometric $Pb_xSn_{1-x}Te$ thermoelectric alloys by Bi_2Te_3 doping. Journal of Applied Physics. 2015; 118:065102.

[6] Cohen I, Kaller M, Komisarchik G, Fuks D, Gelbstein Y. Enhancement of the thermoelectric properties of n-type PbTe by Na and Cl co-doping. Journal of Materials Chemistry C. 2015; 3:9559–9564.

[7] Gelbstein Y, Dashevsky Z, Dariel MP. In-doped $Pb_{0.5}Sn_{0.5}Te$ p-type samples prepared by powder metallurgical processing for thermoelectric applications. Physica B. 2007; 396:16–21.

[8] Hazan E, Ben-Yehuda O, Madar N, Gelbstein Y. Functional graded germanium-lead chalcogenides-based thermoelectric module for renewable energy applications. Advanced Energy Materials. 2015; 5(11):1500272.

[9] Hazan E, Madar N, Parag M, Casian V, Ben-Yehuda O, Gelbstein Y. Effective electronic mechanisms for optimizing the thermoelectric properties of GeTe rich alloys. Advanced Electronic Materials. 2015; 1:1500228.

[10] Gelbstein Y, Davidow J. Highly-efficient functional $Ge_xPb_{1-x}Te$ based thermoelectric alloys. Physical Chemistry and Chemical Physics. 2014; 16:20120.

[11] Davidow J, Gelbstein Y. A comparison between the mechanical and thermoelectric properties of the highly efficient p-type GeTe rich compositions – TAGS-80, TAGS-85 and 3 % Bi_2Te_3 doped $Ge_{0.87}Pb_{0.13}Te$. Journal of Electronic Materials. 2013; 42(7):1542–1549.

[12] Sadia Y, Aminov Z, Mogilyansky D, Gelbstein Y. Texture anisotropy of higher manganese silicide following arc-melting and hot-pressing. Intermetallics. 2016; 68:71–77.

[13] Sadia Y, Madar N, Kaler I, Gelbstein Y. Thermoelectric properties in the quasi-binary $MnSi_{1.73}$-$FeSi_2$ system. Journal of Electronic Materials. 2015; 44(6):1637.

[14] Sadia Y, Elegrably M, Ben-Nun O, Marciano Y, Gelbstein Y. Sub-micron features in higher manganese silicide. Journal of Nanomaterials. 2013; 701268.

[15] Appel O, Zilber T, Kalabukhov S, Beeri O, Gelbstein Y. Morphological effects on the thermoelectric properties of $Ti_{0.3}Zr_{0.35}Hf_{0.35}Ni_{1+\delta}Sn$ alloys following phase separation. Journal of Materials Chemistry C. 2015; 3:11653–11659.

[16] Kirievsky K, Sclimovich M, Fuks D, Gelbstein Y. *Ab initio* study of the thermoelectric enhancement potential in nano grained TiNiSn. Physical Chemistry and Chemical Physics. 2014; 16:20023.

[17] Appel O, Gelbstein Y. A comparison between the effects of Sb and Bi doping on the thermoelectric properties of the $Ti_{0.3}Zr_{0.35}Hf_{0.35}NiSn$ half-Heusler alloy. Journal of Electronic Materials. 2014; 43(6):1976–1982.

[18] Kirievsky K, Gelbstein Y, Fuks D. Phase separation and antisite defects possibilities for enhancement the thermoelectric efficiency in TiNiSn half-Heusler alloys. Journal of Solid State Chemistry. 2013; 203:247–254.

[19] Appel O, Schwall M, Kohne M, Balke B, Gelbstein Y. Microstructural evolution effects of spark plasma sintered $Ti_{0.3}Zr_{0.35}Hf_{0.35}NiSn$ half-Heusler compound on the thermoelectric properties. Journal of Electronic Materials. 2013; 42(7):1340–1345.

[20] Gelbstein Y, Tal N, Yarmek A, Rosenberg Y, Dariel MP, Ouardi S, Balke B, Felser C, Köhne MM. Thermoelectric properties of spark plasma sintered composites based on TiNiSn half Heusler alloys. Journal of Materials Research. 2011; 26(15):1919–1924.

[21] Biswas K, He J, Blum ID, Chun-I-Wu, Hogan TP, Seidman DN, Dravid VP, Kanatzidis M. High-performance bulk thermoelectrics with all-scale hierarchial architectures. Nature. 2012; 489:414.

[22] Biswas K, He J, Zhang Q, Wang G, Uher C, Dravid VP, Kanatzidis MG. Strained endotaxial nanostructures with high thermoelectric figure of merit. Nature Chemistry. 2011; 3:160.

[23] Ohta M, Biswas K, Lo S-H, He J, Chung DY, Dravid VP, Kanatazidis MG. Enhancement of thermoelectric figure of merit by the insertion of MgTe nanostructures in p-type PbTe doped with Na_2Te. Advanced Energy Materials. 2012; 2:1117–1123.

[24] Pei Y, LaLonde AD, Heinz NA, Snyder GJ. High thermoelectric figure of merit in PbTe alloys demonstrated in PbTe-CdTe. Advanced Energy Materials. 2012; 2:670–675.

[25] Zhao L-D, Lo S-H, Zhang Y, Sun H, Tan G, Uher C, Wolverton C, Dravid VP, Kanatzidis MG. Ultralow thermal conductivity and high thermoelectric figure of merit in SnSe crystals. Nature. 2014; 508:373.

[26] Gelbstein Y, Davidow J, Girard SN, Chung DY, Kanatzidis M. Controlling metallurgical phase separation reactions of the $Ge_{0.87}Pb_{0.13}Te$ alloy for high thermoelectric performance. Advanced Energy Materials. 2013; 3:815–820.

[27] Gelbstein Y, Dashevsky Z, Dariel MP. Highly efficient bismuth telluride doped p-type $Pb_{0.13}Ge_{0.87}Te$ for thermoelectric applications. Physica Status Solidi (RRL). 2007; 1(6): 232–234.

[28] Gelbstein Y, Dado B, Ben-Yehuda O, Sadia Y, Dashevsky Z, Dariel MP. Highly efficient Ge-rich $Ge_xPb_{1-x}Te$ thermoelectric alloys. Journal of Electronic Materials. 2010; 39(9): 2049.

[29] Gelbstein Y, Rosenberg Y, Sadia Y, Dariel MP. Thermoelectric properties evolution of spark plasma sintered $(Ge_{0.6}Pb_{0.3}Sn_{0.1})Te$ following a spinodal decomposition. Journal of Physical Chemistry C. 2010; 114:13126–13131.

[30] Rosenberg Y, Gelbstein Y, Dariel MP. Phase separation and thermoelectric properties of the $Pb_{0.25}Sn_{0.25}Ge_{0.5}Te$ compound. Journal of Alloys and Compounds. 2012; 526:31–38.

[31] Ben-Yehuda O, Gelbstein Y, Dashevsky Z, Shuker R, Dariel MP. Highly textured Bi_2Te_3-based materials for thermoelectric energy conversion. Journal of Applied Physics. 2007; 101:113707.

[32] Gelbstein Y, Dashevsky Z, Dariel MP. The search for mechanically stable PbTe based thermoelectric materials. Journal of Applied Physics. 2008; 104:033702.

[33] Webman I, Jortner J, Cohen MH. Thermoelectric power in inhomogeneous materials. Physical Review B. 1977; 16(6):2959–2964.

[34] Bergman DJ, Levy O. Thermoelectric properties of a composite medium. Journal of Applied Physics. 1991; 70:6821.

[35] McLachlan DS, Blaszkiwicz M, Newnham RE. Electrical resistivity of composites. Journal of American Ceramic Society. 1990; 73(8):2187–2203.

[36] Gelbstein Y, Haim Y, Kalabukhov S, Kasiyan V, Hartman S, Rothe S, Frage N. Correlation between thermal and electrical properties of spark plasma sintered (SPS) porous copper. In: Lakshmanan A, editor. Sintering, InTech Publisher, Rijeka, Croatia. 2014. ISBN 978-953-51-4113-6

[37] Gelbstein Y. Phase morphology effects on the thermoelectric properties of $Pb_{0.25}Sn_{0.25}Ge_{0.5}Te$. Acta Materialia. 2013; 61(5):1499–1507.

[38] Gelbstein Y. Morphological effects on the electronic transport properties of three-phase thermoelectric materials. Journal of Applied Physics . 2012; 112:113721.

Chapter 13

Thermal Conductivity and Non-Newtonian Behavior of Complex Plasma Liquids

Aamir Shahzad and Maogang He

Additional information is available at the end of the chapter

http://dx.doi.org/10.5772/65563

Abstract

Understanding of thermophysical properties of complex liquids under various conditions is of practical interest in the field of science and technology. Thermal conductivity of nonideal complex (dusty) plasmas (NICDPs) is investigated by using homogeneous nonequilibrium molecular dynamics (HNEMD) simulation method. New investigations have shown, for the first time, that Yukawa dusty plasma liquids (YDPLs) exhibit a non-Newtonian behavior expressed with the increase of plasma conductivity with increasing external force field strength F^{ext}. The observations for lattice correlation functions $\Psi\ (t)$ show, that our YDPL system remains in strongly coupled regime for a complete range of plasma states of (Γ, κ), where (Γ) Coulomb coupling and (κ) Debye screening length. It is demonstrated, that the present NICDP system follows a simple scaling law of thermal conductivity. It has been shown, that our new simulations extend the range of F^{ext} used in the earlier studies in order to find out the size of the linear ranges. It has been shown that obtained results at near equilibrium ($F^{ext} = 0.005$) are in satisfactory agreement with the earlier simulation results and with the presented reference set of data showed deviations within less than ±15% for most of the present data points and generally overpredicted thermal conductivity by 3–22%, depending on (Γ, κ).

Keywords: thermophysical properties, thermal conductivity, nonlinear effects, lattice correlation functions, nonideal complex (dusty) plasmas, nonequilibrium molecular dynamics

1. Introduction

The exact numerical investigation of transport properties of complex liquids is a fundamental research task in the field of thermophysics, as various transport data are closely related with setup and the confirmation of equations of state. A reliable knowledge of transport data is also important for optimization of technological processes and apparatus design in various engineering and science fields (incl. thermoelectric devices) and, in particular, when provision of precise data for parameters of heat, mass, and momentum transport is required [1–3]. In thermophysical properties of fluids, chemical properties remain unaffected, but physical properties of material are changed by variable temperature, composition, and pressure. These properties of simple and complex liquids explain the phase transition [4]. These fluids can be examined experimentally, theoretically, and by simulation techniques. Thermophysical properties (thermodynamics and transport coefficients) include thermal conductivity, thermal expansion, thermal radiative properties, thermal diffusivity, enthalpy internal energy, Joule-Thomson coefficients, and heat capacity, as well as, thermal diffusion coefficients, mass coefficients, viscosity, speed of sound, and interfacial and surface tension. Thermophysical properties of gases and liquids, such as hydrogen, H_2; oxygen, O_2; nitrogen, N_2; and water, H_2O are different from ideal gas at high pressure and low temperature. Specific models are required for the calculation of these properties in the widest range of pressure and temperature. Different fluids, such as gases and liquids are used as a power generation source in different power plants. For example, heavy water, steam, air, and different gases are used for power generation in nuclear power plants, gas turbine plants, and internal combustion plants. Also, for cooling in refrigerators and fast nuclear reactors, ammonia and sodium in liquid phase are used as a cooling agent.

1.2. Dusty plasma

Nowadays dusty plasma refers to as complex plasma in analogy to the condensed matter field of "complex liquids" in soft matter (colloidal suspensions, polymers, surfactants, etc.). The dust particles combine physics of nonideal plasmas and condensed matter, and this field has played an important role in both newly system designs and advance development micro- and nanotechnology. This complex plasma system has four components, i.e., ions, electrons, neutral atoms, and dust particles with high charges, as compared to other species, which are responsible for the extraordinary plasma properties. The study of complex (dusty) plasmas reveals rich variety of interesting phenomena and extends knowledge on fundamental aspects of plasma physics at the microscopic level. Among these, the freezing (gaseous-liquid-solid) phase transition is of particular interest. Complex plasma is called strongly coupled plasma, in which thermal energy (kinetic energy) of nearest neighbors is much smaller than their Coulomb interaction potential energy, whereas plasma is called weakly coupled when Coulomb interparticle potential energy of nearest particles is much smaller than their kinetic energy [5–7].

Plasma is the fourth state of matter, and usually, it is said, that there are three states of matter, but another state was also found to exist, named as plasma. Irving Langmuir (American physicist) defined plasma as "it is a quasi-neutral gas of charged and neutral particles, which exhibits collective behavior," and he got the Nobel Prize in 1927 because firstly he was using the term plasma [8]. In this definition, quasi-neutral means that plasma is electrically neutral and has approximately equal ion and electron density ($n_i \approx n_e$). The term collective behavior shows, that due to Coulomb potential or electric field, plasma's particles collide with each other. Simply, plasma is an ionized gas. In a gas, sufficient energy is given to eject free electrons from atoms or molecules, and, as a result, ions and electrons, both species, coexist. There are several ways to convert a gas into plasma, but all include pumping the gas with energy. For instance, plasma can be created due to a spark in a gas. Usually, neutral plasmas are relatively hot, such as the solar corona (1000000 K), a candle flame (1000 K), or the ionosphere around our planet (300 K). In this universe, the main source of plasma is sun in which the large number of electrons of hydrogen and helium molecules is removed. So, the sun is a great big ball of plasma like other stars.

1.2.1. Types of plasma

There are different types of plasma that are described by many characteristics, such as temperature, degree of ionization, and density.

1.2.1.1. Cold plasma

In laboratory, in the positive column of a glow discharge tube, there exists plasma in which the same number of ions and electrons is present. When gas pressure is low, collision between electrons and gas molecules is not frequent. So, nonthermal equilibrium between energy of electrons and gas molecules does not exist. So, energy of electrons is very high as compared to gas molecules and the motion of gas molecules can be ignored. We have $T_e \gg T_i \gg T_g$, where T_e, T_i, and T_g represent temperature of electron, ion, and gas molecules and such type of plasma is known as cold plasma. In cold plasma, the magnetic force can be ignored and only the electric force is considered to act on the particles. In cold plasma technique, cold gases are used to disinfect surfaces of packaging or food products. Vegetative microorganisms and spores on packaging materials can be inactivate at temperature below 40°C. This process can have a clear advantage compared to heat treatment for temperature-sensitive products. Also it can reduce the amount of water used for disinfection of packaging materials. Because cold plasma is in the form of gas, so, the irregularly shaped packages, such as bottles can be treated easily as compared to UV or pulsed light where shadowing occurs.

1.2.1.2. Hot plasma

When gas pressure is high in the discharge tube, then electrons collide with gas molecules very frequently and thermal equilibrium exists between electrons and gas molecules. We have $T_e \cong T_i$. Such type of plasma is known as hot plasma and it is also known as thermal plasma. Hot plasma is one which approaches to a local thermodynamic equilibrium (LTE). Atmospheric arcs, sparks, and flames are used for the production of such type of plasmas.

1.2.1.3. Ultracold plasma

If plasma occurs at temperature as low as 1 K, then such type of plasma is known as ultracold plasma, and it can be formed by photoionizing laser-cooled atoms and pulsed lasers. In ultracold plasmas, the particles are strongly interacting because their thermal energy is less than Coulomb energy between neighboring particles [9].

1.2.1.4. Ideal plasma

There are mainly two types of plasma according to plasmas' ideality and properties study, nonideal plasmas (weakly coupled and strongly coupled plasmas) and ideal plasmas (very weakly coupled plasmas). Whenever the kinetic energy of plasma is much larger than the potential energy and plasma has a low temperature and high density, then such type of plasma is known as ideal plasma. Ideal plasma is one in which Coulomb collisions are negligible. If the average distance among the interacting particles is large, then the interaction potential can be ignored due to this large-distance ideal plasma that does not have any arrangement of particles [2].

1.2.1.5. Nonideal (complex) plasma

Nonideal plasmas are often found in nature, as well as, in technological services. They can be shown as electron plasma in solid and liquid metals and electrolytes, the superdense plasma of the matter of white dwarfs, the sun and the interiors (deep layers) of the giant planets of the solar system, and astrophysical objects, whose structure and evolution are defined by plasma characteristics [2]. Further examples of nonideal plasmas are brown dwarfs, laser-generated plasmas, capillary discharges, plasma-opening switches, high-power electrical fuses, exploding wires, etc. On the bases of Coulomb coupling, nonideal plasma can be divided into two families: weakly coupled plasma (WCP, $\Gamma < 1$) and strongly coupled plasma (SCP, $\Gamma \geq 1$).

Nonideal complex plasmas are found in daily life and can be found in processing industries to manufacture many products that we deal in our everyday life directly or indirectly at moderate temperature, such as plastic bags, automobile bumpers, airplane turbine blades, artificial joints, and, most importantly, in semiconductor circuits. Moreover, nonideal plasmas (terrestrial plasmas) are not hard to find. They occur in gas-discharge lighting, such as neon lighting used for commercial purposes and fluorescent lamps, for instance, compact fluorescence light sources, which have a higher performance than the traditional incandescent light sources, a variety of laboratory experiments, and a growing array of industrial processes. Modern display methods contain plasma screens, in which small plasma discharges are used to stimulate a phosphor layer, which then emits light [2, 7].

1.2.2. Complex (dusty) plasma

Dust is present everywhere in the universe and mostly it is present in solid form. It is also present in gaseous form, which is often ionized, and thus the dust coexists with plasma and forms "dusty plasma." In dusty plasma, dust particles are immersed in plasma, in which ions, electrons, and neutrals are present. These dust particles are charged and then affected by electric or magnetic fields and can cause different changes in the properties of plasma. The presence of dust component gives rise to new plasma phenomena and allows study of fundamental aspects of plasma physics at the microscopic level. Dust particles are charged due to the interaction between dust particles and the surrounding plasmas. Due to this interaction, grains are charged very rapidly. The charge on grains depends on the flow of ions and electrons. These charged grains enhance plasma environments, for example, setting up space charges. Also, to determine the charge on dust grain, it is assumed, that a spherically symmetric isolated dust grain is injected in plasma and only the effect of ion and electron is considered. Moreover, there are many other charging processes, such as secondary emission, electron emission, thermionic emission, field emission, radioactivity, and impact ionization. Complex plasma is condensed plasma characterized by strong interaction between existing molecules and atoms; it is also called strongly coupled complex plasma. Dusty plasma is complex plasma which includes many components: ions, electrons, neutral particle, and dust particles. Last 20–25 years, strongly coupled plasmas were mainly studied theoretically, due to lack of suitable laboratory tools and equipment. However, experimental strongly coupled plasma studies became more common with the discovery of ways to find dusty plasma [10], laser-cooled ion plasmas in a penning trap [11], and ultracold neutral plasmas [12]. Plasma systems can be treated theoretically in a straightforward way in the extreme limits of both weakly coupled and strongly coupled plasmas [2].

The main goals of this chapter are to study thermal conductivity (λ) along with lattice correlation (long-range crystalline order) at the corresponding plasma states and to extend the set of plasma states (Γ, κ) by using the same method as introduced by Shahzad and He [3] with different system sizes (N). The effect of external force field strengths on $\Psi(t)$ and corresponding λ values of complex Yukawa liquids of dust particles is another interesting task that is calculated under near-equilibrium condition.

2. HNEMD model and simulation approach

In this section, we will introduce theoretical background needed in this work. We start by introducing the model system, which is used in our HNEMD simulations. We consider a cubic box of edge length L and have N number of particles or millions of atoms. In MD technique, the range of particle number is chosen as N = 500–1000 [13]. Periodic boundary conditions (PBCs) are used for selection of the size or dimension of the box. Practically, PBCs avoid the surface size effects. The particles present in the box interact with each other with known interaction potential. This potential may be Yukawa, Coulomb, and Lennard-Jones potential depends on the type of the system considered.

The particles interact through screened Coulomb potential, which depends on the physical parameters and the background plasmas. Average interparticle interaction is frequently considered to be isotropic and basically repulsive and approximated by Yukawa interaction potential [1–3]. Yukawa model has been employed in many physical and chemical systems (for instance, biological and pharmaceutical sciences, colloidal and ionic systems, space and environment sciences, physics and chemistry of polymers and materials, etc.) [1–8]. In the present case, the interaction of potential energy of particles is in Yukawa form:

$$\phi_Y(|\mathbf{r}|) = \frac{Q_d^2}{4\pi\varepsilon_0}\frac{e^{-|\mathbf{r}|/\lambda_D}}{|\mathbf{r}|},$$

(1)

where charge on dust particle is Q_d, magnitude of interparticle division is r, and Debye length λ_D accounts for the screening of interaction by other plasma species. Due to the long-range interaction between the particles, Yukawa potential energy cannot be solved directly. Ewald sum method is used with periodic boundary conditions to calculate Yukawa potential energy, force, and heat energy current. Improved nonequilibrium molecular dynamics (NEMD) method proposed by Evans has been employed to estimate thermal conductivity of strongly coupled complex (dusty) plasmas. During the simulation of Yukawa systems, sufficient number of particles N is to be selected to study the size effect of system. It comes to know, that there is no effect of system size on thermal conductivity or on any other properties under limited statistical uncertainties. Negative divergence of Yukawa potential $F = (-\nabla\phi)$ and Newton's equation of motion integrated by predictor-corrector algorithm are used for calculation of force exerted by the particles on each other or in minimum image convection [14].

In HNEMD technique, in order to measure thermal conductivity and nonlinearity of NICDPs, the system will be perturbed by applying the external field along z-axis. In three-dimensional systems, we use standard Green-Kubo relations (GKR) for calculation of thermal conductivity coefficient of uncharged particles [15]:

$$\lambda = \frac{1}{3k_B V T^2}\int_0^\infty \left\langle \mathbf{J}_Q(t).\mathbf{J}_Q(0)\right\rangle dt.$$

(2)

Here, $\mathbf{J}_Q$ is vector of heat flux, V is the volume, T is the temperature of system, and k_B is Boltzmann constant. At microscopic level, heat flux vector has value:

$$\mathbf{J}_Q V = \sum_{i=1}^{N} E_i \frac{\mathbf{p}_i}{m} - \frac{1}{2}\sum_{i\neq j}\mathbf{r}_{ij}\left(\frac{\mathbf{p}_i}{m}.\mathbf{F}_{ij}\right).$$

(3)

In this equation, $\mathbf{r}_{ij} = \mathbf{r}_i - \mathbf{r}_j$ is the position vector, $\mathbf{F}_{ij}$ is the force of interaction on particle i due to j, and $\mathbf{p}_i$ represents the momentum vector of the ith particle. The energy E_i of ith particle is given by:

$$E_i = \frac{\mathbf{p}_i^2}{2m} + \frac{1}{2}\sum_{i\neq j}\phi_{ij},$$

(4)

where $\varnothing_{ij}$ is Yukawa pair potential between particles i and j. According to non-Hamiltonian dynamics, generalization of linear response theory, proposed by Evans, for system moving with equations of motion and recently detailed understanding of Ewald-Yukawa sums [1–3] allows to present thermal conductivity as:

$$\lambda = \lim_{F_z \to 0} \lim_{t \to \infty} \frac{-\langle \mathbf{J}_{Q_z}(t) \rangle}{TF_z},$$

(5)

where J_{Qz} is z-component of heat energy flux vector for strongly coupled complex (dusty) plasma liquids. Thermal conductivity coefficient for charged particle of plasma according to GKR is given by Eq. (5), and further detail is provided in Ref. [3]. Plasma states of Yukawa systems can be illustrated fully by three reduced parameters: plasma coupling parameter $\Gamma = (Q^2/4\pi\varepsilon_0)(a_{ws}k_BT)$, screening parameter $\kappa \equiv a_{ws}/\lambda_D$, and reduced external force F^{ext} for HNEMD model, where a_{ws} is Wigner-Seitz radius, Q is charge on dust particle, and ε_0 is permittivity of free space [1–3]. Gaussian thermostat is used to control temperature of systems [14]. Simulation time step is $dt = 0.001/\omega_p$, where $\omega_p = (nQ^2/\varepsilon_0 m)^{1/2}$ is dust plasma frequency with m is dust particle's mass and n is number density. Reported simulations are performed between 3.0 × $10^5/\omega_p$ and 1.5 × $10^5/\omega_p$ time units in the series of data recording of thermal conductivity (λ) [16, 17].

3. Computer simulation outcomes

3.1. Particle lattice correlation

The structural information of Yukawa system is given by lattice correlation. For the calculation of lattice correlation, density of given material at point r can be calculated as:

$$\rho(r) = \sum_{j=1}^{N} \delta(r - r_j),$$

(6)

where ρ is density of system, N is number of particles, δ represents distribution of particles, and r_j is the position of particle corresponding to particle at position r. Eq. (6) gives information about the system being in ordered state and then it may be in solid or crystal form depending on ρ. The lattice correlation equation according to Fourier transform is:

$$\Psi = \frac{1}{N} \sum_{i=1}^{N} \exp(-i\mathbf{k}.\mathbf{r}_j).$$

(7)

System arrangement (ordered or disordered) is calculated by simulation based on Eq. (7). When value of lattice correlation approaches to $|\Psi| \approx 1$, then the system will be in ordered state, and if value becomes $|\Psi| \approx 0$, then the system will be in liquid or gas (nonideal gas) state. In Eq. (7), k is lattice correlation vector for ordered state, and its value is different for different lattice structures. Its value for face-centered cubic (FCC) is $k = 2\pi/(1,-1,1)l$, for body-centered cubic (BCC) is $k = 2\pi/(1,0,1)l$, and for simple cubic (SC) is $k = 2\pi/(1,0,0)l$; here, l is edge length [14].

Lattice correlation was examined in 3D NICDPs in the limit of appropriate constant near equilibrium external force field strength $F^{ext} = 0.005$. **Figure 1** illustrates lattice correlation in NICDPs versus simulation time at normalized $F^{ext} = 0.005$. In this case, additional parameter includes the heat energy flux J_Q and the external force field strength $F^{ext}(t) = (0,0,F_z)$ is selected along z-axis, in the limit of $t \rightarrow \infty$, and its normalized value $F^{ext} = (F_z)(a_{ws}/J_Q)$ [17].

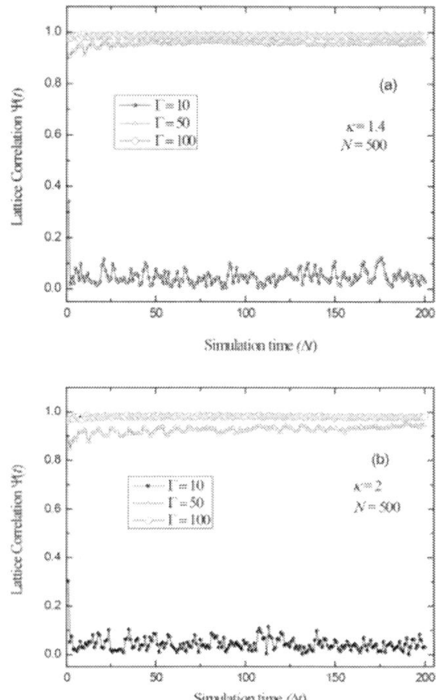

Figure 1. Dependences of lattice correlation $|\Psi\,(t)|$ on simulation time (Δt) at external force field strength $F^{ext} = 0.005$ imposed to Yukawa systems, for three values of coupling states $\Gamma = 10$, $\Gamma = 50$, and $\Gamma = 100$ and system size $N = 500$, (a) at $\kappa = 1.4$ and (b) at $\kappa = 2$.

3.2. Normalized thermal conductivity

We now turn attention to the key results obtained through HNEMD simulations. Obtained computer-simulated data confirm, that thermal conductivity of Yukawa system can be calculated with satisfactory statistics by an extended HNEMD approach. **Figures 2–5** display the main results calculated from HNEMD method for various plasma states for Yukawa liquids at $\kappa = 1.4$ and $\kappa = 2$ and $\kappa = 4$, respectively. HNEMD simulation is used to compute the thermal conductivity normalized by plasma frequency (ω_p) as $\lambda_0 = \lambda/nk_B\omega_p a_{ws}$, or by Einstein frequency (ω_E) as $\lambda^* = \lambda/\sqrt{3}nk_B\omega_E a_{ws}$ of YDPLs, at the normalized external field strength. These normalizations of transport properties, including λ_0, were widely used in earlier studies of one-component complex plasma (OCCP) [18] and NICDPs [1–3, 19–21]. HNEMD method is employed to investigate λ_0 of 3D NICDPs at reduced external force field $F^{ext} = 0.005$ over suitable domain of plasma parameters of coupling ($1 \leq \Gamma \leq 300$) and screening ($1 \leq \kappa \leq 4$).

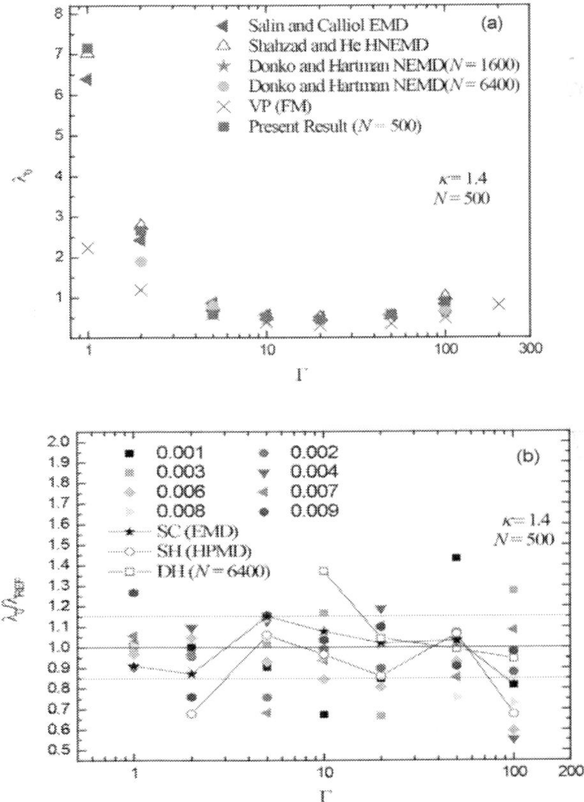

Figure 2. (a) Data for thermal conductivity λ_0 normalized by plasma frequency, calculated by different MD methods for Coulomb coupling parameter ($1 \leq \Gamma \leq 300$) at $N = 500$ and $\kappa = 1.4$. Data obtained by Shahzad and He for homogeneous perturbed MD (HPMD)-SH (HPMD) [1], Salin and Caillol for equilibrium MD (EMD)-SC (EMD) [19], Donko and Hartmann for inhomogeneous NEMD-DH NEMD [20], and Faussurier and Murillo for variance procedure (VP)-FM VP [21]. (b) Dependences of λ_0 normalized by λ_{REF} at $F^{ext} = 0.005$ on Coulomb coupling parameter ($1 \leq \Gamma \leq 100$) at $N = 500$ and $\kappa = 1.4$. Our present and earlier normalized results calculated at different F^{ext}. Results obtained by Shahzad and He [3] with $N = 13500$ at $F^{ext} = 0.005$ are taken as reference set of data λ_{REF}. Dashed lines represent spreading ±15 ranges around the reference data.

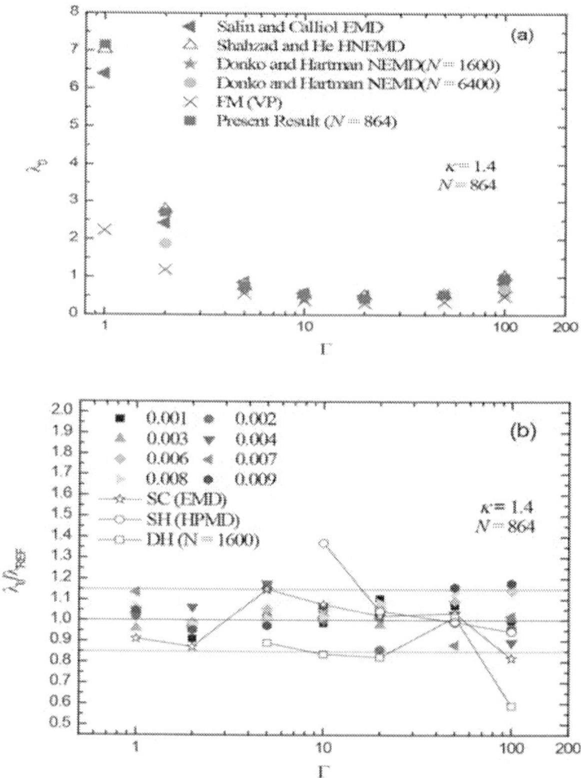

Figure 3. (a) Data for thermal conductivity λ_0 normalized by plasma frequency, calculated by different MD methods for Coulomb coupling parameter ($1 \leq \Gamma \leq 300$) at $N = 864$ and $\kappa = 1.4$. Data obtained by Shahzad and He for homogeneous perturbed MD (HPMD)-SH (HPMD) [1], Salin and Caillol for equilibrium MD (EMD)-SC (EMD) [19], Donko and Hartmann for inhomogeneous NEMD-DH NEMD [20], and Faussurier and Murillo for variance procedure (VP)-FM VP [21]. (b) Dependences of λ_0 reduced by λ_{REF} at $F^{ext} = 0.005$ on Coulomb coupling parameter ($1 \leq \Gamma \leq 100$) at $N = 864$ and $\kappa = 1.4$. Our present and earlier reduced results calculated at different F^{ext}. Results obtained by Shahzad and He [3] with $N = 13500$ at $F^{ext} = 0.005$ are taken as reference set of data λ_{REF}. Dashed lines represent spreading ±15 ranges around the reference data.

Different sequences of λ_0 corresponding to decreasing sequence of F^{ext} are calculated to determine the linear regime of YDPLs under the action of reduced force field strength. The earlier data of thermal conductivities of YDPLs has been limited to the small values of plasma parameters at different F^{ext}. The present HNEMD simulation enables study over the whole domain (Γ, κ) of plasma-state parameters with variation of F^{ext}. In this case, possible low value of the force field strength $F^{ext} = 0.005$ for determination of the near-equilibrium values of Yukawa thermal conductivity is to be chosen, for small reasonable system size. This possible reasonable external force field gives the near-equilibrium thermal conductivity measurements, which are acceptable for the whole domain of plasma parameters (Γ, κ).

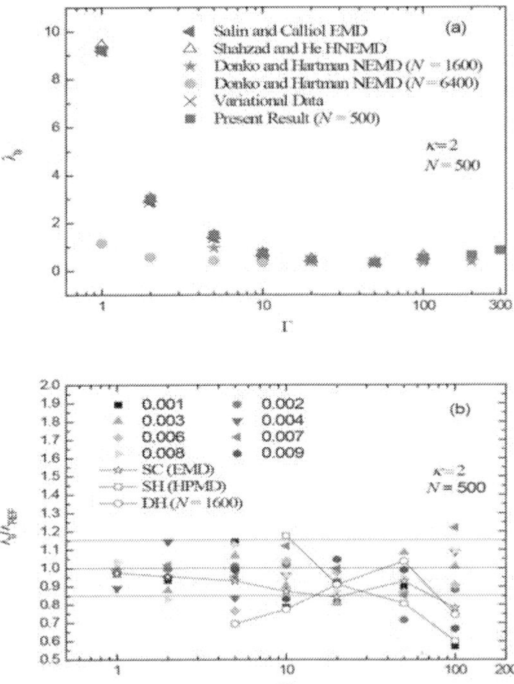

Figure 4. (a) Data for thermal conductivity λ_0 normalized by plasma frequency, calculated by different MD methods for Coulomb coupling parameter ($1 \leq \Gamma \leq 300$) at $N = 500$ and $\kappa = 2$. Data obtained by Shahzad and He for homogeneous perturbed MD (HPMD)-SH (HPMD) [1], Salin and Caillol for equilibrium MD (EMD)-SC (EMD) [19], Donko and Hartmann for inhomogeneous NEMD-DH NEMD [20], and Faussurier and Murillo for variance procedure (VP)-FM VP [21]. (b) Dependences of λ_0 reduced by λ_{REF} at $F^{ext} = 0.005$ on Coulomb coupling parameter ($1 \leq \Gamma \leq 100$) at $N = 500$ and $\kappa = 2$. Our present and earlier reduced results calculated at different F^{ext}. Results obtained by Shahzad and He [3] with $N = 13500$ at $F^{ext} = 0.005$ are taken as reference set of data λ_{REF}. Dashed lines represent spreading ±15 ranges around the reference data.

Figures 2–5 show, that measured thermal conductivity is in satisfactory agreement with earlier HPMD simulations by Shahzad and He [1], inhomogeneous NEMD computations by Donko and Hartmann [20], and EMD measurements by Salin and Caillol [19]. The present results are also higher than Salin and Caillol [19] results at lower $\Gamma = 1$ and 2, for $\kappa = 1.4$ ($N = 500$ and 864). The minimum value of λ_0 is $\lambda_{min} \approx 0.4533$ at $\Gamma = 20$ and $\kappa = 1.4$. Deviation of the present data from the earlier calculated results based on different techniques of EMD, HPMD, and NEMD is also calculated. It is observed, that the results of λ_0 are within the range of ~7–50% for EMD, ~10–16% for NEMD, and ~10–40% for HPMD. Moreover, **Figure 3(a)** shows, that obtained thermal conductivity at $F^{ext} = 0.005$ for $N = 864$ is in good agreement with HPMD simulation of Shahzad and He [1], but it is noted, that our results are slightly greater than EMD of Salin and Caillol [19], NEMD of Donko and Hartmann [20], and VP of Faussurier and Murillo [21] at lower values of Γ. Deviation of the present data from EMD, HPMD, and inhomogeneous

NEMD is within the range of ~4–14%, ~3–16%, and ~4–30%. The minimum value of $\lambda_0 = 0.4331$ at $\Gamma = 20$ and $\kappa = 1.4$.

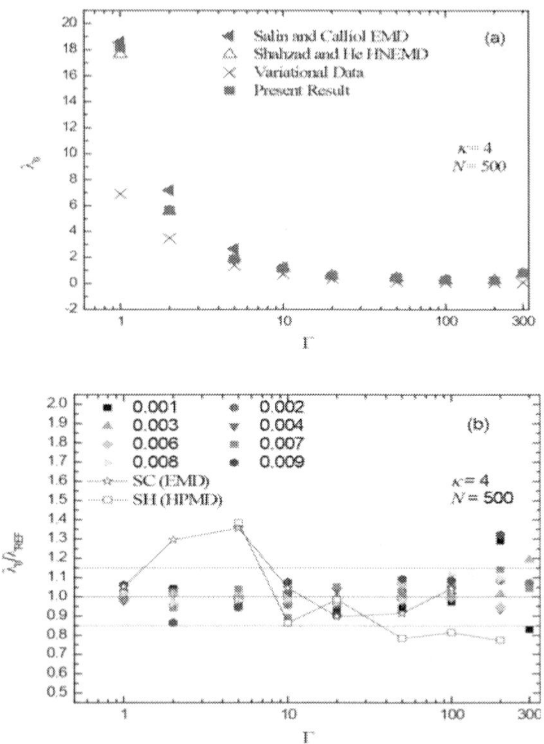

Figure 5. (a) Data for thermal conductivity (λ_0) normalized by plasma frequency, calculated by different MD methods for Coulomb coupling parameter ($1 \leq \Gamma \leq 300$) at $N = 500$ and $\kappa = 4$. Data obtained by Shahzad and He for homogeneous perturbed MD (HPMD)-SH (HPMD) [1], Salin and Caillol for equilibrium MD (EMD)-SC (EMD) [19], Donko and Hartmann for inhomogeneous NEMD-DH NEMD [20], and Faussurier and Murillo for variance procedure (VP)-FM VP [21]. (b) Dependences of λ_0 reduced by λ_{REF} at $F^{ext} = 0.005$ on Coulomb coupling parameter ($1 \leq \Gamma \leq 100$) at $N = 500$ and $\kappa = 4$. Our present and earlier reduced results calculated at different F^{ext}. Results obtained by Shahzad and He [3] with $N = 13500$ at $F^{ext} = 0.005$ are taken as reference set of data λ_{REF}. Dashed lines represent spreading ±15 ranges around the reference data.

This comparison shows, that our data remain within the limited statistical uncertainty range. Panels (b) of **Figures 2–5** compare the present simulation results of thermal conductivity, normalized by reference data, calculated here from HNEMD approach for different sets of external force field strengths with reference set of data and earlier known simulation data of HPMD, EMD, inhomogeneous NEMD, and VP techniques [1, 19–21]. A series of different sequences of HNEMD simulations are performed with $N = 500$ and 864 at various normalized F^{ext} values for varying screening parameters. In our case, results measured by Shahzad and

He [3] with $N = 13500$ at $F^{\text{ext}} = 0.005$ are taken as reference set of data λ_{REF} and are shown panels (b) of the respective figures. It is important, that computationally noted linear regime is traced between $0.001 \leq F^{\text{ext}}$ and ≤ 0.009, which depends on the plasma state points (Γ, κ). It observed, that differences between the simulation data calculated by different authors are large at some state points and differences with their own present data sets are much smaller. Plasma's thermal conductivity is generally overpredicted within ~3–17% (~8–15%), ~5–30% (~5–38%), and ~5–40% (~3–17%) relative to the data of EMD by Salin and Caillol [19], inhomogeneous NEMD by Donko and Hartman [20], and HPMD by Shahzad and He [1], respectively, for $N = 500$ (864). It is concluded, that most of data points of presented results fall under ±15 range around the reference data.

For the cases of $\kappa = 2$ and $\kappa = 4$, the minimum value of λ_0 is $\lambda_{\text{min}} \approx 0.3413$ (for $\kappa = 2$) at $\Gamma = 50$ and $\lambda_{\text{min}} \approx 0.2218$ (for $\kappa = 4$) at $\Gamma = 200$. The comparison of the present data with the earlier investigated results obtained from different techniques of EMD, HPMD, and NEMD is also shown in respective figure panels (a) for $\kappa = 2$ and $\kappa = 4$. It is observed, that the results of λ_0 are within the range ~1–15% (~2–13%, for $\kappa = 4$) for EMD, ~2–20% (~3–13%, for $\kappa = 4$) for HPMD, ~15–22% for inhomogeneous NEMD, and ~1–28% for HPMD. It is observed, that for $\kappa = 4$, thermal conductivity agrees relatively well with the simulation data in general within ~5–35% for EMD and ~2–23% for HPMD.

It is concluded from figures, that obtained results agree well with earlier results at intermediate and high Γ values; however, some data points deviate at the lower Γ values. **Figures 2–5** depict, that extended HNEMD approach can accurately predict thermal conductivity of Yukawa system (dusty plasma). We have used the present developed homogeneous NEMD method, which has an excellent performance; its accuracy is comparable to that of EMD and inhomogeneous NEMD techniques. The first conclusion from above **Figure 5** is that thermal conductivity depends on plasma parameters Γ and κ, confirming the earlier numerical results. Furthermore, it is examined, that the position of minimum value of λ_{min} shifts toward higher Γ with increase in κ, as expected. The minimum value of λ_0 decreases with increasing κ, as $\lambda_0 = 0.4533$ at $\kappa = 1$ to $\lambda_0 = 0.2218$ at $\kappa = 4$ and $\lambda_0 = 0.4331$ at $\kappa = 1$ to $\lambda_0 = 0.2099$ at $\kappa = 4$ for $N = 500$ and 864, respectively. Our numerical data are considerably more comprehensive covering the full range of coupling strengths from nearly nonideal gaseous state to strongly coupled liquid (SCL) as shown in **Figure 1**. The present approach has shown excellent results of λ_0 at steady-state value of F^{ext} at lower, intermediate, and higher coupling values, and it also gives more comparable performance of normalized thermal conductivity at lower N. This approach yields the practical accuracy, compared to EMD, for relatively small total simulation times due to the act of finite nonzero external force field; the signal-to-noise ratio of thermal response is high. It is significant from our simulation data, that the external force field strength of F^{ext} increases with increase in κ.

4. Summary

Thermal conductivity of NICDP system was investigated for wide range of Coulomb coupling parameter ($1 \leq \Gamma \leq 300$) and screening parameter ($1 \leq \kappa \leq 4$), applying external force field, by using HNEMD method. This HNEMD simulation method reveals, that our present results are in good agreement with the earlier results obtained by equilibrium MD and homogeneous and inhomogeneous NEMD simulations for NICDPs. It is confirmed, that lattice correlation is not affected by system size, while lattice correlation decreases with increment of κ and at high temperature ($1/\Gamma$). It is confirmed from presented HNEMD simulation results, in which normalized thermal conductivities follow simple universal (temperature) scaling law. Observations show, that the minimum value of thermal conductivity shifts toward higher Coulomb couplings with increasing screening strength, confirming earlier numerical results. It has been shown, that thermal conductivity depends on both temperature (plasma coupling) and density (screening) in 3D Yukawa systems, which shows previous data for NICDPs. Presented HNEMD approach was particularly a powerful numerical technique, which involves fast calculation of thermal conductivity, on small and intermediate system sizes, in contexts, applicable to the study of matter at microscopic level for 3D NICDPs. The second major contribution of this chapter is that it provides, for the first time, understanding and determination of non-Newtonian behavior of Yukawa liquid. It is important to note, that thermal conductivity is in increasing behavior at higher values of external force field. These indications show, that increasing behavior of field dependence of thermal conductivity decreases with increasing κ and decreasing Γ. Described simulation technique can be helpful for estimation of thermal conductivity regularities in novel liquid, organic (polymer), and multiphase solid-state thermoelectric materials.

Acknowledgements

This work was sponsored by the National Natural Science Fund for Distinguished Young Scholars of China (NSFC no. 51525604) and partially sponsored by the Higher Education Commission (HEC) of Pakistan (no. IPFP/HRD/HEC/2014/916). The authors thank Z. Donkó (Hungarian Academy of Sciences) for providing his thermal conductivity data of Yukawa liquids for the comparisons of our simulation results and useful discussions. We are grateful to the National High Performance Computing Center of Xi'an Jiaotong University and National Advanced Computing Center of the National Centre for Physics (NCP), Pakistan, for allocating computer time to test and run our MD code.

Abbreviations

NICDP	nonideal complex (dusty) plasma
HNEMD	homogeneous nonequilibrium molecular dynamics
Γ	Coulomb coupling
κ	Debye screening length
F^{ext}	external force field strength

YDPL	Yukawa dusty plasma liquid
$\Psi(t)$	lattice correlation functions
MD	molecular dynamics
LTE	local thermodynamic equilibrium
WCP	weakly coupled plasma
SCP	strongly coupled plasma
λ	thermal conductivity
λ_0	normalized thermal conductivity
PBCs	periodic boundary conditions
NEMD	nonequilibrium molecular dynamics
GKR	Green-Kubo relations
HPMD	homogeneous perturbed MD
EMD	equilibrium MD
VP	variance procedure

Author details

Aamir Shahzad[1,2*] and Maogang He[2]

*Address all correspondence to: aamirshahzad_8@hotmail.com; aamir.awan@gcuf.edu.pk

1 Department of Physics, Government College University Faisalabad (GCUF), Faisalabad, Pakistan

2 Key Laboratory of Thermo-Fluid Science and Engineering, Ministry of Education (MOE), Xi'an Jiaotong University, Xi'an, PR China

References

[1] Shahzad A, He M-G. Thermal conductivity calculation of complex (dusty) plasmas. Phys. Plasmas. 2012;19(8):083707. DOI: 10.1063/1.4748526

[2] Shahzad A, He M-G. Computer Simulation of Complex Plasmas: Molecular Modeling and Elementary Processes in Complex Plasmas. 1st ed. Saarbrücken, Germany: Scholar's Press; 2014. 170 p.

[3] Shahzad A, He M-G. Thermal conductivity of three-dimensional Yukawa liquids (dusty plasmas). Contrib. Plasma Phys. 2012;52(8):667. DOI: 10.1002/ctpp.201200002

[4] Fortov VE, Vaulina OS, Lisin EA, Gavrikov AV, Petrov OF. Analysis of pair interparticle interaction in nonideal dissipative systems. J. Exp. Theor. Phys. 2010;110:662–674.DOI: 10.1134/S1063776110040138

[5] Shahzad A, He M-G. Thermodynamic characteristics of dusty plasma studied by using molecular dynamics simulation. Plasma Sci.Technol. 2012;14(9):771–777. DOI: 10.1088/1009-0630/14/9/01

[6] Shahzad A, He M-G. Interaction contributions in thermal conductivity of three-dimensional complex liquids. In: Liejin GUO, editor. AIP Conference Proceedings; 26–30 October; Xi'an, China. USA: AIP; 2013. p. 173–180. DOI: 10.1063/1.4816866

[7] Wigner E. Quasiparticle treatment of quantum-mechanical perturbation theory in the free electron gas. Phys. Rev. 1934; 46: 1002. DOI: 10.1007/BF01029223

[8] Chen FF. Introduction to Plasma Physics and Controlled Fusion. 2nd ed. New York: Spring verlag; 2006. 200 p.

[9] Killian T, Pattard T, Pohl T, Rost J. Ultracold neutral plasmas. Phys. Rep. 2007; 449:77. DOI:10.1016/j.physrep.2007.04.007

[10] Kalman GJ, Rommel JM, Blagoev K. Strongly Coupled Coulomb Systems. New York: Plenum; 1998. 198 p.

[11] Jensen MJ, Hasegawa T, Bollinger JJ, Dubin DHE, et al. Rapid heating of a strongly coupled plasma near the solid-liquid phase transition. Phys. Rev. Lett. 2005;94:025001. DOI: 10.1103/PhysRevLett.94.025001

[12] Robicheaux F, Hanson James D. Ultra cold neutral plasmas. Phys. Plasmas 2003;10 (2217). DOI: 10.1063/1.1573213

[13] Toukmaji AY, Board Jr, John A. Comput. Phys. Commun. 1996;95:73–92.

[14] Rapaport DC. The Art of Molecular Dynamics Simulation. New York: Cambridge University Press; 2004. 250 p.

[15] Evans DJ, Morriss GP. Statistical Mechanics of Non-Equilibrium Liquids. London: Academic; 1990. 20–300 p.

[16] Shahzad A, He M-G. Homogeneous nonequilibrium molecular dynamics evaluations of thermal conductivity 2D Yukawa liquids. Int. J. Thermophys. 2015;36(10–11):2565. DOI: 10.1007/s10765-014-1671-8

[17] Shahzad A, He M-G. Calculations of thermal conductivity of complex (dusty) plasmas using homogeneous nonequilibrium molecular simulations. Radiat. Eff. Defect. S. 2015;170(9):758–770. DOI: 10.1080/10420150.2015.1108316

[18]   Pierleoni C, Ciccotti G, Bernu B. Thermal conductivity of the classical one-component plasma by nonequilibrium molecular dynamics. Europhys. Lett. 1987;4:1115.

[19]   Salin G, Caillol JM. Equilibrium molecular dynamics simulations of the transport coefficients of the Yukawa one component plasma. Phys. Plasmas. 2003;10(5):1220. DOI: DOI: 10.1063/1.1566749

[20]   Donko Z, Hartmann P. Thermal conductivity of strongly coupled Yukawa liquids. Phys. Rev. E. 2004;69(1):016405. DOI: 10.1103/PhysRevE.69.016405

[21]   Faussurier G, Murillo MS. Gibbs-Bogolyubov inequality and transport properties for strongly coupled Yukawa fluids. Phys. Rev. E. 2003;67(4):046404. DOI: 10.1103/PhysRevE.67.046404

Constructional Nanomaterials

Nitrogen-Doped Carbon Nanotube/Polymer Nanocomposites Towards Thermoelectric Applications

Mohammad Arjmand and Soheil Sadeghi

Additional information is available at the end of the chapter

http://dx.doi.org/10.5772/65675

Abstract

This study investigates the impact of nitrogen doping on the performance of carbon nanotube (CNT)/polymer nanocomposites for thermoelectric applications; this was performed through measurement of conductivity of the generated nanocomposites. Three different catalysts (Co, Fe, and Ni) were used to synthesize nitrogen-doped CNTs (N-CNTs) by chemical vapor deposition technique. Synthesized N-CNTs were melt-mixed with a polyvinylidene fluoride (PVDF) matrix with a small-scale mixer at a broad range of loadings from 0.3 to 3.5 wt.% and then compression molded. Measurement of electrical conductivity of the generated nanocomposites showed superior properties in the following order of the synthesis catalyst: Co > Fe > Ni. We employed various characterization techniques to figure out the reasons behind dissimilar electrical conductivity of the generated nanocomposites, i.e., transmission electron microscopy, X-ray photoelectron spectroscopy, Raman spectroscopy, thermogravimetric analysis, light microscopy, and rheometry. It was found out, that the superior electrical conductivity of $(N\text{-}CNT)_{Co}$ nanocomposites was due to a combination of high synthesis yield, high aspect ratio, low nitrogen content, and high crystallinity of N-CNTs coupled with a good state of N-CNT dispersion. Moreover, it was revealed, that nitrogen doping had an adverse impact on electrical conductivity and, thus, on thermoelectric performance of CNT/polymer nanocomposites.

Keywords: carbon nanotube, nitrogen doping, polymer nanocomposites, electrical conductivity, thermoelectric

1. Introduction

1.1. Conductive filler/polymer nanocomposites

Conductive filler/polymer nanocomposites (CPNs) have recently drawn great interest to be employed in various applications due to their unique properties, such as tunable electrical conductivity, light weight, low cost, corrosion resistance, and processability [1, 2]. CPNs are generated by incorporating conductive filler into a polymer matrix. Conventional polymers, such as polycarbonate and polystyrene are insulative; however, incorporating conductive fillers to these polymer matrices can provide them with a broad range of conductivities through the formation of a two- or three-dimensional conductive network (**Figure 1**). Tunable electrical conductivity of CPNs entitles them to be used in a broad spectrum of applications, such as thermoelectric, charge storage, antistatic dissipation, electrostatic discharge (ESD) protection, and electromagnetic interference (EMI) shielding [3–8]. In fact, the level of electrical conductivity defines the applications in which CPNs can be employed. Charge storage and ESD protection are the major applications of CPNs necessitating low and medium electrical conductivity, respectively, whereas thermoelectric and EMI shielding require high electrical conductivity.

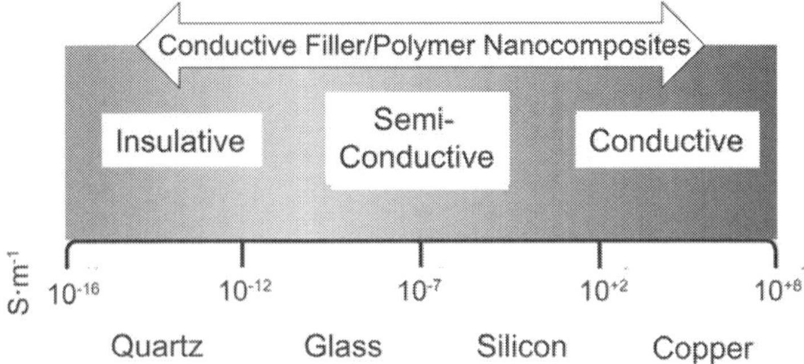

Figure 1. The approximate range of electrical conductivity covered by CPNs [1].

1.2. Conductive filler/polymer nanocomposites for thermoelectric applications

Thermoelectric devices provide an all solid-state means of heat to electricity conversion. These devices feature many advantages in heat pumps and electrical power generators, such as possessing no moving parts, generating zero noise, being easy to maintain, extended lifetime, and being highly reliable [9, 10]. However, their limited efficiency has restricted their usage to specialized applications, where cost and efficiency are of great importance. Thermoelectric efficiency is expressed in terms of dimensionless figure of merit (ZT), defined as $ZT = \dfrac{\alpha^2 \sigma T}{\kappa}$, in which α is Seebeck coefficient, σ is electrical conductivity, κ is thermal conductivity, and T is absolute temperature. The upper limit for thermoelectric energy conversion efficiency is Carnot limit. Several recently developed thermoelectric materials exhibit a ZT near unity, resulting in efficiency equal to 10 % of the Carnot efficiency limit. By reducing the physical dimensionality of the thermoelectric materials (quantum confinement) [11], it is possible to significantly increase Seebeck coefficient and decrease the thermal conductivity. This can lead to ZTs as high as 3 at 550 K in n-type $PbSe_{0.98}Te_{0.02}/PbTe$ quantum-dot superlattices [12]. However, application of these heavy metal thermoelectric materials is associated with high cost of material and production processes, poor processability, and huge adverse environmental impacts [13–15]. Accordingly, polymeric materials have drawn great interest to be used in thermoelectric applications. Improved processability, low cost, low thermal conductivity, and low density of polymeric materials are among key potential benefits stimulating the development of polymer-based thermoelectric materials [16, 17]. It is worth mentioning, that the figure of merit in polymer-based thermoelectric materials is $ZT \sim O(10^{-3})$, which necessitates further investigations in this area [18].

Organic polymers exhibit poor electrical conductivity, which necessitates addition of conductive fillers in order to provide high electrical conductivity and reasonable thermoelectric performance. Recent studies suggested, that carbon-based nanofiller/polymer nanocomposites hold a significant promise in development of lightweight, low-cost thermoelectric materials [19–23]. As an instance, Yu et al. [17] demonstrated, that by forming a segregated network of carbon nanotubes (CNTs), electrical conductivities as high as 48 S cm^{-1} were achievable, while Seebeck coefficient and thermal conductivity were marginally impacted by CNT presence. This resulted in a figure of merit of 0.006 at room temperature. These findings render CNT-based polymer nanocomposite as a basis for development of thermoelectric functional materials for future green energy applications.

1.3. Nitrogen-doped carbon nanotube/polymer nanocomposites for thermoelectric applications

Different types of conductive nanofillers have been employed to develop CPNs, viz., carbonaceous nanofillers and metallic nanowires [4, 5, 24], among which, CNT has appealed remarkable attention due to its large surface area and outstanding electrical, thermal, and mechanical properties [25, 26]. The luminous era of CNTs initiated in 1991 by their discovery from soot using an arc-discharge apparatus [27]. Like fullerene and graphene, CNTs consist of a sp^2 network of carbon atoms. Among these, three carbonaceous poly-types, CNT is the only one produced in large industrial scale. There are two general types of CNTs: multi-walled CNT (MWCNT) and single-walled CNT (SWCNT). MWCNT consists of multiple rolled layers of graphite coaxially arranged around a central hollow core with van der Waals forces between contiguous layers, while SWCNT is made of a single rolled graphene [28]. The global market for CNT primary grades was \$158.6 million in 2014 and is anticipated to reach \$670.6 million in 2019. CNT/polymer nanocomposites represent, by far, the largest segment in the overall market of CNTs [29].

Manipulating the electronic energy gap of CNTs could lead to their superior performance. Since CNTs are sp^2 carbon systems, theoretical [30] and experimental [31] studies showed, that substituting carbon atoms with heteroatoms can result in adjustment of electronic and structural patterns of carbon nanotubes. Nitrogen is the best choice for heteroatom substitution owing to its size proximity to carbon [32]. Bearing in mind, that nitrogen comprises one additional electron as compared to carbon, doping CNTs with nitrogen has emerged as an attractive research topic to improve the electronic properties of CNTs.

Essentially, there are three common nitrogen bonding configurations in nitrogen-doped CNTs (N-CNTs), viz., quaternary, pyridinic, and pyrrolic. As depicted in **Figure 2**, quaternary nitrogen is directly replaced for C atom in the hexagonal network, is sp^2 hybridized, and creates electron-donor state. The pyridinic nitrogen is a part of sixfold ring structure and is sp^2 hybridized, and two of its five electrons are localized sole pair. Pyrrolic nitrogen is a portion of a five-membered ring structure, is sp^3 hybridized, and gives its remaining two electrons to a π orbital, integrating the aromatic ring [33]. Whereas the quaternary and pyridinic nitrogen lead to side-wall defects, the pyrrolic nitrogen is believed to form internal cappings, generating bamboo-like sections [34]. Besides these three types, there is also possibility for N_2 molecules to get trapped inside the tube axis or intercalated into the graphitic layers of N-CNTs.

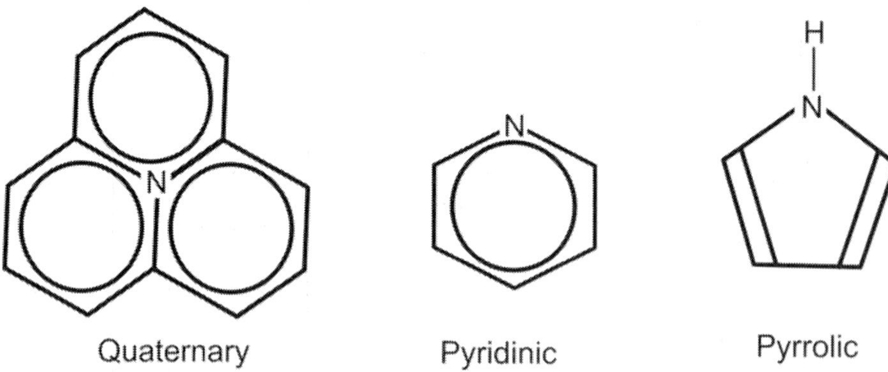

Figure 2. Major types of nitrogen bonding in N-CNTs [35].

Basically, as yet, most of the research studies have investigated the influence of nitrogen doping on electronic properties of CNTs via factors, such as density of states (DOS) and Fermi level; nevertheless, inspecting the electrical properties of polymer nanocomposites containing N-CNTs is still at its infancy [35–38]. Hence, the current study aims to research the impact of nitrogen doping on the performance of N-CNT/polymer nanocomposites for thermoelectric applications by studying its influence on electrical conductivity of N-CNT/polymer nanocomposites. N-CNTs were synthesized with different types of catalyst (Co, Fe, and Ni) to obtain diverse nitrogen contents. Afterward, synthesized N-CNTs were melt mixed into polyvinylidene fluoride (PVDF) and compression molded. We evaluated electrical conductivity of the nanocomposites at different loadings and scrutinized the underlying causes behind dissimilar conductivities of the generated nanocomposites. As high electrical conductivity of CPNs is an important factor on their performance as thermoelectric materials, this study goes significantly beyond the state of the art and gives new insight on the role of nitrogen doping on conductivity and therefore performance of CNT nanocomposites for thermoelectric applications.

2. Experimental

2.1. Materials synthesis

We employed the incipient wetness impregnation technique to produce catalyst precursors. The catalyst precursors were dissolved in water and then impregnated onto aluminum oxide support (Sasol Catalox Sba-200). Thereafter, the developed materials were dried, calcinated, and reduced. Having high solubility and diffusion rate in carbon, Co, Fe, and Ni were chosen as the catalysts [39, 40]. Accordingly, we employed cobalt nitrate hexahydrate, iron (III) nitrate nonahydrate, and nickel (II) sulfate hexahydrate as the catalyst precursors. We set the metal loading at 20 wt.%. The catalyst calcination, reduction, and N-CNT synthesis were performed in CVD setup, detailed in a former study [41]. CVD setup comprised a quartz tubular reactor with an inner diameter of 4.5 cm encapsulated within a furnace. The following steps were implemented for the preparation of the catalysts: first, the catalysts were calcinated under air atmosphere with a flow rate of 100 sccm at 350 °C for 4 h. In this stage, metallic salts were translated into metal oxides. Thereafter, we used a mortar and pestle to achieve a fine powder. Hydrogen gas at a flow rate of 100 sccm at 400 °C for 1 h was utilized to obtain alumina-supported metal catalysts. Afterward, we conveyed a combination of ethane (50 sccm), ammonia (50 sccm), and argon (50 sccm) over the synthesized catalysts. Ethane played the role of carbon source, whereas ammonia and argon were nitrogen source and inert gas carrier, respectively. The synthesis temperature, synthesis time, and catalyst mass were set at 750 °C, 2 h, and 0.6 g, respectively. Catalyst preparation process is elucidated further elsewhere [39].

The polymer matrix utilized for the nanocomposite preparation was semicrystalline PVDF 11008/0001, purchased from 3M Canada, with an average density of 1.78 g/cm^3 and melting point of 160 °C. PVDF was opted as the polymer matrix owing to its ferroelectricity, high dielectric strength (~13 kV mm^{-1}), corrosion resistance, good mechanical properties, thermal stability, good chemical resistance (excellent with acid and alkali), and robust interaction of electrophilic fluorine groups with CNTs [42–44]. The mixing of synthesized N-CNTs with PVDF matrix was carried out with Alberta Polymer Asymmetric Minimixer (APAM) at 240 °C and 235 rpm. PVDF matrix was first masticated within the mixing cup for 3 min, and then N-CNTs were inserted and mixed for an additional 14 min. For each catalyst, the nanocomposites with different N-CNT concentrations, i.e., 0.3, 0.5, 1.0, 2.0, 2.7, and 3.5 wt.%, were prepared. The nanocomposites were molded into circular cavities with 0.5 mm thickness using Carver compression molder (Carver Inc.) at 220 °C under 38 MPa pressure for 10 min. The molded samples were used for electrical, morphological, and rheological characterizations.

2.2. Materials characterization

2.2.1. N-CNT characterization

2.2.1.1. Transmission electron microscopy of N-CNTs

High-resolution transmission electron microscopy (HRTEM) was used to inspect the morphology of synthesized N-CNTs. HRTEM was conducted on Tecnai TF20 G2 FEG-TEM (FEI) at 200 kV acceleration voltage with a standard single-tilt holder. The images were taken with Gatan UltraScan 4000 CCD camera at 2048 × 2048 pixels. For HRTEM, around 1.0 mg of N-CNT powder was dispersed in 10 mL ethanol and bath sonicated for 15 min. A drop of the dispersion was mounted on the carbon side of a standard TEM grid covered with a ~40 nm holey carbon film (EMS). Measurement of the geometrical dimensions of N-CNTs was conducted for over 100 individual ones utilizing MeasureIT software (Olympus Soft Imaging Solutions GmbH).

2.2.1.2. X-ray photoelectron spectroscopy, Raman spectroscopy and thermogravimetric analysis

PHI VersaProbe 5000-XPS was used to obtain X-ray photoelectron spectra. The spectra were achieved employing monochromatic Al source at 1486.6 eV and 49.3 W with a beam diameter of 200.0 μm. The structural defects of N-CNTs were inspected using Raman spectroscopy. Renishaw inVia Raman microscope was used to obtain Raman spectra. Excitation was provided by the radiation of an argon-ion laser beam with 514 nm wavelength. A 5× objective was used to get Raman spectra. The yield of the synthesis process was inspected with Thermogravimetric Analyzer (TA instruments, Model: Q500). The samples were heated under air atmosphere (Praxair AI INDK) from ambient temperature to 950 °C at a rate of 10 °C/min. The samples were kept at 950 °C for 10 min before cooling.

2.2.2. Nanocomposite characterization

2.2.2.1. Light microscopy

The microdispersion state of the nanofillers within PVDF matrix was enumerated using light transmission microscopy (LM) on thin cuts (5 μm thickness) of the compression-molded samples, prepared with Leica Microtome RM 2265 (Leica Microsystems GmbH). Olympus BH2 optical microscope (Olympus Deutschland GmbH) equipped with CCD camera DP71 was used to capture images with dimensions of 600 μm × 800 μm from different cut sections. The software Stream Motion (Olympus) was used to analyze the images. The agglomerate area ratio (in %) was defined by dividing the spotted area of non-dispersed nanofillers (with equivalent circle diameter > 5 μm, area > 19.6 μm^2) over the whole sample area (15 cuts, ca. 7.2 mm^2). Mean value and standard deviation, demonstrating the differences between the cuts and thus heterogeneity, were reckoned. The relative transparency of the cuts provided added information about the amount of dispersed nanofillers in the samples. The relative transparency was quantified by dividing the transparency of the cut over the transparency of the glass slide/cover glass assembly. Ten various areas per sample were used to obtain mean values and standard deviations. Further information on employing LM to evaluate microdispersion state of nanofillers within nanocomposites is presented elsewhere [45, 46].

2.2.2.2. TEM

Ultrathin sections of the samples were cut using ultramicrotome EM UC6/FC6 (Leica) setup with an ultrasonic diamond knife at ambient temperature. The sections were floated off water and thereafter transferred on carbon-filmed TEM copper grids. TEM characterizations were carried out employing TEM LIBRA 120 (Carl Zeiss SMT) with an acceleration voltage of 120 kV.

2.2.2.3. Rheology

Rheological measurements were performed using Anton-Paar MCR 302 rheometer at 240 °C using 25 mm cone-plate geometry with a cone angle of 1° and truncation of 47 μm. The thermal stability of the prepared samples was validated by conducting small-amplitude oscillatory shear measurements prior to and following the long-time exposure of the samples to elevated temperatures. Various rheological properties were measured at 240 °C to characterize the linear and nonlinear response for the neat and nanocomposite samples.

2.2.2.4. Electrical conductivity

Two conductivity meters with 90 V as the applied voltage were employed to measure the electrical conductivity of the generated materials. For nanocomposites with an electrical conductivity higher than 10^{-2} S m^{-1}, the measurements were conducted according to ASTM 257-75 standards employing Loresta GP resistivity meter (MCP-T610 model, Mitsubishi Chemical Co.). An ESP probe was used to avert the effect of contact resistance. For electrical conductivities less than 10^{-2} S m^{-1}, the measurements were carried out with Keithley 6517A electrometer connected to Keithley 8009 test fixture (Keithley Instruments).

3. Results and discussion

3.1. General background

3.1.1. Mechanisms of electrical conductivity

Electrical conductivity derives from ordered movement of charge carriers (electric current). In the absence of an electric field, the conduction electrons are scattered freely in a solid owing to their thermal energy. If an electric field, E, is applied, the force on an electron, e, is $-eE$, and the electron is accelerated in the opposite direction to the electric field because of its negative charge. Accordingly, there is a net velocity and the current density is presented by [47]:

$$J = N_e \times e \times \mu \times E, \tag{1}$$

where J is the current density, N_e is concentration of electrons, e is charge of electron, μ is the electron mobility, and E is the applied electric field. The applied electric field equals to the applied voltage over the thickness of a sample. Hence, the electrical conductivity can be determined as:

$$\sigma = \frac{J}{E},$$

(2)

where σ is electrical conductivity and its SI unit is Siemens per meter (S m^{-1}). Electrical conductivity of materials is an intrinsic property, which spans a very wide range. The conductivity of insulators is typically less than 10^{-10} S m^{-1}, that of semiconductive materials covers the range 10^{-10} to around 10^{-2} S m^{-1}, and for semimetals and metals is more than 10^{-2} S m^{-1}.

Electrical conductivity of materials can be elucidated employing the band theory [48]. In the band theory, the energy level of each electron is reflected as a horizontal line. As any solid possesses a large number of electrons with various energy levels, the sets of energy levels form two continuous energy bands, named valence band and conduction band. The energy gap between the two bands signifies the forbidden region for electrons. Electrons restrained to individual atoms or interatomic bonds are, in the band theory, said to be in valence band. Those electrons, that can move freely in substance upon applying electric field lie in conduction band. **Figure 3** depicts a schematic of the bands in a solid identifying three main types of materials: insulators, semiconductors, and metals. Valence and conduction bands in metals overlap each other; therefore, metals indicate very high conductivity. In intrinsic semiconductors, the valence-conduction band gap is adequately small, so that, electrons in valence band can be excited to conduction band by thermal energy. Among the three types of materials illustrated in **Figure 3**, insulators show the largest valence-conduction band gap, and, therefore, fewer electrons can be excited to their conduction band by thermal energy. This results in a very low conductivity in insulators.

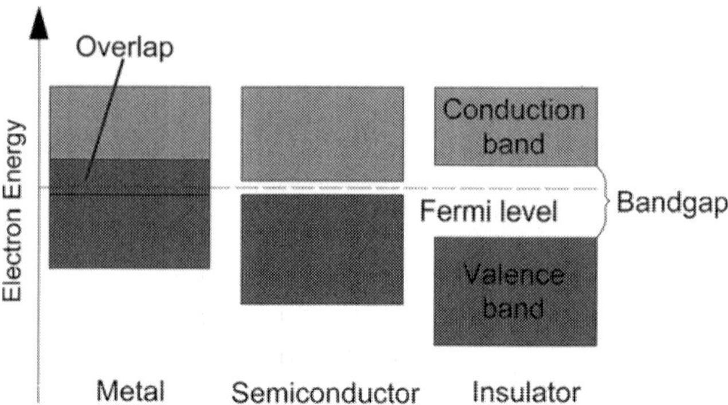

Figure 3. Simplified diagram of the electronic band structure in the band theory, reproduced from [34].

3.1.2. Electrical conductivity in CPNs

High electrical conductivity, i.e., conductive network formation, at very low filler contents has made CPNs distinctive materials for industrial applications [25, 26]. Conductive network formation in CPNs is better understood with the concept of percolation threshold [49, 50]. Percolation means, that at least one conductive pathway forms to allow electrical current to pass across CPNs, thereby transforming CPNs from insulative to conductive. Percolation happens at a narrow filler concentration range, where the electrical conductivity of CPNs drastically increases by several orders of magnitude. Low electrical percolation threshold in CPNs leads to the production of cost-effective composites.

Many statistical, geometric, thermodynamic, and structure-based models have been introduced to anticipate the percolation threshold and electrical conductivity of CPNs [49, 51]. Although the percolation theory is just valid at conductive filler concentrations above the percolation threshold, it is the most acceptable one. Statistical percolation theory estimates the percolation threshold of CPNs as:

$$\sigma = \sigma_0 \cdot \left(V - V_c\right)^t, \tag{3}$$

where σ is electrical conductivity of CPN, σ_0 is electrical conductivity of conductive filler, V is dimensionless volume content of conductive filler, and V_c and t are percolation threshold and critical exponent, respectively [49]. The equation is valid for filler concentration above the percolation threshold, i.e., $V > V_c$. Higher t value and lower percolation threshold correspond to well-dispersed, high-aspect-ratio fillers [52–54].

Figure 4 illustrates a typical percolation curve of CPNs [55]. In general, percolation curve of CPNs can be divided into three regions: (1) region far below the percolation threshold (insulative region), (2) region where percolation occurs (percolation region), and (3) region far above the percolation threshold (conductive region). In the insulative region, the conductive filler loading is very low with the fillers far from each other; thus, polymer matrix dominates the charge transfer. As a matter of fact, at low filler concentrations, the insulating gaps are very large and the chance, that nomadic charge carriers are transferred between conductive fillers is very low.

By enhancing filler loading, the gaps between conductive fillers decrease, and a drastic increase in electrical conductivity is observed over a narrow concentration range (percolation region). In this region, hopping and direct-contact mechanisms become significant. When the mean particle-particle distance reaches below 1.8 nm, the dominant electron transfer mechanism become hopping mechanism [56–58]. It is reported, that the presence of large conductive agglomerates in CPN results in a very high secondary internal electric field between the conductive islands [57, 59]. This high field strength assists free electrons in conductive filler having adequate energy to hop over the insulative gaps. Nevertheless, hopping takes place when an electron receives sufficient energy to pass over distance to nearest free site with lower energy to alter its lattice site. In the percolation region, due to proximity or direct contact of conductive fillers, the nomadic charge carriers in conductive fillers play the dominant role in conduction mechanism. Since these free charge carriers belong to the conduction band, the conductivity of the nanocomposite rises by several orders of magnitude in the percolation region. Next, by adding more filler loading, a well-developed, 3D conductive network initiates to form, but the electrical conductivity increases only marginally. This is due to substantial current dissipation at the contact spots between conductive fillers, i.e., the constriction resistance, leading to a plateau in the percolation curve [26].

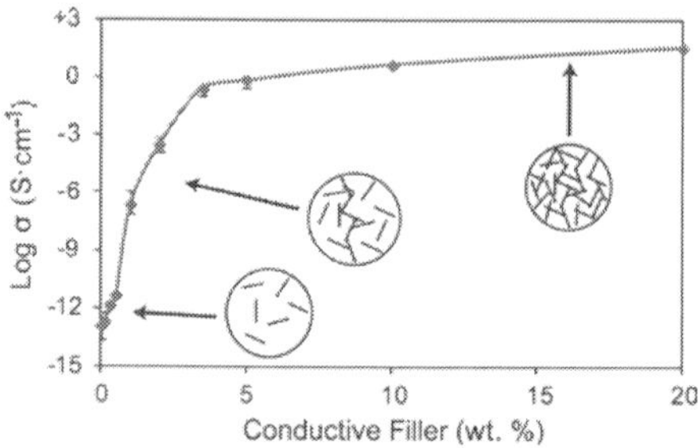

Figure 4. Percolation curve of compression-molded CNT/polystyrene nanocomposite (a typical percolation curve of CPNs) [55].

3.2. Electrical conductivity of N-CNT/PVDF nanocomposites

The percolation curves of N-CNT/PVDF nanocomposites are depicted in **Figure 5**. It was observed, that $(N\text{-}CNT)_{Co}$/PVDF nanocomposites presented the lowest percolation threshold (1.5 wt.%) and highest electrical conductivity (3 S m^{-1} at 3.5 wt.%). However, it was revealed, that Ni-based nanocomposites were insulative up to 2.7 wt.% and experienced a slight increase in electrical conductivity at 3.5 wt.%. The Fe-based nanocomposites presented an increase in electrical conductivity from 1.0 wt.% to 3.5 wt.% with a mild slope.

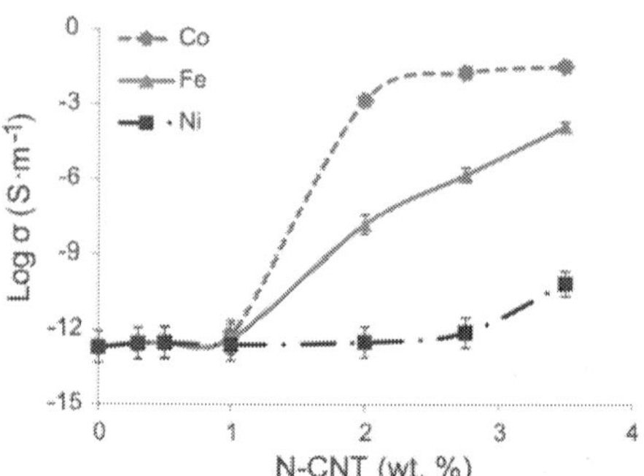

Figure 5. Electrical conductivity of N-CNT/PVDF nanocomposites as a function of N-CNT content. N-CNTs were synthesized over Co, Fe, and Ni catalysts.

There are many factors impacting the electrical conductivity of CPNs, such as loading, intrinsic conductivity, size, and aspect ratio of conductive filler, inherent properties of polymer medium, interfacial properties of CPN constituents, dispersion and distribution of filler, blending method, and crystalline structure of the matrix. The impacts of the aforementioned parameters on electrical conductivity of CNT/polymer nanocomposites have been well reviewed in the literature [1, 60–62]. Accordingly, in succeeding section, we scrutinize structural and morphological features of N-CNTs and their nanocomposites to figure out the reasons behind different electrical behaviors of the generated nanocomposites.

3.3. Morphological and structural characterization of N-CNTs

The morphology and graphitic structure of N-CNTs were analyzed using TEM images. **Figure 6** indicates, that the type of synthesis catalyst played a leading role in creating the final morphology of N-CNTs. As depicted in **Figure 6**, we observed an open-channel morphology for $(N\text{-}CNT)_{Co}$ and a bamboo-like morphology for $(N\text{-}CNT)_{Fe}$ and $(N\text{-}CNT)_{Ni}$. Surface roughness is observed in bamboo-like N-CNTs, deriving from defected bonding of bamboo-like sections. The flawed parts in the wall of N-CNTs are attributed to replacement of nitrogen atoms [63, 64]. Since an open-channel structure was formed for $(N\text{-}CNT)_{Co}$, we can say, that other factors, than nitrogen bonding, are involved in creation of bamboo-like morphology, such as type of catalyst.

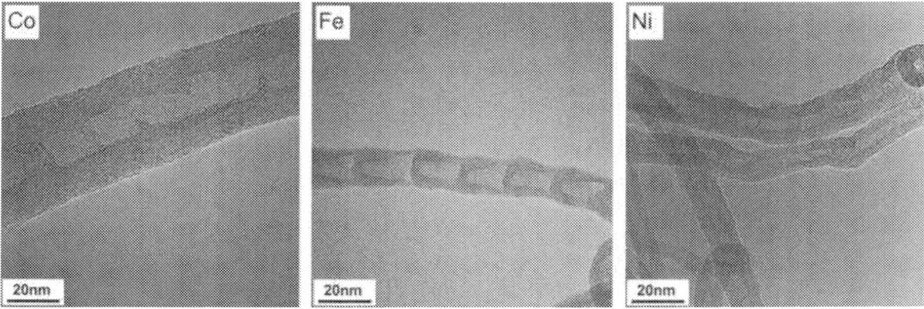

Figure 6. TEM micrographs of N-CNTs synthesized over Co, Fe, and Ni catalysts.

Table 1 tabulates average length and diameter, nitrogen content, Raman feature, and synthesis yield of synthesized N-CNTs. We perceived, that N-CNTs synthesized over Fe catalyst had the largest diameter, over double that of $(N\text{-}CNT)_{Ni}$. Statistical analysis of particle size of the catalysts revealed a good correlation between diameter of N-CNTs and size of catalyst particles. Discrepancies in original size of the catalyst particles and also dissimilar tendencies of the catalyst particles to sinter at synthesis temperatures are among significant parameters affecting the variation in diameter of N-CNTs. It should be noted, that metallic nanoparticles with sizes below 10 nm experience a drastic drop in melting point [65]. High synthesis temperature range (600–1000 °C) coupled with exothermic thermal decomposition of the precursor molecules results in higher temperature than nominal reaction temperature, contributing to metal liquefaction and coalescence of catalyst particles [66, 67].

	Co	Fe	Ni
Length (μm)	2.6	2.6	1.2
Diameter (nm)	25	46	20
Nitrogen content (at.%)	2.2	2.2	3.3
I_D/I_G	0.79	0.73	0.81
Synthesis yield %	89.5	85.1	63.9

Table 1. Physical and structural features of N-CNTs synthesized over Co, Fe, and Ni catalysts.

Table 1 shows, that (N-CNT)$_{Co}$ and (N-CNT)$_{Fe}$ had average length about 2.6 μm, while (N-CNT)$_{Ni}$ exhibited considerably lower average length about 1.2 μm. The shorter length of (N-CNT)$_{Ni}$ can be attributed to either inferior activity of Ni catalyst, as will be shown by TGA, or the presence of larger amount of nitrogen in their structure, as will be exhibited by X-ray photoelectron spectroscopy (XPS) analysis. The presence of nitrogen can be envisaged as an important factor to bend, close, and cap N-CNTs. It is worth noting, that average length and diameter of N-CNTs are of high significance for electrical applications, since CNTs with high aspect ratio provide CPNs with superior electrical performance [68, 69].

The amount of nitrogen content can have a weighty effect on morphological, physical, and electronic properties of N-CNTs. The achieved data revealed, that atomic content of nitrogen incorporated into (N-CNT)$_{Ni}$ was 3.3 at.%, whereas (N-CNT)$_{Co}$ and (N-CNT)$_{Fe}$ had considerably lower nitrogen content, i.e., 2.2 at.%. As nitrogen could have the effect of closing the tube structures and thereby developing more disordered, bent, and capped structures, the larger nitrogen content of N-CNT$_{Ni}$ could be a contributing factor to its lower length.

In Raman spectra of CNTs, tangential mode (*G* band) and defect-active mode (*D* band) offer valuable information about physical and electronic structure of CNTs [70, 71]. Hence, Raman spectroscopy was used to inspect the influence of nitrogen doping on physical and morphological features of N-CNTs. *G* band (~1600 cm^{-1}) derives from the stretching of C—C bond in graphitic materials and is mutual to all sp^2 carbon forms. *D* band (~1400 cm^{-1}) is double-Raman scattering process, which requires lattice distortion to break the basic symmetry of the graphitic structure [72]. Therefore, the presence of structural defects stimulates *D* band feature. Accordingly, the ratio of *D* and *G* band intensities is often used as indicative tool to validate the structural perfection of CNTs [73]. We observed, that (N-CNT)$_{Ni}$ had the uppermost I_D/I_G ratio, signifying the poorest crystallinity. These results are in line with TEM images of (N-CNT)$_{Ni}$, indicating poorer crystalline morphology than the other forms of N-CNTs. Moreover, Villalpando-Paez et al. [74] and Ibrahim et al. [75] reported good correlation between nitrogen concentration and I_D/I_G ratio. This is in agreement with our study and shows the opposing influence of nitrogen doping on the crystalline structure of N-CNTs. We also observed, that

$(N\text{-}CNT)_{Ni}$ went through more breakage during the melt mixing process, ascribed to its poorer crystallinity.

TGA analysis helped investigate the synthesis yield. We obtained residues of 11.5 %, 14.9 %, and 36.1 %, relative to original mass, for $(N\text{-}CNT)_{Co}$, $(N\text{-}CNT)_{Fe}$, and $(N\text{-}CNT)_{Ni}$, respectively. The residue consists of metallic oxide particles and alumina substrate [41, 76]. The higher the yield of the synthesis process, the lower is the amount of the remaining residue. Thus, we can claim, that Ni catalyst had an inferior performance compared to Co and Fe catalysts. The catalyst particles contained 80 wt.% alumina and 20 wt.% metallic particles. Alumina is insulative and metallic particles have much less surface area than synthesized N-CNTs, and their surface area even further reduced due to sintering phenomenon. This justifies the significance of synthesis yield on electrical properties.

3.4. Morphological characterization of N-CNT/PVDF nanocomposites

The dispersion state of conductive filler within polymer matrix is intensely influential on electrical properties. Hence, we inspected the dispersion state at three various scales. Micro-dispersion state of N-CNTs within the polymer medium was investigated via LM. LM talks about the portion of fillers, that appears as big agglomerates and is not disentangled well, and was enumerated as the agglomerate area ratio in our study. Moreover, gray appearance of LM samples helps us quantify the agglomerates with sizes equal to or slightly larger than the wavelength of visible light, ca. 400–700 nm, but smaller than visually identifiable agglomerates. Darker background denotes more nanotubes dispersed in this range. We also employed TEM to obtain information about nanodispersion state of carbon nanotubes, i.e., how well carbon nanotubes disentangle individually.

Figure 7 portrays examples for LM images of three different nanocomposites, corresponding to different synthesis catalysts, with 2.0 wt.% N-CNT content. Quantification of the agglomerate area ratio, as shown in **Table 2**, illustrates the lowest agglomerate area ratio for samples containing Fe-based N-CNTs, followed by Co and Ni. The corresponding relative transparency values indicate the lowest value for Co-based N-CNTs, followed by Fe-based and Ni-based.

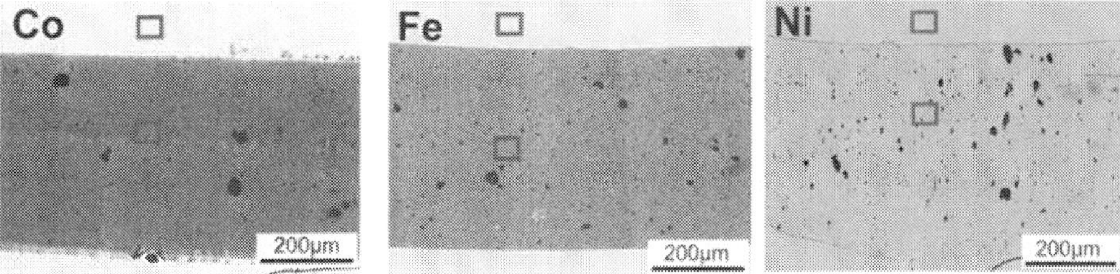

Figure 7. LM images of microtomed sections of 2.0 wt.% N-CNT/PVDF nanocomposites. N-CNTs were synthesized over Co, Fe, and Ni catalysts. The red squares represent areas employed for relative transparency quantifications.

	Co	Fe	Ni
Agglomerate area ratio %	2.3	1.8	2.8
Relative transparency %	37	53	86

Table 2. LM microdispersion parameters of microtomed N-CNT/PVDF nanocomposites with N-CNTs synthesized over different catalysts.

TEM images look into nanodispersion state of N-CNTs in PVDF medium (**Figure 8**). The images clearly show, that (N-CNT)$_{Ni}$ had the worst dispersion state. TEM image of (N-CNT)$_{Ni}$/PVDF nanocomposite shows a few individual nanotubes beside fairly large agglomerates. (N-CNT)$_{Co}$ presented the best state of nanodispersion, while (N-CNT)$_{Fe}$/PVDF nanocomposites held small agglomerates with sizes around 500 nm. In conclusion, microscopy images showed, that (N-CNT)$_{Co}$ and (N-CNT)$_{Fe}$ had better both microdispersion and nanodispersion than their Ni-based counterpart. Co-based and Fe-based N-CNTs indicated only marginal discrepancies in their dispersion state.

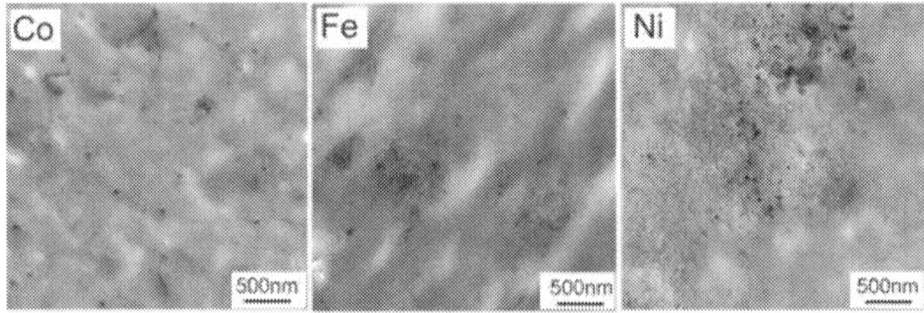

Figure 8. TEM images of 2.0 wt.% N-CNT/PVDF nanocomposites with N-CNTs synthesized with different catalysts, illustrating nanodispersion state of N-CNTs.

3.5. Linear and nonlinear melt-state rheological response of N-CNT/PVDF nanocomposites

Figure 9 depicts storage modulus (G') and loss modulus (G'') of N-CNT/PVDF nanocomposites as a function of frequency under small-amplitude oscillatory shear ($\gamma = 1$ %) for a frequency range from 0.1 rad/s to 625 rad/s at 240 °C.

As shown in **Figure 9**, G' in low frequency region ($\omega \sim 0.1$ rad/s) is significantly larger than G'' (a damping factor smaller than unity) for (N-CNT)$_{Co}$/PVDF and (N-CNT)$_{Ni}$/PVDF nano-composites at concentrations as low as 0.5 wt.%. (N-CNT)$_{Fe}$/PVDF nanocomposite samples, however, exhibited an elastic dominant response ($G' > G''$) only at very high nanofiller con-centrations (~2.0 wt.%). It is noticeable, that all N-CNT/PVDF nanocomposites, regardless of synthesis catalyst, showed a signature for the existence of an ultraslow relaxation proc-ess (a near-zero slope for G' in low frequency region) at concentrations as low as 0.5 wt.%. This indicates, that linear rheological response was affected by the presence of N-CNTs at concentrations, that no significant enhancement in electrical conductivity was observable in N-CNT nanocomposites.

The linear melt-state rheological response is mainly controlled by several factors, such as inter-tube van der Waals interactions, micro- and nanodispersion states, individual CNT stiffness, and CNT network stiffness [77–79]. Individual CNT stiffness is mainly controlled by intra-wall C—C bond strength and graphitic interlayer load transfer [80]. This suggests, that structural imperfections and defects in CNT graphitic walls can deteriorate their elastic properties. Moreover, stiffness of the network structure formed by CNT bundles is mainly determined by the load transfer across CNT/polymer and CNT-CNT interface [79]. In this context, it could be mentioned, that a scenario entirely based on individual CNT stiffness may not be able to fully describe the observations for elastic response of N-CNT nanocomposites in low frequency region. As can be seen in **Figure 9**, Co-, Fe-, and Ni-based N-CNT nanocomposites reached a storage modulus of 9640 Pa, 2290 Pa, and 3860 Pa at 0.1 rad, respectively. The presence of higher amount of structural imperfections may not be responsible for lower elasticity observed for Fe-based N-CNT nanocomposite as $(N\text{-}CNT)_{Fe}$ showed the lowest I_D/I_G. Therefore, the main contributing factors to linear melt-state rheological observations could be considered as the dispersion state and load transfer across the interfacial region.

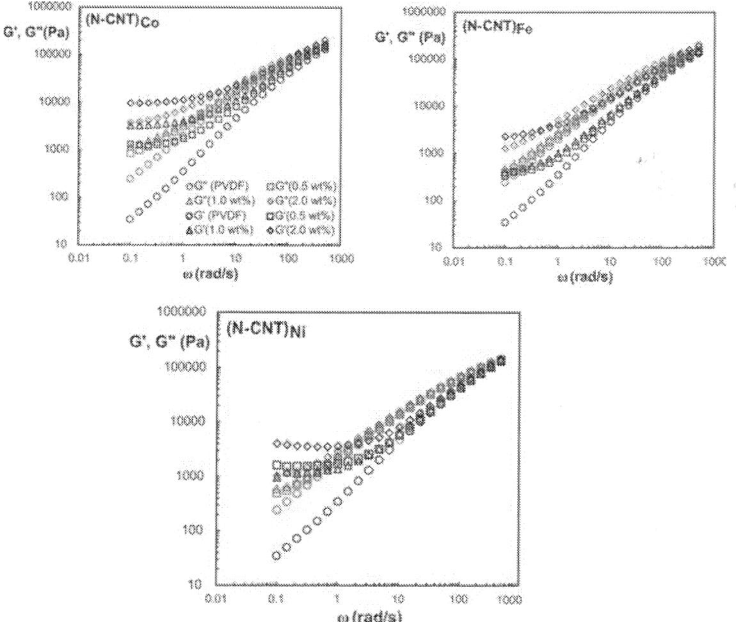

Figure 9. Small amplitude oscillatory shear response at $\gamma = 1$ % and $T = 240$ °C for neat PVDF and N-CNT/PVDF nanocomposites with N-CNTs synthesized over different catalysts.

Figure 10 depicts oscillatory amplitude sweep response of neat PVDF and N-CNT/PVDF nanocomposites containing 3.5 wt.% N-CNTs synthesized over different catalysts over a range of applied strain amplitudes from 0.1 to 1000.0 % at an angular frequency of 0.1 rad/s. The responses observed for neat PVDF and N-CNT/PVDF nanocomposite samples demonstrated a transition from a linear regime to a nonlinear regime and also a drop in G' as strain amplitude increases. It is noticeable, that in low-strain region, all nanocomposite samples exhibited elastic dominant response ($G' > G''$). These results also feature a crossover strain amplitude γ_x ($G' = G''$), which is a measure of N-CNT network sensitivity to deformation-induced microstructural changes. As shown in **Figure 10**, Co-, Fe-, and Ni-based N-CNT nanocomposites exhibited crossover strain amplitudes of 32.0 %, 5.8 %, and 5.6 %, respectively. This implies, that Co-based N-CNTs featured a very resilient behavior toward the applied deformation field and the stress-bearing backbone of the fractal clusters survived up to strain amplitudes one order of magnitude larger than N-CNTs synthesized over Fe and Ni. It is noticeable, that Ni-based N-CNT nanocomposite showed a multistep transition into a nonlinear regime as G' dropped to an intermediate plateau and then significantly decreased. Moreover, the first step decrease in G' in (N-CNT)$_{Ni}$ nanocomposite was accompanied by a dissipation process signified by a weak local peak in G''.

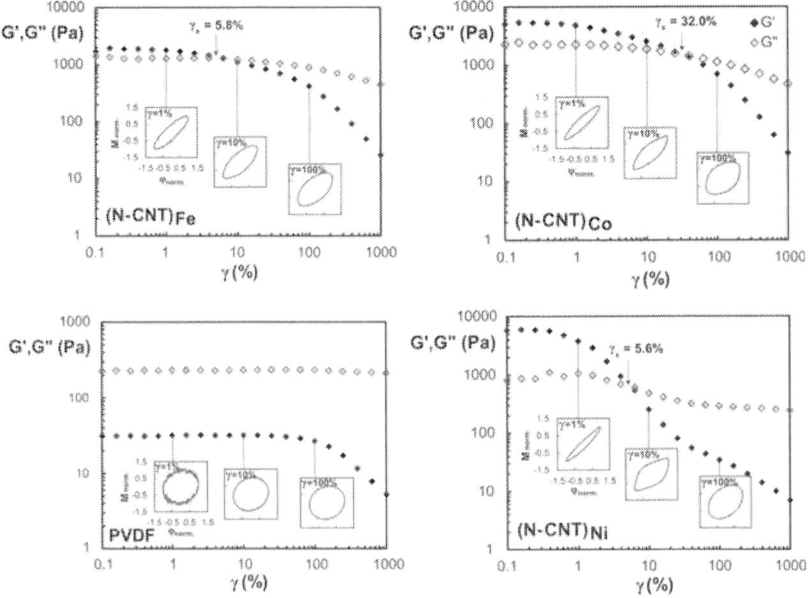

Figure 10. Oscillatory amplitude sweep response of neat PVDF and N-CNT/PVDF nanocomposites containing 3.5 wt.% N-CNTs synthesized over different catalysts for strain amplitudes of $\gamma_0 = 0.1$–1000 % at an angular frequency of $\omega = 0.1$ rad/s. The insets show the non-dimensionalized elastic Lissajous loops for strain amplitudes indicated by solid lines.

Insets in **Figure 10** depict non-dimensionalized elastic Lissajous loops [81–83], in which normalized torque $M_{norm.}$ is plotted as a function of normalized deflection angle $\phi_{norm.}$. At small strain amplitudes ($\gamma \sim 1.0$ %), for neat PVDF and nanocomposites, elastic Lissajous loops were elliptical, corresponding to a linear viscoelastic response. The area enclosed by elastic Lissajous loops is significantly smaller in N-CNT nanocomposites than neat PVDF, indicating an elastic dominant response in this region. As strain amplitude increased, Lissajous loops in N-CNT nanocomposite samples became distorted, indicative of thixotropy and a yielding process in nanocomposite samples. Furthermore, observed patterns for nanocomposites suggest that N-CNT at-rest microstructure partially survived in both weakly ($\gamma \sim \gamma_x$) and strongly ($\gamma > \gamma_x$) nonlinear regimes as the area enclosed by Lissajous loops is relatively smaller, than that observed for neat PVDF. The area enclosed by elastic Lissajous loops in weakly and strongly nonlinear regimes for N-CNT nanocomposites showed the following order: Ni > Fe ~ Co.

As demonstrated by LM observations, Ni-based N-CNT nanocomposite presented the highest agglomerate area and relative transparency, indicating a poor dispersion quality of N-CNTs within PVDF matrix. This could be responsible for observing a multistep transition into a nonlinear regime and strongly nonlinear response at intermediate strain amplitudes in Ni-based N-CNT samples. As explained in the preceding section, N-CNTs synthesized over different catalysts demonstrated fundamentally different dispersion states at different scales. The presence of densely aggregated N-CNT structures in Ni-based N-CNT nanocomposite led to poor load transfer across polymer-aggregate interface, resulting in deformation-induced microstructural changes initiated from aggregate-aggregate boundaries at intermediate strain amplitudes ($\gamma \sim \gamma_x$). This was followed by widespread disintegration of (N-CNT)$_{Ni}$ aggregates, marked by multistep transition into a nonlinear regime. However, in Co-based and Fe-based N-CNT nanocomposites, the transition into nonlinear regime occurred by *stochastic erosion* [84, 85] of network structures formed by individually dispersed N-CNTs bound polymer chains and polymer matrix entanglement network [86, 87].

In this context, it could be added, that Fe-based N-CNT nanocomposites demonstrated a dual nature in a sense, that it showed an almost one-step transition into a nonlinear regime; however, the crossover point occurred at fairly small strain amplitude (γ_x = 5.8 %). This dual behavior can be explained in conjunction with poorer load transfer across interfacial region than Co-based N-CNT nanocomposite as a result of denser N-CNT clusters present in Fe-based N-CNT nanocomposite. Moreover, one-step transition to a nonlinear regime compared to the multistep transition observed for Ni-based N-CNT nanocomposite can be attributed to better nanodispersion state achieved in Fe-based N-CNT nanocomposite (see TEM images in **Figure 8** and relative transparency values in **Table 2**). Overall, it can be expressed, that no direct link between individual N-CNT structural features and rheological response was detectable, and thus N-CNT dispersion state played the main role in determining the melt-state rheological response.

4. Conclusions

In brief, this study revealed, that electrical conductivity of N-CNT/PVDF nanocomposites is highly dependent on N-CNT synthesis catalyst. Measuring electrical conductivity of the generated nanocomposites showed superior electrical conductivity and, thus, thermoelectric performance in the following order of the synthesis catalyst: Co > Fe > Ni. It was observed, that a combination of high synthesis yield, high aspect ratio, low structural defects, enhanced network formation, and good state of N-CNT dispersion can provide N-CNT/PVDF nanocomposites with superior electrical conductivity. Moreover, it was revealed, that nitrogen doping had an adverse impact on electrical conductivity of CNT/polymer nanocomposites and, therefore, their performance as thermoelectric materials.

Acknowledgements

Financial support from the Natural Sciences and Engineering Research Council of Canada (NSERC) is highly appreciated. We would like to thank Prof. Uttandaraman Sundararaj for his supervision to perform this project. We are grateful to Dr. Lars Laurentius for his assistance with Raman spectroscopy. In addition, we thank Dr. Petra Pötschke and Ms. Uta Reuter from IPF Dresden for LM and TEM investigations. Dr. Mohammad Arjmand thanks IPF Dresden for granting a research stay.

Author details

Mohammad Arjmand* and Soheil Sadeghi

*Address all correspondence to: arjmand64@yahoo.com

University of Calgary, Calgary, Canada

References

[1] Arjmand M. Electrical Conductivity, Electromagnetic Interference Shielding and Dielectric Properties of Multi-walled Carbon Nanotube/Polymer Composites [thesis]. Calgary: University of Calgary; 2014.

[2] Al-Saleh MH. Nanostructured Conductive Polymeric Materials [thesis]. Edmonton: University of Alberta; 2009.

[3] McGrail BT, Sehirlioglu A, Pentzer E. Polymer composites for thermoelectric applications. Angewandte Chemie-International Edition. 2015;54:1710–23. DOI: 10.1002/anie. 201408431.

[4] Arjmand M, Moud AA, Li Y, Sundararaj U. Outstanding electromagnetic interference shielding of silver nanowires: comparison with carbon nanotubes. RSC Advances. 2015;5:56590–8. DOI: 10.1039/C5RA08118A.

[5] da Silva AB, Arjmand M, Sundararaj U, Bretas RES. Novel composites of Copper nanowire/PVDF with superior dielectric properties. Polymer. 2014;55:226–34. DOI: 10.1016/j.polymer.2013.11.045.

[6] Pawar SP, Arjmand M, Gandi M, Bose S, Sundararaj U. Critical insights into understanding the effects of synthesis temperature and nitrogen doping towards charge storage capability and microwave shielding in nitrogen-doped carbon nanotube/polymer nanocomposites. RSC Advances. 2016;6:63224–34. DOI: 10.1039/C6RA15037C.

[7] Arjmand M, Mahmoodi M, Park S, Sundararaj U. Impact of foaming on the broadband dielectric properties of multi-walled carbon nanotube/polystyrene composites. Journal of Cellular Plastics. 2014;50:551–62. DOI: 10.1177/0021955X14539778.

[8] Arjmand M, Sundararaj U. Broadband dielectric properties of multiwalled carbon nanotube/polystyrene composites. Polymer Engineering & Science. 2015;55:173–9. DOI: 10.1002/pen.23881.

[9] Liu W, Yan X, Chen G, Ren Z. Recent advances in thermoelectric nanocomposites. Nano Energy. 2012;1:42–56. DOI: 10.1016/j.nanoen.2011.10.001.

[10] Bell LE. Cooling, heating, generating power, and recovering waste heat with thermoelectric systems. Science. 2008;321:1457–61. DOI: 10.1126/science.1158899.

[11] Dresselhaus MS, Chen G, Tang MY, Yang RG, Lee H, Wang DZ, et al. New directions for low-dimensional thermoelectric materials. Advanced Materials. 2007;19:1043–53. DOI: 10.1002/adma.200600527.

[12] Li G, Gadelrab KR, Souier T, Potapov PL, Chen G, Chiesa M. Mechanical properties of $Bi_xSb_{2-x}Te_3$ nanostructured thermoelectric material. Nanotechnology. 2012;23:065703. DOI: 10.1088/0957-4484/23/6/065703.

[13] Li J, Tang X, Li H, Yan Y, Zhang Q. Synthesis and thermoelectric properties of hydrochloric acid-doped polyaniline. Synthetic Metals. 2010;160:1153–8. DOI: 10.1016/j.synthmet.2010.03.001.

[14] Toshima N. Conductive polymers as a new type of thermoelectric material. Macromolecular Symposia. 2002;186:81–6. DOI: 10.1002/1521-3900(200208)186:1<81::AID-MASY81>3.0.CO;2-S.

[15] Yao Q, Chen L, Zhang W, Liufu S, Chen X. Enhanced thermoelectric performance of single-walled carbon nanotubes/polyaniline hybrid nanocomposites. ACS Nano. 2010;4:2445–51. DOI: 10.1021/nn1002562.

[16] Yu B, Zebarjadi M, Wang H, Lukas K, Wang H, Wang D, et al. Enhancement of thermoelectric properties by modulation-doping in silicon germanium alloy nanocomposites. Nano Letters. 2012;12:2077–82. DOI: 10.1021/nl3003045.

[17] Yu C, Kim YS, Kim D, Grunlan JC. Thermoelectric behavior of segregated-network polymer nanocomposites. Nano Letters. 2008;8:4428–32. DOI: 10.1021/nl802345s.

[18] Du Y, Shen SZ, Cai K, Casey PS. Research progress on polymer–inorganic thermo-electric nanocomposite materials. Progress in Polymer Science. 2012;37:820–41. DOI: 10.1016/j.progpolymsci.2011.11.003.

[19] Wang L, Yao Q, Bi H, Huang F, Wang Q, Chen L. PANI/graphene nanocomposite films with high thermoelectric properties by enhanced molecular ordering. Journal of Materials Chemistry A. 2015;3:7086–92. DOI: 10.1039/C4TA06422D.

[20] Song H, Cai K, Wang J, Shen S. Influence of polymerization method on the thermo-electric properties of multi-walled carbon nanotubes/polypyrrole composites. Synthetic Metals. 2016;211:58–65. DOI: 10.1016/j.synthmet.2015.11.013.

[21] Lee W, Kang YH, Lee JY, Jang K-S, Cho SY. Improving the thermoelectric power fac-tor of CNT/PEDOT: PSS nanocomposite films by ethylene glycol treatment. RSC Ad-vances. 2016;6:53339–44. DOI: 10.1039/C6RA08599G.

[22] Gao C, Chen G. Conducting polymer/carbon particle thermoelectric composites: emerging green energy materials. Composites Science and Technology. 2016;124:52–70. DOI: 10.1016/j.compscitech.2016.01.014.

[23] Dey A, Bajpai OP, Sikder AK, Chattopadhyay S, Khan MAS. Recent advances in CNT/graphene based thermoelectric polymer nanocomposite: a proficient move to-wards waste energy harvesting. Renewable and Sustainable Energy Reviews. 2016;53:653–71. DOI: 10.1016/j.rser.2015.09.004.

[24] Qin F, Brosseau C. A review and analysis of microwave absorption in polymer com-posites filled with carbonaceous particles. Journal of Applied Physics. 2012;111:061301. DOI: 10.1063/1.3688435.

[25] Thostenson ET, Ren Z, Chou T-W. Advances in the science and technology of carbon nanotubes and their composites: a review. Composites Science and Technology. 2001;61:1899–912. DOI: 10.1016/S0266-3538(01)00094-X.

[26] Bauhofer W, Kovacs JZ. A review and analysis of electrical percolation in carbon nanotube polymer composites. Composites Science and Technology. 2009;69:1486–98. DOI: 10.1016/j.compscitech.2008.06.018.

[27] Iijima S. Helical microtubules of graphitic carbon. Nature. 1991;354:56–8. DOI: 10.1038/354056a0.

[28] Xie XL, Mai YW, Zhou XP. Dispersion and alignment of carbon nanotubes in polymer matrix: a review. Materials Science & Engineering R-Reports. 2005;49:89–112. DOI: 10.1016/j.mser.2005.04.002.

[29] BCC Research LLC. Global Markets and Technologies for Carbon Nanotubes [Internet]. 2012. Available from: www.bccreserach.com [Accessed: 2016-08-24].

[30] Latil S, Roche S, Mayou D, Charlier J-C. Mesoscopic transport in chemically doped carbon nanotubes. Physical Review Letters. 2004;92:256805. DOI: 10.1103/PhysRevLett.92.256805.

[31] Terrones M, Ajayan PM, Banhart F, Blase X, Carroll DL, Charlier JC, et al. N-doping and coalescence of carbon nanotubes: synthesis and electronic properties. Applied Physics a-Materials Science & Processing. 2002;74:355–61. DOI: 10.1007/s003390201278.

[32] Zhong DY, Liu S, Zhang GY, Wang EG. Large-scale well aligned carbon nitride nanotube films: low temperature growth and electron field emission. Journal of Applied Physics. 2001;89:5939–43. DOI: 10.1063/1.1370114.

[33] Sharifi T, Hu G, Jia XE, Wagberg T. Formation of active sites for oxygen reduction reactions by transformation of nitrogen functionalities in nitrogen-doped carbon nanotubes. ACS Nano. 2012;6:8904–12. DOI: 10.1021/nn302906r.

[34] Pels JR, Kapteijn F, Moulijn JA, Zhu Q, Thomas KM. Evolution of nitrogen functionalities in carbonaceous materials during pyrolysis. Carbon. 1995;33:1641–53. DOI: 10.1016/0008-6223(95)00154-6.

[35] Arjmand M, Sundararaj U. Effects of nitrogen doping on X-band dielectric properties of carbon nanotube/polymer nanocomposites. ACS Applied Materials & Interfaces. 2015;7:17844–50. DOI: 10.1021/acsami.5b04211.

[36] Kanygin MA, Sedelnikova OV, Asanov IP, Bulusheva LG, Okotrub AV, Kuzhir PP, et al. Effect of nitrogen doping on the electromagnetic properties of carbon nanotube-based composites. Journal of Applied Physics. 2013;113:144315. DOI: 10.1063/1.4800897.

[37] Arjmand M, Ameli A, Sundararaj U. Employing nitrogen doping as innovative technique to improve broadband dielectric properties of carbon nanotube/polymer nanocomposites. Macromolecular Materials and Engineering. 2016;301:555-65. DOI: 10.1002/mame.201500365.

[38] Krause B, Ritschel M, Täschner C, Oswald S, Gruner W, Leonhardt A, et al. Comparison of nanotubes produced by fixed bed and aerosol-CVD methods and their electrical percolation behaviour in melt mixed polyamide 6.6 composites. Composites Science and Technology. 2010;70:151–60. DOI: 10.1016/j.compscitech.2009.09.018.

[39] Arjmand M, Chizari K, Krause B, Pötschke P, Sundararaj U. Effect of synthesis catalyst on structure of nitrogen-doped carbon nanotubes and electrical conductivity and electromagnetic interference shielding of their polymeric nanocomposites. Carbon. 2016;98:358–72. DOI: 10.1016/j.carbon.2015.11.024.

[40] Ding F, Larsson P, Larsson JA, Ahuja R, Duan HM, Rosen A, et al. The importance of strong carbon-metal adhesion for catalytic nucleation of single-walled carbon nanotubes. Nano Letters. 2008;8:463–8. DOI: 10.1021/nl072431m.

[41] Chizari K, Vena A, Laureritius L, Sundararaj U. The effect of temperature on the morphology and chemical surface properties of nitrogen-doped carbon nanotubes. Carbon. 2014;68:369–79. DOI: 10.1016/j.carbon.2013.11.013.

[42] Yuan JK, Yao SH, Dang ZM, Sylvestre A, Genestoux M, Bai JB. Giant dielectric permittivity nanocomposites: realizing true potential of pristine carbon nanotubes in polyvinylidene fluoride matrix through an enhanced interfacial interaction. Journal of Physical Chemistry C. 2011;115:5515–21. DOI: 10.1021/jp1117163.

[43] Arjmand M, Sundararaj U. Impact of $BaTiO_3$ as insulative ferroelectric barrier on the broadband dielectric properties of MWCNT/PVDF nanocomposites. Polymer Composites. 2016;37:299–304. DOI: 10.1002/pc.23181.

[44] Dang ZM, Yao SH, Yuan JK, Bai JB. Tailored dielectric properties based on microstructure change in BaTiO(3)-carbon nanotube/polyvinylidene fluoride three-phase nanocomposites. Journal of Physical Chemistry C. 2010;114:13204–9. DOI: 10.1021/jp103411c.

[45] Krause B, Boldt R, Häußler L, Pötschke P. Ultralow percolation threshold in polyamide 6.6/MWCNT composites. Composites Science and Technology. 2015;114:119–25. DOI: 10.1016/j.compscitech.2015.03.014.

[46] Arjmand M, Sundararaj U. Electromagnetic interference shielding of nitrogen-doped and undoped carbon nanotube/polyvinylidene fluoride nanocomposites: a comparative study. Composites Science and Technology. 2015;118:257–63. DOI: 10.1016/j.compscitech.2015.09.012.

[47] Kaiser KL. Electromagnetic Shielding. Boca Raton: CRC Press; 2006. 336 p.

[48] Blythe T, Bloor D. Electrical Properties of Polymers. 2nd ed. New York: Cambridge University Press; 2005. 480 p.

[49] Weber M, Kamal MR. Estimation of the volume resistivity of electrically conductive composites. Polymer Composites. 1997;18:711–25. DOI: 10.1002/pc.10324.

[50] Arjmand M, Mahmoodi M, Gelves GA, Park S, Sundararaj U. Electrical and electromagnetic interference shielding properties of flow-induced oriented carbon nanotubes in polycarbonate. Carbon. 2011;49:3430–40. DOI: 10.1016/j.carbon.2011.04.039.

[51] Lux F. Models proposed to explain the electrical conductivity of mixtures made of conductive and insulating materials. Journal of Materials Science. 1993;28:285–301. DOI: 10.1007/bf00357799.

[52] Li J, Ma PC, Chow WS, To CK, Tang BZ, Kim JK. Correlations between percolation threshold, dispersion state, and aspect ratio of carbon nanotubes. Advanced Functional Materials. 2007;17:3207–15. DOI: 10.1002/adfm.200700065.

[53] Behnam A, Guo J, Ural A. Effects of nanotube alignment and measurement direction on percolation resistivity in single-walled carbon nanotube films. Journal of Applied Physics. 2007;102:044313. DOI: 10.1063/1.2769953.

[54] Arjmand M, Mahmoodi M, Park S, Sundararaj U. An innovative method to reduce the energy loss of conductive filler/polymer composites for charge storage applications. Composites Science and Technology. 2013;78:24–9. DOI: 10.1016/j.compscitech.2013.01.019.

[55] Arjmand M, Apperley T, Okoniewski M, Sundararaj U. Comparative study of electromagnetic interference shielding properties of injection molded versus compression molded multi-walled carbon nanotube/polystyrene composites. Carbon. 2012;50:5126–34. DOI: 10.1016/j.carbon.2012.06.053.

[56] Balberg I. Tunneling and nonuniversal conductivity in composite materials. Physical Review Letters. 1987;59:1305–8. DOI: 10.1103/PhysRevLett.59.1305.

[57] Sichel EK, Gittleman JI, Sheng P. Transport properties of the composite material carbon-poly(vinyl chloride). Physical Review B. 1978;18:5712–6. DOI: 10.1103/PhysRevB.18.5712.

[58] Li C, Thostenson ET, Chou T-W. Dominant role of tunneling resistance in the electrical conductivity of carbon nanotube–based composites. Applied Physics Letters. 2007;91:223114. DOI: 10.1063/1.2819690.

[59] Chekanov Y, Ohnogi R, Asai S, Sumita M. Electrical properties of epoxy resin filled with carbon fibers. Journal of Materials Science. 1999;34:5589–92. DOI: 10.1023/A:1004737217503.

[60] Spitalsky Z, Tasis D, Papagelis K, Galiotis C. Carbon nanotube-polymer composites: chemistry, processing, mechanical and electrical properties. Progress in Polymer Science. 2010;35:357–401. DOI: 10.1016/j.progpolymsci.2009.09.003.

[61] TabkhPaz M, Mahmoodi M, Arjmand M, Sundararaj U, Chu J, Park SS. Investigation of chaotic mixing for MWCNT/polymer composites. Macromolecular Materials and Engineering. 2015;300:482–96. DOI: 10.1002/mame.201400361.

[62] Thomassin J-M, Jérôme C, Pardoen T, Bailly C, Huynen I, Detrembleur C. Polymer/carbon based composites as electromagnetic interference (EMI) shielding materials. Materials Science and Engineering: R: Reports. 2013;74:211–32. DOI: 10.1016/j.mser.2013.06.001.

[63] Sharifi T, Nitze F, Barzegar HR, Tai CW, Mazurkiewicz M, Malolepszy A, et al. Nitrogen doped multi walled carbon nanotubes produced by CVD-correlating XPS and Raman

spectroscopy for the study of nitrogen inclusion. Carbon. 2012;50:3535–41. DOI: 10.1016/j.carbon.2012.03.022.

[64] Barzegar HR, Gracia-Espino E, Sharifi T, Nitze F, Wagberg T. Nitrogen doping mechanism in small diameter single-walled carbon nanotubes: impact on electronic properties and growth selectivity. Journal of Physical Chemistry C. 2013;117:25805–16. DOI: 10.1021/jp409518m.

[65] Moisala A, Nasibulin AG, Kauppinen EI. The role of metal nanoparticles in the catalytic production of single-walled carbon nanotubes—a review. Journal of Physics-Condensed Matter. 2003;15:S3011–35. DOI: 10.1088/0953-8984/15/42/003.

[66] Nasibulin AG, Altman IS, Kauppinen EI. Semiempirical dynamic phase diagrams of nanocrystalline products during copper (II) acetylacetonate vapour decomposition. Chemical Physics Letters. 2003;367:771–7. DOI: 10.1016/s0009-2614(02)01795-5.

[67] Alvarez WE, Kitiyanan B, Borgna A, Resasco DE. Synergism of Co and Mo in the catalytic production of single-wall carbon nanotubes by decomposition of CO. Carbon. 2001;39:547–58. DOI: 10.1016/s0008-6223(00)00173-1.

[68] Huang Y, Li N, Ma YF, Feng D, Li FF, He XB, et al. The influence of single-walled carbon nanotube structure on the electromagnetic interference shielding efficiency of its epoxy composites. Carbon. 2007;45:1614–21. DOI: 10.1016/j.carbon.2007.04.016.

[69] Singh BP, Saini K, Choudhary V, Teotia S, Pande S, Saini P, et al. Effect of length of carbon nanotubes on electromagnetic interference shielding and mechanical properties of their reinforced epoxy composites. Journal of Nanoparticle Research. 2013;16:2161. DOI: 10.1007/s11051-013-2161-9.

[70] Maciel IO, Anderson N, Pimenta MA, Hartschuh A, Qian HH, Terrones M, et al. Electron and phonon renormalization near charged defects in carbon nanotubes. Nature Materials. 2008;7:878–83. DOI: 10.1038/nmat2296.

[71] Rao AM, Richter E, Bandow S, Chase B, Eklund PC, Williams KA, et al. Diameter-selective Raman scattering from vibrational modes in carbon nanotubes. Science. 1997;275:187–91. DOI: 10.1126/science.275.5297.187.

[72] Dresselhaus MS, Dresselhaus G, Jorio A, Souza AG, Saito R. Raman spectroscopy on isolated single wall carbon nanotubes. Carbon. 2002;40:2043–61. DOI: 10.1016/s0008-6223(02)00066-0.

[73] Picher M, Anglaret E, Arenal R, Jourdain V. Processes controlling the diameter distribution of single-walled carbon nanotubes during catalytic chemical vapor deposition. ACS Nano. 2011;5:2118–25. DOI: 10.1021/nn1033086.

[74] Villalpando-Paez F, Zamudio A, Elias AL, Son H, Barros EB, Chou SG, et al. Synthesis and characterization of long strands of nitrogen-doped single-walled carbon

nanotubes. Chemical Physics Letters. 2006;424:345–52. DOI: 10.1016/j.cplett. 2006.04.074.

[75] Ibrahim EMM, Khavrus VO, Leonhardt A, Hampel S, Oswald S, Rummeli MH, et al. Synthesis, characterization, and electrical properties of nitrogen-doped single-walled carbon nanotubes with different nitrogen content. Diamond and Related Materials. 2010;19:1199–206. DOI: 10.1016/j.diamond.2010.05.008.

[76] Ramesh BP, Blau WJ, Tyagi PK, Misra DS, Ali N, Gracio J, et al. Thermogravimetric analysis of cobalt-filled carbon nanotubes deposited by chemical vapour deposition. Thin Solid Films. 2006;494:128–32. DOI: 10.1016/j.tsf.2005.08.220.

[77] Khalkhal F, Carreau PJ, Ausias G. Effect of flow history on linear viscoelastic properties and the evolution of the structure of multiwalled carbon nanotube suspensions in an epoxy. Journal of Rheology (1978-present). 2011;55:153–75. DOI: 10.1122/1.3523628.

[78] Khalkhal F, Carreau PJ. Scaling behavior of the elastic properties of non-dilute MWCNT–epoxy suspensions. Rheologica Acta. 2011;50:717–28. DOI: 10.1007/ s00397-010-0527-9.

[79] Ureña-Benavides EE, Kayatin MJ, Davis VA. Dispersion and rheology of multiwalled carbon nanotubes in unsaturated polyester resin. Macromolecules. 2013;46:1642–50. DOI: 10.1021/ma3017844.

[80] Xiao JR, Gama BA, Gillespie JW. An analytical molecular structural mechanics model for the mechanical properties of carbon nanotubes. International Journal of Solids and Structures. 2005;42:3075–92. DOI: 10.1016/j.ijsolstr.2004.10.031.

[81] Ewoldt RH, Winter P, Maxey J, McKinley GH. Large amplitude oscillatory shear of pseudoplastic and elastoviscoplastic materials. Rheologica Acta. 2010;49:191–212. DOI: 10.1007/s00397-009-0403-7.

[82] Hyun K, Wilhelm M, Klein CO, Cho KS, Nam JG, Ahn KH, et al. A review of nonlinear oscillatory shear tests: analysis and application of large amplitude oscillatory shear (LAOS). Progress in Polymer Science. 2011;36:1697–753. DOI: 10.1016/j.progpolymsci.2011.02.002.

[83] Ewoldt RH, Hosoi AE, McKinley GH. New measures for characterizing nonlinear viscoelasticity in large amplitude oscillatory shear. Journal of Rheology. 2008;52:1427. DOI: 10.1122/1.2970095.

[84] Sprakel J, Lindström SB, Kodger TE, Weitz DA. Stress enhancement in the delayed yielding of colloidal gels. Physical Review Letters. 2011;106:248303. DOI: 10.1103/ PhysRevLett.106.248303.

[85] Sadeghi S, Zehtab Yazdi A, Sundararaj U. Controlling short-range interactions by tuning surface chemistry in HDPE/graphene nanoribbon nanocomposites. The Journal of Physical Chemistry B. 2015;119:11867–78. DOI: 10.1021/acs.jpcb.5b03558.

[86] Pham KN, Petekidis G, Vlassopoulos D, Egelhaaf SU, Pusey PN, Poon WCK. Yielding of colloidal glasses. EPL (Europhysics Letters). 2006;75:624. DOI: 10.1209/epl/i2006-10156-y.

[87] Laurati M, Egelhaaf S, Petekidis G. Nonlinear rheology of colloidal gels with intermediate volume fraction. Journal of Rheology (1978-present). 2011;55:673–706. DOI: 10.1122/1.3571554.

Measurement Techniques for Characterization Materials and Devices

Methods and Apparatus for Measuring Thermopower and Electrical Conductivity of Thermoelectric Materials at High Temperatures

Alexander T. Burkov, Andrey I. Fedotov and Sergey V. Novikov

Additional information is available at the end of the chapter

http://dx.doi.org/10.5772/66290

Abstract

The principles and methods of thermopower and electrical conductivity measurements at high temperatures (100–1000 K) are reviewed. These two properties define the so-called power factor of thermoelectric materials. Moreover, in combination with thermal conductivity, they determine efficiency of thermoelectric conversion. In spite of the principal simplicity of measurement methods of these properties, their practical realization is rather complicated, especially at high temperatures. This leads to large uncertainties in determination of the properties, complicates comparison of the results, obtained by different groups, and hinders realistic estimate of potential thermoelectric efficiency of new materials. The lack of commonly accepted reference material for thermopower measurements exaggerates the problem. Therefore, it is very important to have a clear understanding of capabilities and limitations of the measuring methods and set-ups. The chapter deals with definitions of thermoelectric parameters and principles of their experimental determination. Metrological characteristics of state-of-the-art experimental set-ups for high temperature measurements are analyzed.

Keywords: thermopower, electrical conductivity, high temperature, thermoelectric material, measurement

1. Introduction

Thermoelectric energy conversion is based on two effects discovered in the nineteenth century: Seebeck effect and Peltier effect [1]. Historically, Seebeck effect was the first discovered thermoelectric effect, which consists in appearance of electrical current in the circuit of two different conductors at the presence of temperature difference. In year 1821, Thomas Johann Seebeck discovered, that magnetic field is generated in closed circuit consisting of bismuth (or antimony) and copper in

the presence of temperature difference between two contacts. He first announced this discover in year 1825 in the writings of Berlin Academy of Sciences. Seebeck called this phenomenon thermomagnetism. The term "thermoelectricity" was proposed by Hans Christian Oersted approximately at the same time. There are indications, that this effect was observed and correctly interpreted in years 1794–1795 by Alessandro Volta [2]. Peltier effect was discovered in 1834 by Jean-Charles Peltier. When electrical current is forced to flow through circuit of two conductors, one contact gives out heat, while another absorbs heat. These two physical effects have become the basis for thermoelectric converters. For a long time, their practical application was limited by the use of simple thermoelectric sources for research and metal thermocouples for temperature measurement. The situation changed, when Abram F. Ioffe suggested to use semiconductors instead of metals. Based on PbS and ZnSb compounds, generator for vacuum tube radios was developed.

In the early 1950s, projects on creation of thermoelectric coolers started, and new effective materials based on compounds $(Bi,Sb)_2Te_3$ were discovered. Alloys, based on these compounds, are still the basic materials for thermoelectric refrigeration units. In the 1950–1960s, the complete elementary theory of thermoelectric conversion was created [3–5]. It was shown, that efficiency is determined by parameter $ZT = T\frac{\alpha^2\sigma}{\kappa}$, where T, α, σ and κ are absolute temperature, Seebeck coefficient (or thermopower), electrical and thermal conductivity, respectively. Almost all thermoelectric materials currently used in the industry were discovered, technology for their production was developed: synthesis, crystal growth, metal-ceramic technology (**Figure 1**) [6].

Design and production technology of multi-element assembly of thermocouples, which are called thermoelectric (TE) cells or modules, have been developed. These modules may consist of one or more stages (cascade module); they are used to create different variants of thermoelectric coolers (TEC) and thermoelectric generators (TEG).

After relatively rapid development in years 1950–1960s, further improving of thermoelectric parameters and TE devices progressed more slowly.

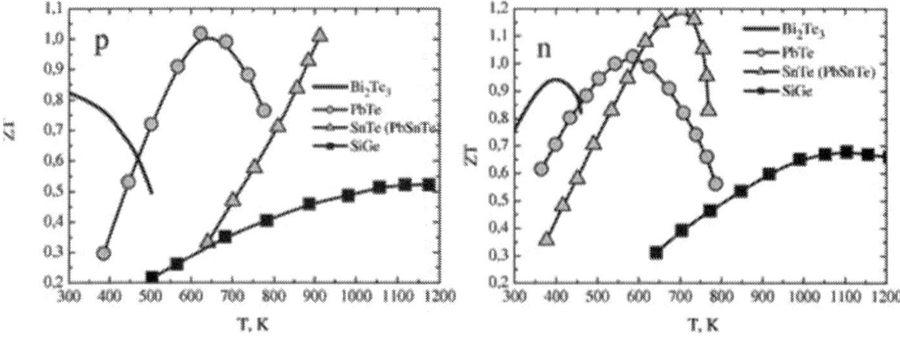

Figure 1. ZT dependence on temperature for the main thermoelectric semiconductor materials according to data of the 1960s: left side is p-type materials; right side is n-type materials.

For a long time, the maximum value of dimensionless parameter ZT does not exceed the value of 1. Areas of application of thermoelectric energy conversion techniques have been largely limited to special applications, such as power sources for spacecraft and military applications, where cost is not a major limiting factor.

In the 1980s, mass application of TE cooling for variety of purposes had started, and the market for thermoelectric cooling continues to expand today. In general, we can say, that method of thermoelectric conversion definitively established itself as one of the high-end technologies, especially for cooling purposes. This is due to its technical advantages. However, its wider application is constrained by insufficiently high efficiency of thermoelectric conversion of modern thermoelectric materials, that make the method economically ineffective. Therefore, the ultimate goal of basic research in physics and chemistry of thermoelectric materials is the development of more efficient thermoelectric materials for TEC and TEG. In view of this problem, precise and reliable measurement of thermoelectric properties (thermoelectric power, electrical and thermal conductivity) of new TE materials plays important role. These measurements must satisfy a number of requirements. Naturally, measurement results must be reliable and sufficiently accurate, measurements must be performed over a wide range of temperatures comparable with a typical range of applications. Despite the relative simplicity of fundamental measuring methods of thermoelectric properties, their practical implementation, accounting of above requirements, is a difficult task. For example, requirements for measurement accuracy are determined by the minimum of practically meaningful change of ZT parameter, which is about 10%. In order to reliably detect such small change of this parameter, measurement accuracy of thermoelectric coefficients α, σ, and κ must be not worse than 3%.

2. Thermoelectric coefficients and principles of experimental determination

2.1. Electrical conductivity

Electrical resistivity ρ, or conductivity $\sigma = \frac{1}{\rho}$, which is inverse value, determines electrical current density $\mathbf{j}$ in conductor, when external electric field $\mathbf{E}$ is applied: $\mathbf{j} = \sigma\mathbf{E}$ (Ohm's law). Coefficient σ does not depend on current value. In general, σ is a second rank tensor, the number of independent components of this tensor depends on the sample material crystallographic symmetry. For crystals with cubic symmetry, tensor σ has only diagonal components, and they are all identical. Thus, it degenerates into scalar in this case. Detailed information on tensor structure for crystal lattice of different symmetry can be found in [7].

Figure 2 shows electrical scheme for measuring electrical conductivity. When electrical current I is forced through the uniform conductor under isothermal conditions, an electric field arises. The sample electrical resistance R can be found from potential difference ΔV between two points on the sample surface and electrical current magnitude: $R = \frac{\Delta V}{I}$. Resistance R depends on parameters of the sample material and its geometrical dimensions: $R = \frac{1}{\sigma} \times \frac{l}{a \times b}$, where l, a, b is distance between potential probes, the sample's width and thickness, respectively. Thus, electrical conductivity σ of the sample material can be determined from measured values: its resistance R and geometrical parameters l, a, and b:

$$\sigma = \frac{1}{R} \times \frac{l}{a \times b}. \tag{1}$$

It should be noted, that geometric parameters do not necessarily coincide with dimensions of the sample. Electrical conductivity value is always positive, in linear approximation it does not depend on electric field (or magnitude of electric current), but depends on temperature. Depending on type of material, electrical conductivity value changes over very wide range. For metals at room temperature, σ is in the range of 10^6–10^4 S cm^{-1}, and for good insulators, it falls to 10^{-20} S cm^{-1}. Electrical conductivity of typical conductor at room temperature or above is inversely proportional to temperature and has finite value as temperature approaches absolute zero 0 K (**Figure 3a**). Electrical conductivity of insulators increases exponentially with increasing temperature and vanishes at low temperatures (**Figure 3b**).

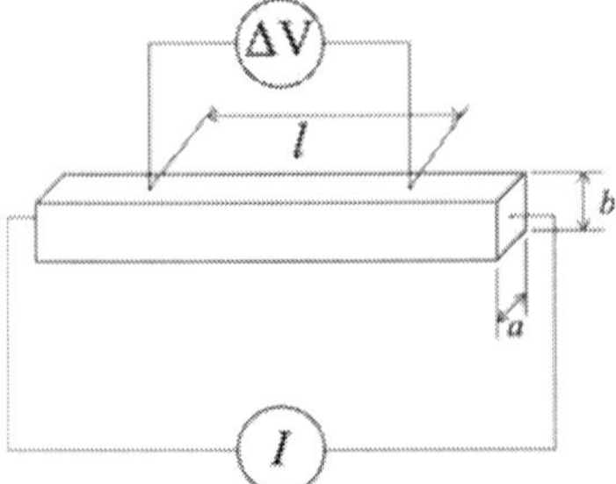

Figure 2. Scheme of electric circuit for measuring electrical conductivity value.

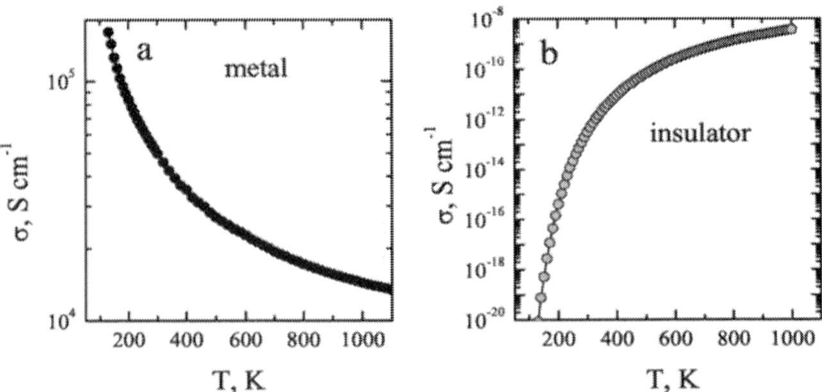

Figure 3. Temperature dependence of electrical conductivity σ of metal (a) and insulator (b).

2.2. Thermoelectric effects

Seebeck effect is occurrence of electromotive force in conductor, which has temperature gradient inside. It can be observed in a simple circuit consisting of two different conductors (x and l), when contacts of these conductors have different temperatures (**Figure 4**). Under these conditions, there will be a potential difference in a circuit: $\Delta V \propto \alpha_{xl}(T_2-T_1)$, where T_2-T_1 is a temperature difference between contacts and coefficient α is known as Seebeck coefficient or thermoelectric power. Seebeck coefficient is formally defined as follows: $\mathbf{E} = \alpha \nabla T$, here $\mathbf{E}$ – electric field induced in conductor in the presence of temperature gradient ∇T. Seebeck coefficient α is a second rank tensor. In contrast to electrical conductivity, Seebeck coefficient can be either positive or negative. Potential difference measured by voltmeter in the circuit shown in **Figure 4**, $\Delta V = \varphi_2-\varphi_1$, where φ_2 and φ_1 are input voltmeter potentials "1" and "2" at the same temperature T_0, equals to $\Delta V = \varphi_2-\varphi_1 = \int_1^2 \nabla \varphi dl$.

Since $\nabla \varphi = -E$, then: $\Delta V = \int_1^2 -Edl = -\int_1^2 \alpha \nabla T dl$.

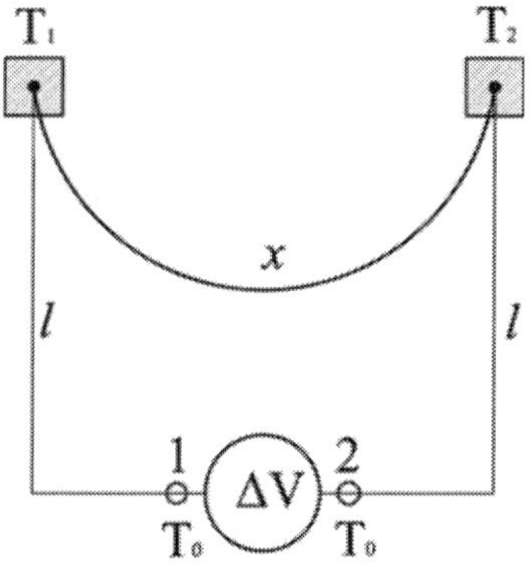

Figure 4. Thermoelectric circuit consisting of two conductors connected in series. Contacts of conductors are maintained at temperatures T_1 and T_2.

The circuit shown in **Figure 4** consists of two different conductors, x ("sample") and l (wires connecting the sample with voltmeter). We assume, that both conductors are uniform. Seebeck coefficient of the sample and wires are denoted as α_x and α_l, respectively. For homogeneous and isotropic conductors, coefficient α is independent on position along the wire and direction of temperature gradient, but usually it depends on temperature. Therefore:

$$\Delta V = -\int_1^{T_1} \alpha_l \nabla T dl - \int_{T_1}^{T_2} \alpha_x \nabla T dl - \int_{T_2}^{2} \alpha_l \nabla T dl \qquad (2)$$

or

$$\Delta V = -\int_{T_0}^{T_1} \alpha_l dT + \int_{T_0}^{T_1} \alpha_l dT - \int_{T_1}^{T_2} \alpha_x dT + \int_{T_1}^{T_2} \alpha_l dT = -\int_{T_1}^{T_2} (\alpha_x - \alpha_l) dT. \tag{3}$$

When temperature difference $T_2 - T_1$ is small compared to average temperature $(T_2 + T_1)/2$, then:

$$\Delta V = -(T_2 - T_1) \times (\alpha_x - \alpha_l). \tag{4}$$

Hence, experimentally measured potential difference is proportional to temperature difference between the sample and probe contacts and Seebeck coefficient difference of the sample material and probes. It means, that in this kind of experiment only difference $\alpha_x - \alpha_l$ can be measured, and it is called relative thermoelectric power of "x" and "l" conductors α_{xl}. In order to determine absolute thermopower of the sample α_x, it is necessary to know thermopower of probe α_l (usually called as reference electrode probe). The magnitude of thermopower of metals ranges from $\pm 10^{-6}$ to $\pm 5 \times 10^{-5}$ V/K (at room temperature), while thermopower of thermoelectric semiconductors can reach $\pm 10^{-3}$ V/K.

Peltier effect can be observed in a similar circuit by replacing voltmeter to current source. When electrical current flows through the circuit, then at one contact, heat is emitted and at another heat is absorbed. Quantity of heat (Q) emitted or absorbed per unit time at contact of two materials is given by formula: $Q = \Pi_{lx} \times I$, here Π_{lx} – Peltier coefficient of materials l and x, I – current flowing through the contacts. Similar to thermopower, Peltier coefficient of each material can be determined: $\Pi_{lx} = \Pi_l - \Pi_x$. Thermopower and Peltier coefficient are interrelated by Thompson relation [8, 9]:

$$\Pi = T \times \alpha. \tag{5}$$

Another important thermoelectric effect is Thompson effect. When electrical current passes through homogeneous conductor in the presence of temperature gradient, then some heat energy is released or absorbed depending on mutual orientation of current and temperature gradient. In contrast to Joule heat, in this effect, heat can be emitted, leading to additional heating of conductor or absorbed, leading to cooling. When electrical current with density j flows through conductor, then quantity of heat (q), emitted in unit volume of conductor per unit time, equals to [8, 9]: $q = -\tau_T \times j \times \nabla T$. In contrast to Seebeck and Peltier coefficients, Thompson coefficient τ_T can be measured for individual conductor. Thompson coefficient is interrelated to two other thermoelectric coefficients by second Thompson relation [8, 9]:

$$\tau_T = T \frac{d\alpha}{dT}. \tag{6}$$

This important relation allows to determine thermopower:

$$\alpha(T) = \int_0^T \frac{\tau_T}{T} dT \tag{7}$$

and to build the *absolute thermoelectric scale*, which we will discuss further.

3. Measurement principles

3.1. Electrical conductivity

If the sample is homogeneous (electrical conductivity is the same everywhere inside the sample), then under uniform electrical current distribution inside the sample, electrical conductivity of the material can be determined by formula (1) on the base of experimentally determined values R, l, a and b. In conductivity measurements, some heat energy is always generated in the sample volume due to Joule heating. Amount of heat generated in unit volume of the sample is determined by Joule-Lenz's law: $q_j = j \times E = j^2 \times \rho$. This heat energy can affect the accuracy of conductivity measurement, changing sample temperature, and inducing thermoelectric contribution to measured potential difference (ΔV). In order to reduce influence of Joule heat on conductivity measurement, one has to use lower current density and provide good thermal contact of the sample with environment.

Eq. (1) is applicable, if the sample is in isothermal conditions. In actual practice, this condition is almost never fulfilled. Moreover, electrical conductivity is often measured simultaneously with thermoelectric coefficient, which requires temperature gradient. Under these conditions, potential difference, measured in the circuit shown in **Figure 2**, will include two components: $\Delta V = R \times I + \alpha_{lx}\Delta T = \frac{1}{\sigma} \times \frac{l}{a \times b}I + \alpha_{lx}\Delta T$, where α_{lx} and ΔT – thermopower of the sample and temperature difference between potential probes, respectively. For thermoelectric materials, both contributions can be of the same order of magnitude. To eliminate the influence of thermal gradient in the sample on electrical conductivity, two methods are used:

1. DC measurements: two measurements of ΔV must be performed. One is carried out with electrical current flow, and another is carried out either without current flow or with current flow in opposite direction: $\Delta V^- = -R \times I + \alpha_{lx}\Delta T$, subtracting the results of these measurements, we get: $\Delta V_R = \Delta V - \Delta V^- = 2R \times I$, σ now can be found from measured R. It was assumed here, that thermoelectric contribution to ΔV does not depend on electrical current. In fact, this is not true, and in measurement of σ with direct current, dependence of thermoelectric contribution on electrical current should be considered. We will discuss this issue in the analysis of electrical conductivity measurement errors.

2. AC measurements: typically, AC current with frequency from several tens to thousands of hertz is used. Thus, due to thermal inertia, thermopower contribution does not contain frequency-dependent components, and signal, proportional to electrical resistance, can be measured at AC current frequency. However, it should be remembered, that for materials with high magnetic permeability, such as ferromagnets, thickness of skin layer, even at low current frequencies, may be equal to unit or even fractions of millimeter. If thickness of skin layer is comparable to or less than thickness of the sample, then determination of σ may contain significant errors.

In the following sections, different methods of measurement σ are described. To exclude the contribution of thermoelectric effects in all of them, AC or DC measurements may be used.

3.1.1. Classic measurement scheme

Classic measurement scheme of electrical conductivity is presented in **Figure 5**. In this method, the sample should be prepared in the form of long thin and uniform wire with diameter d. Potential difference ΔV_{12}, is measured between points "1" and "2" separated by distance "l", when current I passes through the wire. Electrical conductivity is determined by formula:

$\sigma = \frac{I}{\Delta V_{12}} \times \frac{4l}{\pi d^2}.$

The wire must be placed into electrically non-conductive medium having sufficiently high thermal conductivity, which absorbs heat generated in the sample, and minimizes temperature gradient in it. When the sample is prepared in the form of long thin wire, then low measuring current density j can be used. In this case, along with reducing quantity of Joule heat, it is possible to maintain large enough potential difference ΔV_{12} by increasing distance l between potential probes, which improves measurements accuracy. However, this method is rarely used in practice.

First, long wire samples are inconvenient, when measurements are performed in chambers with limited volume, such as cryostat for measurements at low temperatures or vacuum chamber at high temperature measurements. Second, majority of materials is difficult or impossible to prepare in the form of thin homogeneous wire. Therefore, usually, short samples are used in the form of cylinder or parallelepiped, thin plate or film. The accuracy of resistance measurement of such samples is lower than in classic configuration.

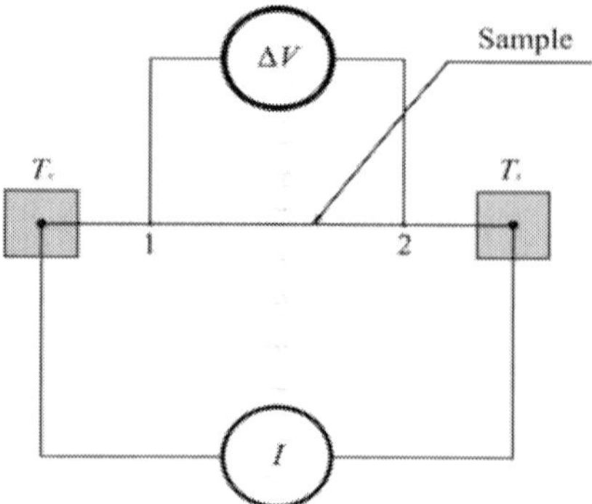

Figure 5. Classic method of electrical resistance measurement. The sample is prepared as a homogeneous wire, contacts with current leads are maintained at temperature T_x.

3.1.2. Samples of regular geometric shape

Figure 6 shows the scheme for measuring electrical conductivity of short samples with a regular geometric shape. The sample for such measurements must have simple geometric shape allowing accurate determination of electrical current density and potential gradient in the sample, which have to be uniform. Current contacts should provide uniform current distribution in the sample.

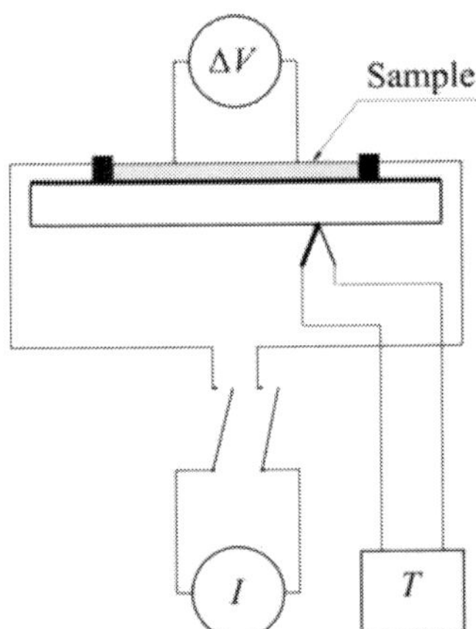

Figure 6. Resistance measuring scheme of short samples.

Since electrical conductivity depends on temperature, then, during measuring it, temperature of the sample must be set and determine precisely. Electrical conductivity is determined by formula: $\sigma = \frac{I}{\Delta V} \times \frac{l}{A}$, here, A is cross-section area of the sample in plane, perpendicular to electric current direction.

3.1.3. Four-probe method of electrical resistivity measurement

In both schemes of measuring electrical conductivity σ, described above, geometric parameters coincide with cross-section of the sample $A = a \times b$ and distance between potential probes (l). There are modifications of these schemes, in which geometric parameters do not match sizes of the sample. They include four-probe method [10–12] and van der Pauw method [12, 13]. Note that all methods of measuring electrical conductivity described here are essentially four-probe methods in the sense, that potential probes are separated from electrical current leads. However, this term is also used as the name of specific embodiment of methods for measuring electrical conductivity. In the most common variant of this method, all four electrodes are arranged along straight line on flat surface of the sample (**Figure 7**). If electrodes are arranged symmetrically, and thickness (d) and minimum distance from electrodes to the edge of the sample is much greater than distance between electrodes (l) (semi-infinite space approximation), then electrical conductivity is determined by simple expression [10–12]:

$\sigma = \frac{2I}{\pi \times \Delta V} \times \left[\frac{1}{S-l} - \frac{1}{S+l} \right]$, here I – electrical current flowing through the sample, S – distance between outermost electrodes (current contacts), l – distance between potential probes. Practical criterion of applicability of this approximation is $S/d < 5$. If electrodes are arranged at the same distance from each other, that is, $S = 3l$, we get $\sigma = \frac{I}{2\pi \times l \times \Delta V}$. In another limit case $d \ll l$, expression for determining electrical conductivity takes the form [10–12]: $\sigma = \frac{I}{\pi \cdot d \cdot \Delta V} \ln\left(\frac{S+l}{S-l}\right)$.

If $S = 3l$, we obtain: $\sigma = \frac{I}{\pi \times d \times \Delta V} \ln 2$. This formula is applicable for $S/d > 5$.

For arbitrary thickness of the sample, expression for σ is as follows [10–12]:

$$\sigma = \frac{2I}{\pi \times \Delta V} \left\{ \frac{1}{S-l} - \frac{1}{S+l} + 2 \times \sum_{n=1}^{\infty} \left[\frac{1}{\sqrt{(S-l)^2 + (4nd)^2}} - \frac{1}{\sqrt{(S+l)^2 + (4nd)^2}} \right] \right\}. \tag{8}$$

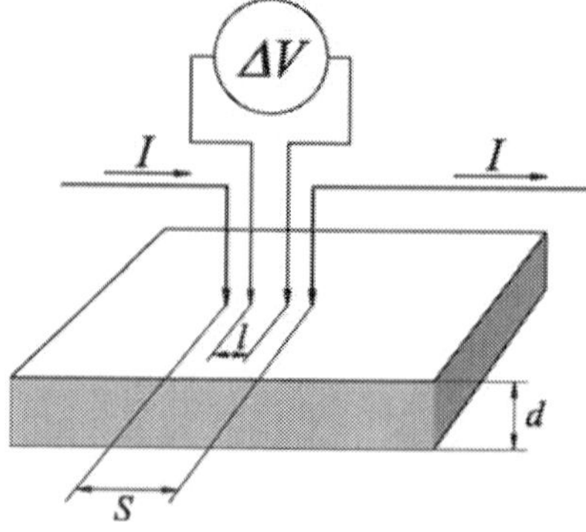

Figure 7. Four-probe method to measure electrical conductivity.

Four-probe method is a convenient way to determine quickly and accurately electrical conductivity and does not require preparation of samples with regular geometric shape. It requires one flat surface only. However, the sample surface area needs to be large enough to satisfy condition $L_{min} > 10S$ for any distance (L) from measuring probes to the edge of the sample. Otherwise, measured potential difference ΔV will depend on type and shape of the sample boundaries.

3.1.4. Van der Pauw method

Van der Pauw method is applied for measuring electrical conductivity of the samples with irregular shape [12–15]. To measure electrical conductivity by van der Pauw method, it is necessary to form four contacts at arbitrary points A, B, C and D on the edge of flat sample (**Figure 8**).

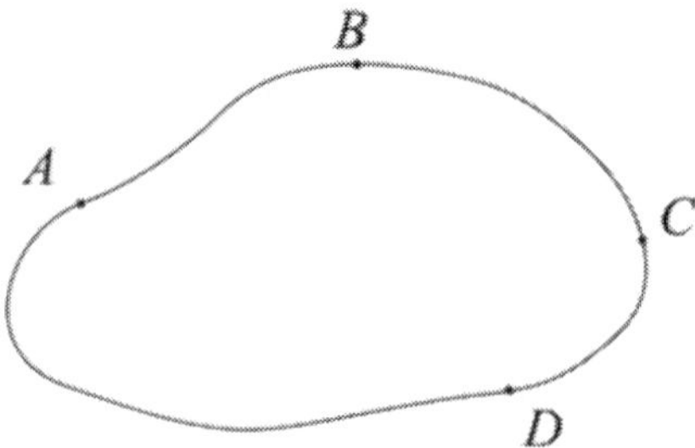

Figure 8. Schematic view of arbitrary shape flat plate (sample) with four contacts A, B, C, D for measuring electrical conductivity by van der Pauw method.

By passing electrical current I_{AB} between contacts A and B, one can determine resistance $R_{AB,CD}$ as follows: $R_{AB,CD} = \frac{\Delta V_{CD}}{I_{AB}}$, where ΔV_{CD} is potential difference between contacts C and D. Similarly: $R_{BC,DA} = \frac{\Delta V_{DA}}{I_{BC}}$.

If the following conditions are fulfilled:

- sample is parallel-sided free form plate,

- sample does not have isolated holes,

- sample is homogeneous (σ is the same everywhere) and isotropic,

- all four contacts are located on the edge of the sample and contact area is negligible compared to dimensions of the sample,

then $R_{AB,CD}$ and $R_{BC,DA}$ satisfy to equation [13, 14]:

$$\exp\left(-\pi d\sigma R_{AB,CD}\right) + \exp\left(-\pi d\sigma R_{BC,DA}\right) = 1, \tag{9}$$

where d is thickness of the sample plate. Since resistances $R_{AB,CD}$, $R_{BC,DA}$ and d are known, σ is the only one unknown quantity in Eq. (9), and it can be found by solving this equation.

Solution of Eq. (9) can be written in the form [13]:

$$\sigma = \frac{2\ln2}{\pi d(R_{AB,CD} + R_{BC,DA})} \frac{1}{f\left(\frac{R_{AB,CD}}{R_{BC,DA}}\right)}, \tag{10}$$

where f is function depending only on the ratio $\frac{R_{AB,CD}}{R_{BC,DA}}$. Graph of this function is shown in **Figure 9**.

When $\frac{R_{AB,CD}}{R_{BC,DA}} \approx 1$, then f can be approximated by expression:

$$f \approx 1 - \left(\frac{R_{AB,CD} - R_{BC,DA}}{R_{AB,CD} + R_{BC,DA}}\right)^2 \frac{\ln2}{2} - \left(\frac{R_{AB,CD} - R_{BC,DA}}{R_{AB,CD} + R_{BC,DA}}\right)^4 \left[\frac{(\ln2)^2}{4} - \frac{(\ln2)^3}{12}\right]. \tag{11}$$

Situation is considerably simplified, if the sample has symmetry axis [14]. Assume that contacts A and C are located on the symmetry axis and contacts B and D are placed symmetrically relative to this axis (**Figure 10**). Then $R_{AB,CD} = R_{AD,CB}$. According to the theorem of reciprocity for passive four poles [16], we have $R_{AD,CB} = R_{CB,AD} = R_{BC,DA}$, and it follows from Eq. (9):

$$\sigma = \frac{\ln2}{\pi d\, R_{AB,CD}}.$$

3.2. Thermopower

Figure 11 shows principle of measuring thermoelectric power. There are two direct methods of thermopower measurement: the integral method is historically the first and is conceptually simpler (Eq. (3)); and the differential method, which is practically the most used (Eq. (4)).

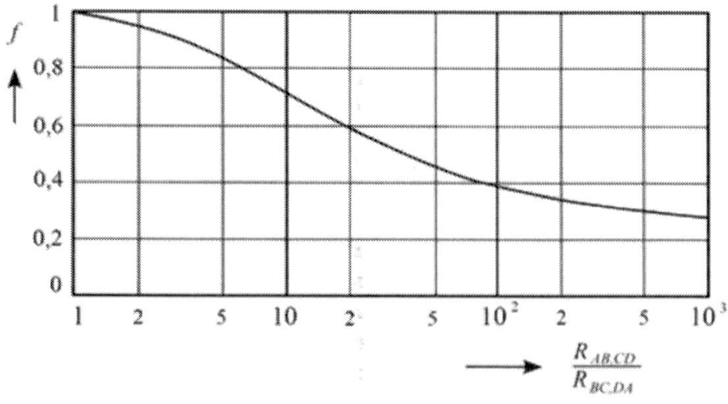

Figure 9. Dependence of function f on ratio $\frac{R_{AB,CD}}{R_{BC,DA}}$ [14].

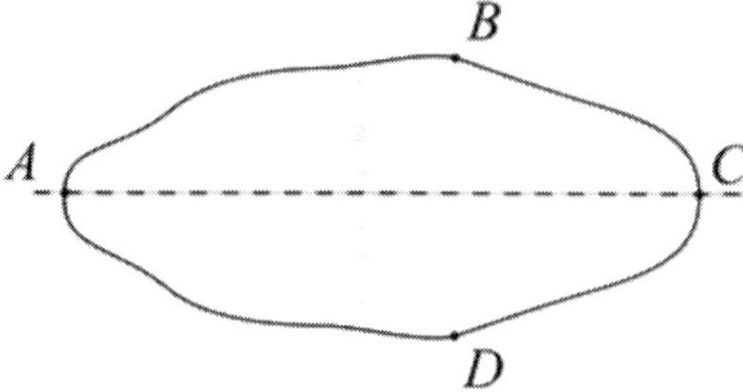

Figure 10. Contacts configuration for the sample having axis of symmetry.

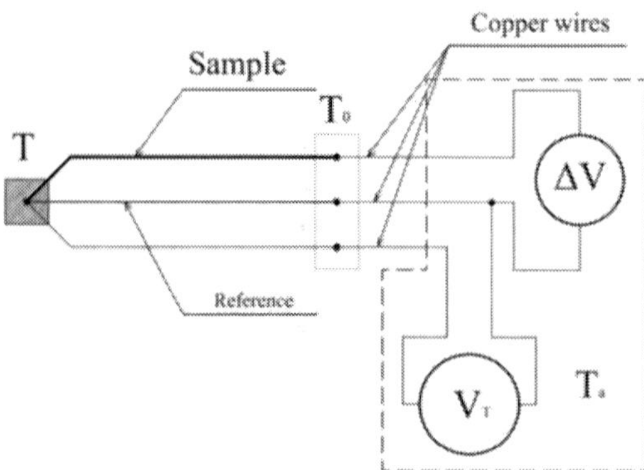

Figure 11. Scheme of integral method for thermopower measurements: voltage of thermocouple, consisting of sample and reference wires, is measured as function of thermocouple junction temperature T by voltmeter ΔV. Temperature is measured with help of another thermocouple by voltmeter V_T. Junctions of the sample, reference and thermocouple wires with cupper wires, connected to voltmeters are kept at fixed temperature T_0. T_a is ambient temperature.

3.2.1. Integral method

Figure 11 shows electrical circuit for measuring thermopower by integral method. Voltage of thermocouple, consisting of the sample and reference electrode wires, is measured as function

of temperature: $\Delta V = -\int_{T_0}^{T}(\alpha_x - \alpha_l)dT$. Hence, $\alpha_x(T) - a_l(T) = -\frac{d\Delta V}{dT}$.

In this method, besides ΔV, it is necessary to measure temperature T of contacts of material under study and reference electrode. This can be done by using additional electrode with known thermopower, which forms thermocouple with reference electrode. Note, that serious disadvantage of this method is that samples must be prepared in the form of homogeneous wires. But, many materials, which are considered as prospective thermoelectrics, are very difficult or impossible to prepare in form of wire. Therefore, integral method of thermopower measurement is used very rarely now.

3.2.2. Differential method

In contrast to integral method, differential one is designed for measuring thermopower of short samples of any shape, including thin films. Therefore, the vast majority of thermopower measurements have been performed by this method. **Figure 12** shows a scheme of differential method. Temperature difference between two points on the sample is measured with two thermocouples (or other temperature sensors), and thermopower signal ΔV can be measured by the same branches of thermocouples. Using Eq. (4), expression for determining absolute thermopower of the sample can be written as follows:

$$\alpha_x = -\frac{\Delta V}{\Delta T} + \alpha_l. \tag{12}$$

3.3. Absolute thermoelectric scale

In order to determine absolute thermopower of the material α_x, it is necessary to know absolute thermopower of reference electrode α_l. This is a key point in thermopower measurements. There is no direct method for measuring absolute thermoelectric power. Determination of absolute thermopower is based on two physical phenomena:

Thomson's relationship between Seebeck (α) and Thomson coefficients (τ_T) [9].

Property of superconductors: electric field $E = 0$ inside superconductor. Hence, it follows, that thermopower of superconductor is zero.

Based on these two phenomena, absolute thermopower of some materials was determined. Currently, lead, copper, and platinum are the main materials of reference electrodes. Dataset of absolute thermoelectric power of these metals establish *absolute thermoelectric scale*. This scale is based on experimental data of Thomson coefficient τ_T. Absolute thermoelectric power can be calculated according to second Thompson relation (Eqs. (6) and (7)).

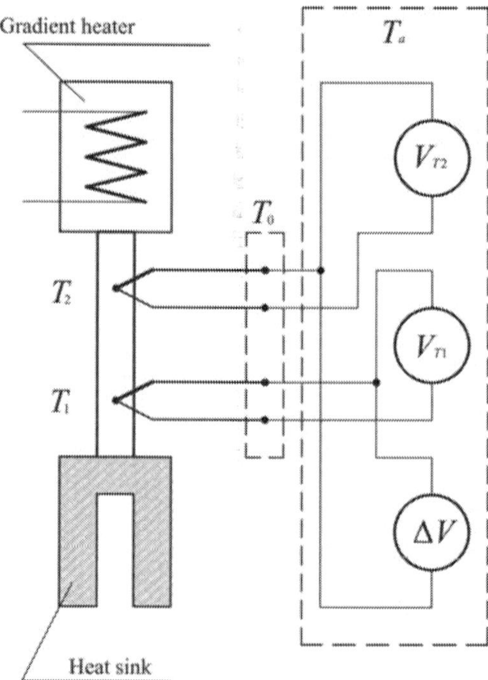

Figure 12. Scheme of differential method for thermopower measurements. Heat flow generated by gradient heater passes through the sample and creates temperature gradient in it. Temperature difference between two points on surface of the sample is measured using thermocouples. The same thermocouple branches are used to measure potential difference between points on the sample.

However, in practice, we cannot determine by Eq. (7) absolute thermopower of studied material, since it requires information about Thomson coefficient in temperature range from absolute zero to T, which is fundamentally impossible. This problem can be solved with superconducting materials. In superconducting state, that is, at $T < T_c$, thermopower $\alpha = 0$. Hence, for such materials, it is sufficient to know Thomson coefficient value at temperature $T > T_c$ only.

Nyström [17] created first absolute thermoelectric scale, which was based on his measurements of Thomson coefficient of copper in temperature range from 723 to 1023 K and Borelius's low-temperature data [18, 19]. Using data of absolute thermoelectric power of copper, Nyström determined absolute thermoelectric power of platinum. Later, Rudnitskii [20] has extrapolated Nyström's data for platinum up to 1473 K. Cusack and Kendall [21] have processed Thompson coefficient data and calculated absolute thermoelectric power of number of metals in wide temperature range, including platinum up to 2000 K, and molybdenum and tungsten up to 2400 K (using Thomson coefficient data obtained by Lander [22]). The most accurate thermoelectric scale was created by Roberts, who carried out measurements of Thomson coefficient of lead, copper, and platinum [23–25]. Thomson coefficient was measured for lead in temperature range from 7 K (i.e., from superconducting transition temperature) to 600 K (up to nearly melting temperature). Thomson coefficient of copper was measured up to 873 K, and for platinum and tungsten up to 1600 K. On the basis of these data, thermoelectric scale overlapping temperature range between 0 and 1600 K was created. According to Roberts estimations, his thermoelectric scale has error not more than ±0.01 μV/K at room temperature, ±0.02 μV/K at 600 K, ±0.05 μV/K at 900 K, and ±0.2 μV/K at 1600 K. At higher temperatures, absolute thermopower data is much less precise. Accuracy of the data at 2000 K is about ±2 μV/K. The results of these studies are summarized in **Table 1**.

T (K)	α_{Pb} (μV/K) Roberts [23, 24]	α_{Cu} (μV/K) Roberts [24]	α_{Pt} (μV/K) Roberts [25]	α_{Pt} (μV/K) Rudnitskii [20]
80	−0.544			
120	−0.631			
160	−0.734			
200	−0.834			
250	−0.948			
300	−1.05	1.94	−4.92	−4.2
350	−1.16	2.22	−6.33	−2.87
400	−1.28	2.5	−7.53	−7.33
450	−1.41	2.78	−8.59	−8.61
500	−1.56	3.07	−9.53	−9.68
550	−1.73	3.35	−10.41	−10.54
600		3.62	−11.22	−11.33
650		3.89	−11.98	−12.05
700		4.16	−12.71	−12.78
750		4.43	−13.42	−13.50
800		4.7	−14.14	−14.23
900		5.23	−15.66	−15.68
1000			−17.21	−17.13
1100			−18.77	−18.58
1200			−20.29	−20.03
1300			−21.78	−21.45
1400			−23.18	−22.93
1500			−24.49	
1600			−25.67	

Table 1. Thermopower of lead (α_{Pb}), copper (α_{Cu}), and platinum (α_{Pt}).

In practice, for measuring thermopower at high temperature (above 100 K) are used thermocouples copper-constantan and platinum-platinum/rhodium and reference electrode of platinum or copper, respectively. For both, platinum and copper, the absolute thermopower was accurately determined by Roberts only above room temperature. Therefore, it was necessary to expand temperature range of accurate determination of absolute thermopower of these metals to lower temperature region. Absolute thermoelectric power of platinum in temperature range from 25 to 1600 K was determined in [26] using Roberts's data and Moore's and Grave's low temperature data [27], which were adjusted using Roberts's data for lead [23], so that, corrected data are consistent with Roberts' high temperature data. These data and experimental results for platinum are shown in **Figure 13** and summarized in **Table 1** [26].

By using combined experimental data obtained in temperature range 70–1500 K, thermopower of platinum can be described by empirical interpolation formula $\alpha_{Pt}(T)$:

$$\alpha_{Pt}(T) = 0.186T\left[\exp\left(-\frac{T}{88}\right) - 0.0786 + \frac{0.43}{1+\left(\frac{T}{84.3}\right)^4}\right] - 2.57. \tag{13}$$

This function and its deviation from experimental points are shown in **Figure 13**.

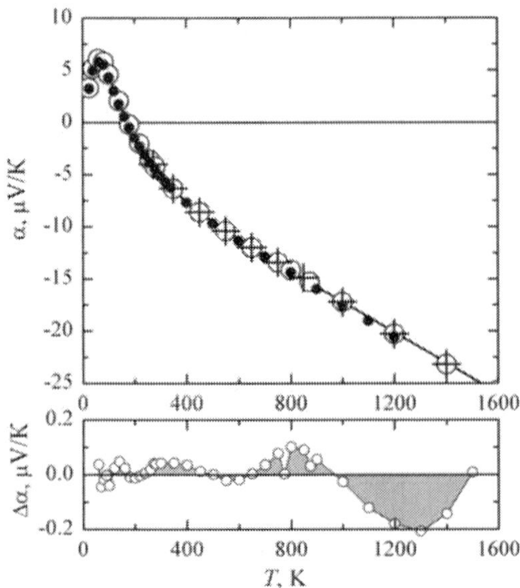

Figure 13. The top panel shows absolute thermoelectric power of platinum: • – Moore's data [27]; + –Roberts's data [24, 25]; ○ – combined data (not all data points are depicted). Solid line shows interpolation function. The bottom panel presents deviation of interpolation function from experimental data $\Delta\alpha = \alpha_{exp\,er} - \alpha_{Pt}(T)$.

Absolute thermoelectric power of copper was determined in [26] using Roberts' data for temperature range 273–900 K [24], and at temperatures below 273 K using Cusack's and Kendall's results [21]. Small correction was introduced in the data, so that, this low-temperature dependence smoothly joints with Roberts' high-temperature data. Adjusted and original experimental data and empirical interpolation formula α_{Cu} for temperature range 70–1000 K are shown in **Figure 14** and **Table 1** [26]. Interpolation function α_{Cu} is given by:

$$\alpha_{Cu}(T) = 0.041\, T \left[\exp\left(-\frac{T}{93}\right) - 0.123 + \frac{0.442}{1 + \left(\frac{T}{172.4}\right)^3} \right] + 0.804. \tag{14}$$

The error of this practical thermoelectric scale (considering the interpolation error) is estimated as follows [26]:

In temperature range 70–900 K: ±0.1 µV/K and 1000–1500 K: ±0.5 µV/K.

In formulas (13) and (14), thermopower is expressed in µV/K, and temperature is expressed in Kelvin degree.

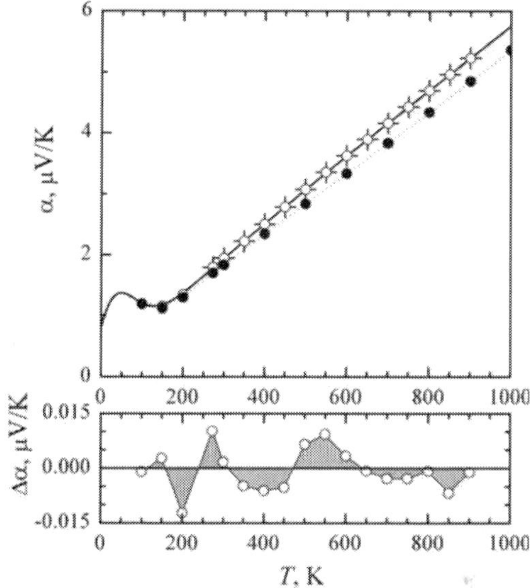

Figure 14. The top panel shows absolute thermoelectric power of copper: • – Cusack's data [21]; + – Roberts's data [24]; ○ – adjusted data. Solid line is interpolation function. The bottom panel shows deviation of interpolation function from experimental data $\Delta\alpha = \alpha_{exper} - \alpha_{Cu}(T)$.

4. Error analysis

4.1. Electrical conductivity

Errors in measurements of electrical conductivity can be divided into three categories. First, it is electrical signal measurement errors, that is, potential difference and current magnitude. Second, it is errors associated with shape of the sample and of measuring electrodes. And third, there are errors associated with change in temperature of the sample during measurement process.

The first kind of errors is common to all measurements of electrical signals and are not specific for measuring electron transport properties. When modern measuring equipment is used and proper organization of measuring system and procedure are applied, then these errors generally are not a factor limiting the accuracy of measurements. Possible exceptions are measurements of electrical conductivity of high pure metals at very low temperatures. However, these cases are not typical for high-temperature measurements of thermoelectric materials, and not analyzed here.

4.1.1. Errors associated with shape

Errors associated with sample's and electrode's shape are, perhaps, the main problem in most cases. When measuring conductivity, actual measured value is a total resistance of the sample between potential probes $R = \frac{\Delta V}{I}$. In order to obtain electrical conductivity of the sample, it is necessary to know cross-section of the sample (A) and distance between potential probes (l) $\sigma = \frac{1}{R} \times \frac{l}{A}$. There are four sources of errors associated with geometric factor. The easiest is inaccuracy in determining size and shape of the sample. Assume, that the sample has parallelepiped shape with typical dimensions $2 \times 2 \times 10$ mm^3. In ordinary methods of sample machining and measurement of lengths, typical error of size determination is of the order 0.01 mm. This error includes distance measurement inaccuracy, and shape and surface imperfections of the sample as well. This error causes error of determining the section $\Delta A/A$ equals to 1%. The error of determining distance l, which includes both error in measurement of length and finite size of potential contact, is of the order 0.1 mm. Thus, total error in determining geometric factor is equal to $\frac{\Delta A}{A} + \frac{\Delta l}{l} = 0.02$, that is, 2%. It is accuracy limit of measuring resistance by four-probe method using bulk samples. Of course, accuracy can be improved by using a special high-precision technology for manufacturing of the sample and measuring its dimensions. However, these methods are not applicable for mass measurements.

A second important factor, determining accuracy of resistivity measurements, is electrical current distribution in the sample. Ideally, electrical current distribution in the sample must be uniform (**Figure 15a**). In this case, electrical current lines are parallel to axis of the sample and potential distribution on sample surface, where it can be measured, is the same as in the bulk. However, in most cases, point current contacts are used for measuring resistance and, in such case, current distribution is not uniform in the sample (**Figure 15b**).

As a result, potential distribution on surface of the sample may differ significantly from distribution in volume. To minimize this error, distance between the nearest current and potential probes must be (for highly conductive samples) more, than the maximum transverse dimension of the sample. With increasing resistance of the sample material, this distance must be also increased. Potential probes must be arranged along electrical current lines. If potential probes are arranged along line directed at angle ψ with respect to current lines, then effective length is $l^* = l \cos \psi$. For small angles ψ, error can be expressed as follows: $\Delta l = |l - l^*| = l(1 - \cos \psi) \approx l \times \psi^2$ and $\frac{\Delta l}{l} = \psi^2$. The probe position error of 6° results in resistance error of 1%.

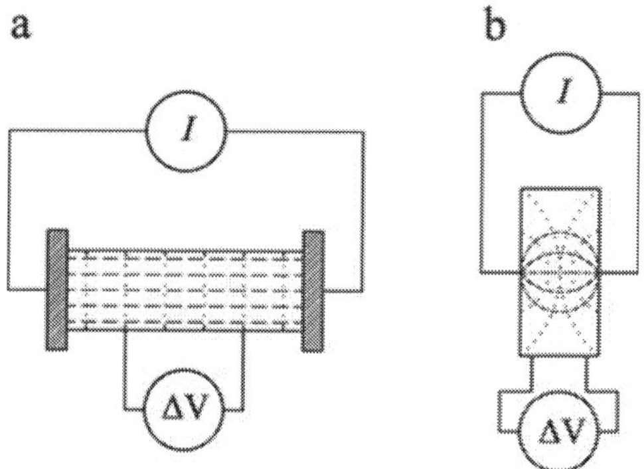

Figure 15. Errors associated with geometry of the sample and current leads. Dashed lines indicate electrical current flow; doted lines represent equipotential surfaces: a. ideal current lead contacts-homogeneous current distribution; b. point current lead contacts-nonhomogeneous current distribution.

Errors related to mechanical imperfection of samples are the most common and bring the greatest trouble. This may be pores, cracks, non-uniformity in composition, and so on. There is no general recipe to minimize such errors. Errors associated with the presence of pores can be reduced in part by corrections proportional to deviation of actual density of the sample from theoretical, calculated on the basis of structural data. It should be noted, that geometrical factor leads also to errors in determining temperature coefficient of electrical conductivity $\frac{d\sigma}{dT}$.

4.1.2. Errors associated with changes in thermal regime of the sample during measurement

Two types of phenomena leading to such kind of errors can be distinguished: sample temperature changes due to Joule heating and changes of temperature distribution in the sample due to thermoelectric effects.

Since Joule heat released in the sample is equal to $I^2 R$, then measuring at lower current and improving heat transfer from sample to the environment can effectively solve the problem of temperature changes. More difficult task is to eliminate the influence of thermoelectric effects, namely, Peltier and Seebeck effects. This influence arises, when measurement of electrical conductivity is performed with direct current (DC), whereas in measurements of electrical conductivity with alternative current (AC), thermoelectric effects do not affect measurement accuracy.

Since the sample and connected to it electrical current leads represent nonuniform electrical circuit, Peltier heat will be released at one contact of current lead with the sample, while at another contact it will be absorbed. This will produce temperature difference across the sample. **Figure 16** shows time diagram of potential difference across the sample during measurement of resistance with taking into account Peltier effect. We assume, that in the initial state, when electrical current is turned off, there is no temperature gradient in the sample, so the potential difference $\Delta V_0 = 0$. Due to finite sample heat capacity, immediately after electrical current is switch-on, temperature of the contact is not changed and measured voltage is equal to $\Delta V^+ = R \times I^+$. However, due to Peltier effect, heat flow from one contact to another creates in the sample temperature gradient. Therefore, additional potential difference arises, so that the total potential difference between probes is $\Delta V^+ = R \times I^+ + \alpha \times \Delta T(t)$, here α is relative thermoelectric power of the pair "sample-potential probe," ΔT is temperature difference between potential probes. This difference increases with time at rate depending on heat capacity of the system and rate of release and absorption of heat on the contacts due to Peltier effect.

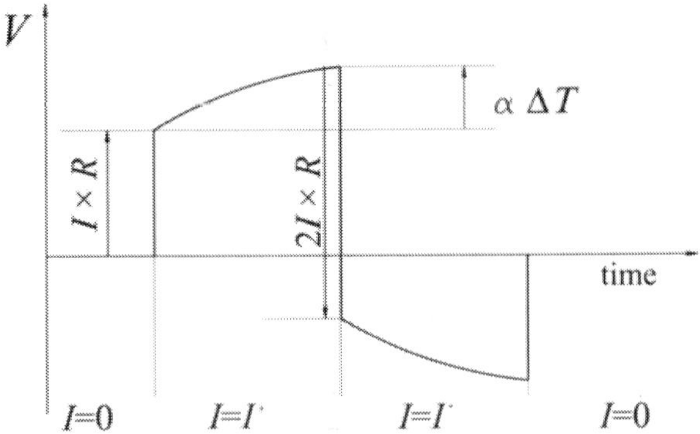

Figure 16. Timing diagram of potential difference across the sample, when measuring electrical conductivity with DC current.

ΔT is stabilized at level, which is determined by balance between rate of heat generation at the contacts, thermal conductivity of the sample and conditions of heat exchange between sample and the environment. To estimate the maximum value of the effect, we assume, that there is no heat exchange between sample and the environment. When electrical current with density j flows, then the amount of Peltier heat q_p, generated at contact between current lead and the sample, is equal to: $q_p = \Pi \times j$, where Π is Peltier coefficient of the pair "sample-current lead". At stationary conditions and in the absence of heat exchange with the environment, whole heat flow passes through the sample due to thermal conductivity: $q = -\kappa \nabla T$, here κ is a thermal conductivity of the sample. The flow balance $q_p = -q$ determines temperature gradient: $\nabla T = \frac{\Pi \times j}{\kappa}$. The effect of this temperature gradient on conductivity measurement precision depends on ratio of voltage drop across the sample $\Delta V_\rho = I \times R = j \times A \times \rho \times \frac{l}{A} = j \times l \times \rho$, which occurs when electrical current passes, to potential difference related to temperature gradient $\Delta V_{thermo} = \alpha \times l \times \nabla T$:

$$\frac{\Delta V_{thermo}}{\Delta V_\rho} = \frac{\Pi \times j \times \alpha \times l}{\rho \times j \times \kappa \times l} = T\frac{\alpha^2 \sigma}{\kappa}. \tag{15}$$

In deriving the latter expression, Thompson relation (5) was used. As we can see, $\frac{\Delta V_{thermo}}{\Delta V_\rho}$ is determined by dimensionless figure of merit $ZT = T\frac{\alpha^2\sigma}{\kappa}$. For good thermoelectric materials, this value can be of the order of unity. It is important, that error related to Peltier effect does not depend on direction or magnitude of electrical current or sample geometry. Therefore, it cannot be eliminated by changing these parameters of experiment. The error can be significantly reduced in two ways:

1. Proper design of the sample holder should provide good thermal contact between sample and environment, that assures absorption of heat released in the sample and provides uniform temperature distribution in the sample. Very effective is gas environment with high thermal conductivity. Taking into account all the properties, helium is the best environment.

2. The signals measurement must be properly organized. From **Figure 16**, it is clear, that measurement of voltage drop should be performed as soon as possible, immediately before and immediately after electrical current switch (on, off or change direction).

4.1.3. Measurement errors in four-probe method

When conditions of applicability of four-probe method are fulfilled, then errors of electrical conductivity measurements will be caused by inaccuracy of determining the distance between potential contacts and sample thickness. Distance between potential contacts is limited by conditions of method applicability, and it should be much less than linear dimensions of the sample. For typical sample having flat surface area 10×10 mm^2, the distance between the contacts must be less than 1 mm. Typically, the diameter of contact area of potential probe is of order of 0.01 mm, therefore, the error in determining distance between contacts will be $\Delta l/l \geq 1\%$. The error in determining of average thickness of the sample is of the same order. Thus, accuracy of determining electrical conductivity with four-probe method will usually be at least 2%.

4.1.4. Error estimation: van der Pauw method

Measurement errors in van der Pauw method are associated with non-ideal contacts, that is, with their finite size and offset from the edge of the sample. Estimation of errors has been done for three typical cases of non-ideal contacts and is shown in **Figure 17** [14]. For simplicity, let us consider circle shape sample with diameter D, electrical contacts to which are arranged at equal distance from each other. Assume, that only one contact is imperfect. In practice, there are no ideal contacts. To the first approximation, the total error is the sum of errors on each contact. Advantage of van der Pauw method is applicability to samples of different (including irregular) forms, because in many cases test material is available in the form of small plates. Such samples do not require further processing and can be used for other purposes after van der Pauw measurement. However, in cases, where high measurement accuracy is required, the samples of special form should be used [12]. They can be divided into two groups. The first group includes the samples having the shape of cloverleaf. Such form allows to increase the length of the border, so that imperfect contacts make negligible error in measurement results. The second group includes samples, having symmetrical shape and extended contacts, which respective correction functions have already been calculated for.

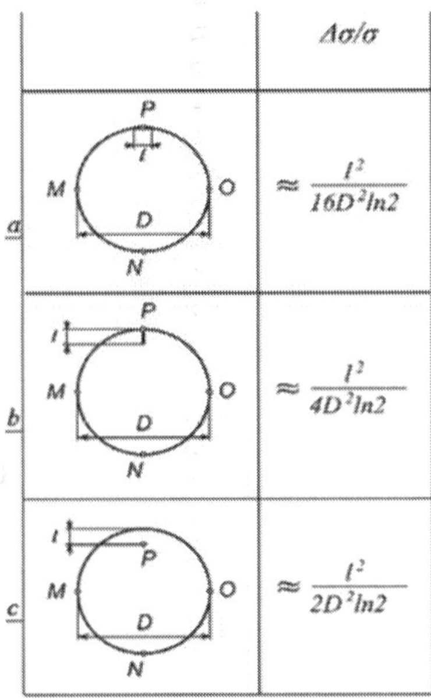

Figure 17. The relative errors $\Delta\sigma/\sigma$ when measuring electrical conductivity of circle shape sample [14]: a. One of contacts has length l along the edge of the sample; b. One of contacts has length l perpendicular to the edge of the sample; c. One of contacts is point contact located at distance l from the sample edge.

4.2. Measurement errors of thermopower

Measurement errors of thermopower by differential method are mainly related to incorrect determination of temperature difference ΔT. We can distinguish two sources of errors in determination of ΔT:

1. Temperature sensors and calibration are non-ideal. Thermocouples are almost exclusively used as temperature sensors at high-temperature thermopower measurements. To ensure precise determination of ΔT, thermocouple must satisfy very rigid requirements, such as the branches homogeneity and stability of their properties. Typical value of ΔT is about 10 K. At sample temperature of about 1000 K, just 0.1% difference in average thermopower of two thermocouples will lead to errors in determining ΔT of 10%. For example, average thermopower of one of the most commonly used thermocouple, consisting of platinum wire and wire of alloy Pt+10% Rh, is about 10 µV/K in temperature range 300–1000 K. Deviation of average thermopower of one thermocouple on another of the order of 0.01 µV/K will result in error in ΔT of 10%. Therefore, for measurements of thermopower, high-

quality thermocouple wires should be used only and their homogeneity should be monitored during operation.

2. The main source of errors in thermopower measurements associated with mismatch between the points, where ΔT and ΔV are measured (see, e.g., [28]). Junction of thermocouple used for measuring temperature at the point of electrical contact of reference electrode (which is usually one of used thermocouple branches) has finite dimensions. In real conditions of high temperature measurements, significant heat flow may occur along thermocouple branches. Combination of these factors leads to the fact, that average temperature of the junction and real temperature of electrical contact of reference electrode with the sample differ, that leads to error in determining of thermopower. For this type of errors, it is difficult to make general numerical estimate, because errors depend on several factors, which are difficult to control: size of thermocouple junction, cross-section and thermal conductivity of thermocouple branches, the value of thermal resistance at contact of thermocouple with the sample, temperature distribution in contact area. Error evaluation can be done by measuring thermopower of well-known materials, which have stable properties. Unfortunately, as we have already noted, so far, there is no standard for thermopower at high temperatures. Some metals can be used as reference samples. Due to the combination of the properties, platinum and nickel are the most suitable for high temperatures. It should be noted, that if platinum is used as a reference electrode for thermopower measurements, platinum sample is not suitable as a reference for evaluation of measurement error. In this case, as follows from Eq. (4) $\Delta V = 0$ (since $\alpha_x = \alpha_l$). Thermopower $\alpha_x = \frac{\Delta V}{\Delta T} + \alpha_l$, determined in such measurements, will have correct value, regardless of the accuracy of determining ΔT.

Specifications analysis of set-ups for thermopower measurements and experience allow to state, that accuracy of determination of thermopower at high temperatures is limited by about ±5%. This estimate includes also uncertainty of modern absolute thermoelectric scale, which at high temperatures reaches ±0.5 µV/K. However, for thermoelectric materials, in which thermopower value is of the order of 100 µV/K or more, this uncertainty is not significant. Note also, that errors associated with inhomogeneity of thermocouple wires may be partially removed, when using alternating temperature difference [29–32]. At the same time, the second-type errors cannot be eliminated with alternating temperature difference and/or by use of differential thermocouple for measuring temperature difference, as it is sometimes assumed [30].

5. Devices for measuring thermopower and electrical conductivity

Devices realizing differential thermopower measurement technique can be divided into two classes: with variable (modulated) and static temperature difference. Measurements with variable temperature difference allow to eliminate or significantly reduce errors associated with inhomogeneity of branches of thermocouples, with slow instrumental drift or constant voltages caused by inhomogeneity of electrical circuits due to thermoelectric effects. This method has an advantage comparing to measurements with static temperature difference at low temperatures, when amplitude of ΔT is very small, because condition, which must be satisfied is $\Delta T \ll T$. Therefore, there have been numerous variants of its implementation, designed for measuring thermopower at low temperatures [31–36]. At high temperatures, gradient modulation does not bring significant increase in accuracy, and implementation of this method is more difficult. Nevertheless, variable gradient method has been used at high temperatures as well [37–39].

Further, we describe in detail two experimental set-ups for measuring thermopower and electrical conductivity in temperature range from 80 to 2000 K [26, 40] and give brief overview of other devices for measurement of these properties.

5.1. Set-up for thermopower and electrical conductivity measurements at 80–1300 K

General view of measuring apparatus shown in photograph (**Figure 18**). Set-up was built to provide the fast and high quality electrical conductivity and Seebeck coefficient measurements using samples of any shape, including thin films. These objectives were fully achieved [26, 40].

Figure 18. Experimental set-up for thermopower and electrical conductivity measurements at 80–1300 K.

The set-up consists of four main parts:

1. Sample holder.

2. Main heater and temperature control system.

3. Vacuum system.

4. Data collection and processing system.

Sample holder is located inside vacuum chamber, which can be pumped out using turbo molecular pump to residual pressure down to 10^{-4} Pa. Typically, chamber is filled with helium gas to pressure slightly above atmospheric. Measurements can be performed in vacuum, but in this case, accuracy of measurement of thermoelectric power decreases. Moreover, it must be borne in mind, that metallization of isolators may occur at high temperatures due to vaporization of metals.

General view of sample holder is shown in **Figure 19** [26]. The basis of the holder is two coaxial tubes made of high-temperature steel, which are mounted on vacuum flange (19). Inner tube (16) is mounted on top of the flange (19). Gradient heater (11), supporting plate (8) and heat sink (4) mounted on other end of inner tube. Outer tube (15) is centered relative to inner tube with steel disks (14), which are mounted on inner tube at distance of 50 mm from each other. This system of two coaxial tubes is rigid and stable, which is especially important at high temperatures. All current and thermocouple wires are arranged in the space between inner and outer tubes and, therefore, they are well protected from mechanical damage and contamination. Outer tube can be easily removed, allowing access to the wires in case of repair. Sample supporting plate (8) is located between gradient heater (11) and heat sink (4) made of molybdenum. Selection of molybdenum as material for heater and heat sink is motivated by its high thermal conductivity and mechanical stability at high temperatures. The sample (5) is pressed against supporting plate (8) by press arm (10), pressure plate (9) and steel spring (13). These parts are made of special high-temperature steel. Cold junctions of thermocouples are made in the form of copper block (17), inside of which is made connection of thermocouple branches with copper wires, connecting thermocouple with the measuring equipment.

In this case, two conditions should be fulfilled:

1. All junctions of thermocouple branches with copper wires should be at the same temperature T_0.

2. All junctions should just be electrically isolated from each other. Quality of isolation must satisfy condition: $R_{ij} > 10^3 \times R_{max}$, where R_{ij} – resistance between any two junctions of open thermocouple, R_{max} – maximum resistance of the samples.

Connection of sample holder with measuring equipment is carried out by means of connector made of conductors with low thermopower relative to thermopower of copper. Temperature of reference point (17) is measured by thermistor (18).

Selection of thermocouple is mainly determined by temperature measurement interval. For temperature range from 80 to 600 K, the best choice is thermocouple copper-constantan, it has good sensitivity, stable enough, thermopower of copper is well-known and it is rather low. For temperatures from 300 K to ≈1600 K Pt-Pt/Rh thermocouples are the best choice, where the second branch is alloy of platinum and rhodium. Usually, as the second branch of these thermocouples, alloys of platinum with 10 and 13% rhodium are used. Thermopower of platinum, which is normally used here as reference electrode in measurements of thermopower, is also well known.

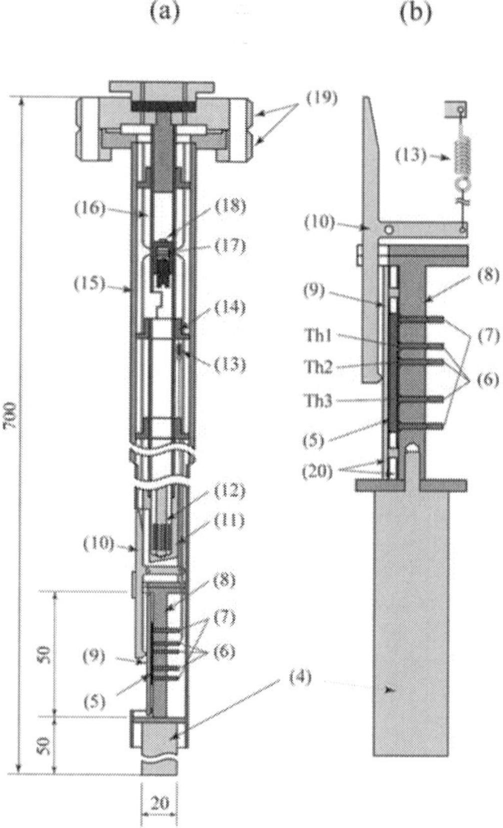

Figure 19. General view of sample holder (a), and sample supporting plate (b). The dimensions are given in millimeter.

Figures 19b and **20** show details of mounting and pressing mechanism of thermocouples on the supporting plate. The basis for mounting thermocouples (6) and current contacts (7) are two-channel tube (21) made of Al_2O_3 of 1 mm diameter. Tubes are pressed against the sample (5) using small springs (22) made of iridium wire. The springs are welded to the supporting plate (8). Such system provides reliable contact of thermocouples and current contacts with the sample within the whole operating temperature range. The choice of material for the springs (22) is important for providing reliable and stable contacts. The most important condition is to maintain elasticity of the material up to about 1300 K, as well as, mechanical and chemical stability. Iridium satisfies in full these requirements. Other good materials are tungsten-rhenium alloys; however, they cannot be used in oxidizing atmosphere. For electrical isolation of the sample from supporting (8) and pressing (9) plate, thin mica sheets (20) are used. Gradient heater (12) (**Figure 19a**) is used to regulate temperature gradient in the sample. Temperature gradient is mainly formed by slightly asymmetric sample's position relative to the center of the heater (**Figure 18**). Typical value of temperature gradient between measuring thermocouples is in the range from 5 to 20 K (depends on temperature).

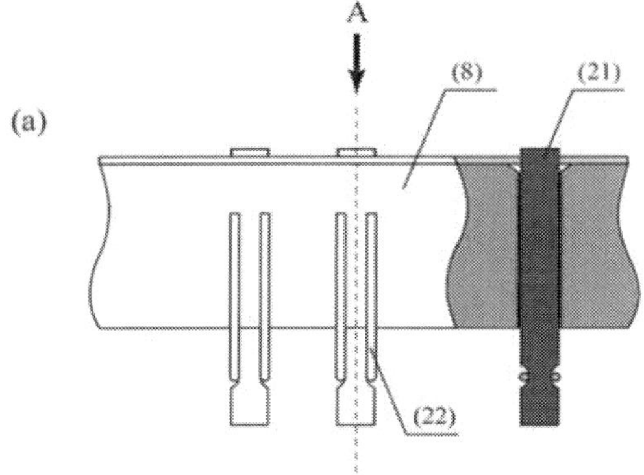

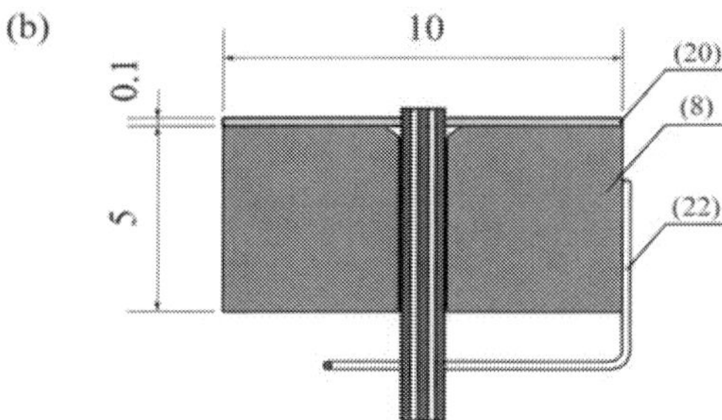

Figure 20. Detailed view of mechanical contact mechanism: (a) side view of the sample supporting plate with ceramic tubes (21) and springs (22); distance between tubes is not in scale with their diameter; (b) cross-sectional view of the sample supporting plate.

Five electrodes are used in the sample holder: 3 – for thermocouple contacts and 2 – for current contacts. The distance between Th1 and Th2 equals to 3 mm, between Th1 and Th3 – 10 mm (**Figure 19**). Such configuration allows measuring properties of the samples of various sizes with optimal accuracy. The sample holder allows measurements with both bulk samples and thin films as well.

5.1.1. Measurement procedure

Standard four-probe DC current method is used for measurements of electrical conductivity. Differential method with constant temperature gradient is utilized for thermopower measurement. Performing reliable measurements of thermopower requires accurate temperature measuring and availability of precise and detailed information about thermoelectric power of reference electrodes depending on temperature. As shown above, thermopower in differential measurement method is given by Eq. (12). To determine ΔT, precise calibration data for thermocouples must be used: $T = F(V)$. For standard thermocouples, calibration dependences are usually presented in the form of tables or dependencies of $V(T)$. If measurement is automated, it is more convenient to have calibration dependence in the form of analytic functions. In this case, it is important to choose the most natural analytic representation. In rough approximation, metal thermopower is linear function of temperature (generally, this is incorrect statement, but for metals and alloys used in thermocouples it is true), and then thermopower of thermocouple can be expressed as follows:

$$V(T) = \int_{T_0}^{T} \alpha_{12}(T)dT \propto \int_{T_0}^{T} kTdT = \tfrac{1}{2}k(T^2 - T_0^2)$$

and, hence, $T \propto \left(\tfrac{2}{k}\right)^{1/2} \times \sqrt{V + \tfrac{k}{2}T_0^2}$. Therefore, we represent thermocouple calibration dependence in the form of:

$$T = \sum_{i=0}^{n} b_i \left(\sqrt{V + a}\right)^i.$$

Coefficients b_i of interpolating polynomials for four standard thermocouples are shown in **Table 2**.

	Pt–Pt+13% Rh	Pt–Pt+10% Rh	Chromel-alumel	Copper-constantan
a	0.1676	0.2045	6.4	6.1
b_0	237.54	230.43	28.5	34.4
b_1	0	0	0	0
b_2	273.7	269.884	135.5	116.9675
b_3	−179.79	−162.308	−90.6	−73
b_4	85.454	65.836	33.363	29.5
b_5	−22.35955	−9.851	−6.6509	−7.02
b_6	2.9002	−1.207	0.7315	1.488
b_7	−0.13503	0.524	−0.04125	−0.9712
b_8	−0.0015	−0.04238	−0.00092	0.0347

Table 2. Coefficients of interpolating polynomials for thermocouples [26].

These polynomials can be used in the following temperature ranges:

Pt-Pt+13% Rh from 273 to 1873 K;

Pt-Pt+10% Rh from 233 to 1883 K;

chromel-alumel from 43 to 1543 K;

copper-constantan from 53 to 673 K.

Polynomial coefficients were obtained by fitting polynomials to calibration tables recommended by International Electrotechnical Commission for standard thermocouples. Deviation from calibration tables in specified temperature ranges does not exceed 0.1 K for copper-constantan and Pt-Pt+13% Rh thermocouples; 0.15 K for thermocouple Pt-Pt+10% Rh; and 1.5 K for chromel-alumel thermocouple. Additional error in determining temperature difference across the sample due to these deviations is within ±1% for copper-constantan thermocouple and both platinum thermocouples, and ±3% for chromel-alumel thermocouple.

Measurements of both properties are performed simultaneously. When measuring temperature dependence of the parameters, it is not required to establish steady temperature at each point. Measurements are performed with a continuous change in temperature at rate up to 10 K/min.

5.2. Set-up for measuring thermopower and electrical conductivity at 300–2000 K

Thermopower measurements at very high temperatures, particularly above 1500 K, are rather difficult due to several factors:

1. Structural materials lose their strength and stability. Cycling of temperature between room temperature and high temperatures leads to deformation of structures.

2. Almost all electrical insulating materials have significant electrical conductivity at these temperatures.

3. At temperatures approaching to 2000 K, difficulties with heating of the sample and maintaining stable temperature and temperature gradient in the sample arises. Therefore, measurement of transport properties, in particular thermopower, at temperatures above 1000 K is quite rare. Typically, measurements at higher temperatures are less accurate compared with measurements at low temperatures.

Described apparatus allows measurements of thermopower and resistance of bulk samples of conductors at temperatures from 300 K to temperature slightly above 2000 K with good accuracy. This system allows to work with samples of various shapes and sizes. Perhaps, this is the most high-temperature experimental device for direct measurement of thermopower described in the literature. The exception is the device used Lander [22] for measurement of Thompson coefficient of some metals up to 2400 K.

The main original part of this set-up is sample holder [41]; scheme of this holder is shown in **Figure 21**. The basis of holder is molybdenum tube (2), in the lower part of which is fixed molybdenum massive heat sink (1), where replaceable molybdenum bottom sample support (4) is installed. At the top of the tube (working position of the holder is vertical), molybdenum pusher (8) is located, which is isolated from tube by ceramic (Al_2O_3) rings (7). The lower ring is held by molybdenum stop (6), which also protects ceramic ring from metallization by metal vapors from the hot zone at the bottom. The sample (3) is clamped between the upper (5) and lower (4) supports under weight of gravity transmitted through molybdenum (8) and stainless steel (9) pushers. The holder is mounted in vacuum chamber with heater through ceramic insulating tube (10).

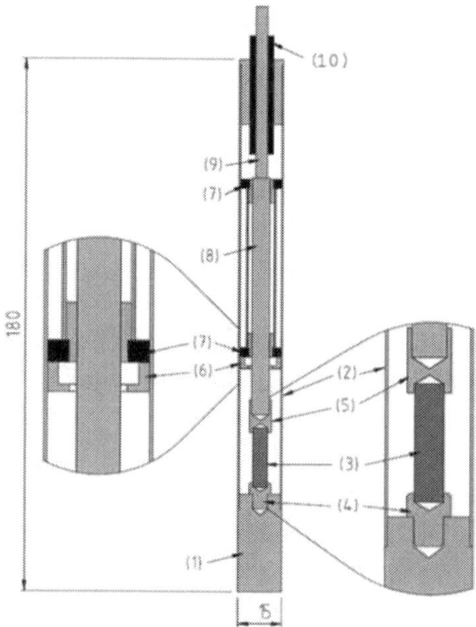

Figure 21. Sample holder for measuring thermopower and resistance at 300–2000 K: (1) – heat sink; (2) – outer molybdenum tube; (3) – sample; (4) – bottom sample support; (5) – upper sample support; (6) – molybdenum support for insulation; (7) – insulating ring made of Al_2O_3; (8) – molybdenum pusher; (9) – stainless steel pusher; (10) – ceramic insulation.

To measure temperature and thermopower in this set-up, thermocouples of tungsten-rhenium alloys are used: WR10-WR20. These are alloys of W + 10% Re and W + 20% Re, respectively. WR20 alloy is used as the reference electrode in measuring thermopower. For thermocouple WR10-WR20, there is standard calibration, however, the absolute thermopower of the branches is not known. The absolute thermoelectric power of WR20 alloy was determined by measuring thermopower of reference metal samples. As standards were used: platinum in temperature range 300–1700 K, and molybdenum at 1700–2100 K. Thermopower of high-purity molybdenum sample was beforehand accurately measured in temperature range from 80 to 1600 K relative to copper and platinum. Cusack and Kendall data [21] were used at higher temperatures. However, in order to provide a smooth joining of low-temperature data with Cusack's data, it must be entered temperature-independent correction of 2 µV/K in these data. A possible reason for this difference is insufficient purity of metal, which was used by Lander [22] in measurement of Thomson coefficient of molybdenum. Thermopower of molybdenum and WR20 alloy are shown in **Figure 22**. At temperatures from 100 to 2000 K, thermopower of WR20 can be calculated using interpolation polynomial:

$$\alpha_{WR20} = 1.6337 \times 10^{-12} \times T^4 - 1.2669 \times 10^{-8} \times T^3 + 2.6192 \times 10^{-5} \times T^2 - 1.6889 \times 10^{-2} \times T + 3.111.$$

$$(16)$$

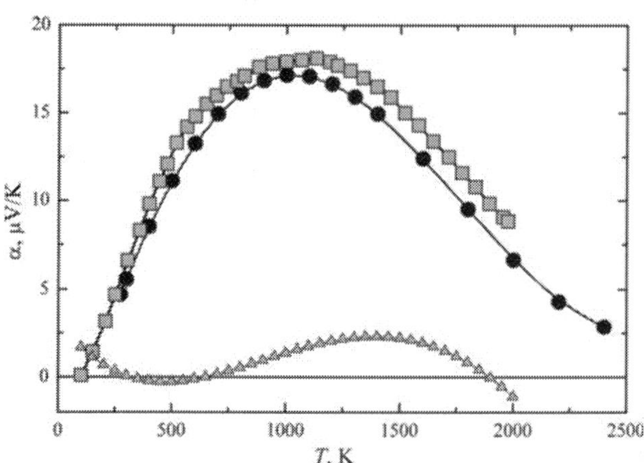

Figure 22. Thermopower of molybdenum and WR20 alloy: • – molybdenum thermopower according Cusack [21]; ■ – adjusted molybdenum thermopower; ▲ – thermopower of WR20 alloy.

5.3. Other techniques

Petrov [42] built set-up for simultaneous measurement of thermopower, thermal conductivity and electrical conductivity of thermoelectric materials (i.e., materials with very low thermal conductivity), at temperatures from 100 to 1300 K, which operates successfully (in upgraded form) up to nowadays. In this device, method of electrical conductivity measurement with DC current, differential method of thermopower measurement and classic steady-state method of thermal conductivity measurement are used. The measurements at each value of temperature must be carried out in stationary temperature conditions. Since, achievement of thermal equilibrium, especially at low temperatures, is slow, measurements over whole temperature range takes several days. To suppress the heat loss by radiation, active heat shield and special ceramic filling with very low and known thermal conductivity are used. This system allows to determine parameter ZT as a result of simultaneous measurement of α, σ, and κ with accuracy of ±5%.

In contrast to electrical conductivity measurements, thermopower measurement is difficult to automate using analog methods. Therefore, before the advent of personal computers, these measurements were very time-consuming. There are several original analog automated devices for measuring thermoelectric power [30, 33]; however, they were not widely used.

Interesting device for measuring thermopower at high temperatures has been developed by Wood et al. [37]. This device uses a differential method for measuring thermoelectric power with modulation of temperature difference over the sample. Interchangeable heating the ends of the sample by light flash lamps was used for the modulation of temperature difference. Light beam energy was applied to the sample by means of sapphire optical fibers, between which the sample was clamped. The device allowed to measure thermopower up to 1900 K, with amplitude of temperature difference modulation of a few degrees. Author estimates measurement error of thermopower as ±1%, but does not specify experimental evidence of stated accuracy.

In apparatus for measuring thermal conductivity and thermoelectric power at temperatures 300–750 K, described in Ref. [43], stationary method of measuring thermal conductivity and differential method of thermopower measurement are used. Measurement of thermal conductivity is based on the comparison between temperature difference of heat source and heat sink in the presence of the sample and without the sample. At each temperature, after thermal stabilization, measurements of ΔT with the sample in contact with heat source and heat sink are performed. Then, heat source is disconnected from the sample and ΔT is measured again. Assumed, that heat losses in the system are the same in both states, and losses due to radiation from the sample are not considered. This put in question the correctness of the measurement. Thermopower is measured by differential method with constant temperature gradient.

In set-up for measuring electrical conductivity and thermoelectric power at 300–1300 K [44], electrical conductivity is measured with AC current at frequency 16 Hz, and for measuring thermopower, differential method with constant temperature difference is used. Thermocouples, which are used for the measurement of temperature gradient and thermopower, are mounted in holes drilled in the sample by using graphite paste. After installation of the sample, paste must be heat treated to ensure proper contact. This, as well as, current leads design, which cannot provide stable electrical contact, is a serious disadvantage of the system. Extremely small thermopower measurement error 0.3%, stated by authors, has not been experimentally confirmed.

AC electrical conductivity measurement procedure and differential method with temperature gradient modulation for measuring thermopower are utilized in set-up for thermopower and electrical conductivity measurements at 300–1273 K [38]. The publication, however, contained only measurement principles, which are not original. No details of measuring device were presented.

Interesting sample holder design for measuring thermoelectric power at temperatures up to 1200 K was suggested in Ref. [45]. This is further development of Wood's system [37], but with significant changes. Distinctive feature of the design is axial location of thermocouples. Thermocouples, supported by four-channel thin tubes, extend along the central axis of gradient heaters, between which is clamped the sample. Working junctions of thermocouples are pressed against the ends of the sample by springs. Therefore, sample does not require special preparation for measurement. Thermopower is measured by differential method with temperature gradient modulation; amplitude of modulation is up to 20 K. The article provides fairly detailed analysis of measurement errors of thermopower.

A feature of the holder for measuring thermopower and electrical conductivity proposed in Ref. [46] is the material: the main parts of this device are made of ceramics (Al_2O_3). Therefore, this device can be used for high temperature (1200 K) measurements in oxidizing atmosphere in the case of using platinum thermocouples. Thermoelectric power is measured by differential method with variable temperature gradient.

Relatively detailed overview of methods and devices for measurement of thermopower and electrical conductivity was published by Martin et al. [47].

Apart from temperature, pressure and magnetic field are accessible experimental parameters affecting the material properties. Dependences of electrical conductivity and thermoelectric power on magnetic field and pressure provide important information about electronic structure and conductivity mechanisms. Generally, studies of thermoelectric and conductivity dependencies on pressure and magnetic field are carried out at low temperatures. However, for thermoelectric materials, dependence of their properties on pressure and magnetic field at high temperatures is of considerable interest. Therefore, considerable effort has been directed toward the study of these dependences and development of devices for such measurements [48–52].

6. Conclusion

Research and successful development of novel effective materials for thermoelectric energy converters is critically dependent on obtaining accurate and reliable information about properties of these materials. The most important characteristics of thermoelectric materials are thermopower and electrical conductivity. They determine potential effectiveness of thermoelectric material and provide important information on its electronic structure. Measurements of these properties must meet a number of requirements. Measurement results must be reliable and sufficiently accurate. Measurements must be performed over a wide range of temperatures comparable with a typical range of applications. In experimental research for new thermoelectric materials, the versatility of measurement set-ups is especially important. They should make affordable measurements of samples of different shapes and dimensions in a wide range of temperatures. Despite relative simplicity of fundamental methods of measuring thermoelectric properties of materials, their practical implementation is a difficult task. Additional difficulty is the lack of commonly accepted reference materials for measuring thermopower at high temperatures, making it difficult to compare the results obtained by independent groups. In such circumstances, it is crucial to understand clearly possibilities and limitations of different methods for measuring thermoelectric properties and unconditional implementation of some basic requirements by researchers. When measuring thermopower, the most important points are: (1) thermoelectric signal and temperature difference must be measured between the same points of the sample; (2) potential contacts and temperature sensors must be in good thermal and electrical contact with the sample; (3) when using thermocouples, special attention must be given to thermoelectric homogeneity of their branches.

Author details

Alexander T. Burkov*, Andrey I. Fedotov and Sergey V. Novikov

*Address all correspondence to: a.burkov@mail.ioffe.ru

Ioffe Institute, Saint Petersburg, Russian Federation, Russia

References

[1] Stilbans LS. Physics of Semiconductors. Moscow: Sovetskoe Radio; 1967. 451 p. (in Russian).

[2] Anatychuk LI. On the discovery of thermoelectricity by Volta. Journal of Thermoelectricity. 2004; **N2**:5–10.

[3] Ioffe AF, Stilbans LS, Iordanishvili EK, Stavitskaya TS. Thermoelectric Cooling. Moscow, Leningrad: USSR Academy of Sciences Publishing; 1956. 114 p. (in Russian).

[4] Ioffe AF. Physics of Semiconductors. London: Infosearch; 1960. 436 p.

[5] Ioffe AF. Semiconductor Thermoelements and Thermoelectric Cooling. London: Infosearch; 1957. 184 p.

[6] Manasian YuG. Sudovye termoelektricheskie ustroystva i ustanocki. Leningrad: Sudostroenie; 1968. 283 p. (in Russian).

[7] Nye JF. Physical Properties of Crystals: Their Representation by Tensors and Matrices. Oxford: Clarendon Press; 1985. 329 p.

[8] Landau LD, Lifshitz EM. Course of Theoretical Physics, v. 10: Pitaevskii LP, Lifshitz EM, Physical Kinetics. Butterworth-Heinemann; 1981. 452 p.

[9] Barnard RD. Thermoelectricity in Metals and Alloys. London: Taylor & Francis; 1972. 259 p.

[10] Bowler N. Theory of Four-Point Direct-Current Potential Drop Measurements on a Metal Plate. Research in Nondestructive Evaluation. 2006;**17**:29–48. DOI: 10.1080/09349840600582092.

[11] Bowler N.: Four-point potential drop measurements for materials characterization. Measurement Science and Technology. 2011;**22**:012001-1-11. DOI:10.1088/0957-0233/22/1/012001.

[12] Pavlov LP. Methods for measurement of parameters of semiconducting materials. Vysshaya skola; 1987. 239 p. (in Russian).

[13] van der Pauw LJ. A method of measuring specific resistivity and Hall effect of disks of arbitrary shape. Philips Research Reports. 1958;**13**:1–9.

[14] van der Pauw LJ.: A method of measuring specific resistivity and Hall effect of lamellae of arbitrary shape. Philips Technical Review. 1958/1959;**20**:220–224.

[15] Webster JG, Halit Eren. The Measurement, Instrumentation, and Sensors Handbook, 2nd ed., CRC Press; 2014, 1921 p. ISBN-13: 978-1-4398-4893-7.

[16] Kontorovich M. Operational calculus and non-stationary phenomena in electrical circuits. 2nd ed. GITTL; 1955. 230 p. (in Russian).

[17] Nystrom J. Themopower. Landolt-Bornstein: Numerical data and functions. Vol. 2. Berlin: Springer; 1959.

[18] Borelius G, Keesom WH, Johansson CH. Communication from the Physical Laboratory at Leiden. 1928;**31**:no. 196a.

[19] Borelius G, Keesom WH, Johansson CH, Linde JO. Establishment of an absolute scale for the thermo-electric force. Proc. Kon. Akad. Amsterdam. 1932;**35**:no. 10.

[20] Rudnitskii AA. Thermoelectric properties of noble metals and their alloys. Moscow: USSR Academy of Sciences Publishing; 1956 (in Russian).

[21] Cusack N, Kendall P. The Absolute Scale of Thermoelectric Power at High Temperatures. Proceedings of Physical Society. 1958;**72**:898–901.

[22] Lander JJ. Measurements of Thomson coefficients for metals at high temperatures and of peltier coefficients for solid–liquid interfaces of metals. Physical Review. 1948;**74**:479–488.

[23] Roberts RB. The absolute scale of thermoelectricity. Philosophical Magazine. 1977;**36**:91–107.

[24] Roberts RB. The absolute scale of thermoelectricity II. Philosophical Magazine. 1981;**43**:1125–1135.

[25] Roberts RB, Righini F, Compton RC. The absolute scale of thermoelectricity III. Philosophical Magazine. 1985;**52**:1147–1163.

[26] Burkov AT, Heinrich A, Konstantinov PP. et al. Experimental set-up for thermopower and resistivity measurements at 100–1300 K. Measurement Science and Technology. 2001;**12**:264–272.

[27] Moore JP, Graves RS. Absolute Seebeck coefficient of platinum from 80 to 340 K and the thermal and electrical conductivities of lead from 80 to 400 K. Journal of Applied Physics. 1973;**44**:1171–1178. DOI: 10.1063/1.1662324.

[28] Horne RA. Errors associated with thermoelectric power measurements using small temperature differences. Review of Scientific Instruments. 1960;**31**:459–460. DOI: 10.1063/1.1717013.

[29] Testardi LR, McConnell GK.: Measurement of the seebeck coefficient with small temperature differences. Review of Scientific Instruments. 1961;**32**:1067–1068. DOI: 10.1063/1.1717624.

[30] Berglund CN, Beairsto RC. An automatic technique for accurate measurements of Seebeck coefficient. Review of Scientific Instruments. 1967;**38**:66–68. DOI: 10.1063/1.1720530.

[31] Aubin M, Ghamlouch H, Fournier P. Measurement of the Seebeck coefficient by an ac technique: application to high-temperature superconductors. Review of Scientific Instruments. 1993;**64**:2938–2941. DOI: 10.1063/1.1144387.

[32] Resel R, Gratz E, Burkov AT. et al. Thermopower measurements in magnetic fields up to 17 tesla using the toggled heating method. Review of Scientific Instruments. 1996;**67**:1970–1075. DOI: 10.1063/1.1146953.

[33] Caskey GR, Sellmver DJ, Rubin LG. A technique for the rapid measurement of thermoelectric power. Review of Scientific Instruments. 1969;**40**:1280–1282. DOI: 10.1063/1.1683764.

[34] Chaikin PM, Kwak JF. Apparatus for thermopower measurements on organic conductors. Review of Scientific Instruments. 1975;**46**:218–220. DOI: 10.1063/1.1134171.

[35] Putti M, Cimberle MR, Canesi A, Foglia C, Siri AS. Thermopower measurements of high-temperature superconductors: experimental artifacts due to applied thermal gradient and a technique for avoiding them. Physical Review B. 1998; **58**:12344–12349.

[36] Chen F, Cooley JC, Hults WL, Smith JL. Low-frequency ac measurement of the Seebeck coefficient. Review of Scientific Instruments. 2001;**72**:4201–4206. DOI: 10.1063/1.1406930.

[37] Wood C, Zoltan D, Stapfer G.: Measurement of Seebeck coefficient using a light pulse. Review of Scientific Instruments. 1985;**56**:719–722. DOI: 10.1063/1.1138213.

[38] D'Angelo J, Downey A, Hogan T. Temperature dependent thermoelectric material power factor measurement system. Review of Scientific Instruments. 2010;**81**:075107-1-4. DOI: 10.1063/1.3465326.

[39] Ravichandran J, Kardel JT, Scullin ML, Bahk J-H, Heijmerikx H, Bowers JE, Majumdar A. An apparatus for simultaneous measurement of electrical conductivity and thermopower of thin films in the temperature range of 300–750 K. Review of Scientific Instruments. 2011;**82**:015108-1-4. DOI: 10.1063/1.3529438.

[40] Burkov AT. Measurements of resistivity and thermopower: principles and practical real-
      ization. In: M. Rowe, editor. Thermoelectrics Handbook: Macro to Nano. Boca Raton:
      CRC Press; 2006. p. 22-1-12.

[41] Burkov AT, Dvunitkin VG. Simple sample-holder for high temperature measurements of
      thermopower and electrical resistivity. Pribory i technika eksperimenta. 1985; **5**:210–211
      (in Russian).

[42] Petrov AV. Method for measurement of thermal conductivity of semiconductors at high
      temperatures. In: Thermoelectric properties of semiconductors. Moskva: Akademia Nauk
      SSSR; 1963. pp. 27–35 (in Russian).

[43] Dasgupta T, Umarji AM. Apparatus to measure high-temperature thermal conductivity
      and thermoelectric power of small specimens. Review of Scientific Instruments.
      2005;**76**:094901-1-5. DOI: 10.1063/1.2018547.

[44] Zhou Z, Uher C. Apparatus for Seebeck coefficient and electrical resistivity measure-
      ments of bulk thermoelectric materials at high temperature. Review of Scientific Instru-
      ments. 2005;**76**:023901-1-5.

[45] Iwanaga S, Toberer ES, LaLonde A, Snyder GJ. A high temperature apparatus for mea-
      surement of the Seebeck coefficient. Review of Scientific Instruments. 2011;**82**:063905-1-6.
      DOI: 10.1063/1.3601358.

[46] Byl C, Berardan D, Dragoe N.: Experimental setup for measurements of transport prop-
      erties at high temperature and under controlled atmosphere. Measurement Science and
      Technology. 2012;**23**:035603-1-8. DOI: 10.1088/0957-0233/23/3/035603.

[47] Martin J, Tritt T, Uher C. High temperature Seebeck coefficient metrology. Journal of
      Applied Physics. 2010;**108**:121101–1–12. DOI: 10.1063/1.3503505.

[48] Polvani DA, Meng JF, Hasegawa M, Badding JV. Measurement of the thermoelectric
      power of very small samples at ambient and high pressures. Review of Scientific Instru-
      ments. 1999;**70**:3586–3589. DOI: 10.1063/1.1149964.

[49] Polvani DA, Fei Y, Meng JF, Badding JV. A technique for thermoelectric power measure-
      ments at high pressure in an octahedral multianvil press. Review of Scientific Instru-
      ments. 2000;**71**:3138–3140. DOI: 10.1063/1.1305517.

[50] Choi ES, Kang H, Jo YJ, Kang W. Thermoelectric power measurement under hydrostatic
      pressure using a self-clamped pressure cell. Review of Scientific Instruments.
      2002;**73**:2999–3002. DOI: 10.1063/1.1489076.

[51] Mun E, Budko SL, Torikachvili MS, Canfield PC. Experimental setup for the measure-
      ment of the thermoelectric power in zero and applied magnetic field. Measurement
      Science and Technology. 2010;**21**:055104-1-8. DOI: 10.1088/0957-0233/21/5/055104.

[52] Yuan B, Tao Q, Zhao X. et al. In situ measurement of electrical resistivity and Seebeck
      coefficient simultaneously at high temperature and high pressure. Review of Scientific
      Instruments. 2014;**85**:013904-1-4. DOI: 10.1063/1.4862654.

Chapter 16

Novel Measurement Methods for Thermoelectric Power Generator Materials and Devices

Patrick J. Taylor, Adam Wilson, Jay R. Maddux,
Theodorian Borca-Tasciuc, Samuel P. Moran,
Eduardo Castillo and Diana Borca-Tasciuc

Additional information is available at the end of the chapter

http://dx.doi.org/10.5772/65443

Abstract

Thermoelectric measurements are notoriously challenging. In this work, we outline new thermoelectric characterization methods that are experimentally more straightforward and provide much higher accuracy, reducing error by at least a factor of 2. Specifically, three novel measurement methodologies for thermal conductivity are detailed: steady-state isothermal measurements, scanning hot probe, and lock-in transient Harman technique. These three new measurement methodologies are validated using experimental measurement results from standards, as well as candidate materials for thermoelectric power generation. We review thermal conductivity measurement results from new half-Heusler (ZrNiSn-based) materials, as well as commercial $(Bi,Sb)_2(Te,Se)_3$ and mature PbTe samples. For devices, we show characterization of commercial $(Bi,Sb)_2(Te,Se)_3$ modules, precommercial PbTe/TAGS modules, and new high accuracy numerical device simulation of Skutterudite devices. Measurements are validated by comparison to well-established standard reference materials, as well as evaluation of device performance, and comparison to theoretical prediction obtained using measurements of individual properties. The new measurement methodologies presented here provide a new, compelling, simple, and more accurate means of material characterization, providing better agreement with theory.

Keywords: thermal conductivity, Seebeck coefficient, electrical resistivity, ZT, device efficiency

1. Introduction

The efficiency with which a thermoelectric (TE) power generator can convert heat energy to electricity is determined, in part, by thermal conductivity, κ, of the materials used for fabricating TE devices. Experimental measurement of that property usually results in surprisingly

significant error, $>\pm$ 10% [1]. There are many causes of that error, and in this chapter, we provide new solutions which are experimentally faster, yield results which are more consistent with physical devices, and address several sources of experimental error, which may reduce uncertainty by a factor of 2 or more.

In this work, we describe several new, more accurate techniques to measure thermal conductivity, κ, of thermoelectric materials. The first is based on detailed control of the heat flows within a sample under steady-state conditions; the so-called *steady-state isothermal technique*. The second is nondestructive microscale analysis technique called *scanning hot-probe (or scanning thermal microscopy)*. And the third is *lock-in transient Harman* method, which is a comprehensive modification of transient Harman technique, employing a lock-in procedure and considers detailed contact effects. A new interesting follow-on is frequency-dependent Nyquist analysis, which presages a different perspective on the material analysis.

The truest test of the accuracy of measurement is comparison with fabricated devices. To support the validation of measurements of individual material properties, we outline a new device metrics, which allows comparison between theoretical and measured device efficiency. We outline a new *slope-efficiency method*, which can be used to determine informative index $ZT_{maximum}$ of any device. The second method of device evaluation is a numerical device model called the *discretized heat balance model*, which considers a piecewise continuous collection of discrete layers within a device, where boundary heat flows have energy and current continuity relationships, and enable incredibly easy determination of device efficiency.

2. Novel measurements of thermal conductivity

2.1. Steady-state isothermal technique

This new measurement of κ leverages Peltier heat, $\pm Q_\Pi$, an electronically controlled internal heat source unique to thermoelectric materials. Peltier heat causes either heating ($+Q_\Pi$) or cooling ($-Q_\Pi$) at the junction between a thermoelectric material and a metal by passing the proper polarity of electric current, $\pm I$.

Q_Π was first employed to roughly estimate κ by Putley [2]. In Putley's experiment, convective, parasitic, and nonsymmetric heat flows required correction factors larger than 20%. Harman dramatically improved upon Putley's demonstration by performing the measurements in vacuum, and by reducing other parasitic heat flows. Despite these improvements, error is still obtained, because parasitic heat flows are nonzero [3].

In these past studies, principal parasitic heat flows causing error include conduction along lead wires, conduction along thermocouples, Joule heating within lead wires, and radiation. The magnitudes of these parasitic heat flows can be as large as 30% of Peltier heat. Penn quantified the significance of parasitic heat flows in Harman's technique, and showed, that they induced error of more than 10% [4]. Bowley and Goldsmid [5], as well as Buist [6] reported, that parasitic heat flows cause error, usually larger than 20%.

The focus of the present work is description of a new, correctionless method to measure κ by balancing two independently controlled heat sources: Q_Π, and a radiatively coupled input heat as per the Stefan-Boltzmann law, Q_{SB}. In this new method, Q_Π and Q_{SB} can be independently balanced. A finite temperature difference across the sample imposed by Q_{SB} can be cancelled and even inverted by application of Q_Π. When exactly cancelled, there is no temperature difference across the sample (i.e., $\Delta T = 0$) and a steady-state isothermal condition is obtained leaving the steady-state temperature of the sample exactly equal to that of the surroundings, T_e. Because there is no ΔT, parasitic heat flows, such as those along lead-wires/thermocouples (Q_{wires}) and radiative heat loss, ($Q_{radiation-error}$), which would otherwise cause significant error [7], converge exactly to zero. Analysis of this technique using a Peltier cooler demonstrates error of less than $\pm 1\%$, an improvement of over an order of magnitude [8]. When considering thermoelectric power generators, other considerations must be taken into account to determine experimental uncertainty (such as view angle for radiative heat flows from the environment, material emissivity, etc.), which are more complicated and beyond the scope of the work presented here. However, as is demonstrated qualitatively here, experimental uncertainty is reduced using the steady-state isothermal technique by significantly more than a factor of 2.

Assume a sample having the temperature of one end anchored to the temperature of the environment, T_e, by a large heat-sink, and the opposite end, T_{top}, having small thermal mass and capable of temperature diversion by the application of Q_{SB}. Vacuum is used to obviate convective heat flow, and Q_{SB} is applied by small heater having approximately the same subtended area as that of the cross-sectional area (A) of the sample, such that Q_{SB} is localized to the top and there is no direct line of sight along the length of the sample (ℓ). Thermocouples are attached to each end to determine the temperature difference across the sample, and electrical leads are used for passing $\pm I$ to control Q_Π. **Figure 1** depicts this experimental setup.

When Q_{SB} is applied, the temperature of the heated contact will increase with respect to T_e and some magnitude of Q_κ will be conducted through the sample. At the heated contact, contributing heat flows include Q_{SB}, Q_κ, $Q_{radiation-error}$, and Q_{wires}. Radiation-error flow $Q_{radiation-error}$ is governed by the emissivity (ε), Stefan-Boltzmann constant (σ), sidewall temperature ($T_{sidewall}$), and sidewall area ($A_{sidewall}$) of the sample. Q_{wires} follows the usual Fourier's law description where, for simplicity, aspect ratio (A_{wires}/ℓ_{wires}) and thermal conductivity (κ_{wires}) of all wires and thermocouples are combined into one lumped parasitic term. When electrical current flows through the sample, Q_Π is absorbed at the heated contact, where the first Kelvin relation gives $Q_\Pi = (\alpha_{sum}IT)$ and α_{sum} is the sum of Seebeck coefficients of the sample and the contact metal, and T is the temperature of the heated contact. An equal and opposite value of Q_Π is liberated at the contact between the sample and large heat-sink, but is too small to cause any measurable temperature change of the heat-sink, and is therefore negligible. Including Q_Π at the heated contact yields Eq. (1), which, under steady-state conditions sums to zero:

$$\sum Q = Q_{SB} - Q_\Pi - Q_\kappa - Q_{radiation-error} - Q_{wires} = 0. \tag{1}$$

To quantify the magnitude of Q_κ, a range of electrical currents can be passed, which enables Q_Π to absorb a corresponding range of Q_{SB} at the contact, so we get:

$$Q_{SB} = (\alpha_{sum}IT) + \left[\kappa\left(\frac{A}{\ell}\right)\Delta T\right] + \varepsilon\sigma A_{sidewall}(T^4_{sidewall}\,T_e^{\,4}) + \left[\kappa_{wires}\left(\frac{A_{wires}}{\ell_{wires}}\right)\Delta T\right]. \qquad (2)$$

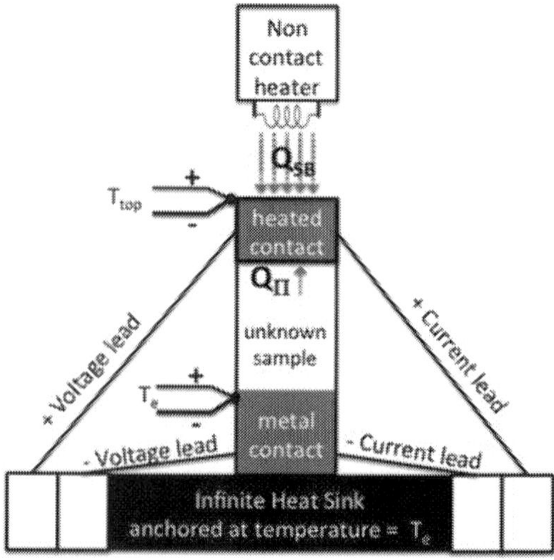

Figure 1. Experimental schematic representation of steady-state isothermal technique.

For progressively larger I, Q_Π absorbs increasingly more of Q_{SB} and temperature of the heated contact begins to converge to T_e, such that the overall ΔT across the sample goes to zero. As ΔT becomes smaller with increasing I, the only relevant heat flows are Q_{SB}, Q_Π, and Q_κ because $Q_{radiation\text{-}error}$ and Q_{wires} are only statistically significant [7] for larger ΔT, say >10 K, and all parasitic heat flows converge to zero. Therefore, under these conditions at any given I, ΔT across the sample is required to satisfy Eq. (3):

$$Q_{SB} = (\alpha_{sum}IT) + \left[\kappa\left(\frac{A}{\ell}\right)\Delta T\right]. \qquad (3)$$

From the requirement imposed by Eq. (3), a new method for measuring κ is obtained. Eq. (3) is solved to show the dependence of ΔT on electrical current. By taking the derivative of Eq. (3), prior knowledge of Q_{SB} is not required, because it is a constant, and the analysis yields the following:

$$\frac{\partial\Delta T}{\partial I} = -\left[\frac{\alpha_{sum}T}{\kappa\left(\frac{A}{\ell}\right)}\right]. \qquad (4)$$

To determine κ, Eq. (4) is solved using the slope at $\Delta T = 0$ of steady-state ΔT as function of I and that slope is combined with α_{sum}, as well as geometrical aspect ratio (A/ℓ) of the sample [8]. **Figure 2** shows dependence of the temperature difference between ends of the thermoelement sample on value of passing current (left panel) and real picture measurement configuration (right panel).

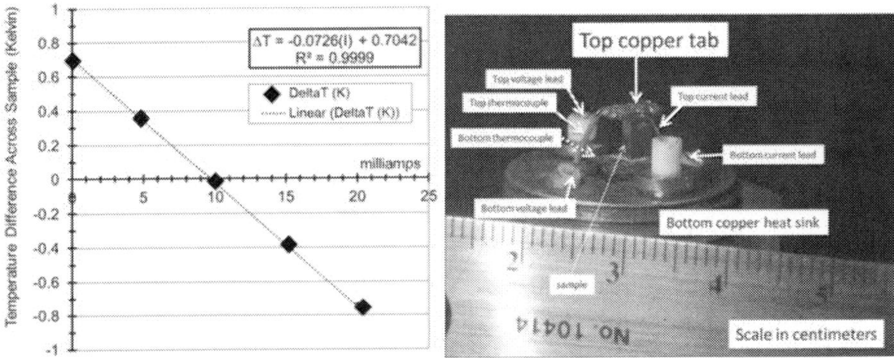

Figure 2. (Left) Linear ΔT decrease across sample by application of milliamps of current (I) for Peltier cooling and (right) picture showing a fully connected sample.

2.1.1. Thermal conductivity of n-type half-Heusler

Thermal conductivity of a half-Heusler alloy (**Figure 3**) was collected (triangles) and is presented with respect to previously published data (squares, circles, and diamonds) measured by the laser-flash thermal diffusivity technique and reported by researchers from GMZ corporation [9]. Because of the speed, ease, and simplicity of the new technique presented, there is opportunity for significantly more collected data. One data point can be collected in seconds. However, as can be seen in **Figure 3**, it is consistent to within experimental error and falls within the bounds of the published laser-flash data.

2.1.2. Thermal conductivity of PbTe

One category of high performance thermoelectric materials is near-degenerate semiconductors. Such materials do not directly obey the Wiedemann-Franz relationship between electrical resistivity and thermal conductivity, due to the significant contribution of lattice thermal conduction to the total thermal conductivity. However, utilizing a modified Wiedemann-Franz relationship to find the thermal conductivity due to electron flow allows direct, real-time deconvolution of lattice thermal conductivity (κ_{lattice}) from electronic contribution ($\kappa_{\text{electronic}}$). For charge carriers concentrations near degeneracy, and random scattering of charge carriers, Rosi et al. [10] describe how $\kappa_{\text{electronic}}$ can be determined using electrical resistivity, ρ, by:

$$\kappa_{\text{electronic}} = \left(\frac{\pi^2}{3}\right)\left(\frac{k_B}{q}\right)^2\left(\frac{T}{\rho}\right). \tag{5}$$

If thermal conductivity and electrical resistivity are measured, then κ_{lattice} can be determined by $\kappa_{\text{lattice}} = [\kappa - \kappa_{\text{electronic}}]$.

Figure 4 shows temperature dependences of PbTe thermal conductivity (including measured data) and deconvolution of electronic and lattice contributions to total thermal conductivity.

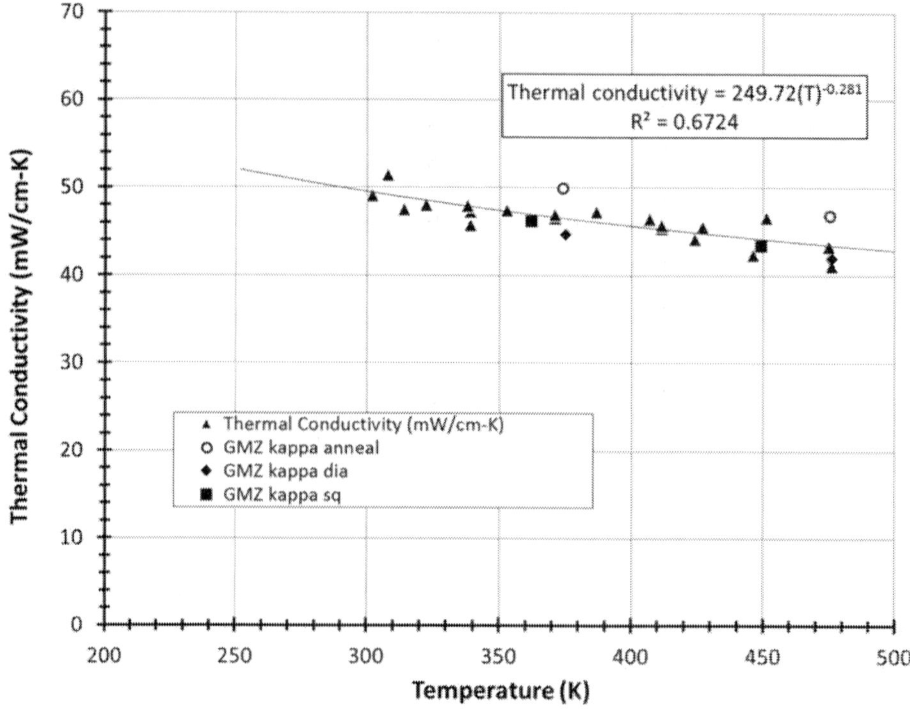

Figure 3. Thermal conductivity of n-type half-Heusler measured by steady-state isothermal technique as compared to published data [9].

2.2. Scanning hot probe

The scanning hot probe technique provides measurement of local thermal conductivity and Seebeck coefficient of a sample by measuring average probe temperature when probe tip and sample are in "thermal contact", i.e., when probe tip is in physical contact with the sample or is at known distance near enough to the sample to induce measurable heat exchange between probe and sample. Average probe temperature is also measured far from the sample, to account for the amount of heat lost to the surroundings and through the probe contacts. Difference in average probe temperature between these two cases is due to the heat transferred to the sample, which may be quantified through the following analytical derivation. **Figure 5** depicts the thermal exchange between probe, sample, and surroundings, as well as the series thermal resistance network between probe and sample.

For steady-state probe heating using DC current, (or AC current at low frequency, when the heat capacity effects are negligible, and temperature rise amplitude is frequency independent and equivalent to DC temperature rise to good approximation) the governing equation describing amplitude of the temperature profile of the probe shown in **Figure 5** is given by [12]:

$$\frac{d^2 T^*}{dx^2} - \left(\frac{2h_{\text{eff}}}{\kappa_P r} - \frac{I^2 \rho_0 \text{TCR}}{\kappa_P \pi^2 r^4}\right) T^* + \frac{I^2 \rho_0}{\kappa_P \pi^2 r^4} = 0, \tag{6}$$

where $T^* = T(x) - T_0$, $h_{\text{eff}} = h + 4\epsilon\sigma T_0^3$ (here, h is the convective heat transfer coefficient, ϵ is the probe's emissivity, T_0^3 is an approximation for the exact $(T^4 - T_o^4)$ term and σ is Stefan-Boltzmann constant. ρ_0 and κ_P are the probe's electrical resistivity and thermal conductivity, respectively, TCR is the probe's temperature coefficient of resistance, I is root-mean-square electrical current passed through the probe, and r is the radius of the probe. Contribution from radiation is negligible for $\Delta T < 100$ K [13].

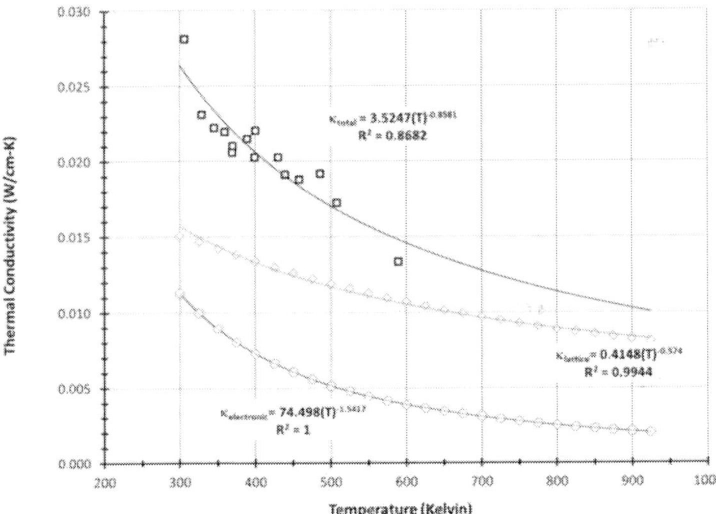

Figure 4. Measurement of thermal conductivity of PbTe, and deconvolution of electronic and lattice contributions to the total thermal conductivity.

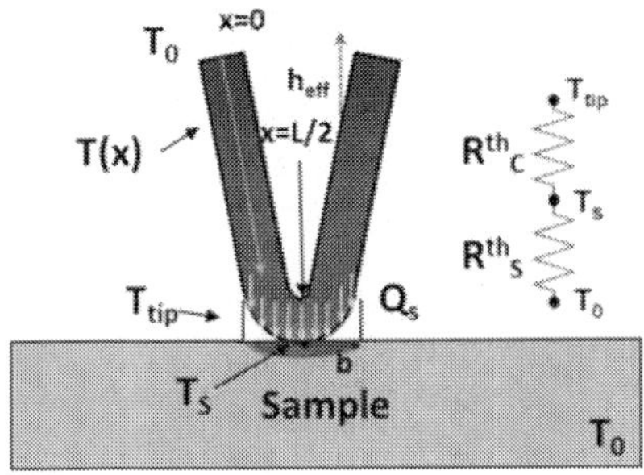

Figure 5. Diagram showing thermal phenomenology around probe tip. Reproduced from [11] with permission from The Royal Society of Chemistry.

To obtain an analytical solution to the second order differential equation, two boundary conditions are employed. The first assumption is that the ends of the probe are at ambient temperature (i.e., $T(0) = T_0$). The second assumption is that the tip region of the probe of length $2b$ is of uniform temperature, and by energy balance at the probe tip region, we get:

$$-\kappa_P A \frac{dT^*}{dx}\Big|_{x=L/2-b} + I^2 \rho_0 (1 + TCR \times T^*|_{x=L/2-b}) \frac{b}{A} = \frac{Q_s}{2}, \tag{7}$$

where the left-hand side is heat conduction and Joule heating of the probe, $A = \pi r^2$ is the probe's cross-sectional area, and L is the length of the probe, and where the right-hand side is heat transfer through one leg of the probe (thus half the total heat transfer to the sample, by symmetry). Finally, heat transfer rate between probe and sample, Q_s, is:

$$Q_s = \frac{\Delta T_{tip}}{R_C^{th} + R_S^{th}} = \frac{\Delta T_S}{R_S^{th}}, \tag{8}$$

where $\Delta T_{tip} = T_{tip} - T_0$ and $\Delta T_S = T_S - T_0$ are temperature of the probe and sample, respectively, at the tip region, and R_S^{th} is samples thermal resistance. Solving Eq. (6) to obtain temperature profile along the probe for a given value of Q_s yields the following expression:

$$\Delta T_P(x) = C_1 e^{\lambda x} - C_2 e^{-\lambda x} + \frac{\Gamma}{\lambda^2}, \tag{9}$$

where $\lambda = \frac{I^2 \rho_0}{\kappa_P \pi^2 r^4}$, $\Gamma = \frac{2h_{eff}}{\kappa_P r} - \frac{I^2 \rho_0 TCR}{\kappa_P \pi^2 r^4}$, and constants C_1 and C_2 are easily obtained by applying boundary condition $T(0) = T_0$.

If the sample is bulk, or has bulk-like thickness, thermal conductivity is found from R_S^{th} by employing semiinfinite medium assumption and 2D bulk sample assumption [14]:

$$R_S^{th} = \frac{1}{4\kappa_{sample} b}. \tag{10}$$

If the sample is a thin film of thickness l on substrate, and is thin enough, that there is negligible heat spreading in the in-plane directions of the sample, then thermal conductivity is found by solving the expression for the series thermal resistance across substrate and film, with 1D heat transfer across the thickness of the film [14]:

$$R_S^{th} = \frac{1}{4\kappa_{substrate} b} + \frac{l}{\pi \kappa_{film} b^2}. \tag{11}$$

When heat transfer may be multidimensional and anisotropic, models developed by Son et al. [15] for laser heating may be used to predict thermal resistance of the sample, based on the respective values of thermal conductivity for the film and substrate.

Data collected from scanning hot thermoelectric probe experiment are probe voltage, voltage across a reference resistor, Seebeck voltage, and photodetector voltage for position sensing. The value of current passing through the system is obtained by dividing voltage across the reference resistor by known electrical resistance of that resistor. Probe resistance is then found by dividing probe voltage by that value of current. Often, instead of single reference resistor, Wheatstone bridge is utilized. **Figure 6** depicts the difference in circuit between measurement taken (a) with and (b) without Wheatstone bridge.

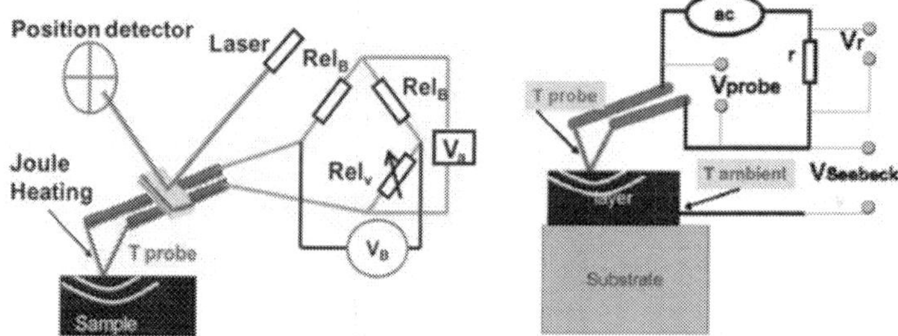

Figure 6. Schematic scanning thermal microscopy circuits (left) with Wheatstone bridge and (right) with a reference resistor.

In DC mode, with resistor wired in series with the probe, measured probe voltage, V_P, may be expressed in terms of voltage across reference resistor (V_r) as:

$$V_P = R_P I = \frac{R_P V_r}{R_r}, \tag{12}$$

where I is DC current passing through the circuit, R_P is electrical resistance of the probe, and R_r is known electrical resistance of reference resistor. Probe's electrical resistance is proportional to temperature rise above ambient, when the probe undergoes Joule heating. It may be expressed as:

$$R_P = \Delta T_P (R_0 \text{TCR}) + R_0 + R_c, \tag{13}$$

where ΔT_P is average probe temperature rise, R_0 is nominal probe electrical resistance at 19.9° C, (not including electrical contacts to the circuit), when the probe is not being heated, R_c is electrical resistance arise from contacts and circuit's wiring, and TCR is probe's temperature coefficient of resistance, in terms of 1/°C.

Defining average probe thermal resistance as average probe temperature rise divided by Joule heating power, we can write average probe thermal resistance as:

$$R_P^{\text{th}} = \frac{\Delta T_P}{I^2 R_0 (1 + (\text{TCR})\Delta T_P)}. \tag{14}$$

Eq. (14) allows determination of R_P^{th} by the slope of probe temperature rise with power applied, reducing the overall experimental uncertainty compared with a single value of temperature at a given power. If the circuit uses Wheatstone bridge, equations differ only by the method of finding electrical resistance of the probe. In this case, R_P reduces to:

$$R_P = R_0 + \frac{V_B}{V_A} \frac{(R_B + R_0)^2}{R_B} + R_c, \tag{15}$$

where V_B and V_A are voltages across bridge side and probe side, respectively. R_B is the total resistance of bridge side of the circuit, and R_0 is no heating resistance of the probe side of the bridge. With the probe resistance obtained, remaining equations are left unchanged. When AC current of amplitude I_0 is passed through the circuit, then measured probe resistance can be expressed as:

$$R_P = \underbrace{R_0 \left(1 + (TCR)\Delta T_{P,DC}\right)}_{DC\ Component} + \underbrace{R_0(TCR)\Delta T_{P,2\omega}cos(2\omega t + \phi)}_{AC\ Component}. \tag{16}$$

The probe tip voltage is expressed as:

$$V = I_0 R_P \left(1 + (TCR)\Delta T_{P,DC}\right) cos\,(\omega t) + \frac{I_0 R_P (TCR)\Delta T_{P,2\omega}}{2} \left[\,cos\,(3\omega t + \phi) + cos(\omega t + \phi)\right]. \tag{17}$$

Thus, temperature amplitude is determined to be:

$$\Delta T_{ave} = \frac{2V_{3\omega}}{(TCR)V_{1\omega}}. \tag{18}$$

To obtain sample thermal conductivity, the probe must be calibrated. Quantities in Eqs. (6)–(11), which are not determined directly from experimental measurement are: h_{eff}, TCR, k_P, A, L, ρ_0, b, and R_C^{th}. To be fully calibrated, these quantities must be known. The probe manufacturer specifies values for TCR, k_P and ρ_0, and these values are used in this work. Values A and L may be found by determining probe's geometry (typically from SEM or microscope images, but may also be determined by measuring R_P^{th} in a vacuum and in air). h_{eff} is determined by measuring R_P^{th} far from contact, and matching the value predicted by the analytical model by adjusting h_{eff} and integrating Eq. (9) from $x = 0$ to L and dividing by L, with $Q_s = 0$ to obtain the average probe temperature when no heat is transferred to the sample. Finally, probe-to-sample thermal exchange parameters, b and R_C^{th}, must be determined. Typically, these values have been assumed to be sample-independent for given probe-to-sample contact force or probe-to-sample distance. As such, calibration strategies utilize measurements on two samples. However, these parameters are now shown to change with sample thermal conductivity. **Figure 7** demonstrates change in b and R_C^{th} with sample thermal conductivity. Alternatively, if the sample is electrically grounded and probe tip is capable of making good electrical contact with the sample, then sample with known thermal conductivity and Seebeck coefficient may be used to determine both b and R_C^{th} simultaneously. **Figure 7** demonstrates this calibration strategy, together with the typical "intersection method" using two or more samples. Care must be taken to calibrate in the correct range of thermal conductivity, as samples with thermal conductivity of higher than 1.1 W/mK yield a different pair of b and R_C^{th} values compared with samples with thermal conductivity of 1.1 W/mK and lower. **Table 1** presents the results of measurements taken with properly calibrated Wollaston probe tips, showing good agreement with independent measurements.

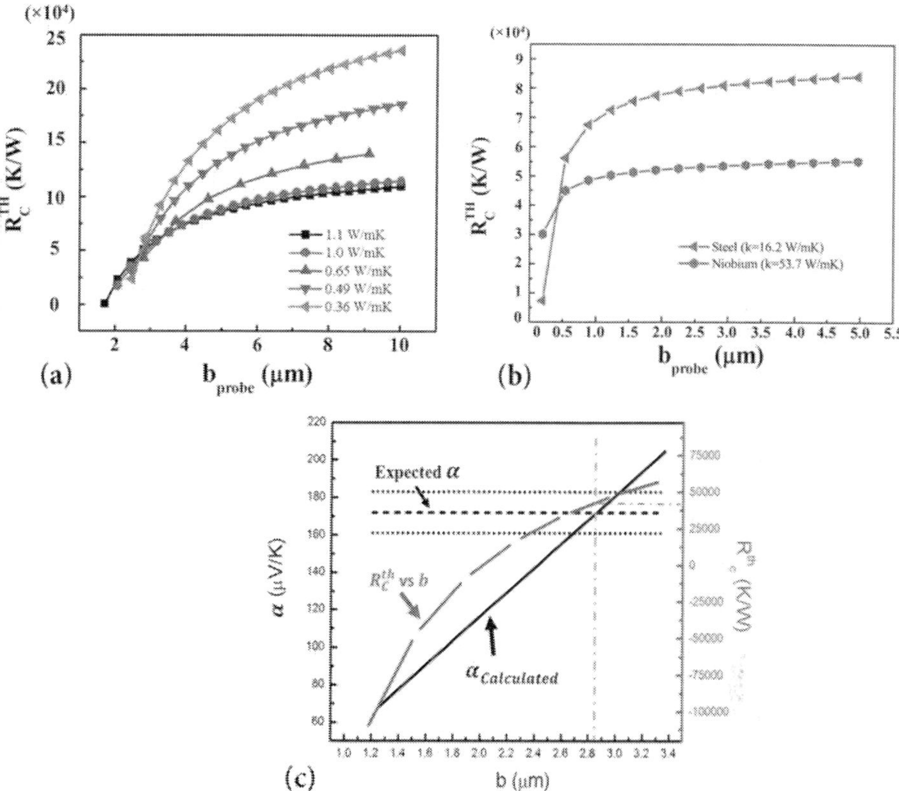

Figure 7. Empirical probe calibration strategies for determining b and R_C^{th} using (top) intersections of curves taken from samples with known κ for (left) low thermal conductivity and (right) high thermal conductivity and (bottom) determining b and R_C^{th} using a single sample with known κ and α. Adapted from [11] with permission from The Royal Society of Chemistry.

2.3. Transient and lock-in Harman techniques to decouple material ZT and thermoelectric properties

Finally, a new method of measuring material thermal conductivity by simultaneously measuring thermoelectric figure of merit (ZT), electrical resistivity (ρ), and Seebeck coefficient (α) is proposed. ZT is dimensionless measure of the efficiency of material at converting thermal into electrical energy, or vice versa, at a given temperature, T, and may be expressed as:

$$ZT = \alpha^2 T / \rho\kappa. \tag{19}$$

Thus, thermal conductivity is obtained if the other terms in Eq. (19) are known. This new method also allows for measuring intrinsic ZT with reduced experimental error by accounting for losses through nonideal contacts and geometry.

Sample	l	b	R_P^{th}	R_C^{th}	R_S^{th}	κ_{film}, this work // κ_{film}, expected
SiGe film on glass substrate	1.8 μm	2.8 ± 0.3 μm	14.948 ± 54 K/W	44.927±7820 K/W	76.134 ± 9494 K/W	1.22 ± 0.21 W/K·m // 1.23 ± 0.12 W/K·m[16]
Fe-doped PCDTBT (1:1 doping concentration)	3.0 μm	2.8 ± 0.3 μm	15.220 ± 155 K/W	44.927±7820 K/W	87.022 ± 14.631 K/W	1.03 ± 0.15 W/K·m [17]
PCDTBT (non-doped)	3.0 μm	2.8 ± 0.3 μm	17.866 ± 204 K/W	44.927±7820 K/W	358.859 ± 66.204 K/W	0.25 ± 0.04 W/K·m //0.20 ± 0.02 W/K·m [17]
Tellurium Film	2.74 μm	2.8 ± 0.3 μm	15.749 ± 75.5 K/W	44.927±7820 K/W	112.476 ± 6.480 K/W	0.79 ± 0.04 W/K·m //0.78 ± 0.08 W/K·m [18]
Au film on silicon substrate	150 nm	428 ± 24 nm	11.624 ± 157 K/W	40.191±1532 K/W	5505 ± 253 K/W	104.2 ± 67.4W/K·m //110 ± 2 W/ K·m [19]
PEDOT [CAL]	Bulk	2.8 ± 0.3 μm	17.429 ± 217 K/W	44.927 ± 7820 K/W	241.732 ± 37.672 K/W	0.37 ± 0.05 W/K·m // 0.36 W/K·m
PANI-5 % GNP [CAL]	Bulk	2.8 ± 0.3 μm	17.018 ± 115 K/W	44.927 ± 7820 K/W	188.595 ± 27.836 K/W	0.47 ± 0.06 W/K·m // 0.49 W/K·m [20]
PANI-7 % GNP [CAL]	Bulk	2.8 ± 0.3 μm	16.314 ± 118 K/W	44.927 ± 7820 K/W	131.760 ± 25.913 K/W	0.68 ± 0.08 W/K·m //0.65 W/ K·m [20]
p-typeBi$_2$Te$_3$ [CAL]	Bulk	2.8 ± 0.3 μm	15.700 ± 145 K/W	44.927 ± 7820 K/W	92.113 ± 11.911 K/W	0.97 ± 0.11 W/K·m // 1.0 W/K·m
Borosilicate Glass [CAL]	Bulk	2.8 ± 0.3 μm	15.516 ± 134 K/W	44.927 ± 7820 K/W	82.313 ± 9787 K/W	1.08 ± 0.11 W/K·m // 1.1 W/K·m
AISI 304 Steel [CAL]	Bulk	428 ± 24 nm	13.811 ± 119 K/W	40.191 ± 1532 K/W	37.511 ± 3511 K/W	15.6 ± 2.2 W/K·m // 16.2 W/K·m
Goodfellow®99.9 % pure Niobium [CAL]	Bulk	428 ± 24 nm	12.194 ± 140 K/W	40.191 ± 1532 K/W	10.632 ± 2329 K/W	54.9 ± 8.9 W/K·m // 53.7 W/K·m

Table 1. Tabular results for a range of materials [16–20]. Reproduced from [11] with permission from The Royal Society of Chemistry.

Conventional application of Harman method uses four probes–two to pass current, and two to measure the voltage response of the sample. Harman demonstrated that, while electrical response of the sample was nearly instantaneous, voltage generated by Seebeck effect, which is thermally driven, is much slower. By taking advantage of this fact, thermal signal could be determined from voltage response of the sample to a sudden change in voltage over time, or response to an AC current passed, locking into thermally driven signal. It was shown, by letting $\alpha = V_\alpha/\Delta T$, $\kappa = -\Delta T A/(\alpha T I)$, and $\rho = V_\rho A/(LI)$, that ZT could be reduced to the ratio of resistive voltage to Seebeck voltage (*i.e.*, $ZT = V_S/V_Q$). This assumes ideal contacts (negligible thermal and electrical losses through the contact leads) and that temperature rise is due only to Peltier heating, neglecting effects of Joule heating. However, these effects are often difficult to mitigate, and may be accounted for by appropriate modeling (see **Figure 8**).

From resistive voltage, one may be able to determine electrical resistivity of the material; however, Seebeck coefficient and thermal conductivity remain coupled in equation for ZT. To decouple them, Seebeck coefficient may be simultaneously determined by adding a pair of thermocouple wires at the top and bottom surfaces of the sample as per **Figure 9**. If we label electric potential in each corner of the sample $E_1 - E_4$, respectively, using Ivory technique [17], we may find Seebeck coefficient from taking voltage measurements across the sample. If voltage is measured at opposite corners, then voltage values measured are E_{13} and E_{24}. Value α is determined from expression below, where m is the slope of E_{13} vs E_{24}:

$$\alpha_{\text{sample}} = \lim_{E_{13}\to E_{24}} \frac{1}{1-m}\alpha_{ba} + \alpha_b. \tag{20}$$

This technique for measuring Seebeck coefficient reduces the required number of voltage measurements to determine α from three to two and mitigates mismatch in thermocouples, since DC offsets are removed by using a slope. It also allows for AC measurements of the total voltage, and determination of ZT from Nyquist diagrams.

2.3.1. Transient Harman technique–analytical model

Several experimental setups used by the research community for thermoelectric characterization of thin films employ clean room microfabrication techniques to pattern a metallic electrode on top of the sample, while others use bonded wires or micromanipulated probes to make electrical contact with the top surface of the film sample [3, 21–24]. Configuration modeled in this work is similar to these situations, as shown in **Figure 8**. Thermoelectric film (3) with cross-sectional area, A_3, is deposited on substrate; metallic electrode (2) covers the top film's surface; and electrically conductive probe wire (1) of diameter, d_1, is brought in contact with the top surface of the sample. Substrate electrode (4) situated at the interface between thermoelectric film and substrate is used to close the loop and pass current into the film. The substrate electrode is assumed to have negligible electrical and thermal resistance. Its contribution to thermoelectric transport is therefore neglected, with exception of Peltier effect. Substrate electrode temperature is assumed to be the same as the top surface of the substrate (T_b). Electrical and thermal contact resistances expressed as specific values $R_{C\_i\text{-}j}$ and $R_{\text{th}\_i\text{-}j}$ are assumed at interfaces between adjacent layers indexed by i and j, with ($j = i + 1$). Classical thermoelectric transport model, which neglects electron-phonon nonequilibrium effects, is developed by assuming, that the thickness of thermoelectric film is much larger than phonon-electron thermalization length [25]. Under these conditions, thermoelectric transport in the probe, electrode, and sample is considered one-dimensional. In each layer i, x is the spatial coordinate; h, κ, and ρ are convection heat transfer coefficient, thermal conductivity and electrical resistivity, respectively; P is perimeter, T is absolute temperature, α is Seebeck coefficient, A_3 is area perpendicular to thermoelectric transport direction, and J is current density.

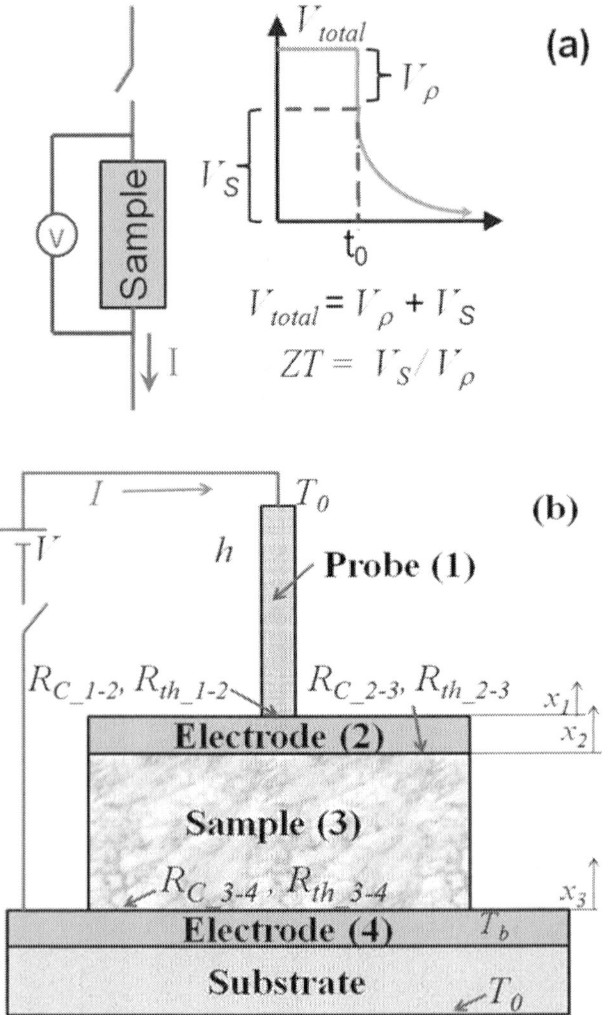

Figure 8. Schematic representation of lock-in Harman technique, under assumptions of (top) ideal contacts and (bottom) considering nonnegligible thermal and electrical resistances arise from contacts. Reprinted from [21] with the permission of AIP Publishing.

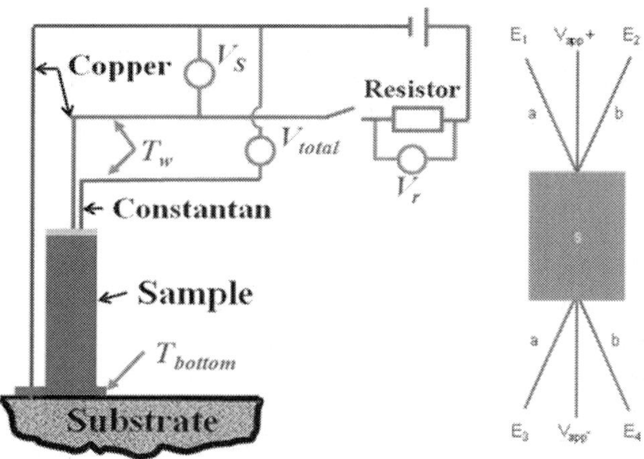

Figure 9. Schematic representation of sample measurement, including thermocouple contacts (ivory technique). (Left) sectional view (reprinted from [21] with the permission of AIP Publishing) and (right) view of sample from above.

Eq. (21) represents steady state energy balance in layers 1–3; the first term on the left side represents heat conduction, the second represents lateral convection and the third represents Joule heating. Thermal radiation and temperature dependence of thermoelectric properties have been neglected as small temperature differences are assumed to occur during the experiments.

$$\frac{d^2 T_i}{dx_i^2} - \frac{h_i P_i}{\kappa_i A_i}(T_i - T_0) + \frac{J_i^2 \rho_i}{\kappa_i} = 0; \quad i = 1, 2, 3. \tag{21}$$

General solution of Eq. (21) for temperature profile in the layer i is of the form,

$$T_i(x_i) = T_0 + c_{1,i} e^{\sqrt{\frac{h_i P_i}{\kappa_i A_i}} x_i} + c_{2,i} e^{-\sqrt{\frac{h_i P_i}{\kappa_i A_i}} x_i} + \frac{J_i^2 \rho_i A_i}{\kappa_i h_i}; \quad i = 1\text{–}3. \tag{22}$$

Integration constants $c_{1,\,i}$ and $c_{2,\,i}$ are determined using six boundary conditions. First, the temperature at free end of the probe of length l_1 is assumed to be ambient temperature as stated by:

$$T_1 = T_0; x_1 = l_1. \tag{23}$$

Second, the temperature of the end of the probe in contact with the surface is assumed to be constant temperature (see expression in the ensuing discussion, Eq. (31)).

The third boundary condition considers energy conservation at the interface between the film and the substrate:

$$-\kappa_3 A_3 \frac{dT_3}{dx_3}\Big|_{x_3=0} + J_3 \alpha_3 A_3 T_3\Big|_{x_3=0} = \frac{T_b|-T_3|_{x_3=0}}{R_{th\_3\text{–}4}/A_3} + \frac{1}{2} J_4^2 A_4 R_{C\_3\text{–}4} =$$
$$\frac{T_0 - T_b}{\Theta_{subst}} + J_4^2 A_4 R_{C\_3\text{–}4} + J_4 \alpha_4 A_4 T_b. \tag{24}$$

In Eq. (24), the left side represents heat transfer rate out of the interface. It includes heat conduction and Peltier terms in layer 3, respectively. The middle section of Eq. (24) represents one way to express heat transfer rate entering the interface. It is written as the sum of heat conduction across the interface thermal contact resistance and contact Joule heating term deposited at the interface in layer 3. It is assumed, that the total contact Joule heating is split equally on both sides of the interface. The right side of Eq. (24) represents the second way to express heat transfer rate entering the interface and includes: (1) substrate heat conduction transfer rate written as the temperature difference across the substrate divided by substrate's thermal conduction resistance; (2) total Joule heating due to electrical contact resistance; and (3) Peltier contribution due to electric current flowing through the substrate electrode. It is assumed, that the bottom surface of the substrate is at ambient temperature. Thermal conduction resistance of the substrate, Θ_{subst}, can be determined by conduction shape factor. For instance, for sample of diameter d_3 on semiinfinite substrate with thermal conductivity κ_3 shape factor is $0.5\ \kappa_3^{-1} d_3^{-1}$.

Similar to Eq. (23), the fourth boundary condition is energy balance at the interface between probe and electrode:

$$-\kappa_1 A_1 \tfrac{dT_1}{dx_1}\big|_{x_1=0} + J_1\alpha_1 A_1 T_1\big|_{x_1=0} + h_2(A_2-A_1)(T_2\big|_{x_2=l_2}-T_0) =$$
$$\frac{T_2\big|_{x_2=l_2}-T_1\big|_{x_3=0}}{R_{th\_1-2}/A_1} + \frac{1}{2}J_2^2 A_2 R_{C\_1-2} = -\kappa_2 A_2\frac{dT_2}{dx_2}\big|_{x_2=l_2} + J_2\alpha_2 A_2 T_2\big|_{x_2=l_2} + J_2^2 A_2 R_{C\_1-2}. \tag{25}$$

Here, heat transfer rate exiting the interface (the left side of the equation) also includes convection from the top surface of the electrode to the ambient. Similar analysis is performed at the electrode-sample interface, as stated in (26):

$$-\kappa_2 A_2\tfrac{dT_2}{dx_2}\big|_{x_2=0} + J_2\alpha_2 A_2 T_2\big|_{x_2=0} = \frac{T_3\big|_{x_3=l_3}-T_2\big|_{x_2=0}}{R_{th\_1-2}/A_1} + \frac{1}{2}J_3^2 A_3 R_{C\_2-3} =$$
$$-\kappa_3 A_3\tfrac{dT_3}{dx_3}\big|_{x_3=l_3} + J_3\alpha_3 A_3 T_3\big|_{x_3=l_3} + J_3^2 A_3 R_{C\_2-3}. \tag{26}$$

Finally, continuity of electrical current in the layers of the sample requires:

$$J_1 A_1 = J_2 A_2 = J_3 A_3 = J_4 A_4. \tag{27}$$

Modeling approach discussed above can be used to study in detail effects on temperature profile due to thermal and electrical properties of individual layers and contacts.

Thermal conductivity of the thermoelectric film is typically determined from relationship between temperature rise (usually the measured surface temperature) and dissipated power. In addition, difference between the surface and substrate temperature together with Seebeck voltage developed across the film is used to calculate Seebeck coefficient of the film. Practitioners in thermoelectric field need a way to evaluate steady state surface temperature before electrical current is switched off for transient Harman method under nonideal boundary conditions.

The main strategy pursued here is to use the superposition principle to calculate the total temperature rise by solving separately for temperature solutions under Joule heating and Peltier effects. Rather than using full set of Eqs. (21)–(27), several assumptions are made in this section in order to arrive at an easy to use expression for the surface temperature, which still reflects the main thermoelectric transport mechanisms in many practical situations. These assumptions are: (1) electrode's contributions (layer 2) to thermoelectric transport are neglected because metallic electrode layers typically have low Seebeck coefficient similar to the probe and much lower electrical and thermal resistances compared to thermoelectric films; (2) Seebeck coefficient of the current probe is neglected; (3) convection terms on the film surfaces are neglected; (4) substrate thermal resistance and film-substrate electrode thermal contact resistances are neglected when compared to film thermal resistance, since thermoelectric film samples are typically low thermal conductivity films on high thermal conductivity substrates; and (5) Joule heating at film-substrate electrode contact is neglected because under assumption (4) the substrate acts as a heat sink.

Total temperature, T, in thermoelectric film is then divided into two components as: $T = T' + T^*$, where T' is linear temperature component, LTC, that is independent of Joule heating terms and includes Peltier effects, while T^* is nonlinear component, NLTC, and takes into account Joule heating effects, including electrical contact resistance heating. Then the set of Eqs. (21)–(27) for T' becomes:

$$-\kappa_3 A_3 \frac{d^2 T_3'}{dx_3^2} = 0, \tag{28}$$

$$T_3' = T_0; x_3 = 0, \tag{29}$$

$$-\kappa_3 A_3 \frac{dT_3'}{dx_3}\Big|_{x_3=l_3} + J_3 A_3 \alpha_3 T_s' = \frac{T_s' - T_w'}{R_{th\_1-3}/A_1} = q_w'; x_3 = l_3, \tag{30}$$

where T_s' and q_w' are temperature of the top surface of the sample and heat transfer rate through the probe, respectively. Heat transfer rate through the probe is calculated using a fin model with ambient temperature at free end and constant temperature T_w' at its base [16]:

$$q_w' = a(T_w' - T_0), \tag{31}$$

where the constant a is defined as:

$$a = \frac{\sqrt{h_1 P_1 \kappa_1 A_1}}{\tanh\left(\frac{l_1^2 h_1 P_1}{\kappa_1 A_1}\right)^{0.5}}. \tag{32}$$

Then, the solution for LTC of the top surface of the sample can be calculated as:

$$T_s' = \frac{(a_{th} l_3/A_3 + \kappa_3) T_0}{a_{th} l_3/A_3 + \kappa_3 - J_3 l_3 \alpha_3}, \tag{33}$$

where a_{th} is the total heat conductance through the contact and the probe defined as:

$$a_{th} = \frac{a}{1 + a R_{th\_1-3}/A_1}. \tag{34}$$

Next, T^* is calculated from the following equations:

$$\frac{d^2 T_3^*}{dx_3^2} + \frac{J_3^2 \rho_3}{\kappa_3} = 0, \tag{35}$$

$$T_3^* = 0; x_3 = 0, \tag{36}$$

$$-\kappa_3 A_3 \frac{dT_3^*}{dx_3}\Big|_{x_3=l_3} + J_3 A_3 \alpha_3 T_3^*\Big|_{x_3=l_3} + J_3^2 A_3 R_{C\_1-3} = \\ \frac{T_S^* - T_w^*}{R_{th\_1-3}/A_1} + \frac{1}{2} J_1^2 A_1 R_{C\_1-2} = q_w^*; x_3 = l_3. \tag{37}$$

Heat transfer along the probe is calculated by solving the fin model with volumetric Joule heating and a temperature rise equal to zero (relative to the ambient) at free end of the fin (away from the sample). The equation for the probe heat transfer is:

$$q_w^* = a T_w^* - b, \tag{38}$$

where the constant b is expressed as:

$$b = -a \frac{J_3^2 A_3^2 \rho_1}{h_1 P_1 A_1} \left[1 - \frac{1}{\cosh[(l_1^2 h_1 P_1 / \kappa_1 A_1)^{0.5}]} \right]. \tag{39}$$

Then, the solution for NLTC is given by:

$$T_S^* = \frac{J_3^2 l_3 \left(R_{C\_1-3} \frac{A_3}{A_1} + \frac{\rho_3 l_3}{2} \right) - \frac{l_3 b_{th} b_{cont}}{A_3}}{a_{th} l_3 / A_3 + \kappa_3 - J_3 l_3 \alpha_3}, \tag{40}$$

where,

$$b_{th} = \frac{b}{1 + a R_{th\_1-3}/A_1} \qquad b_{cont} = 1 + \frac{\frac{1}{2} J_3^2 \frac{A_3^2}{A_1} R_{C\_1-3}}{b} \frac{a R_{th\_1-3}}{A_1}. \tag{41}$$

Finally, total temperature of the top surface of the sample, T_s, is:

$$T_S = \frac{1}{a_{th} l_3 / A_3 + \kappa_3 - J_3 l_3 \alpha_3} \left[\begin{array}{l} (a_{th} l_3 / A_3 + \kappa_3) T_0 + \\ J_3^2 l_3 \left(R_{C\_1-3} \frac{A_3}{A_1} + \frac{\rho_3 l_3}{2} \right) - b_{th} b_{cont} \frac{l_3}{A_3} \end{array} \right], \tag{42}$$

where the first term contains Peltier effect's induced contributions to the surface temperature, the second term includes Joule heating effects from the sample and contact, and the third term includes Joule heating contribution from the probe wire.

Temperature of the probe at junction with the sample surface is then calculated as:

$$T_w = \frac{T_S + \frac{R_{th\_1-3}}{A_1} \left(\frac{1}{2} J_3^2 \frac{A_3^2}{A_1} R_{C\_1-3} + a T_0 - b \right)}{1 + a R_{th\_1-3}/A_1}. \tag{43}$$

Understanding how to eliminate or reduce the effects due to heat loss and electrical and thermal contact resistances is critical in designing test structures amenable for accurate thermoelectric transport measurements. Parasitic effects are expected to be different for macroscale versus microscale samples, and this section focuses on microscale samples. To illustrate these effects, the surface temperature predictions as a function of current density are discussed for thermoelectric sample of $10 \times 10 \times 10$ μm^3 in contact with copper probe of 5 μm diameter and 1.3 mm length. Thermoelectric properties of thermoelectric film are similar to n-type $Bi_2Te_{2.7}Se_{0.3}$ and are listed in **Table 2**.

Material	Probe	Sample	Electrode
	Cu	$Bi_2Te_{2.7}Se_{0.3}$	In
Lateral dimension (μm)	76.2[a]	1014[b]	1014[b]
Length (mm)	20	1.8	0.05
Seebeck coefficient ($\mu V/K$)	1.84	−212	1.68
Thermal conductivity (W/m K)	401	1.5	82
Convection heat transfer coefficient (W/m^{2}K)	200[c]	10	10
Electrical resistivity (nOhm m)	17.1	10500	84

a = Diameter, b = Width, c = The convection heat transfer coefficient for 5 μm probe is 3000 (W/m^{2}K).

Table 2. Sample parameters and thermal/thermoelectric properties. Reprinted from [21] with the permission of AIP Publishing.

Figure 10 shows rise of surface temperature with respect to ambient temperature, calculated from Eq. (42) for a range of specific thermal contact resistances. Electrical resistance of contact was assumed to be equal to theoretical limit predicted for electrical boundary resistance between Bi_2Te_3 and metal electrode [25]. Direction of electrical current was chosen such, that the sample surface undergoes Peltier cooling. At low current densities, Peltier cooling term dominates over Joule heating terms and temperature of the top surface of the sample decreases linearly as electrical current density increases. After reaching the maximum cooling temperature at optimum current density, Joule heating terms start to dominate over Peltier terms. Parasitic conduction heat transfer effect is apparent even at very low current densities, as shown by inset in **Figure 10**. It leads to reduction of temperature difference across the sample as compared to predictions of an ideal Harman model. On the other hand, as thermal contact resistance increases, the sample cooling is stronger because the thermal barrier created at the contact reduces heat transfer rate with the probe. Importance of thermal barrier effect is gauged by comparison between thermal resistances of the probe, probe-sample contact, and the sample itself. The modeled probe has thermal resistance of ~5 × 10^4 K/W, which is similar to 6 × 10^4 K/W thermal resistance of the sample; therefore, a significant heat transfer occurs through the probe. As the thermal contact resistance increases, the probe heat transfer is reduced, particularly after contact thermal resistance becomes of the same order as thermal resistance of the probe. Alternative way to minimize the probe heat transfer rate could be realized by reducing diameter of the probe. However, besides practical challenges, this may have a negative impact associated with increase in resistance of electrical contact, as shown in **Figure 10**.

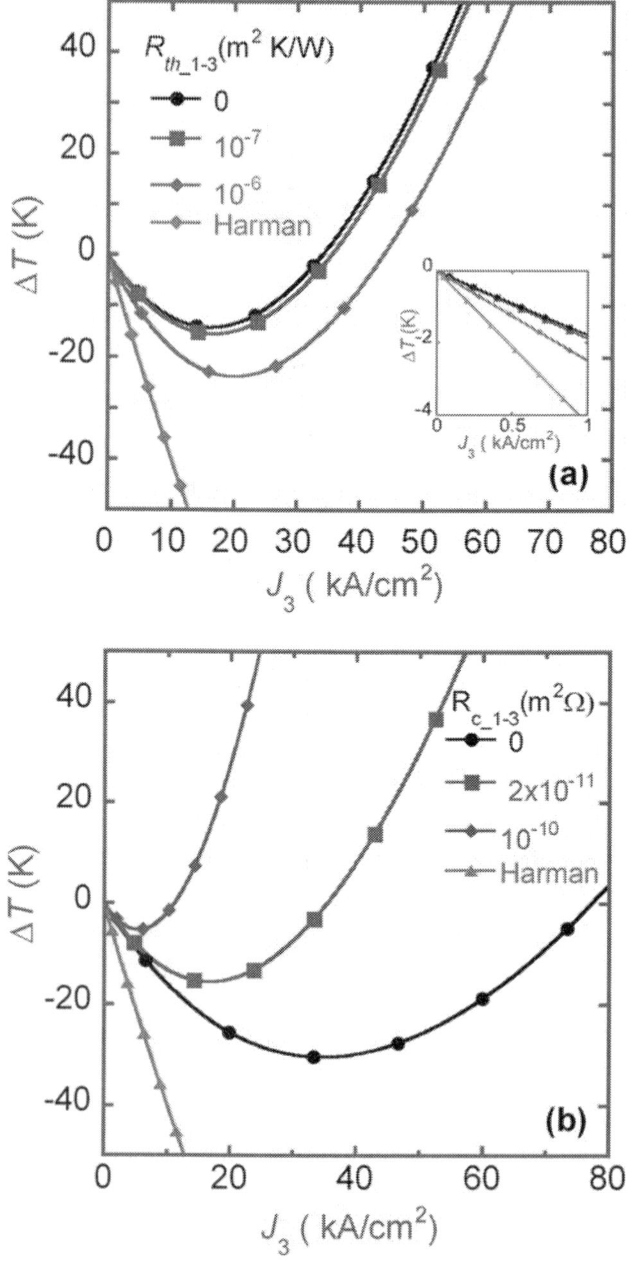

Figure 10. Calculated temperature response for specified R_{th} and R_c. Reprinted from [21] with the permission of AIP Publishing.

Heat transfer through the substrate could also play a major role in establishing the surface temperature. For large current densities, the strength of heat transfer through the substrate is indicated by large positive temperature difference measured across the sample. This difference is in contrast with the predictions of an ideal Harman model, where Joule heating effects never generate a temperature difference across the sample.

The effect of electrical contact resistance in absence of thermal contact resistance is investigated in **Figure 10**. The modeled probe has electrical resistance of ~1 Ohm, which is similar to electrical resistance of the sample, therefore Joule heating effects can occur simultaneously in the sample and the probe. In addition, even for very low specific electrical contact resistance of 1×10^{-11} Ohm $\times$ m^2, electrical contact resistance for the modeled probe is considerable (0.5 Ohm). Electrical contact resistance increases by two orders of magnitude if the probe diameter is reduced by a factor of 10. This illustrates strong requirements to control the probe and contact electrical resistances in transient Harman experiments performed on film on substrate samples. This is because good thermoelectric samples have low electrical resistances, so Joule heating effect in contacts and probe can easily become dominant. Inspection of the probe and wire Joule heating terms suggests mitigation of the electrical contact resistance problem may be achieved by preparing thick film samples (large l_3) for measurements, but with small cross-sectional area (A_3) in order to increase the relative importance of electrical resistance of the sample. Since this strategy leads to an increase in thermal resistance of the sample, one must simultaneously address the need to mitigate parasitic probe heat conduction effects by choosing a probe with thermal resistance larger than that of the sample. The probe's thermal resistance may be calculated by knowing the material properties and diameter of the probe.

One proposed strategy for determining the properties under nonideal conditions is the bipolar method, where transient Harman experiments are performed using direct and reversed current directions and where measured Seebeck and resistive voltages across the sample are averaged. This is believed to eliminate Joule heating effects and reveal the intrinsic Peltier effects in the sample [21, 23]. However, as demonstrated in this work, when nonideal boundary conditions are present, parasitic effects cannot be always completely eliminated by this strategy. Nevertheless, the analysis below demonstrates the ability to exploit this behavior to determine both thermal and electrical transport properties of the samples and their contacts.

Bipolar resistive voltage difference measured across the sample $\Delta V_{\rho\pm}$ is related to the total electrical resistance through the expression:

$$\frac{V_{\rho+}-V_{\rho-}}{J_3 A_3} = \frac{\Delta V_{\rho\pm}}{J_3 A_3} = 2\left(R_{C\_1\text{-}3}/A_1 + R_{C\_3\text{-}4}/A_3 + \frac{\rho_3 l_3}{A_3}\right). \tag{44}$$

To find expression for bipolar Seebeck voltage, first Seebeck coefficient, α_s, is expressed as measured V_S as a function of the surface temperature of the sample as:

$$\alpha_S = \frac{-V_S}{T_S - T_0}, \tag{45}$$

where V_s is experimental Seebeck voltage measured between the probe and electrode situated at the bottom of the sample. If the probe temperature is measured, then:

$$\alpha_S = \frac{-V_S}{(T_w-T_0)\left(1 + \frac{aR_{th\_1-3}}{A_1}\right) + \frac{R_{th\_1-3}}{A_1}\left(b - \frac{1}{2}J_3^2 A_3^2 \frac{R_{C\_1-3}}{A_1}\right)}. \tag{46}$$

When bipolar method is used, then α_s can be extracted from bipolar Seebeck voltage difference $\Delta V_{S\pm}$ and temperature difference $\Delta T_{w\pm}$ using expression:

$$\alpha_S = \frac{-\Delta V_{S\pm}}{\Delta T_{S\pm}} = \frac{-\Delta V_{S\pm}}{\Delta T_{w\pm}\left(1 + \frac{aR_{th\_1-3}}{A_1}\right)}. \tag{47}$$

Next, under small current approximations:

$$\kappa_3 + a_{th}\frac{l_3}{A_3} \gg J_3 l_3 \alpha_3 \quad \& \quad T_0\left(\kappa_3 + a_{th}\frac{l_3}{A_3}\right) \gg J_3^2 l_3 (R_{C\_1-3}\frac{A_3}{A_1} + \frac{\rho_3 l_3}{2}) - \frac{l_3 b_{th} b_{cont}}{A_3}, \tag{48}$$

a simplified expression for bipolar surface temperature difference is obtained as:

$$\Delta T_{S\pm} = \frac{2T_0}{\kappa_3 + a_{th}l_3/A_3} J_3 l_3 \alpha_3. \tag{49}$$

Expression for ZT obtained through bipolar technique is found as:

$$ZT_0 = \frac{-\Delta V_{S\pm}}{\Delta V_{\rho\pm}}\left(1 + \frac{a_{th}l_3}{\kappa_3 A_3}\right) \times \left(1 + \frac{R_{C\_1-3}l_3}{\rho_3 A_3 A_1} + \frac{R_{C\_3-4}l_3}{\rho_3 A_3{}^2}\right). \tag{50}$$

Another strategy is to perform experiments over a range of currents and use differential changes in V_S and V_ρ with current I. Under small current conditions, the following expression is then obtained for the figure of merit:

$$ZT_0 = \frac{-dV_S/dI}{dV_\rho/dI}\left(1 + \frac{a_{th}l_3}{\kappa_3 A_3}\right) \times \left(1 + \frac{R_{C\_1-3}l_3}{\rho_3 A_3 A_1} + \frac{R_{C\_3-4}l_3}{\rho_3 A_3{}^2}\right). \tag{51}$$

Neither bipolar nor differential current methods alone are able to account for all parasitic effects. As a result, these effects must be considered in data reduction or otherwise minimized. A variable thickness method [21, 23, 24] is used to account for electrical contact resistance effects, while heat losses and thermal resistance effects are neglected. A different method to determine all thermoelectric properties without the need for extensive sample preparations is outlined below.

The strategy explored here is to use bipolar experiments performed over a wide range of currents rather, than small current regime required by above methods. It is expected, that at large currents, experimental Seebeck voltage and temperature signals become sensitive to electrical transport properties of the sample and contacts and could be used to determine the sample and contact thermoelectric properties. In addition to Seebeck and resistive voltage drops, method requires measurement of the sample surface temperature or the probe temperature (at the contact with the samples surface).

Proposed strategy takes into consideration selective sensitivity of thermal signals to Peltier and combined Peltier and Joule heating effects under low and high current regimes, respectively. Under small current approximations, temperatures of the probe and sample surface are linear with current, and thermal conductivity can be expressed as a function of experimentally measured slope of the probe temperature as:

$$\kappa_3 = \frac{l_3}{A_3}\left[\frac{\alpha_3 T_0}{\left(1 + \frac{aR_{th\_1\text{-}3}}{A_1}\right)dT_w/dI} - a_{th}\right].$$
(52)

In Eq. (52), value of Seebeck coefficient is substituted from Eq. (47), which is valid at any current. Next, Eqs. (44), (47), and (52) are substituted in Eq. (43). For the sake of discussion it is assumed, that specific electrical contact resistance is similar at the top and bottom contacts (other assumptions are discussed in Section 3). After the above substitutions, predicted probe temperature and Seebeck coefficient become a function of two unknowns, specific electrical and thermal contact resistances, which are then used to fit experimental signals under large current regime for both direct and reverse currents. This strategy allows the unique determination of all thermoelectric properties of the sample and electrical and thermal contact resistances. Details of the fitting procedure are presented in experimental validation section.

2.3.2. Lock-in Harman technique–analytical model

To find *frequency-dependent* temperature solution in the sample, governing equation and boundary conditions for the problem were first expressed as a function of time and then transformed to frequency domain. Governing thermal transport equation for the sample with attached wires was derived by balancing the energy in infinitesimal length *dx* of wire or sample domain treated as Joule heated fin and neglecting Thompson effects. Approach is similar to the steady state model [25]. The governing equation is:

$$\frac{\partial^2 T_i(x,t)}{\partial x_i^2} - m_i^2(T_i - T_0) + \frac{J^2 \rho_i}{2\kappa_i} = \frac{1}{\Delta_i}\frac{\partial T_i(x,t)}{\partial t}.$$
(53)

The first term on the left side accounts for conduction through the wire/sample, where T_i is temperature along length of the wire/sample as a function of position, x, and time, t. The second term represents convective heating or cooling from the environment, where $m_i^2 = h\,p/(\kappa\,A)$, with h being heat transfer coefficient, p wire/samples circumference, A its cross-sectional area, and T_0 ambient temperature. The third term accounts for Joule heating, and the right side of equation is transient heat storage term. J is electrical current density, and Δ is thermal diffusivity.

Figure 11 represents the same configuration of sample and wires as considered before, but here shows heat transfer domains used in the model. 1D heat transfer was modeled in each of seven domains, one each for the six wires plus another for the sample. Details of the boundary conditions are given below. Temperature solution requires a total of 14 boundary conditions, two for the sample and four for each set of wires. The wire boundary conditions are as follows: two boundary conditions per wire (or six total) were defined by assuming, that the end of each wire was at room temperature, since the wires in the experiment were relatively long compared to their width and measurements were conducted in ambient conditions. The remaining two boundary conditions per wire (summing to twelve in total) are that the ends of each wire in contact with the sample are at a fixed temperature. The two boundary conditions across the sample come from the fact, that heat transfer at the interfaces must be balanced. For the interface between the first set of wires (domains 1–3) and the sample (domain 4), the energy balance yields Eq. (54):

$$q_{1\text{-}3} + J_4^2 A_4 R_{14} + J_4 \alpha_{34} A_4 T_4(0, t) = -\kappa_4 A_4 \frac{\partial T_4(0, t)}{\partial x_4} \tag{54}$$

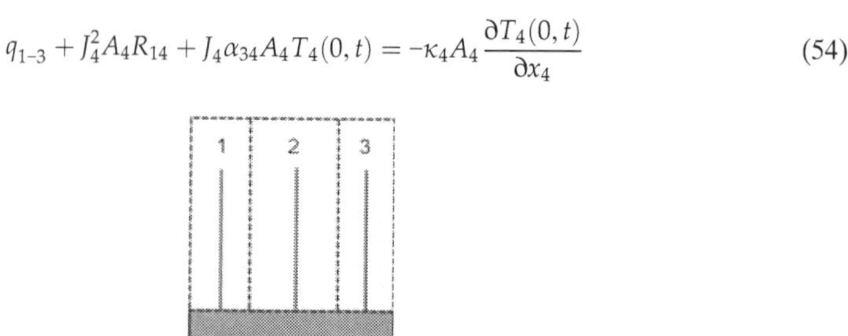

Figure 11. Schematic representation of the sample showing heat transfer domains.

The first term is the rate of heat conduction to and from the wires and is a function of temperature gradient at the wire-sample interface and thermal conductivity of the wires. This is calculated using the pin fin equation below for the experimental results. The second term accounts for Joule heating due to electrical contact resistivity between the current lead and the sample, R_{14}. The third term is Peltier heating at the interface, where $\alpha_{34} = \alpha_3 - \alpha_4$, relative Seebeck coefficient between the wire (3) and the sample (4). The right side of the equation is the heat conducted through the sample. Form of boundary condition for the other wire-sample interface is identical. To transform the problem from a partial differential equation in time to an ordinary differential equation in frequency, ω, we used Fourier transform. Before applying Fourier transformation, it was convenient to represent the sample temperature as a Fourier series. This was possible since temperature is a function of periodic excitation signal and is therefore itself periodic. Temperature as the sum of its DC component ($\eta = 0$) and all of the harmonics (all other values of η) of fundamental frequency, ω_0, is given by Eq. (55):

$$T_i(x,t) = \sum_{-\infty}^{\infty} T_i(x, n\omega_0) e^{i\eta\omega_0 t}. \tag{55}$$

Substituting this into governing equation and applying Fourier transform gives transformed governing equation:

$$\frac{\partial^2 T_i(x,\omega)}{\partial x_i^2} - m_i^2 T_i(x,\omega) + m_i^2 T_0 \sqrt{2\pi}\delta(\omega) + \frac{J^2 \rho_i}{8\kappa_i}\sqrt{2\pi}\left(2\delta(\omega) + \delta(\omega-2\omega_0) + \delta(\omega+2\omega_0)\right) = \frac{i\omega}{\Delta_i} T_i(x,\omega). \tag{56}$$

Dirac delta function, δ, is employed because signals of constant frequency in the time domain become delta functions in frequency domain. Conduction, convection, and heat storage terms are present at each harmonic with additional convection term present at $\omega = 0$. Joule heating occurs only at $\omega = 0$ and $\omega = 2{\cdot}\omega_0$. The transformation was next applied to the boundary conditions. The transformed boundary condition for the first interface is given by Eq. (57):

$$q_{1-3} + \frac{J_4 A_4 R_{14}}{2}\sqrt{2\pi}\left(2\delta(\omega) + \delta(\omega-2\omega_0) + \delta(\omega+2\omega_0)\right)$$
$$+ \frac{J_4 S_{34} A_4}{2}\sum_{-\infty}^{\infty} T_4(0, n\omega_0)\sqrt{2\pi}\left(\delta\left(\omega-\omega_0(n+1)\right) - \delta\left(\omega-\omega_0(n-1)\right)\right) = -\kappa_4 A_4 \frac{\partial T_4(0,\omega)}{\partial x_4}. \tag{57}$$

Joule heating term is again present at $\omega = 0$ and $2{\cdot}\omega_0$, while Peltier heating term occurs only at the fundamental frequency. This demonstrates mathematically how Joule and Peltier components of heat transfer are separated by measuring the harmonics of temperature. All measurements in this work use Peltier component. Joule component is not used in measurements described here. Since this work focuses on the first harmonic measurements, the solution for the first harmonic and its derivative are given by Eq. (58) and Eq. (59), where R is the root of homogenous form of governing equation and is given by Eq. (60):

$$T_i(x_i, \omega) = c_{i1,1\omega} e^{R_{1\omega}x} + c_{i2,1\omega} e^{-R_{1\omega}x}, \tag{58}$$

$$\frac{\partial T_i(x_i, \omega)}{\partial x_i} = c_{i1,1\omega} R_{1\omega} e^{R_{1\omega}x} - c_{i2,1\omega} R_{1\omega} e^{-R_{1\omega}x}, \tag{59}$$

$$R = \sqrt{m^2 + \frac{i\omega}{\Delta}}. \tag{60}$$

The undetermined coefficients were solved numerically using transformed boundary conditions.

2.3.3. Experimental results–transient Harman

Figure 12 shows experimentally measured total and Seebeck voltages as a function of current.

Resistive voltage drop obtained after subtracting Seebeck voltage from total voltage includes contributions from the sample, probe-sample contact, and sample-substrate electrode contact. Inset in **Figure 12** shows an example of measured voltage as a function of time during an experiment at 163 mA. The figure of merit calculated according to classical Harman method yields an average value of 0.11, much smaller than the manufacturer value of 0.85. This discrepancy is due to parasitic effects neglected in classical technique.

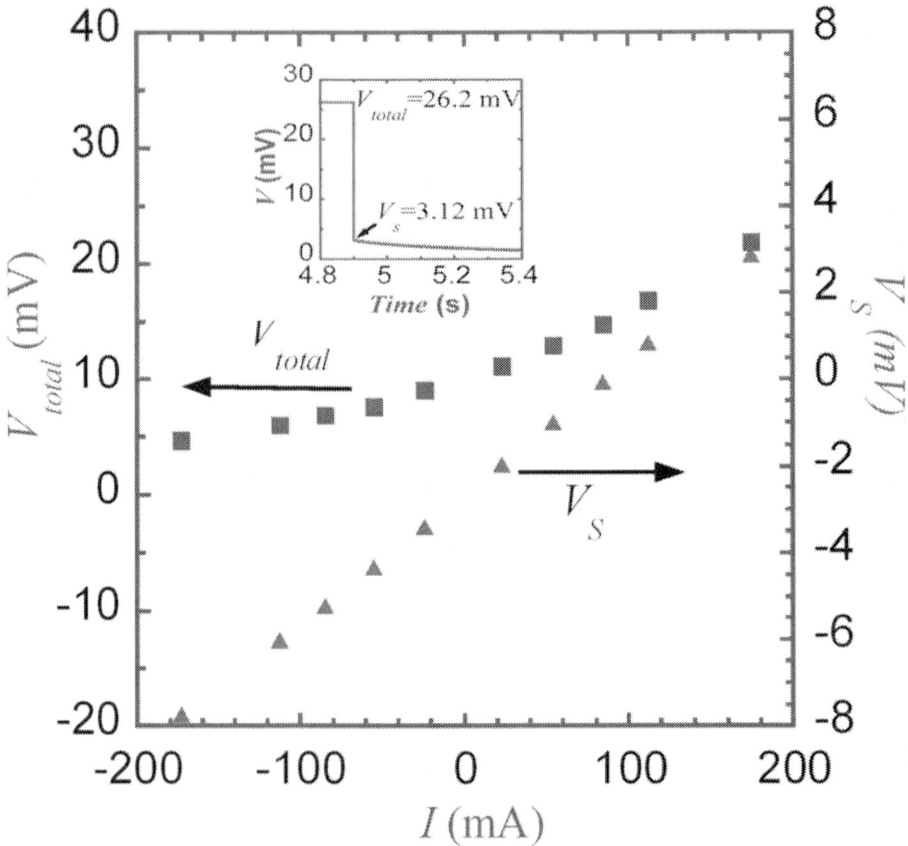

Figure 12. The deconvolution of resistive and Seebeck voltage contributions. Reprinted from [21] with the permission of AIP Publishing.

Measured probe temperature is linear with current for the smallest two bipolar currents and yields a slope of −55.1 K/A and was substituted in Eq. (52). Originally, only one probe was modeled in contact with the sample, the energy loss across constantan wire was evaluated and found to be in an order of magnitude smaller than for copper wire, therefore its effect is expected to be negligible.

Transient Harman experiments performed under the highest direct and reverse current conditions were used in the fitting of sample's thermoelectric properties and contact resistances. In the fitting procedure, thermal contact resistance $R_{th\_1\text{-}3}$ was varied over a wide range, typically between 1.0×10^{-8} m^2 × K/W and 1.0×10^{-5} m^2 × K/W with step of 1.0×10^{-8} m^2 K/W. For each $R_{\text{th\_1-3}}$ value, sample's Seebeck coefficient was determined from Eq. (47) and sample's thermal conductivity was determined from Eq. (52). In addition, two sets of solutions for electrical contact resistance as function of $R_{\text{th\_1-3}}$ are generated by fitting experimental Seebeck coefficient and the probe temperatures. The first set of solutions is obtained by minimizing the mean square deviation between Seebeck coefficient values calculated from Eqs. (46)–(47). The second set of solutions is obtained by minimizing the mean square deviation between experimental temperatures of the probe and the predictions of Eq. (43). In Eq. (43), T_S was substituted from Eq. (42), thermal conductivity was substituted from Eq. (52), Seebeck coefficient from Eq. (47) and resistivity of the sample from Eq. (41). It was found, that the first set of solutions is more sensitive to thermal contact resistance, while the second set was more sensitive to electrical contact resistance. The intersection of the two sets of solutions leads to the unique solution for thermal and electrical contact resistances and thermoelectric properties of the sample.

Since the original sample-substrate interface from commercial Peltier device has a negligible electrical contact resistance compared with the sample, $R_{C,3-4}$ was taken as zero in Eq. (41). It was also assumed, that $R_{c,1-3}$ and $R_{\text{th},1-3}$ represent respectively lump electrical and lump thermal contact resistance contributions due to the probe-indium interface, indium layer (negligible contribution), and indium-sample interfaces. However, measured resistive voltage drop through the constantan probe does not include electrical resistance contribution due to the probe-indium interface. Therefore, the fitting procedure was repeated several times under four different assumptions regarding contact electrical resistance. The assumptions and the fitting results are presented and discussed below.

The thermoelectric properties arise from each of the following four cases are summarized in **Table 3**. In case 1, electrical contact resistance R_{c_1-3} was assumed to originate only from the probe-indium interface. In case 2, electrical contact resistance R_{c_1-3} was assumed to originate only from the indium-sample interface. In case 3, electrical contact resistance R_{c_1-3} was assumed to split equally between probe-indium and indium-sample interfaces. Finally, in case 4, manufacturer's value for sample resistivity $\rho_3 = 1 \times 10^{-5}$ Ohm × m, was used in the fitting process, which allowed the exact determination of the split of probe-sample electrical contact resistance between two contributions. In this case, $R_{c,1-2} = 4.9 \times 10^{-10}$ Ohm × m^2 may be split as 1.1×10^{-10} Ohm × m^2 due to the probe-indium interface and 3.8×10^{-10} Ohm × m^2 due to indium-sample interface. Relatively large values of measured electrical and thermal contact resistances are due to the imperfections of the mechanical contact under the small contact load used in this proof-of-concept experiment. The indium-pellet interface dominates the contact resistance.

The highest deviations between measured sample's Seebeck and thermal conductivity as compared with manufacture's values are respectively 6 µV/K (3%) and 0.12 W/(m × K) (8 %). These deviations are smaller than the experimental uncertainty. The uncertainty in thermal conductivity due to propagation of the uncertainty in temperature and voltage measurements was calculated to be 0.26 W/(m × K). Similarly, for Seebeck coefficient, the uncertainty is equal to 9.9 µV/K. To accurately determine the sample resistivity, resistive voltage drop should be measured through the copper probe, which is used to pass electrical current, at the same time as through the constantan wire, so total electrical contact resistance and sample resistivity can be accurately determined. When correct resistivity of the sample was employed in the fitting, the sample thermal conductivity was within 3 % of manufacture's values.

Case #	$R_{th,1\text{-}3}$ (K/W)	κ_3 (W/mK)	α_3 (μV/K)	$R_{c,1\text{-}3}$ (m^{2}Ohm)	ρ_3 (Ohm m)
1	1.1×10^{-6}	1.38	−212	2.9×10^{-10}	5.7×10^{-5}
2	6.6×10^{-7}	1.48	−218	6.2×10^{-10}	Nonphysical (negative)
3	9.4×10^{-7}	1.43	−215	3.9×10^{-10}	3.3×10^{-5}
4	7.8×10^{-7}	1.46	−217	4.9×10^{-10}	1×10^{-5} (manufacturer specified)

Table 3. Summary of cases simulated to explore effect of contacts in transient Harman measurements.

Figure 13 shows comparison between measured and calculated temperature of the probe as a function of electrical current passed through it.

The theoretical predictions use the fitted thermoelectric properties and employ Eq. (43) with either all terms or only Peltier terms. The theoretical predictions were performed for all cases 1–4 and, since they superpose along the same line, they are not individually distinguishable in **Figure 13**. Joule heating effects are important at large currents in tested sample, as demonstrated by the discrepancy between Peltier heating only predictions and combined Peltier and Joule heating model. Predictions based on solving Eq. (20)–(27), that also include the convection on the sample surface and the contributions from indium electrode, show no significant difference with prediction based on Eq. (43). There is an excellent agreement between experimental and modeling data over entire electrical current range. Data sets for intermediate current values (~85 mA) show also excellent agreement, although they have not been used in the fitting.

2.3.4. Experimental results–lock-in Harman

ZT and individual thermoelectric properties may also be characterized experimentally using a Nyquist plot (plotting imaginary vs real parts of the complex voltage signal or sample temperature rise). Nyquist analysis of voltage measurements across the sample allows for direct calculation of the slope m, which, in turn, yields α. ZT is obtained from finding each of V_R and V_S from the different regimes represented in the plot (see **Figure 14**). V_R is obtained from the value of the real part of the voltage response when the imaginary part is equal to zero, and V_S is the radius of the circular portion of the Nyquist plot.

The samples measured were bulk bismuth telluride alloys with dimensions of 4.5 × 3.8 × 3.8 mm^3. Thin layer of gold was deposited on either end to improve adhesion and current spreading between the sample and lead wires. One lead wire and one thermocouple were soldered to either end of the sample. Current was applied through un-insulated 50.8 μm diameter copper wire, and voltage was measured using 50.8 μm E-type thermocouples. Two sets of voltage measurements were made across the sample using each set of thermocouple wires for excitation frequencies between 10 mHz and 10 Hz. Amplitudes of resulting voltages, E_{13} and E_{24}, are plotted in **Figure 14**. When the signal is applied at low frequencies, then measured voltage is sum of total voltage across the sample and Seebeck voltage in thermocouple's wires. As frequency is increased, thermal component in the sample and wires decays and voltage approaches resistive voltage of the sample, and E_{13} approaches E_{24}.

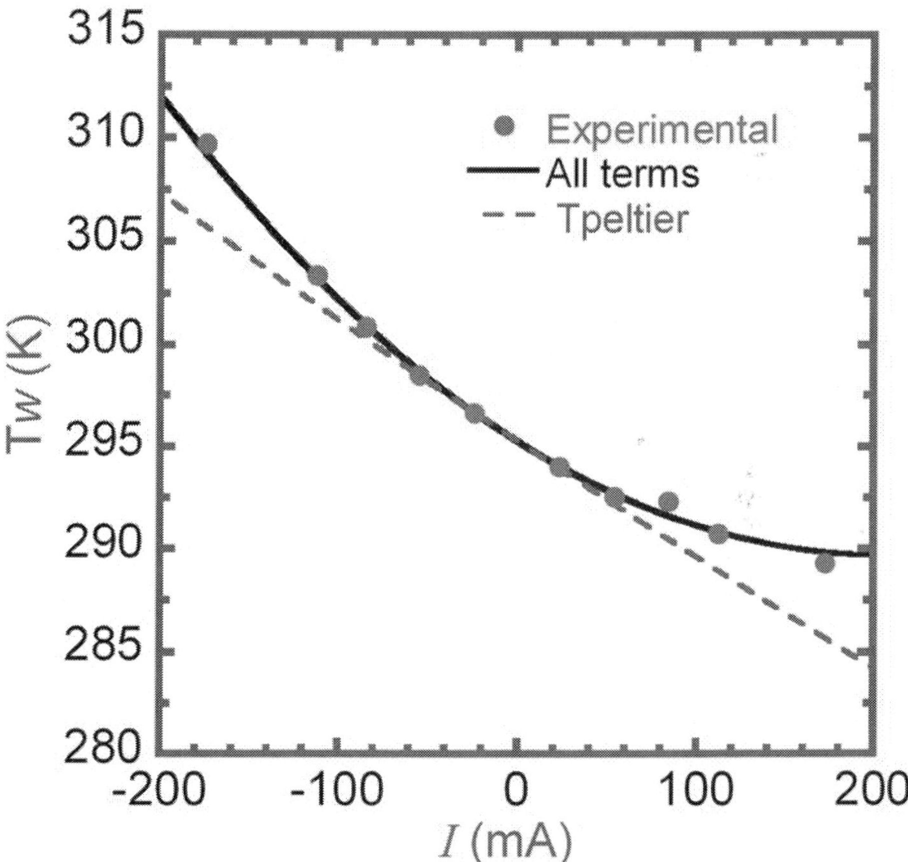

Figure 13. Measured versus calculated probe temperature. Reprinted from [16] with the permission of AIP Publishing.

These two voltages were used to find Seebeck coefficient by Ivory's technique [22] using Eq. (21). Non-imaginary values of two voltages are plotted against each other in **Figure 15** and resulting in Seebeck coefficient α_{sample} = 202.6 ± 1.4 μV/K. Real parts of signals are used, as these are components, that are in phase with excitation signal and as a result are in phase with each other. Amplitudes may be out of phase with each other and imaginary part is much smaller. Advantage of Ivory technique is that the magnitudes of measured voltages are greater than in traditional technique, if α_{sample} is larger than average value of α_a and α_b, which is often the case, when measuring thermoelectric materials, because voltage measured across the sample is equal to $\Delta T(\alpha_a-\alpha_{sample})$, whereas that measured in thermocouples is $\Delta T(\alpha_a-\alpha_b)/2$. This assumes, that the sample is symmetric and that ΔT is total temperature gradient across the sample. Thus, temperature gradient measured by one set of thermocouples is $\Delta T/2$. Since temperatures on the two sides of the sample are 180° out of phase, the total temperature difference is twice the temperature amplitude registered on one side. The larger signal results in better signal-to-noise ratios and less error in the final calculation.

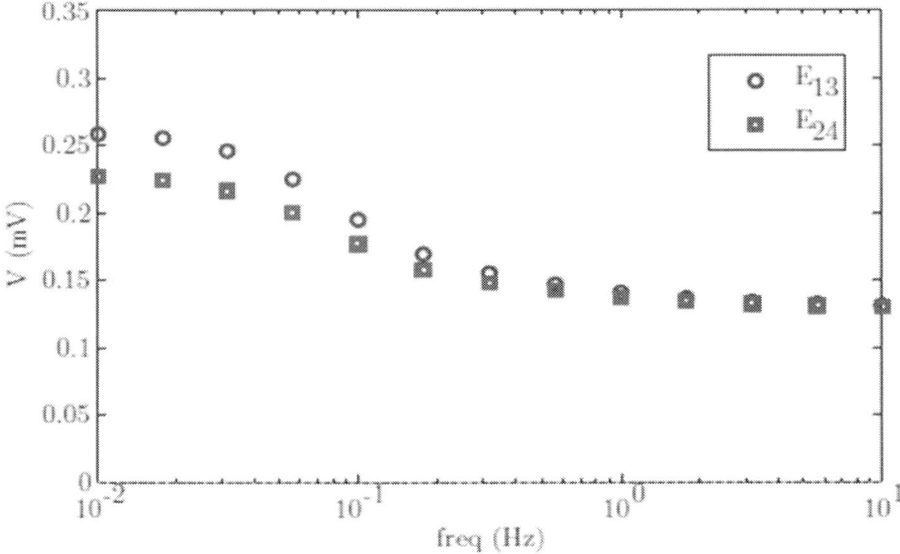

Figure 14. E_{13} and E_{24} voltages as a function of frequency.

Total voltage in the sample V_{sample} was calculated using Eq. (61) and then plotted on Nyquist diagram, shown in **Figure 16**:

$$V_{sample} = \frac{\alpha_b E_{13} - \alpha_a E_{24}}{\alpha_b - \alpha_a}. \tag{61}$$

Obtained data is again shown as superposition of resistive V_R and Seebeck V_S voltages, and V_{sample} and V_R can be found by extrapolating the data to the real axis as described in introduction. High frequency behavior of real devices may not obey the −45° assumption, if contacts have significant heat capacity [25]. As seen in **Figure 16**, behavior of the sample deviates somewhat from −45°, which can be attributed to heat capacity of the solder between the wires and the sample. From **Figure 16**, V_{sample} is 0.17 mV and V_R is 0.088 mV. Extrinsic ZT of the device is 0.93 for this measurement. To find intrinsic ZT of the material, then nonidealities in measurement system must be accounted for.

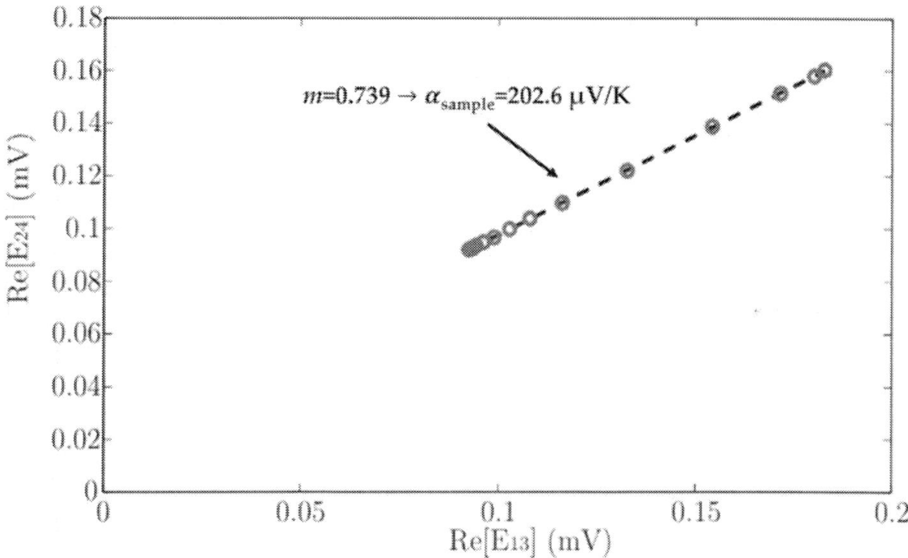

Figure 15. Real parts of E_{24} and E_{13} to find the slope, m, and Seebeck coefficient.

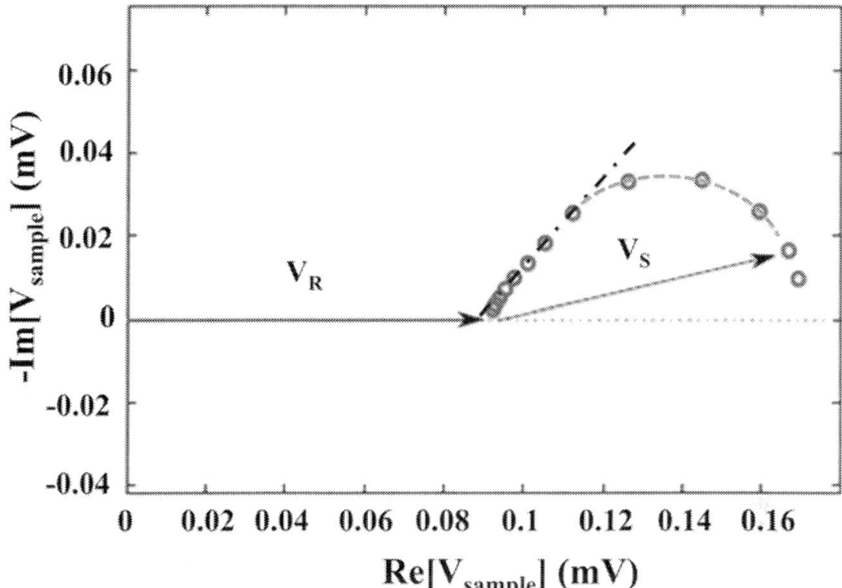

Figure 16. Voltage measurements plotted on Nyquist diagram.

To find equation for intrinsic ZT, the derivation of ZT performed by Harman can be repeated to include terms for heat loss and contact resistance. This adds two correction factors as shown in Eq. (62). The first is the ratio of heat lost from the sample to Peltier heat generation at the wire-sample interface. This heat loss may be due to either convection or radiation from the end of the sample or conduction through the contacts. The second correction factor is the ratio of voltage drop across the contacts to that in the sample, which is equal to resistance of the contacts divided by that of the sample:

$$\mathrm{ZT} = \frac{V_S}{V_R}\left[\left(1-\frac{q_{\mathrm{loss}}}{q_{\mathrm{Peltier}}}\right)\left(1-\frac{V_C}{V_R}\right)\right]^{-1}. \tag{62}$$

For this experiment, radiation and convection from the sample itself were negligible compared to heat loss through the contacts and only the latter was considered. Since ZT was calculated using voltages approximating DC, where heat transfer is in steady state, and high frequency AC, which is not affected by heat losses, steady-state equation can be used to account for heat loss. Conduction through the wires is described by Eq. (63), which is the fin equation, where the base temperature is equal to that of the wire-sample interface, T_s, and T_0 is the ambient temperature:

$$q_w = \sqrt{hp\kappa A}(T_S-T_0)\tanh\left(\sqrt{\frac{hP}{\kappa A}}L\right). \tag{63}$$

For long thin wires, the hyperbolic tangent goes to one and may be neglected. We considered losses by convection through the wire and temperature at the end of the wire away from the sample was assumed to have reached ambient (since wires were long and thin). Even though, the sample temperature was not measured directly, the temperature gradient across the sample was found from measured Seebeck coefficient value, α_{sample}, and Seebeck voltage, V_S. Once T_s was found this way and substituted in Eq. (63), q_w was determined for each of two wires, summed up and the value used in Eq. (62). In the equation, $q_{\mathrm{Peltier}} = \alpha\mathrm{IT}$. After all these substitutions, intrinsic ZT of the material was calculated as 1.04.

Samples resistivity was determined as 7.0×10^{-6} Ohm $\times$ m based on resistive voltage, V_R, and neglecting the contribution of the contacts. This was confirmed in independent measurement probing resistive voltage profile along the length of the sample, which resulted in value of 7.25×10^{-6} Ohm $\times$ m. Using intrinsic ZT, measured Seebeck coefficient and the first value of resistivity, thermal conductivity of 1.55 W/(m $\times$ K) is obtained, while if the second resistivity value is used κ is determined as 1.6 W/(m $\times$ K).

The same material properties were also found by fitting the predictions based on the numerical model described above in Eqs. (58)–(60) to experimental data. The fitting is shown in **Figure 17**, where Seebeck voltage data was converted to temperature amplitude using measured Seebeck coefficient. Adjusting thermal conductivity in the model and using the least squares fit, thermal conductivity was 1.55 W/(m $\times$ K) and thermal diffusivity was 9.5×10^{-3} cm^2/s.

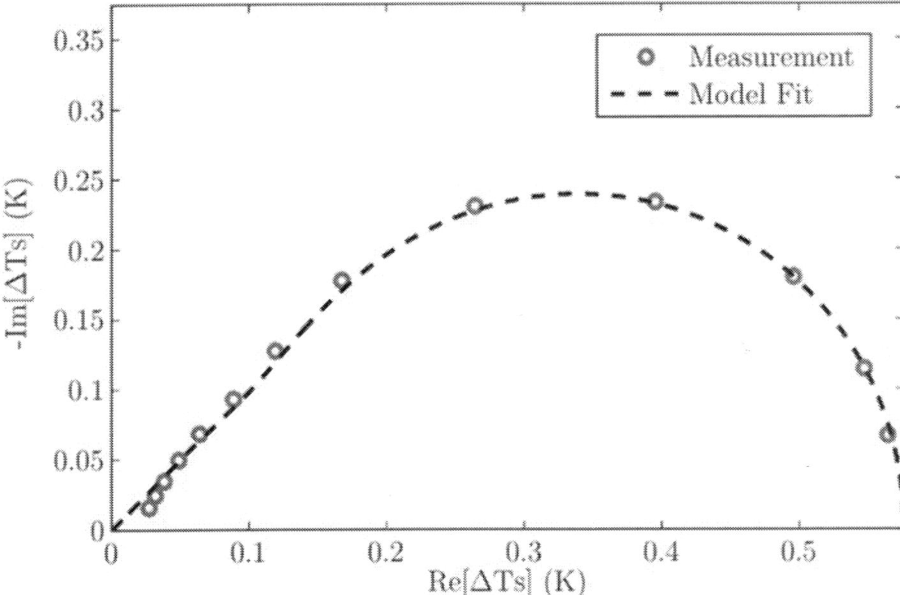

Figure 17. Sample temperature and fitted prediction plotted on Nyquist diagram.

Overall, calculated uncertainty in this experiment was small with that in Seebeck coefficient and extrinsic ZT measurements being lower, than that of calculated intrinsic ZT. The latter is due to the uncertainty in calculating the heat loss through the wires, specifically calculating the heat transfer coefficient between the wires and the air. The heat transfer coefficient was calculated assuming a horizontal cylinder in air. While the uncertainty for the heat transfer coefficient was determined to be about 2%, its uncertainty was assumed to be closer to 10%. This was done because there is some additional uncertainty surrounding the assumptions of free convection and due to its dependence on lab conditions. The uncertainty could be improved by testing the sample in an evacuated chamber, eliminating entirely the need to calculate heat transfer coefficient. The uncertainty of extrinsic ZT is due to that of voltage measurements, and was assumed to be 1% of the measured value, and temperature, assumed to be 296 ± 2 K. The uncertainty in voltage measurement was assumed 1% as conservative estimate. The error of the device was lower, but noise in the system and variation between measurements was closer to 1%. Some error in Seebeck coefficient measurement will be present due to the assumption, that temperature gradients across all the wires in each thermocouple are identical. Since the junctions of thermocouples were somewhat embedded in solder, there may be slight temperature gradient between two wire-solder interfaces. However, this difference was assumed to be negligible compared to temperature gradient along lengths of wires, and the uncertainty in Seebeck coefficient measurement was calculated as less than 1%. The uncertainty in V_{sample} was similarly low, while that in determining V_R from the sample voltage was calculated to be about 2–3%.

These were calculated using error propagation from the uncertainty of the measurements. For extrinsic value of ZT, the uncertainty was calculated as 2.7%. With addition of the uncertainty in the heat loss calculation, that for intrinsic ZT increased to 5.9%.

3. Verification strategies for measurements

3.1. Slope-efficiency method: rapid measurement of device $ZT_{maximum}$.

Maximum electrical power output, P_{max}, of any thermoelectric generator (TEG) depends on open-circuit voltage, V_{oc}, and occurs when internal device resistance, R_{int}, exactly equals to resistance of external load. When $R_{int} = R_{load}$, then total system resistance $= 2R_{int}$, and V_{oc} drops exactly by half leading to:

$$P_{max} = \frac{V_{oc}^2}{4R_{int}}. \tag{64}$$

For TEG consisting of some number "i" of individual "thermocouples" connected in series and each having n-type thermoelement and p-type thermoelement, Seebeck effect relates V_{oc} to the temperature difference, ΔT, induced by the heat source as described:

$$V_{oc} = \sum_0^i (\alpha_n + \alpha_p)_i \Delta T, \tag{65}$$

where α_n and α_p are values of n-type and p-type Seebeck coefficients from each individual thermoelement, respectively. Thus, the sum of Seebeck coefficients from i thermocouples is ensemble-average proportionality between V_{oc} and ΔT. Likewise, R_{int} is the sum of resistances from i thermocouples, and it is the ensemble-average electrical resistivity of n-type (ρ_n) and p-type (ρ_p) thermoelements times their respective area (A)-to-length (ℓ) values:

$$R_{int} = \sum_0^i \left(\rho_n \frac{\ell}{A} + \rho_p \frac{\ell}{A} \right)_i. \tag{66}$$

P_{max} can be expressed in terms of Seebeck coefficients:

$$P_{max} = \frac{\left(\sum_0^i (\alpha_n + \alpha_p)_i \right)^2 \Delta T^2}{4R_{int}}. \tag{67}$$

This expression highlights first important point: P_{max} increases with ΔT^2. So, for large electrical power output, the largest possible ΔT is desired.

The efficiency, Φ, with which TEG can convert heat flow, Q, to electrical power is also important because the most electrical power possible from a given amount of heat flow is desirable. A new expression for the efficiency of a TEG can be obtained starting with expression for P_{max}.

The ratio of electrical energy generated per given amount of input heat energy is the definition of efficiency:

$$\Phi = \frac{P_{max}}{Q}.$$ (68)

Eq. (68) can be rewritten, assuming for simplicity a unicouple (i = 1), as:

$$\Phi = \frac{(\alpha_n + \alpha_p)^2 \Delta T^2}{4R_{int}Q}.$$ (69)

The flow of heat is dominated by thermal conductivity of the materials from which TEG is constructed, so Fourier's law can be used to express Q:

$$\Phi = \frac{(\alpha_n + \alpha_p)^2 \Delta T^2}{4R_{int}\left([\kappa_n + \kappa_p]\frac{A}{\ell}\Delta T\right)}.$$ (70)

Then expressing R_{int} as described earlier:

$$\Phi = \frac{(\alpha_n + \alpha_p)^2 \Delta T^2}{4\left(\rho_n\frac{\ell}{A} + \rho_p\frac{\ell}{A}\right)\left([\kappa_n + \kappa_p]\frac{A}{\ell}\Delta T\right)}.$$ (71)

For planar TEG devices, the values of ℓ of both n-type and p-type thermoelements are equal; however, cross-sectional areas of n-type and p-type may be quite different. Identifying cross-sectional area of n-type as A_n and that of p-type as A_p allows a simplification, yielding Φ in terms of measurable materials properties and temperature difference:

$$\Phi = \frac{1}{4}\left(\frac{(\alpha_n + \alpha_p)^2}{\left(\frac{\rho_n}{A_n} + \frac{\rho_p}{A_p}\right)(\kappa_n A_n + \kappa_p A_p)}\right)\Delta T.$$ (72)

The proportionality between Φ and ΔT will be termed "Z_{device}":

$$Z_{device} = \left(\frac{(\alpha_n + \alpha_p)^2}{\left(\frac{\rho_n}{A_n} + \frac{\rho_p}{A_p}\right)(\kappa_n A_n + \kappa_p A_p)}\right).$$ (73)

Note, that when area-to-length ratios are optimized for maximum efficiency, this relationship reduces to the common, well-known expression for device ZT:

$$Z_{max} = \left(\frac{(\alpha_n + \alpha_p)}{\left(\sqrt{\kappa_n\rho_p} + \sqrt{\kappa_p\rho_n}\right)}\right)^2.$$ (74)

TEG efficiency can be measured as function of ΔT, and the slope of that data should be equal to:

$$\frac{\partial \Phi}{\partial \Delta T} = \frac{1}{4}\left(\frac{(\alpha_n + \alpha_p)^2}{\left(\frac{\rho_n}{A_n} + \frac{\rho_p}{A_p}\right)(\kappa_n A_n + \kappa_p A_p)}\right) = \frac{Z_{device}}{4}. \tag{75}$$

This expression highlights a second important point, that Φ should linearly increase as a function of ΔT according to the slope indicated by ¼ of the quantity in parentheses. This makes sense, because Φ increases linearly with ΔT, and P_{max} increases as ΔT^2. Taking the ratio yields a simple linear dependence on ΔT. Note, that the material properties are all temperature dependent, so taking the derivative would necessarily yield higher-order terms. However, we make use of the following assumptions: (1) the temperature dependence of the electrical component of thermal conductivity depends on the mobility of charge carriers, and the electrical resistivity depends on the inverse of that mobility, so these dependencies can be assumed to be first-order to cancel completely. (2) Seebeck coefficient does have a relatively small, but finite temperature dependence; however the derivative of Seebeck coefficient should yield a temperature dependence of T^1 which approximately cancels with the temperature dependence of lattice thermal conductivity in the denominator. So, to first order the linearity of the slope would be expected, and is in fact experimentally observed as will be shown.

A new index to determine maximum ZT of TEG device can be obtained by measuring the slope of TEG efficiency. To calculate maximum ZT, four times the slope of TEG efficiency multiplied by maximum temperature, under which TEG displays linear behavior with respect to ΔT. Outside the linear regime of TEGs, the basic properties can no longer be described by these functions. Therefore, measure of maximum ZT of TEG device can be determined by:

$$(Maximum)Z_{device}T = 4\left(\frac{(\alpha_n + \alpha_p)^2}{\left(\frac{\rho_n}{A_n} + \frac{\rho_p}{A_p}\right)(\kappa_n A_n + \kappa_p A_p)}\right)T_{maximum}. \tag{76}$$

The significance of this analysis is that it allows unique means to rapidly obtain $ZT_{maximum}$ and confirm properties and individual measurements. Measurements can be confirmed by measuring slope of efficiency as function of ΔT and ZT can be obtained and compared to theoretical ZT as calculated by individual measurements.

3.1.1. Analysis of commercial (Bi,Sb)$_2$(Te,Se)$_3$ module

Efficiency of commercial (Bi,Sb)$_2$(Te,Se)$_3$ device is presented in **Figure 18**. This device is designed for high thermal impedance and has optimum performance window from nearly room-temperature to roughly 425 K. Slope of efficiency was determined and is shown in inset of **Figure 18**, and is given as 0.0004/K. As expected, the slope is highly linear function of ΔT until deviation from non-linearity begins at 405 K. The maximum ZT can be obtained by the simple relationship, and observed maximum temperature of roughly 405 K:

$$Z_{device}T_{maximum} = 4\left(\frac{\partial \Phi}{\partial \Delta T}\right)T_{maximum}. \tag{77}$$

Obtained value of $ZT_{maximum}$ is equal to 0.7, which is consistent with established values for commercial devices.

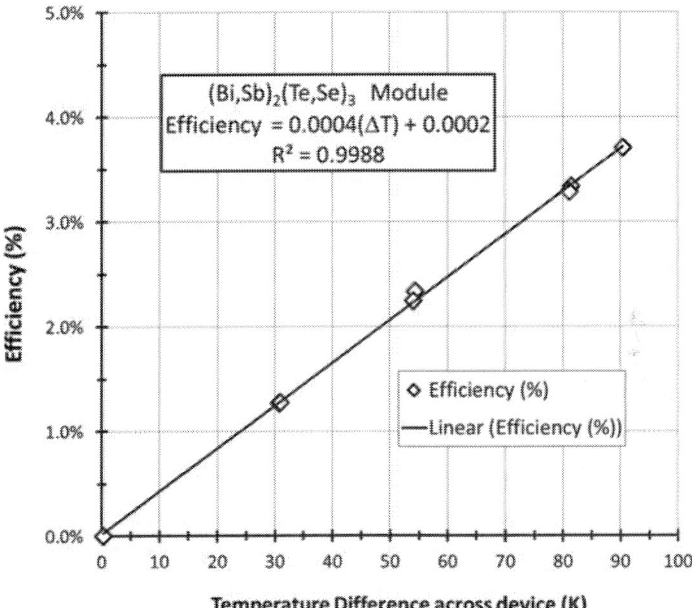

Figure 18. Slope of efficiency from (Bi,Sb)$_2$(Te,Se)$_3$ to determine ZT_{max}.

3.1.2. Analysis of PbTe/TAGS module

Efficiency of PbTe/TAGS device is presented below. **Figure 19** shows temperature dependence of PbTe/TAGS module efficiency. At low temperature, the slope is somewhat a nonlinear function of ΔT because, it is well known, that properties of these materials are uninteresting at low-temperature, but optimum at elevated temperature. So, slope of efficiency is measured in temperature range >500 K, where properties are linear. The maximum ZT can, therefore, be obtained by Eq. (77), and observed maximum temperature for linear device behavior, which for the device being measured is equal to 873 K. Obtained slope is 0.0002/K resulting in value $ZT_{maximum}$ = 0.7, which is consistent with established values for well-known PbTe/TAGS modules.

3.2. Discretized heat-balance model and analysis

More detailed device analysis and performance modeling including effects of temperature-dependent material properties may be accomplished through the use of numerical methods. One technique for performing numerical analysis on TEG was reported by Lau and Buist [26] and later confirmed and expanded upon by Hogan and Shih [27]. It involves partitioning the legs of TEG into virtual segments for computational purposes, where each segment is taken to be isothermal. Neighboring segments then vary in temperature such, that governing thermo-electric heat balance equations based on constant parameter theory are satisfied [28, 29]. This process is illustrated in **Figure 20**.

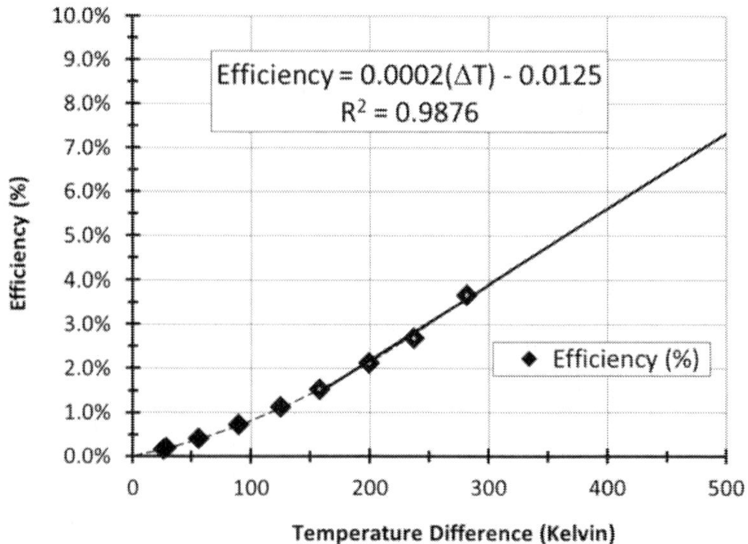

Figure 19. Slope of efficiency from pre-commercial PbTe/TAGS to determine ZT_{max}.

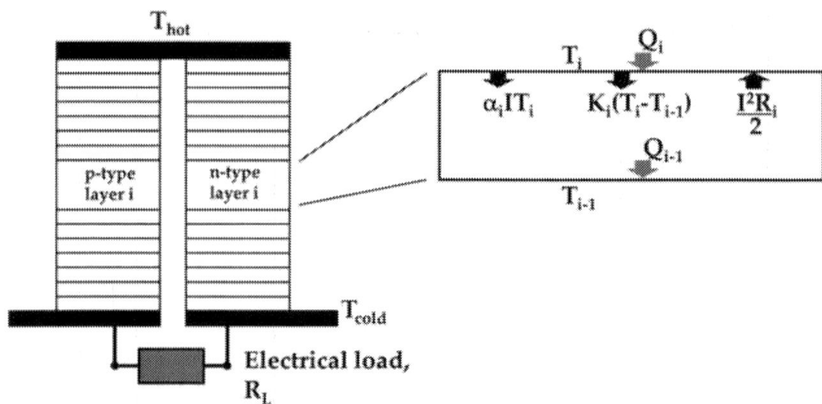

Figure 20. Discrete communicating layers having thermal and electrical flux continuity. Expanded view of ith segment explicitly showing the heat flows that must be balanced to maintain continuity through the bulk of the TEG device.

Based on constant-parameter theory [29] and heat balance at the top surface of the ith segment, we have:

$$Q_i = a_i I T_i - I^2 R_i/2 + K_i(T_i - T_{i-1}), \qquad (78)$$

where the total heat flux Q_i into segment is the sum of individual components indicated.

Here α_i is Seebeck coefficient of the ith segment at temperature T, R_i is the electrical resistance of the ith segment at temperature T, K_i is thermal conductance of the ith segment at temperature T. Last, I, and T_i are the electrical current and temperature of the ith segment, respectively. The electrical power, P, is determined by:

$$Q_{i-1} = Q_i - P, \tag{79}$$

where power P is delivered to the external load resistor, R_L:

$$P = I^2 R_L. \tag{80}$$

The discrete heat balance equations (Eqs. (78) and (79)) derived in this manner can be easily solved for single leg of TEG with an iterative technique [26, 27]. For a given T_C, an initial estimate is made for the heat delivered to the cold junction, Q_C, and temperature and heat flow in each segment is determined sequentially, ending with a numerical solution for heat absorbed at the hot junction Q_H and the hot-side temperature T_H. If calculated T_H is not equal to the desired T_H boundary condition, then Q_C is adjusted accordingly and the process is repeated until the desired T_H is achieved.

The initial hot-side temperature of segment is taken as a uniform temperature for the entire segment and its thermoelectric properties are then determined from a curve fit to measured data. Adjacent segments attain different temperatures as the system is solved according to the energy balance requirements. Thus, temperature-dependent effects are fully incorporated into the model. In fact, Hogan and Shih [27] were able to demonstrate excellent agreement using the discrete approximation as compared with an exact analysis of temperature-dependent TEG performance by Sherman et.al. [30].

Figure 21 shows temperature profiles calculated for n-type Skutterudite material with temperature-dependent properties, operating at the indicated boundary conditions. For simplicity, electrical current is treated here as though there were an external load resistance matched to the internal resistance of the thermoelectric material leg, thus producing maximum output power.

It is also instructive to examine calculated heat flow through the leg as this helps to illustrate thermal-to-electrical conversion process. **Figure 22** shows heat flow corresponding to temperature profiles depicted in **Figure 21**. From the hot side to the cold side of the leg (or right to left on the plot), heat flow is reduced as thermal energy is converted to electrical power and delivered to the load. Examining the specific case of T_H = 750 K and T_C = 300 K, there is approximately 5.7 W of thermal power incident on the hot side and 5.0 W rejected at the cold side, leaving 0.7 W which is delivered as electrical power to the load. So the conversion efficiency within just one of the legs of the TEG itself is simply 0.7 W/5.7 W = 12.3%. The same methodology may be applied to the companion p-type leg to complete the analysis of a full TEG at a given set of temperature boundary conditions. Proper temperature profiles of n-type and p-type legs together are shown in **Figure 23**. Taking T_H = 800 K and T_C = 300 K, the combined incident thermal power is equal to 6.42 Watts$_{thermal}$ (n-type) + 7.85 Watts$_{thermal}$ (p-type) = 14.27 Watts$_{thermal}$ (TEG). And that rejected to the cold side is equal to 5.58 Watts$_{thermal}$ (n-type) + 7.16 Watts$_{thermal}$ (p-type) = 12.74 Watts$_{thermal}$ (TEG). This calculation finds the overall efficiency = 1.53 W/14.27 W = 10.7%.

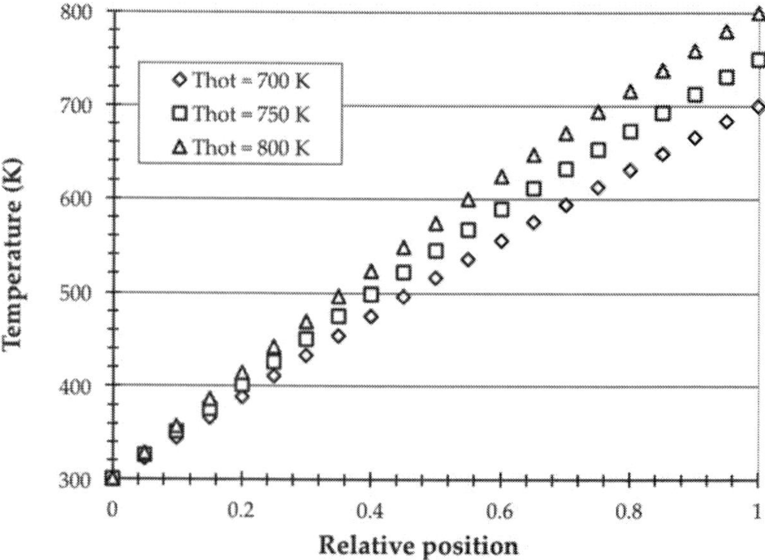

Figure 21. Temperature profile in n-type Skutterudite at maximum power. T_C = 300 K.

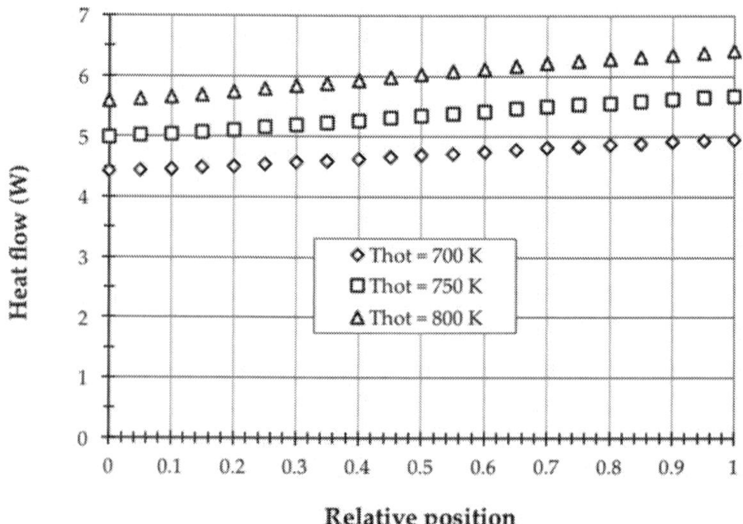

Figure 22. Heat flow profile in n-type Skutterudite sample at maximum output power, corresponding to temperature profiles depicted in **Figure 21.** T_C = 300 K.

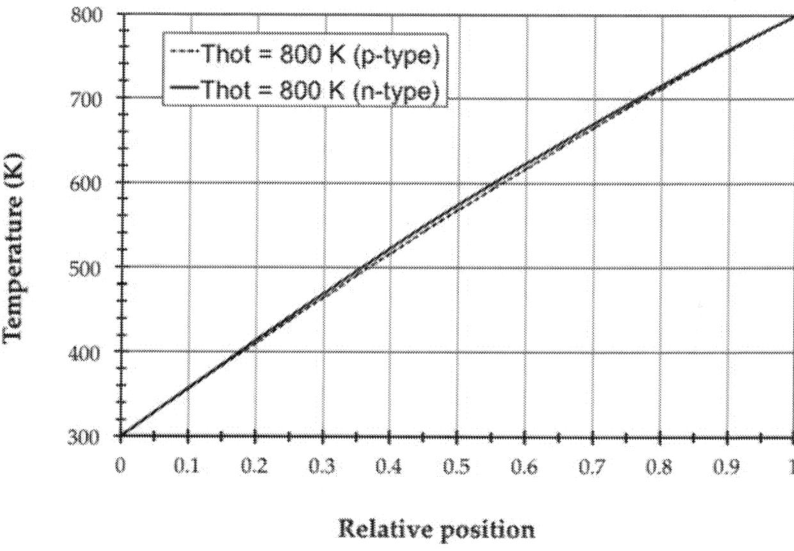

Figure 23. Temperature profiles in both n- and p-type legs of TEG. T_C = 300 K.

This discussion has highlighted a simple, but powerful temperature-dependent phenomenological model for precisely calculating temperature profiles, heat flows, power outputs, and efficiencies in a single leg of TEG. Full TEG device modeling is accomplished by simultaneously solving the discrete heat balances as described for each leg (subject to hot side and cold side boundary conditions) along with the simultaneous energy balance relationship required for electrical power being delivered to the load.

Figure 24 shows calculated efficiency using the *discretized heat balance* theory of idealized Skutterudite n-type and Skutterudite p-type devices.

Electrical and thermal contact resistivities are defined to be zero, but could easily be included as finite quantities, which would add penalties to the efficiency. Slope of efficiency identified in the best-fit is quantified first using unitless efficiency data. The equation is then re-included on the plot after converting to percent. This is so that $ZT_{maximum}$ can be calculated using **slope/ efficiency method** described in Section 3.1, and for overall clarity in the final plot. Slope of efficiency is 0.0002/K, and the upper-limit temperature is 800 K, so, following Eq. (77), value $ZT_{maximum}$= 0.64 is obtained for the device.

Therefore, to confirm measurements for device fabricated using materials, from which measurements were collected, it could be assembled and the efficiency is measured. If measurements are accurate and not overestimated, then performance should be consistent with $ZT_{maximum}$ value equals to 0.64. The slope of the data should be roughly 0.0002/K.

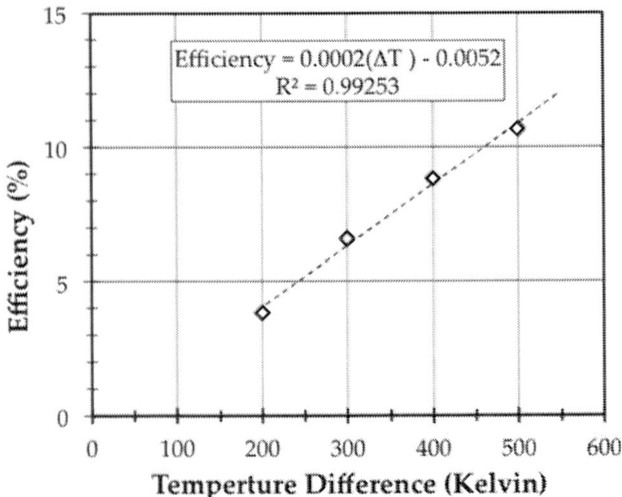

Figure 24. Calculated efficiency of single p-n couple TEG. T_C = 300 K and T_h(max) = 800 K.

4. Conclusions

In conclusion, we have presented several novel approaches to the significant challenge of accurately determining the thermal conductivity of thermoelectric materials. The new solutions can be much faster experimentally, and they successfully address several sources of experimental error. The overall result is significantly reduced error, which may reduce uncertainty by a factor of 2 or more. Further, we introduce new approaches to compare device performance with physical property measurements as a novel means of confirming measurements. Using this approach, the new measurements can be clearly seen to yield physical property measurements which are more consistent with physical device performance.

The first new thermal conductivity measurement method, *steady-state isothermal technique*, improves accuracy by collecting data under conditions where thermal losses and errors are unimportant. The validity was confirmed by comparing the thermal conductivity extracted from a Peltier cooling device with the lab measured value: the error was ~2% [8]. The second is nondestructive micro-scale analysis technique called the *scanning hot-probe*, and the third is *lock-in transient Harman* method, which is a comprehensive modification of transient Harman technique. The second and third methods reduce error by highly detailed treatment of interfacial contact effects including electrical contact resistance and thermal contact effects. The high accuracy for both of these methods is obtained by comparison with established standard reference materials whose properties are well-known and accepted. A new interesting follow-on is frequency-dependent Nyquist analysis, which presages a different perspective on the materials analysis, and even further simplified measurement.

The truest test of the accuracy of measurement is comparison with fabricated devices. To support the validation of measurements of individual material properties, we have outlined new device metrics, which allows comparison between theoretical and measured device efficiency. We outline a new *slope-efficiency method*, which can be used to determine informative index $ZT_{maximum}$ of any device. The second method of device evaluation is a numerical device model called the *discretized heat balance model*. Using this modeling approach, we showed, that a piecewise continuous collection of discrete layers within a device, where boundary heat flows have energy and current continuity relationships, can yield an incredibly easy theoretical determination of device efficiency, which can be compared with experimental values.

Author details

Patrick J. Taylor[1]*, Adam Wilson[1,2], Jay R. Maddux[1,3], Theodorian Borca-Tasciuc[2], Samuel P. Moran[2], Eduardo Castillo[2] and Diana Borca-Tasciuc[2]

*Address all correspondence to: patrick.j.taylor36.civ@mail.mil

1 US Army Research Laboratory, Sensors and Electron Devices Directorate, Adelphi, MD, USA

2 Department of Mechanical, Aerospace and Nuclear Engineering, Rensselaer Polytechnic Institute, Troy, NY, USA

3 General Technical Services, NJ, USA

References

[1] Wang H, Bai S, Böttner S, Chen L, Harris F, Kiss L, Kleinke H, Konig J, Lo J, Mayolet A, Porter W, Sharp J, Smith C, Tritt T. Transport properties of bulk thermoelectrics-an international round-robin study, Part I: Seebeck coefficient and electrical resistivity. Journal of Electric Materials. 2013:**42**:4 p.654–664. DOI: 10.1007/s11664-012-2396-8

[2] Putley E. Thermoelectric and galvanomagnetic effects in lead selenide and telluride. Proceedings of the Physical Society. Section B. 1955:**68** p.35. DOI: 10.1088/0370-1301/68/1/306

[3] Harman T, Cahn J, Logan M. Measurement of thermal conductivity by utilization of the Peltier effect. Journal of Applied Physics. 1959:**30**:9 p.1351. DOI: 10.1063/1.1735334

[4] Penn A. The correction used in the adiabatic measurement of thermal conductivity using the Peltier effect. Journal of Scientific Instruments. 1964:**41**:10 p.626. DOI: 10.1088/0950-7671/41/10/311

[5] Bowley A, Cowles L, Williams G, Goldsmid H. Measurement of the figure of merit of a thermoelectric material. Journal of Scientific Instruments. 1961:**38**:11 p.433. DOI: 10.1088/0950-7671/38/11/309

[6] Buist R. A new method for testing thermoelectric materials and devices. In: Proceedings of the 11th International Conference on Thermoelectrics (ICT '92); 7–9 October 1992; Arlington, Texas. Available from: https://www.tetech.com/wp-content/uploads/2013/10/ICT92RJB.tif [accessed 2016-08-22]

[7] Taylor P, Jesser W, Rosi F, Derzko Z. A model for the non-steady-state temperature behaviour of thermoelectric cooling semiconductor devices. Semiconductor Science and Technolology. 1997:**12**:4 p.443. DOI: 10.1088/0268-1242/12/4/018

[8] Taylor P, Maddux J, Uppal P. Measurement of thermal conductivity using steady-state isothermal conditions and validation by comparison with thermoelectric device performance. Journal of Electronic Materials. 2012:**41**:9 p.2307. DOI: 10.1007/s11664-012-2178-3

[9] Joshi G, Yan X, Wang H, Liu W, Chen G, Ren. Enhancement in thermoelectric figure-of-merit of an N-type half-Heusler compound by the nanocomposite approach. Advanced Energy Materials. 2011:**1**:4 p.643. DOI: 10.1002/aenm.201100126

[10] Rosi F, Hockings E, Lindenblad N. Semiconducting materials for thermoelectric power generation. RCA (Radio Corporation of America) Review. 1961:**22**. OSTI: 4838510

[11] Wilson A, Munoz M, Abad B, Perez J, Maiz J, Schomacker J, Martin-Gonzalez M, Borca-Tasciuc D, Borca-Tasciuc T. Thermal conductivity measurements of high and low thermal conductivity films using a scanning hot probe method in the 3ω mode and novel calibration strategies. Nanoscale. 2015:**7**: p.15404–15412. DOI: 10.1039/C5NR03274A

[12] Borca-Tasciuc T. Scanning probe methods for thermal and thermoelectric property measurements. Annual Review of Heat Transfer. 2013:**16**:80 p.211–258. DOI: 10.1615/AnnualRevHeatTransfer.v16.80

[13] Lefèvre S, Volz S, Chapuis P. Nanoscale heat transfer at contact between a hot tip and a substrate. International Journal of Heat and Mass Transfer. 2006:**49**:1–2 p.251. DOI: 10.1016/j.ijheatmasstransfer.2005.07.010

[14] Incropera F, DeWitt D. Fundamentals of heat and mass transfer. 5th ed. New York: Wiley; 2002. ISBN: 0471386502

[15] Son Y, Pal S, Borca-Tasciuc T, Ajayan P, Siegel R. Thermal resistance of the native interface between vertically aligned multiwalled carbon nanotube arrays and their SiO2/Si substrate. Journal of Applied Physics. 2008:**103**:2 p.024911. DOI: 10.1063/1.2832405

[16] Taborda J,. Romero J, Abad B, Muñoz-Rojo M, Mello A, Briones F, and Gonzalez M. Low thermal conductivity and improved thermoelectric performance of nanocrystalline silicon germanium films by sputtering. Nanotechnology. 2016:**27**:17. p.175401. DOI: 10.1088/0957-4484/27/17/175401

[17] Maiz J, Rojo M, Abad B, Wilson A, Nogales A, Borca-Tasciuc D, Borca-Tasciuc T, Martín-González M. Enhancement of thermoelectric efficiency of doped PCDTBT polymer films. RSC Advances. 2015:**82**:5 p.66687–66694. DOI: 10.1039/C5RA13452H

[18] Abad B, Rull-Bravo M, Hodson S, Xu X, Martin-Gonzalez M. Thermoelectric properties of electrodeposited tellurium films and the sodium lignosulfonate effect. Electrochimica Acta. 2015:**169**: p.37. DOI: 10.1016/j.electacta.2015.04.063

[19] Langer G, Hartmann J, Reichling M. Thermal conductivity of thin metallic films measured by photothermal profile analysis. Review of Scientific Instruments. 1997:**68**: p.1510. DOI: 10.1063/1.1147638

[20] Abad B, Alda I, Díaz-Chao P, Kawakami H, Almarza A, Amantia D, Gutierrez D, Aubouy L, Martín-González M. Improved power factor of polyaniline nanocomposites with exfoliated graphene nanoplatelets (GNPs). Journal of Materials Chemistry A. 2013:**1**:35 p.10450. DOI: 10.1039/C3TA12105D

[21] Castillo E, Hapenciuc C, Borca-Tasciuc T. Thermoelectric characterization by transient Harman method under non-ideal contact and boundary conditions. Review of Scientific Instruments. 2010:**81**:4 p.044902. DOI: 10.1063/1.3374120

[22] Ivory J. Rapid method for measuring Seebeck coefficient as ΔT approaches zero. Review of Scientific Instruments. 1962:**33**:9 p.992. DOI:10.1063/1.1718048

[23] Sharp J, Bierschenk J, Lyon Jr. H. Overview of solid-state thermoelectric refrigerators and possible applications to on-chip thermal management. Proceedings of the IEEE. 2006:**94**:8 p.1602. DOI: 10.1109/JPROC.2006.879795

[24] De Marchi A, Giaretto V. An accurate new method to measure the dimensionless figure of merit of thermoelectric devices based on the complex impedance porcupine diagram. Review of Scientific Instruments. 2011:**82**:10 p.104904. DOI: 10.1063/1.3656074

[25] Efremov M, Olson E, Zhang M, Lai S, Schiettekatte F, Zhang Z, Allen L. Thin-filmdifferential scanning nanocalorimetry: Heat capacity analysis. Thermochimica Acta.2004:**412**:1 p.13. DOI: 10.1016/j.tca.2003.08.019

[26] Lau P, Buist R. Calculation of thermoelectric power generation performance using finite element analysis. In: Proceedings of the 16th International Conference on Thermoelectrics (ICT '97); 26–29 August 1997; Dresden, Germany. New York: IEEE; 1997. p.563.

[27] Hogan T, Shih T. Modeling and Characterization of Power Generation Modules Based on Bulk Materials. In: Rowe D, editor. Thermoelectrics Handbook: Macro to Nano. 1st ed. Washington DC: CRC Press; 2005. p.12–21. DOI: 10.1201/9781420038903.ch12

[28] Buist R. Calculation of Peltier Device Performance. In: Rowe D, editor. CRC Handbook of Thermoelectrics. 1st ed. Washington DC: CRC Press; 1995. p.143–156. DOI: 10.1201/9781420049718.ch14

[29] Cobble M. Calculation of Generator Performance. In: Rowe D, editor. CRC Handbook of Thermoelectrics. 1st ed. Washington DC: CRC Press; 1995. p.489. DOI: 10.1201/9781420049718.ch39

[30] Sherman B, Heikes R, Ure R. Calculation of efficiency of thermoelectric devices. Journal of Applied Physics. 1960:31: p.1. DOI: 10.1063/1.1735380

Thermoelectric Generators Simulation, Modeling and Design

Chapter 17

Thermoelectric Power Generation Optimization by Thermal Design Means

Patricia Aranguren and David Astrain

Additional information is available at the end of the chapter

http://dx.doi.org/10.5772/65849

Abstract

One of the biggest challenges of the twenty-first century is to satisfy the demand for electrical energy in an environmentally speaking clean way. Thus, it is very important to search for new alternative energy sources along with increasing the efficiency of current processes. Thermoelectric power generation, by means of harvesting waste heat and converting it into electricity, can help to achieve above-mentioned goal. Nowadays, efficiency of thermoelectric power generators limits them to become key technology in electric power generation, but their performance has potential of being optimized, if thermal design of such generators is optimized. Heat exchangers located on both sides of thermoelectric modules (TEMs), mass flow of refrigerants and occupancy ratio (the area covered by TEMs related to base area), among others, need to be fine-tuned in order to obtain the maximum net power generation (thermoelectric power generation minus consumption of auxiliary equipment). Finned dissipator, cold plate, heat pipe and thermosiphon are experimentally tested to maximize net thermoelectric generation on real-working furnace based on computational model. Maximum generation of 137 MWh/year using thermosiphons is achieved with 32% of area covered by TEMs.

Keywords: thermoelectric generator, optimization, computational model, heat exchanger, occupancy ratio

1. Introduction

The excessive use of fossil fuels has lead into severe environmental issues. Consequently, global warming, greenhouse gases emissions, climate change, acid rain and ozone depletion are commonly heard on the media. Moreover, combustible resources are limited, and more

restrict environmental regulations are arising. Hence, one of the biggest challenges of the twenty-first century is to satisfy energetic demand in environmentally friendly manner.

In order to fulfill the previous aim, new tendencies are springing, such as smart utilization of energy throughout boosting savings, avoiding waste and developing more efficient, so as less fuel consuming, equipment and through the development of renewable energies. Thermo-electric generation contributes to diminish the impact that fossil fuels generate. A better exploitation of fossil fuels is possible due to their potential to harvest waste heat and convert it into electricity, improving efficiency of energy generating systems.

Nowadays, thermoelectrics is an emerging technology, which converts waste heat into electricity. Solid-state operation of thermoelectric generators (TEGs) eliminates the presence of moving parts and/or chemical reactions, and thus the maintenance is reduced to minimum. It cancels greenhouse gases emissions to environment, and long lives are achieved due to safe operation of thermoelectric generators.

Waste heat is defined as by-product heat of a process, which is not exploited afterward, but it is emitted to the ambient. Nowadays, great amount of produced energy is lavished as waste heat, and at least 40% of the primary energy utilized in industrialized countries is emitted to the ambient as waste heat [1]. Nevertheless, most of this waste heat presents low temperature levels (low temperature grade heat), as **Figure 1** presents, explaining the most studied use up to the moment, heating of fluids for heating or other purposes [2–4]. It has been estimated, that double the heating needs of the United States, the 16.4 % of the primary energy consumed worldwide, could be supplied with waste heat [1].

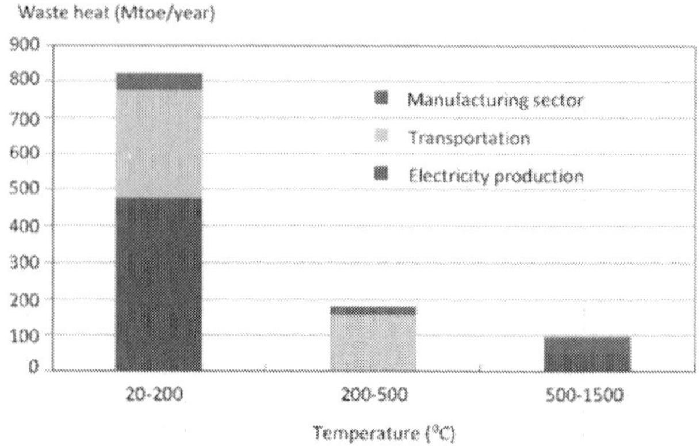

Figure 1. Temperature grade of waste heat [1].

Particular temperature grade, that waste heat presents, restricts applicable technologies to harvest it with effective conversion to electricity. However, thermoelectricity is a promising technology to recover low temperature grade waste heat [5]. Several studies have ratified promising future, that TEGs demonstrate ability to produce electric energy from waste heat of different applications. Some of them are introduced here: Bi_2Te_3-PbTe TEG obtains 211 kW electrical power from waste heat of Portland Cement Rotary Kilns [6]; study conducted in Japan presents potential of recovering radiant heat from steelmaking processes with 10-kW-class grid-connected TEG system [7]; thermoelectric power density of approximately 193.1 W/m^2 is obtained from waste heat of biomass gasifier [8], while power density nearly 100 W/m^2 is obtained from combustion chamber [9]; thermoelectric generator integrated within photovoltaic/thermal absorber improves total efficiency of generating system [10]; nearly 5 kWh/year-m^2 can be produced from solar ponds [11]; and the most common and studied TEGs, which recover waste heat from exhaust gas of vehicles in order to improve their efficiency [12–14].

Efficiency, that normally TEGs present between 5 and 10% [15, 16], is deterrent to make these systems attractive enough to pass the thin line between laboratory experimentation and simulation and commercialization and expansion of this technology. Nowadays, the two issues that are the main objectives are to improve efficiency of thermoelectric generation systems: the first objective is development and improvement of thermoelectric materials through modification of conventional materials with new technologies, such as introducing nanostructures into conventional semiconductors [17, 18] or creating novel thermoelectric materials, such as polymers [19], oxides [20], half-heusler [21] or skutterudites [22]; the second objective is to optimize thermal design of the system. To achieve the latter objective, different approaches can be studied and implemented, for example, heat exchangers located on both sides of thermoelectric modules can be optimized through many different approaches that will be detailed afterwards, and also the number of thermoelectric modules (TEMs) has to be properly selected to reach the maximum thermoelectric generation. Occupancy ratio δ, parameter that includes number of used TEMs M_{TEM}, that is, ratio between area covered by TEMs A_{TEM} and base area A_b of heat exchangers (Eq. (1)), is crucial parameter to optimize thermoelectric generation:

$$\delta = \frac{M_{TEM} A_{TEM}}{A_b} . \tag{1}$$

Although it seems, that higher number of thermoelectric modules would mean higher thermoelectric power generation, thermal resistance per thermoelectric module of heat exchangers worsens, if occupancy ratio rises, resulting in reduction in thermoelectric power generation per TEM. Each application presents optimum point, where thermoelectric power generation is maximum [23–25]. Moreover, reduction in the number of modules does not only imply increase in thermoelectric power generation, but also decrease in initial investment.

Optimization of heat exchangers attached to hot and cold sides of TEMs is very important to maximize thermoelectric power generation, and improvement in thermal resistances will result in higher temperature difference between hot and cold TEM sides close to temperature difference between heat exchangers, and, hence, will provide higher thermoelectric power generation [26–29]. Optimization of heat dissipation systems can be done by modifying their geometry, such as increasing number, height or spacing of fins of finned dissipator [30, 31] or by properly selecting channel's diameter, internal distribution and/or internal inserts of cold plates [32–35]. Besides, inclusion of novel heat exchangers, such as heat pipes [23, 36] or thermosiphons [37–39], could procure higher thermoelectric power generation. Nevertheless, increase in power generation does not necessarily mean improvement in net generation (usable energy obtained from any application) due to increase in the consumption of auxiliary equipment, coolant pump or fans, in order to optimize the thermal behavior of the systems [9, 33, 34, 40].

In this chapter, computational optimization of real furnace located in Spain is performed giving experimental data of thermal resistances of different kinds of heat exchangers (finned dissipator, cold plate, heat pipe and thermosiphon) as function of occupancy ratio, mass flow of refrigerants and heat power to dissipate. Net power generation, that is, thermoelectric power generation minus power consumption of auxiliary equipment Eq. (2), is computed and maximized by means of previously mentioned parameters:

$$\dot{W}_{net} = \dot{W}_{TEM} - \dot{W}_{aux}. \tag{2}$$

2. Computational methodology

Thermoelectric generators produce electricity when there is temperature gradient between hot and cold sides of TEM. Therefore, harvesting of waste heat to produce electricity by thermoelectric generation is becoming very interesting field of studying. Gratuity of waste heat and its great presence in numerous applications overcome low efficiency values, that TEGs present; however, until to date not many applications have been materialized. Initial investment and payback time (due to low efficiency) are deterrents for the development of this technology. This is the reason why computational models are playing very important role in the development of thermoelectric power generation. Due to complicated physical phenomena, that take place in TEGs, knowledge of TEG-based systems' behavior in different conditions is crucial to evaluate their potential, as well as to improve their performance, basing on both thermoelectric material properties and properties and dimensions of heat exchangers located on both sides of TEMs.

Modeling of each component of TEG is essential to perform accurate simulation of behavior of TEG-based systems. TEG is formed by TEMs (which present thermoelectric material, ceramic plates, joints…), the heat exchangers that are located on both sides of thermoelectric modules, as well as by any elementary component for correct assembly of the whole system; consequently, everything needs to be included into the model [41, 42]. Moreover, each thermoelectric phenomenon (Seebeck effect, Thomson effect, Peltier effect and Joule effect) needs to be taken into account, especially in thermoelectric generation due to significant temperature difference between hot and cold sides of TEMs, to obtain accurate results [43–45]; likewise, thermoelectric properties need to be defined as function of temperature not to commit big errors [46]. Furthermore, resolution has to bear in mind transient state of operation [47, 48], especially if trying to model combustion systems with permanent changes in performance and, thus, permanent changes in temperature and mass flow, as vehicles or combustion stoves. The latest applications are very precious due to gratuity of waste heat.

Computational model developed to optimize any thermoelectric application, especially TEGs, which harvest waste heat to produce electricity, includes each thermoelectric phenomenon, each component of the system, temperature dependence of thermoelectric properties and transient state of operation. Moreover, it includes novel parameters, such as occupancy ratio, that is, the ratio between area covered by thermoelectric modules and dissipative base area (Eq. (1)), mass flow of refrigerants and temperature decrease in flue gases when flowing along TEG. Previously mentioned parameters are determinant of net thermoelectric power generation (Eq. (2)), the main parameter to optimize in any application.

Computational methodology uses finite differences approach to solve behavior of the system. It solves each thermoelectric phenomenon, Seebeck effect Eq. (3), Peltier effect Eq. (4), Thomson effect Eq. (5) and Joule effect Eq. (6), and it includes Fourier law of heat conduction used in one-dimensional form, when heat power generation Eq. (7) takes place:

$$\alpha_{AB} = \frac{dE_t}{dT} = \alpha_A - \alpha_B, \tag{3}$$

$$\dot{Q}_{Peltier} = \pm\pi_{AB}I = \pm IT(\alpha_A - \alpha_B), \tag{4}$$

$$\dot{Q}_{Thomson} = -\sigma\vec{I}(\overrightarrow{\Delta T}), \tag{5}$$

$$\dot{Q}_{Joule} = R_0 I^2, \tag{6}$$

$$\rho c_p \frac{\delta T}{\delta t} = k\left(\frac{\delta^2 T}{\delta x_2}\right) + \overline{q}. \tag{7}$$

Resolution methodology is based on previously published and validated computational model [26, 49].

Temperature decrease in flue gases is achieved by discretizing pipe, where flue gases circulate. Within each block, thermoelectric phenomenon is solved. To that objective, temperature of flue gases must be known. Temperature of heat source in each block T_H^i is selected as the mean temperature between entry T_e^i and exit T_s^i temperatures of each block, $T_H^i = T_m^i = \frac{1}{2}(T_e^i + T_s^i)$. Exit temperature is obtained using heat power extracted from flue gases by TEG in that block, Eq. (8). As blocks are located sequentially, exit temperature of previous block coincides with entry temperature of the following block, $T_e^{i+1} = T_s^i$:

$$T_s^i = T_e^i - \frac{\dot{Q}^i}{\dot{m}_{gas} c_p}. \tag{8}$$

Figure 2 presents block "i" of pipe and discretization of that block in order to apply finite differences method to solve thermoelectric phenomena. There are totally 16 nodes, which represent the whole TEG: node 1 is heat source, while node 16 is heat sink; nodes 2 and 15 are hot side and cold side heat exchangers, respectively; and nodes 3–14 represent TEM, where nodes 3 and 14 are hot and cold sides and from node 4 to node 13 thermoelectric material is represented. Electrical analogy is composed by thermal resistances, thermal capacities and absorbed or generated heat fluxes. R_{HD}^i and R_{CD}^i stand for resistances of hot side and cold side heat dissipators, respectively, R_{cont}^i are contact resistances and R_{per}^i and R_{tor}^i stand for two alternative ways for heat power to reach heat sink. The best scenario would be where the total amount of heat power circulates through thermoelectric modules, but in real application there are parasitic heats that circulate along other elements. In this case, heat power that reaches cold sink directly from hot source is born in mind through R_{per}^i, and heat power that flows through assembling screws attaching cold and hot dissipators is represented by R_{tor}^i.

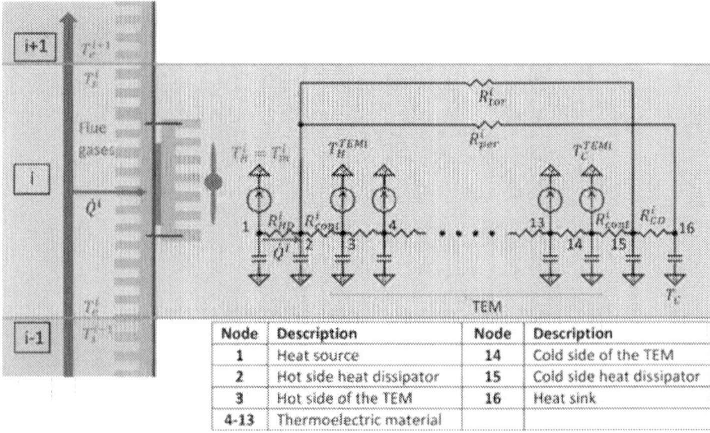

Node	Description	Node	Description
1	Heat source	14	Cold side of the TEM
2	Hot side heat dissipator	15	Cold side heat dissipator
3	Hot side of the TEM	16	Heat sink
4-13	Thermoelectric material		

Figure 2. Thermoelectric generator discretization of block "i".

Particularly, this model considers temperature loss of flue gases, while they circulate along TEG, occupancy ratio and mass flow of refrigerants. Methodology used can be seen in **Figure 3**. The first step is to choose the number of blocks, in which the pipe is discretized, n_{blo}. Once this parameter is selected and information of application is introduced into the model, resolution starts from the first block, where the mean temperature of the block is supposed to be entry temperature of the block, in the case of first block's temperature is that of flue gases. The next step is to suppose heat power that needs to be dissipated by heat exchangers, $\dot{Q}_i^c$, parameter that determines thermal resistance of dissipation systems, as it will be seen in the next section. Thermal resistances of dissipation systems are now determined, so the finite differences method can be used to solve thermoelectric phenomena, obtaining heat power to dissipate and closing the most interior iteration loop. As heat dissipators are function of heat power to dissipate, and at the same time, they define amount of heat that TEG is extracting from flue gases, this issue is solved through an iteration process, which obtains the heat power to dissipate. Once known, the mean temperature of the block needs to be obtained. The mean temperature is computed as the mean value between entry and exit temperatures of flue gases, and exit temperature is obtained using Eq. (8), so new iteration loop solves this situation. Finally, when everything has converged, thermoelectric generation is saved and resolution follows to the next block. This procedure keeps on until each block has been solved and the total power generation has been computed. **Figure 3** presents schematic of the methodology used to obtain thermoelectric power generation.

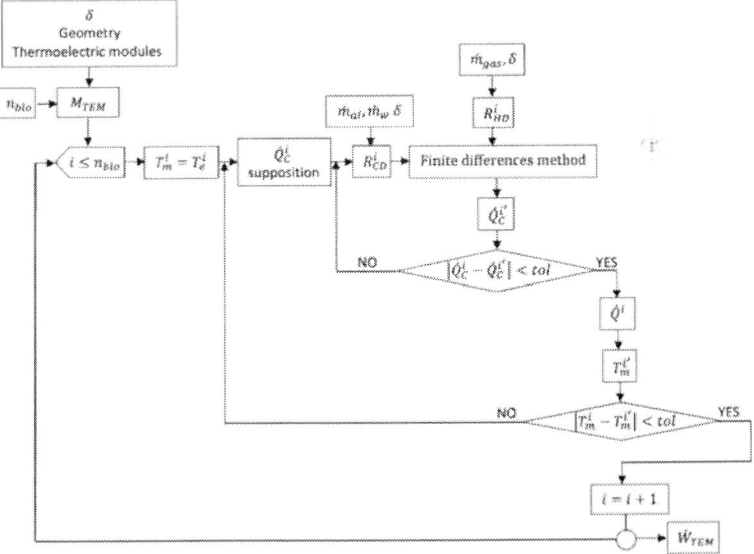

Figure 3. Computational model for thermoelectric power generation.

Net power generation, Eq. (2), is afterward obtained, giving power consumption of auxiliary equipment, determined by the test conducted to thermally characterize the different types of heat exchangers studied, finned dissipator, cold plate, heat pipe and thermosiphon. Experimental thermal characterization of these systems is explained in the next section, very important data that are essential to include into the computational model in order to calculate accurate results about thermoelectric power generation from any application.

3. Thermal characterization of heat exchangers

Thermal characterization of heat dissipation systems is crucial to obtain accurate results using computational model presented in the above section. Four different heat dissipation systems (cold plate, finned dissipator, heat pipe and thermosiphon) have been experimentally tested in order to obtain their thermal resistances as function of influential parameters in thermo-electric power generation: occupancy ratio, mass flow of refrigerants and heat power to dissipate.

Thermal resistances are presented as thermal resistances per thermoelectric module and computed through experimentation as Eq. (9) presents:

$$R^{TEM} = \frac{T_m^{HX} - T_{amb}}{\dfrac{\dot{Q}_C}{M_{TEM}}}. \tag{9}$$

T_m^{HX} stands for average temperature of each heat exchanger, where heat is applied and T_{amb} represents ambient temperature, it is selected constant, 22°C, as heat dissipation systems are placed into climatic chamber ensuring constant temperature during experiments. To test different occupancy ratios, heat plates of the same size of TEMs have been used. Heat power to dissipate corresponds to electric power supplied to heat plates ($\dot{Q}_C = V_{HP}I_{HP}$). One side of heat plate is thermally isolated to assure that total supplied electrical power is transformed into heat power directed to heat exchangers. Finally, the number of TEMs (M_{TEM}) contributes to get medium thermal resistance per thermoelectric module of each heat exchanger.

Variable parameters during experiments are occupancy ratio, heat power to dissipate and mass flow of the refrigerants. Each configuration has been replicated three times ($M_{sample} = 3$) to reduce the random standard uncertainty of the mean ($S_{\bar{R}^{TEM}}$). Expanded uncertainty of measured resistances (Eq. (10)) is composed of previously mentioned uncertainty (Eqs. (11) and (12)), systematic standard uncertainty (Eq. (13)) and level of confidence, in this case chosen to be the 95% [50]:

$$U_{R^{TEM}} = 2\left(b_{R^{TEM}}^2 + s_{\bar{R}^{TEM}}^2\right)^{\frac{1}{2}}, \tag{10}$$

$$s^2_{\overline{R}^{TEM}} = \frac{1}{M_{sample}\left(M_{sample}-1\right)} \sum_{k=1}^{M_{sample}} (R_k^{TEM} - \overline{R}^{TEM})^2, \tag{11}$$

$$\overline{R}^{TEM} = \frac{1}{M_{sample}} \sum_{k=1}^{M_{sample}} R_k^{TEM}, \tag{12}$$

$$b^2_{R^{TEM}} = \left(\frac{\partial R^{TEM}}{\partial T_m^{HX}}\right)^2 b^2_{T_m} + \left(\frac{\partial R^{TEM}}{\partial T_{amb}}\right)^2 b^2_{T_{amb}} + \left(\frac{\partial R^{TEM}}{\partial V_{HP}}\right)^2 b^2_{V_{HP}} + \left(\frac{\partial R^{TEM}}{\partial I_{HP}}\right)^2 b^2_{I_{HP}}. \tag{13}$$

3.1. Cold plate

Use of fluids as heat carrier enhances thermal transfer. In the case of tested system, water has been used in order to characterize heat dissipation system thermally and to analyze results, if net thermoelectric generation increases. Heat dissipation system is formed by cold plate (cold side heat exchanger), fan-coil composed by core and fans to make air circulate through its fins (the secondary heat exchanger in charge of reducing temperature of heat carrier fluid), pump, necessary elements to direct fluid flow and secure safe performance of dissipation system and sensors to obtain the data, as shown in **Figure 4** [51]. Cold plate has 26 transversal channels with diameter 6.2 mm and two manifolds to distribute water coolant along the channels. Plate exterior dimensions are 190 mm × 230 mm. The fan-coil presents core formed by two 8 mm diameter pipes with total number of 12 passes. It is provided with wind tunnel, which presents three fans to make air circulate through its fins. Pump used in the system has been specially chosen, and pumping level can be chosen from one to four using switch.

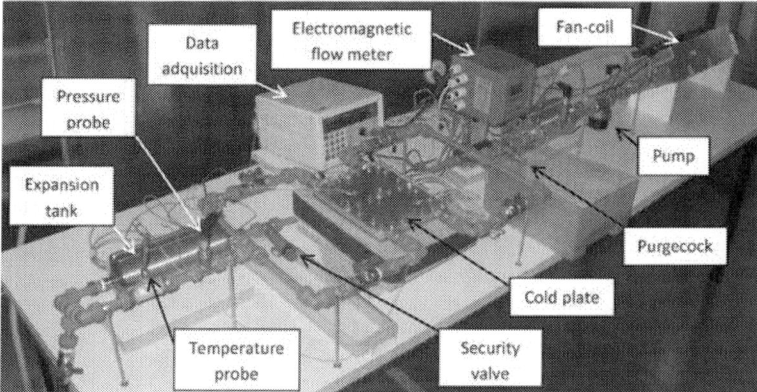

Figure 4. Test bench used to obtain experimentally thermal resistance of cold plate system [51].

Power consumption of pump and fans of fan-coil as function of water and air mass flows, respectively, is shown in **Figure 8**, which represents all relations between mass flows and consumption of auxiliary equipment used for different heat exchangers.

Thermal characterization has been performed using experimental data and validated computational model, enabling to obtain thermal resistances on test bench. Description of the model and validation details can be found in publications, Aranguren et al. [40, 51]. Thermal resistance of cold plate is not function of heat power to dissipate due to small influence of this parameter on temperature of water coolant, the term that could influence on thermal resistance. **Figure 5a** depicts influence of heat power for specific water mass flow. **Figure 5b** presents the influence of occupancy ratio on thermal resistance per thermoelectric module for fixed heat power and water mass flow. As ratio grows, implying that number of TEMs grows, thermal resistance worsens due to the reduction in dissipative area per thermoelectric module. **Figure 5c** presents dependence of thermal resistance air and water mass flows at different occupancy ratio. Occupancy ratio has a great influence on thermal resistance, showing that increasing number of modules harms thermal resistance. Within the same occupancy ratio, water and air mass flow show influence on thermal resistance, most notable for high occupancy ratios, where dissipative area is reduced and any improvement in convective coefficients procures important benefits to thermal resistance.

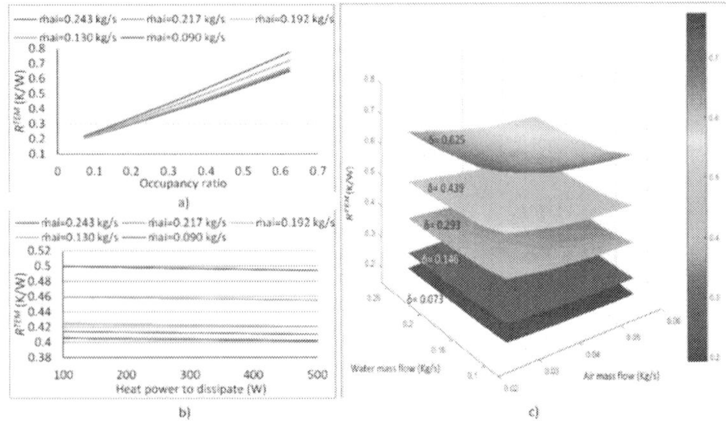

Figure 5. Thermal resistance per thermoelectric module of cold plate. (a) Thermal resistance as function of heat power to dissipate for $\dot{m}_w = 0.055$ kg/s, (b) thermal resistance as function of occupancy ratio for $\dot{m}_w = 0.044$ kg/s, (c) thermal resistance as function of air and water mass flows.

3.2. Finned dissipator

Finned dissipators up to date have been the most used heat exchangers in thermoelectricity due to their simplicity. Studied finned dissipator has external dimensions of 190 mm × 230 mm, base thickness of 14.5 mm and height, thickness and spacing of fins of 39.5, 1.5 and 3.3 mm, respectively. It is provided with wind tunnel, which includes two fans to make air circulate along its fins. The finned dissipator is shown in **Figure 7a**. Relation between power consumption of fans and air mass flow is presented in **Figure 8**.

Figure 6 presents thermal resistance of finned dissipator as function of heat power to dissipate, occupancy ratio and mass flow of air. Heat power to dissipate does not determine thermal resistance, as (shown in **Figure 6a** and **b**). Each panel of **Figure 6** presents thermal resistance of finned dissipator as function of heat power to dissipate, the first one for fixed air mass flow of $\dot{m}_{ai} = 0.024$ kg/s and the second one for $\dot{m}_{ai} = 0.060$ kg/s. Occupancy ratio influences highly thermal resistance, and higher occupancy ratios procure higher thermal resistances per thermoelectric module, due to the reduction in dissipative area per TEM, as presented in **Figure 6**. **Figure 6d** shows the influence of air mass flow, and for high occupancy ratios, influence is more remarkable than for low ones, due to the higher benefits that improvement in convection coefficients has for small convective areas.

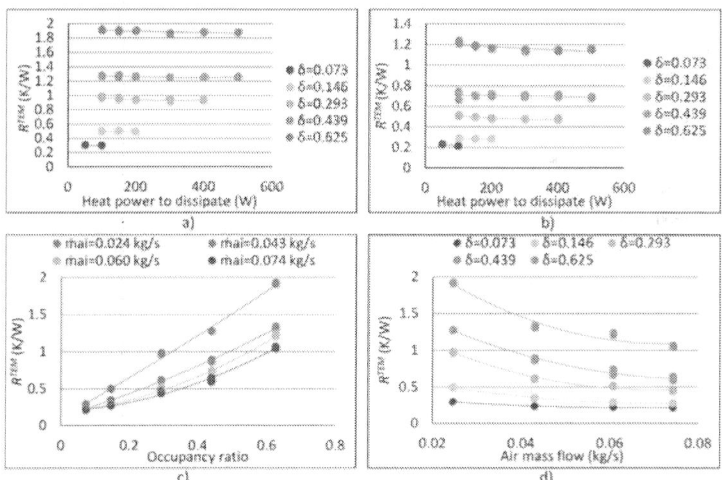

Figure 6. Thermal resistance per thermoelectric module of finned dissipator. (a) Thermal resistance as function of heat power to dissipate for $\dot{m}_{ai} = 0.024$ kg/s, (b) thermal resistance as function of heat power to dissipate for $\dot{m}_{ai} = 0.060$ kg/s, (c) thermal resistance as function of occupancy ratio, (d) thermal resistance as function of air mass flow.

The expanded uncertainty of thermal resistance R^{TEM} is equal to ±10.80%.

3.3. Heat pipe

Heat pipes are passive devices able to transfer great amount of heat with small temperature differences. Heat pipes present sealed volumes provided with porous media and divided into three regions: evaporator, where heat is absorbed; condenser, where heat is emitted; and adiabatic region. Working fluid evaporates due to heat gained and flows into the condenser, where it condensates and returns to evaporator due to capillary lift. Tested heat pipe is composed by 10 8 mm diameter pipes with length of 350 mm and spaced 7 mm. Base external dimension of heat pipe is 90×192.5 mm^2, and pipes are inserted, being the region, where heat arrives. To help condensation of working fluid water, heat pipe includes wind tunnel provided with fan, as **Figure 7b** presents. Air mass flow as function of power consumption is shown in **Figure 8**.

Figure 7. Heat exchanger devices: (a) Finned dissipator; (b) heat pipe.

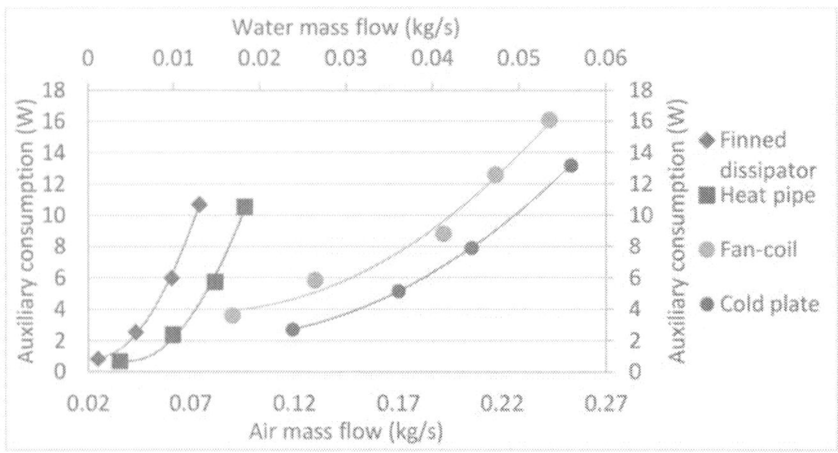

Figure 8. Power consumption of fans of finned dissipator, heat pipe and fan-coil as function of air mass flow and power consumption of the pump of cold plate heat dissipation system as function of water mass flow.

Thermal resistance of heat pipe is function of heat power to dissipate. Condensation and boiling coefficients depend on temperature of fluid and walls, and, therefore, thermal resistance is function of heat power that has to be dissipated, as shown **Figure 9a**. Occupancy ratio and air mass flow present the same tendency as in previous cases, as shown in **Figure 9b, c** and **d**. The expanded uncertainty of thermal resistance R^{TEM} is equal to ±7.88%.

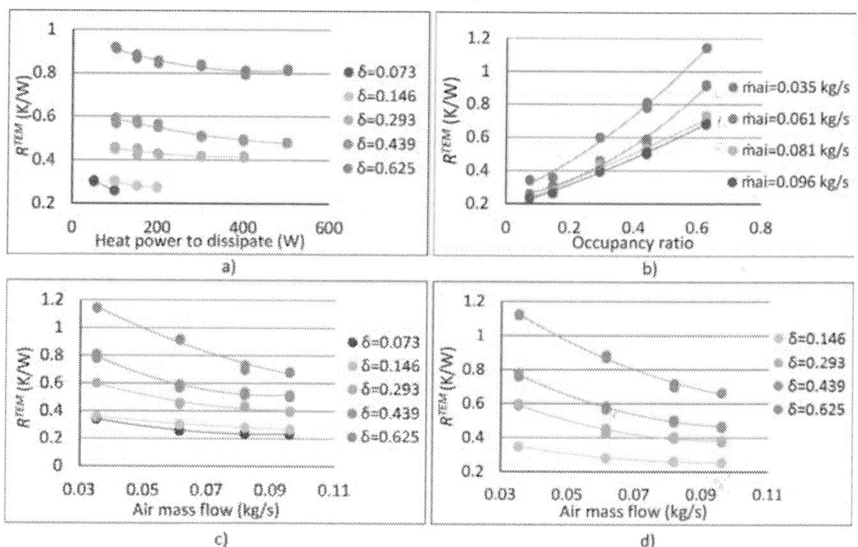

Figure 9. Thermal resistance per thermoelectric module of heat pipe. (a) Thermal resistance as function of heat power to dissipate for $\dot{m}_{ai} = 0.061$ kg/s, (b) thermal resistance as function of occupancy ratio for $\dot{Q}_C = 100$ W, (c) thermal resistance as function of air mass flow for $\dot{Q}_C = 100$ W, and (d) thermal resistance as function of air mass flow for $\dot{Q}_C = 150$ W.

3.4. Thermosiphon

Thermosiphons with phase change present the same physical phenomena, than heat pipes, but they do not present porous media. Hence, they need gravitational forces to ensure that condensate heat carrier returns to evaporator. Tested thermosiphon has vessel of 160×200 mm² and 22 mm diameter pipe that connects the circuit. Pipe is divided into six channels with diameter of 10 mm. Condenser area is composed by seven levels extended along 850 mm with width of 240 mm and depth of 500 mm. This area has 8 mm spaced fins in order to help working fluid, R134a, to condensate. Thermosiphon does not present auxiliary equipment as previous heat exchangers presented. **Figure 10** shows heat dissipation system. Thermosiphon test does not present any fans to help working fluid to condensate, so thermal resistance depends only on heat power to dissipate and occupancy ratio, as displayed in **Figure 11**. Due to natural convection to exterior space and boiling and condensation coefficients, thermal resistance depends on calorific power to a higher extent to dissipate. Higher heat power to dissipate procures higher temperatures, which benefit transfer coefficients involved, and procuring lower thermal resistances, especially high occupancy ratios, is more affected due to high occupation, as presented in **Figure 11b**. The expanded uncertainty of thermal resistance R^{TEM} is equal to ±8.42%.

Figure 10. Tested thermosiphon.

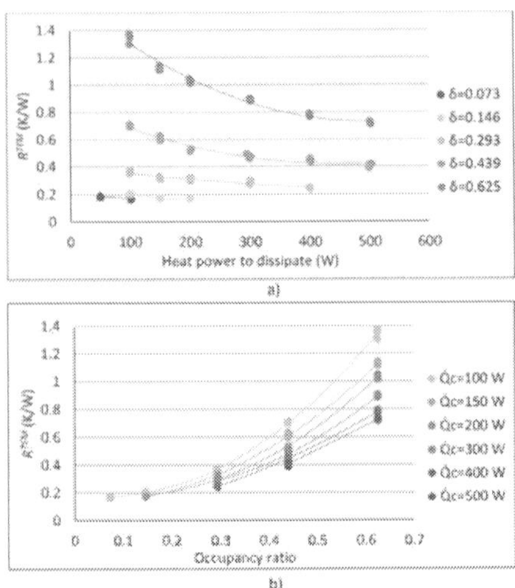

Figure 11. Thermal resistance per thermoelectric module of thermosiphon. (a) Thermal resistance as function of heat power to dissipate, (b) thermal resistance as function of occupancy ratio.

4. Thermoelectric computational optimization of waste heat energy harvesting from real application

Computational model, which enables determining behavior of any TEG and thermal characterization of four different types of studied heat exchangers used to optimize net thermoelectric power generation of real application, is tested on furnace located in Spain. Computational model computes thermoelectric power generation, including calculation power consumption of auxiliary equipment shown in **Figure 8**, and net power generation can be computed (Eq. 2) as well, which is a real target of optimization in any application.

Selected application is furnace, which works 24 h a day, 350 days a year. Temperature of flue gases is 187°C and mass flow is 5.49 kg/s. Chimney has diameter of 0.8 m, transversal area of 0.5 m^2 and height of 12 m. Therefore, available surface to locate TEG is 33.6 m^2. Flue gases emitted to ambient atmosphere are heat source of TEG, while ambient air is heat sink. Temperature of heat sink has been chosen as medium temperature of the year of the location of the furnace, T_{amb} = 17 °C. TEMs simulated are TG12-8-01L from Marlow Industries [52], where hot and cold sides area equals to 40 × 40 mm^2 and TEMs can work up to 250°C on hot side.

Optimization is done for cold side of TEG, where four types of heat exchangers are simulated as function of occupancy ratio and mass flow of refrigerants in order to look for the maximum net power generation. On hot side, that is, interior of chimney, finned dissipator with base thickness of 4 mm and height, thickness and spacing of fins of 50, 6 and 1.5 mm, respectively, have been simulated. Thermal resistance of the latter heat exchanger has been computed as function of occupancy ratio and velocity of flue gases using a Computational Fluid Dynamics program, ANSYS Fluent. Eq. (14) presents thermal resistance of hot side of TEG per thermoelectric module:

$$R^{TEM} = 0.046127 - 0.887591\delta - 0.000251v_{gas} + 0.385376/\ln(v_{gas}) + 0.304593\delta^2 - 0.281665/\ln^2(v_{gas}) + 4.35262\delta/\ln(v_{gas}), \tag{14}$$

Total consumption of auxiliary equipment is essential to obtain net thermoelectric power generation, the goal of this optimization. Consumption of auxiliary equipment is obtained as function of mass flow presented in **Figure 8** and with accounting for number of TEM units necessary to cover the whole available surface of the chimney, totally 769 units. Cold plate, finned dissipator and heat pipe present auxiliary consumption, while thermosyphon does not, as explained in previous section. As it can be seen in **Figures 5, 6** and **9**, increment in mass flow of refrigerants, with simultaneous increment in auxiliary equipment consumption, causes improvement in thermal resistances. This fact leads to higher thermoelectric power generation, due to improvement in heat transfer on both sides of the TEMs, obtaining higher difference of temperature between their sides and consequently higher thermoelectric power generation. Nevertheless, consumption of auxiliary equipment grows, so it is not so clear as in the case, when increasing mass flow of refrigerants leads to increase in net power generation. **Figure 12** shows thermoelectric and net power generation as function of occupancy ratio, when finned dissipators are located on cold side of TEG. It can be seen, that higher air mass flow produces higher thermoelectric power generation; however, net power generation has optimum near the second smallest mass flow simulated, and after this value, net power generation decreases significantly, even obtaining negative values for small occupancy ratios and high mass flow of the air.

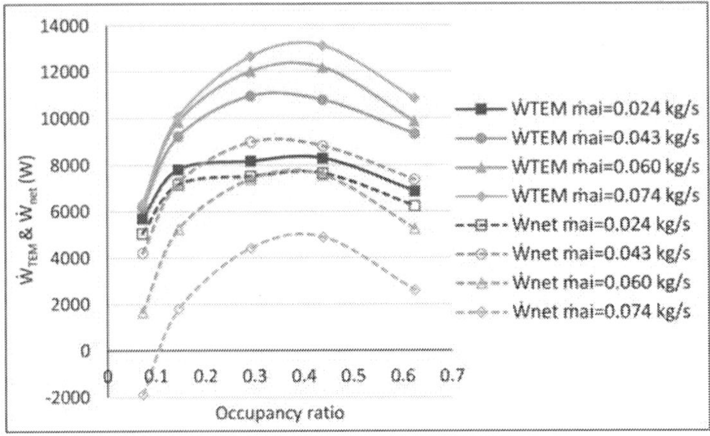

Figure 12. Thermoelectric and net power generation as function of occupancy ratio, when finned dissipators are simulated on cold side of the chimney.

Occupancy ratio is also determined for thermoelectric power generation. Higher occupancy ratio leads to higher thermal resistances per thermoelectric module and, therefore, less power generation per module unit; however, number of units to produce electricity is higher. Once more, it is necessary to elaborate optimization to get the maximum net power generation point. **Figure 12** presents the influence of this parameter, when finned dissipators are simulated on cold side. Occupation ratio value that provides maximum power generation is equal to $\delta \approx 0.4$, that is, optimum is reached, when approximately 40% of available surface is covered by TEMs only. This optimization is crucial to obtain the highest thermoelectric power generation and to optimize initial investment as well, because reduction in number of TEMs, which is necessary to install, reduces the cost of the application.

Figure 13 presents optimum points for net power generation for each occupancy ratio simulated. These points have been obtained by optimizing mass flow of refrigerants of every heat exchanger at each value of occupancy ratio. **Figure 13** shows that the best cold side heat exchanger for this case is thermosiphon, obtaining up to 16280 W from waste heat of the furnace. The optimum occupancy ratio that provides this value equals to $\delta = 0.32$. The number of TEMs necessary to cover 32% of chimney surface is 6720, and the smallest TEMs number is required, if compared with other optimum values as function of occupancy ratio. Therefore, thermosiphons are heat exchangers that not only provide the highest net thermoelectric power generation, but also the ones that require the smallest initial investment. Moreover, these systems have no moving parts, so they are completely robust and lack of maintenance.

Thermosiphons produce 30% more net optimal power than the second best option, heat pipes. Besides, occupancy ratio to maximize net power generation for heat pipes is higher, $\delta \approx 0.42$, so the initial investment has to be approximately 30% higher to obtain the maximum electrical energy. If optimal heat exchangers are compared with cold plates and finned dissipators, electrical energy production is 72 and 86% higher. Cold plates present optimum values of δ higher than thermosiphons, while finned dissipators present approximately the same occupancy ratio, so the same initial investment. Consumption of auxiliary equipment is deterrent, and small thermal resistances that cold plate presents produce higher thermoelectric power generation, but large consumption of auxiliary equipment negatively influences on net power generation, as **Figure 13** shows, even when auxiliary equipment consumption has been optimized to obtain the maximum net power generation for each occupancy ratio.

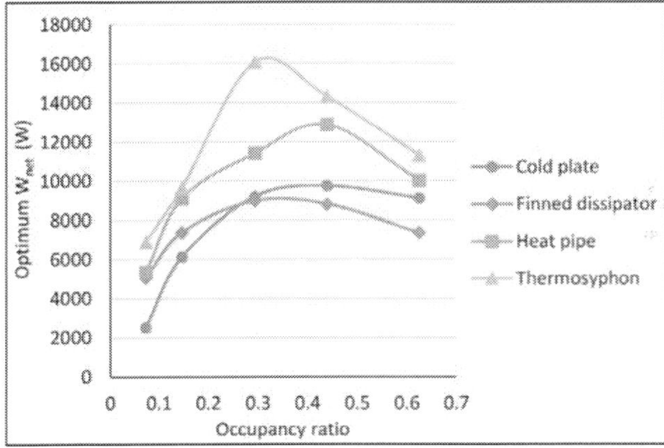

Figure 13. Optimal net thermoelectric power generation for four studied heat exchangers as function of occupancy ratio.

Optimized TEG is able to generate 137 MWh/year, taking into account that furnace works 8400 h a year, with power generation of 484.5 W/m^2 and average of 2.42 W/TEM. Produced electrical energy could supply 40 Spanish dwellings, just harvesting waste heat that furnace emits to the ambient with TEG formed by finned dissipators on hot side and thermosiphons on cold side.

5. Conclusions

Harvesting of waste heat to produce electrical energy via TEGs is a promising technology to help mitigate environmental issues that nowadays society is facing. TEGs are solid-state systems, which barely present moving parts, and, therefore, they are very robust, reliable, silent and long-lasting.

Developed general computational model allows predicting behavior of any TEG. It does not include the most common simplifications that the rest of models from publications have and besides it includes new parameters, such as occupancy ratio, mass flow of refrigerants and temperature reduction in flue gases, while they flow along the system. The latter parameters are determinant for thermoelectric power generation and, therefore, very important to bear in mind for optimization study.

Thermal resistances of different heat exchange systems are function of novel parameter, occupancy ratio, included in computational model. Occupancy ratio has negative influence on thermal resistance per thermoelectric module, if it increases, due to the reduction in available dissipative area per TEM. Calorific power to dissipate influences just heat exchangers, where phase change is involved and mass flow of refrigerants determines thermal resistance, but in greater extent, when occupancy ratio has high values.

Thermosiphon with phase change is a dissipation system that provides the highest net thermoelectric power generation, 137 MWh/year, which is equivalent to supply 40 Spanish dwellings, when 32% of chimney surface is covered by TEMs. This production is 30, 72 and 87% higher than optimal productions of heat pipe, cold plate and finned dissipators, respectively. Moreover, the number of TEMs required for use in TEG with thermosiphons is lower or similar to that for the rest of heat dissipation systems, so not only power generation is optimum, but initial investment also.

The absence of moving parts for TEG built with thermosiphon procures really robust, reliable and silent power generation system that can produce electrical energy from waste heat of any system, improving their efficiency and, therefore, collaborating to satisfy demand for electrical energy in green manner.

Nomenclature

δ	Occupancy ratio
ϱ	Density, kg/m^3
σ	Thomson coefficient, V/K
α	Seebeck coefficient, V/K
π	Peltier coefficient, V
K	Thermal conductivity, W/(m × K)
c_p	Specific heat at constant pressure, J/(kg × K)
A_{TEM}	Area of a TEM, m^2
A_b	Area of the heat exchanger base, m^2
$b_R{}^{TEM}$	Systematic standard uncertainty
I	Current , A

I_{HP}	Current supplied to heat plates, A
M^{TEM}	Number of TEMs
M_{sample}	Number of samples for each configuration
$\dot{m}_{ai}$	Mass flow of air, kg/s
$\dot{m}_{gas}$	Mass flow of flue gases, kg/s
$\dot{m}_{w}$	Mass flow of water, kg/s
n_{blo}	Number of blocks of pipe
$\dot{Q}_C$	Heat power to dissipate, W
$\dot{Q}_{Joule}$	Joule heat power, W
$\dot{Q}_{Peltier}$	Peltier heat power, W
$\dot{Q}_{Thomson}$	Thomson heat power, W
$\dot{Q}^i$	Heat power extracted from flue gases in block "i", W
$\bar{q}$	Volumetric heat generation, W/m^3
R^{TEM}	Thermal resistance per thermoelectric module, K/W
R^i_{CD}	Thermal resistance of cold side heat dissipators of block "i", K/W
R^i_{cont}	Contact thermal resistance of block "i", K/W
R^i_{HD}	Thermal resistance of hot side heat dissipators of block "i", K/W
R^i_{per}	Thermal resistance of heat losses through free surface of block "i", K/W
R^i_{tor}	Thermal resistance of heat losses through bolts of block "i", K/W
R_0	Electrical resistance , Ohm
$s_{\bar{R}TEM}$	Random standard uncertainty of the mean,
T_{amb}	Ambient temperature, K
T_C	Temperature of cold side , K
T_C^{TEMi}	Temperature of cold side of TEMs in block "i", K
T_e^i	Entry temperature of block "i", K
T_H^i	Temperature of heat source in block "i", K
T_H^{TEMi}	Temperature of hot side of TEMs in block "i", K

T_m^i	Mean temperature of block "i", K
T_m^{HX}	Mean temperature of heat exchanger, where heat is applied, K
T_s^i	Exit temperature of block "i", K
$U_{R}TEM$	Expanded uncertainty
v_{gas}	Velocity of flue gases, m/s
V_{HP}	Voltage supplied to heat plates, V
$\dot{W}_{aux}$	Consumption of auxiliary equipment, W
$\dot{W}_{TEM}$	Thermoelectric power generation, W
$\dot{W}_{net}$	Net power generation, W

Author details

Patricia Aranguren[1,2*] and David Astrain[1,2]

*Address all correspondence to: patricia.arangureng@unavarra.es

1 Mechanical, Energy and Materials Engineering Department, Public University of Navarre, Pamplona, Spain

2 Smart Cities Institute, Pamplona, Spain

References

[1] Rattner AS, Garimella S. Energy harvesting, reuse and upgrade to reduce primary energy usage in the USA. Energy. 2011;36(10):6172–6183. doi:10.1016/j.energy.2011.07.047

[2] Torío H, Schmidt D. Development of system concepts for improving the performance of a waste heat district heating network with exergy analysis. Energy and Buildings. 2010;42(10):1601–1609. doi:10.1016/j.enbuild.2010.04.002

[3] Patil A, Ajah A, Herder P. Recycling industrial waste heat for sustainable district heating: A multi-actor perspective. International Journal of Environmental Technology and Management. 2009;10(3–4):412–426. doi:10.1504/IJETM.2009.023743

[4] Law R, Harvey A, Reay D. Opportunities for low-grade heat recovery in the UK food processing industry. Applied Thermal Engineering. 2013;53(2):188–196. doi:10.1016/j.applthermaleng.2012.03.024

[5] Rowe DM, Min G. Evaluation of thermoelectric modules for power generation. Journal of Power Sources. 1998;73(2):193–198. doi:10.1016/S0378-7753(97)02801-2

[6] Luo Q, Li P, Cai L, Zhou P, Tang D, Zhai P, et al. A thermoelectric waste-heat-recovery system for portland cement rotary kilns. Journal of Electronic Materials. 2014;44(6): 1750–1762. doi:10.1007/s11664-014-3543-1

[7] Kuroki T, Murai R, Makino K, Nagano K, Kajihara T, Kaibe H, et al. Research and development for thermoelectric generation technology using waste heat from steel-making process. Journal of Electronic Materials. 2015;44(6):2151–2156. doi:10.1007/s11664-015-3722-8

[8] Ma H, Lin C, Wu H, Peng C, Hsu C. Waste heat recovery using a thermoelectric power generation system in a biomass gasifier. Applied Thermal Engineering. 2015;88:274–279. doi:10.1016/j.applthermaleng.2014.09.070

[9] Aranguren P, Astrain D, Rodríguez A, Martínez A. Experimental investigation of the applicability of a thermoelectric generator to recover waste heat from a combustion chamber. Applied Energy. 2015;152:121–130. doi:10.1016/j.apenergy.2015.04.077

[10] Makki A, Omer S, Su Y, Sabir H. Numerical investigation of heat pipe-based photo-voltaic-thermoelectric generator (HP-PV/TEG) hybrid system. Energy Conversion and Management. 2016;112:274–287. doi:10.1016/j.enconman.2015.12.069

[11] Ding LC, Akbarzadeh A, Date A. Transient model to predict the performance of thermoelectric generators coupled with solar pond. Energy. 2016;103:271–289. doi: 10.1016/j.energy.2016.02.124

[12] Meng J-H, Wang X-D, Chen W-H. Performance investigation and design optimization of a thermoelectric generator applied in automobile exhaust waste heat recovery. Energy Conversion and Management. 2016;120:71–80. doi:10.1016/j.enconman.2016.04.080

[13] Tao C, Chen G, Mu Y, Liu L, Zhai P. Simulation and design of vehicle exhaust power generation systems: The interaction between the heat exchanger and the thermoelectric modules. Journal of Electronic Materials. 2015;44(6):1822–1833. doi:10.1007/s11664-014-3568-5

[14] Baker C, Vuppuluri P, Shi L, Hall M. Model of heat exchangers for waste heat recovery from diesel engine exhaust for thermoelectric power generation. Journal of Electronic Materials. 2012;41(6):1290–1297. doi:10.1007/s11664-012-1915-y

[15] Tritt TM. Thermoelectric phenomena, materials, and applications. Annual Review of Materials Research. 2011;41:433–448.

[16] Riffat SB, Ma X. Thermoelectrics: A review of present and potential applications. Applied Thermal Engineering. 2003;23(8):913–935. doi:10.1016/S1359-4311(03)00012-7

[17] Xie W, Tang X, Yan Y, Zhang Q, Tritt T. High thermoelectric performance BiSbTe alloy with unique low-dimensional structure. Journal of Applied Physics. 2009;105(11): 113713. doi:10.1063/1.3143104

[18] Heremans JP, Jovovic V, Toberer ES, Saramat A, Kurosaki K, Charoenphakdee A, et al. Enhancement of thermoelectric of the electronic density of states. Science. 2008;321 (July):1457–1461. doi:10.1126/science.1159725

[19] Culebras M, Gómez C, Cantarero A. Review on polymers for thermoelectric applications. Materials. 2014;6701–6732. doi:10.3390/ma7096701

[20] Kahraman F, Diez JC, Rasekh S, Madre MA, Torres MA, Sotelo A. The effect of environmental conditions on the mechanical and thermoelectric properties of $Bi_2Ca_2Co_{1.7}O_x$ textured rods. Ceramics International. 2015;41(5):6358–6363. doi:10.1016/j.ceramint.2015.01.070

[21] Zhang H, Wang Y, Dahal K, Mao J, Huang L, Zhang Q, et al. Thermoelectric properties of n-type half-Heusler compounds $(Hf_{0.25}Zr_{0.75})_{1-x}Nb_xNiSn$. Acta Materialia. 2016;113:41–47. doi:10.1016/j.actamat.2016.04.039

[22] Li X, Zhang Q, Kang Y, Chen C, Zhang L, Yu D, et al. High pressure synthesized Ca-filled $CoSb_3$ skutterudites with enhanced thermoelectric properties. Journal of Alloys and Compounds. 2016;677:61–65.doi:10.1016/j.jallcom.2016.03.239

[23] Chen J, Zuo L, Wu Y, Klein J. Modeling, experiments and optimization of an on-pipe thermoelectric generator. Energy Conversion and Management. 2016;122:298–309. doi: 10.1016/j.enconman.2016.05.087

[24] Yang Z, Winward E, Lan S, Stobart R. Optimization of the number of thermoelectric modules in a thermoelectric generator for a specific engine drive cycle. SAE Technical Papers. 2016. doi:10.4271/2016-01-0232

[25] Favarel C, Bédécarrats J-P, Kousksou T, Champier D. Experimental analysis with numerical comparison for different thermoelectric generators configurations. Energy Conversion and Management. 2015;107:114–122. doi:10.1016/j.enconman.2015.06.040

[26] Astrain D, Vián JG, Martínez A, Rodríguez A. Study of the influence of heat exchangers' thermal resistances on a thermoelectric generation system. Energy. 2010;35(2):602–610. doi:10.1016/j.energy.2009.10.031

[27] Martínez A, Vián JG, Astrain D, Rodríguez A, Berrio I. Optimization of the heat exchangers of a thermoelectric generation system. Journal of Electronic Materials. 2010;39(9):1463–1468. doi:10.1007/s11664-010-1291-4

[28] Wang XY, Wang SM, Zhou L, Li XC. Modeling and optimization of heat exchanger in automotive exhaust thermoelectric generator based on fluid kinematics. Advanced

Materials Research. 2014;9863–987:8483–851. doi:10.4028/www.scientific.net/AMR.
986-987.848

[29] Su CQ, Wang WS, Liu X, Deng YD. Simulation and experimental study on thermal
optimization of the heat exchanger for automotive exhaust-based thermoelectric
generators. Case Studies in Thermal Engineering. 2014;4:85–91. doi:10.1016/j.csite.
2014.06.002

[30] Barma MC, Riaz M, Saidur R, Long BD. Estimation of thermoelectric power generation
by recovering waste heat from biomass fired thermal oil heater. Energy Conversion and
Management. 2015;98:303–313. doi:10.1016/j.enconman.2015.03.103

[31] Wang CC, Hung CI, Chen WH. Design of heat sink for improving the performance of
thermoelectric generator using two-stage optimization. Energy. 2012;39(1):236–245.
doi:10.1016/j.energy.2012.01.025

[32] Esarte J, Min G, Rowe DM. Modelling heat exchangers for thermoelectric generators.
Journal of Power Sources. 2001(Feb);93(1–2):72–76. doi:10.1016/S0378-7753(00)00566-8

[33] Amaral C, Brandão C, Sempels ÉV., Lesage FJ. Thermoelectric power enhancement by
way of flow impedance for fixed thermal input conditions. Journal of Power Sources.
2014;272:672–680. doi:10.1016/j.jpowsour.2014.09.003

[34] Lesage FJ, Sempels ÉV., Lalande-Bertrand N. A study on heat transfer enhancement
using flow channel inserts for thermoelectric power generation. Energy Conversion
and Management. 2013;75:532–541. doi:10.1016/j.enconman.2013.07.002

[35] Zhou S, Sammakia BG, White B, Borgesen P, Chen C. Multiscale modeling of thermoelectric generators for conversion performance enhancement. International Journal of
Heat and Mass Transfer. 2015;81:639–645. doi:10.1016/j.ijheatmasstransfer.2014.10.068

[36] Brito FP, Martins J, Hançer E, Antunes N, Gonçalves LM. Thermoelectric exhaust heat
recovery with heat pipe-based thermal control. Journal of Electronic Materials.
2015;44(6):1984–1997. doi:10.1007/s11664-015-3638-3

[37] Jang J, Chi R, Rhi S, Lee K, Hwang H, Lee J, et al. Heat pipe-assisted thermoelectric
power generation technology for waste heat recovery. Journal of Electronic Materials.
2015;44(6):2039–2047. doi:10.1007/s11664-015-3653-4

[38] Singh R, Tundee S, Akbarzadeh A. Electric power generation from solar pond using
combined thermosyphon and thermoelectric modules. Solar Energy. 2011; 85(2):371–
378. doi:10.1016/j.solener.2010.11.012

[39] Kim S, Park S, Kim S, Rhi SH. A thermoelectric generator using engine coolant for light-
duty internal combustion Engine-Powered Vehicles. Journal of Electronic Materials.
2011;812–816. doi:10.1007/s11664-011-1580-6

[40] Aranguren P, Astrain D, Pérez MG. Computational and experimental study of a
complete heat dissipation system using water as heat carrier placed on a thermoelectric
generator. Energy. 2014;74(C):346–358. doi:10.1016/j.energy.2014.06.094

[41] Massaguer E, Massaguer A, Montoro L, Gonzalez JR. Development and validation of a new TRNSYS type for the simulation of thermoelectric generators. Applied Energy. 2014;134:65–74. doi:10.1016/j.apenergy.2014.08.010

[42] Meng JH, Zhang XX, Wang XD. Characteristics analysis and parametric study of a thermoelectric generator by considering variable material properties and heat losses. International Journal of Heat and Mass Transfer. 2015;80:227–235. doi:10.1016/j.ijheat-masstransfer.2014.09.023

[43] Montecucco A, Buckle JR, Knox AR. Solution to the 1-D unsteady heat conduction equation with internal Joule heat generation for thermoelectric devices. Applied Thermal Engineering. 2012;35(1):177–184. doi:10.1016/j.applthermaleng.2011.10.026

[44] Nguyen NQ, Pochiraju KV. Behavior of thermoelectric generators exposed to transient heat sources. Applied Thermal Engineering. 2013;51(1–2):1–9. doi:10.1016/j.applther-maleng.2012.08.050

[45] Meng F, Chen L, Sun F. A numerical model and comparative investigation of a thermoelectric generator with multi-irreversibilities. Energy. 2011;36(5):3513–3522. doi:10.1016/j.energy.2011.03.057

[46] Meng F, Chen L, Sun F. Effects of temperature dependence of thermoelectric properties on the power and efficiency of a multielement thermoelectric generator. International Journal of Energy and Environment. 2012;3(1):137–150.

[47] Lineykin S, Ben-Yaakov S. Modeling and analysis of thermoelectric modules. IEEE Transactions on Industry Applications. 2007;43(2):505–512. doi:10.1109/TIA.2006.889813

[48] Alata M, Al-Nimr MA, Naji M. Transient behavior of a thermoelectric device under the hyperbolic heat conduction model. International Journal of Thermophysics. 2003;24(6):1753–1768. doi:10.1023/B:IJOT.0000004103.26293.0c

[49] Rodríguez A, Vián JG, Astrain D, Martínez A. Study of thermoelectric systems applied to electric power generation. Energy Conversion and Management. 2009;50(5);1236–1243. doi:10.1016/j.enconman.2009.01.036

[50] Coleman HW, Steele WG. Experimentation, Validation, and Uncertainty Analysis for 25 Engineers, 3rd Edition. John Wiley & Sons, Inc., Hoboken, New Jersey.

[51] Aranguren P, Astrain D, Martínez A. Study of complete thermoelectric generator behavior including water-to-ambient heat dissipation on the cold side. Journal of Electronic Materials. 2014;43(6):2320–2330. doi:10.1007/s11664-014-3057-x

[52] TG12-8-01L Power Generators | Generator Modules. http://www.marlow.com/power-generators/standard-generators/tg12-8-01l.html

Chapter 18

Modeling of a Thermoelectric Generator Device

Eurydice Kanimba and Zhiting Tian

Additional information is available at the end of the chapter

http://dx.doi.org/10.5772/65741

Abstract

Thermoelectric generators (TEGs) are devices that employ Seebeck effect in thermopile to convert temperature gradient induced by waste heat into electrical power. Recently, TEGs have enticed increasing attention as green and flexible source of electricity able to meet wide range of power requirements from thermocouple sensors to power generators in satellites. Thermoelectric generators suffer from low-conversion efficiency; however, they could be promising solutions, when they are used to harvest waste heat coming from industry processes or central-heating systems. This chapter covers the working principles behind TEGs, depicts numerous schematics explaining functionality of TEGs, and investigates performance of TEGs. A detailed derivation process, which provides performance expressions dictating operation of TEGs, is exposed in this chapter. In addition, thermal resistance network is shown to explain thermal connection of thermocouples in TEGs in parallel and electrical connection of thermocouples in series. Performance features shown in this chapter are power output, efficiency, and voltage induced within TEG as functions of numerous parameters.

Keywords: Seebeck effect, Peltier effect, Thomson effect, Joule heating, thermal resistance network, electrical resistance network, structure of TEGs, TEGs performance expressions derivation, analytical model, performance analysis of TEGs

1. Introduction

Increase in greenhouse gases emissions in the atmosphere due to burning of fossil fuels for the production of electricity and heat energy has motivated the development of alternative efficient and clean-energy-generation systems including that for the recovery of waste heat into electrical power. Numerous power-generation systems, such as solar panels, wind turbines, and geothermal power plants, which utilize renewable energies, have been designed to reduce dependency on fossil fuels, thus reducing greenhouse gases emissions. However, such power-generation systems require high maintenance and are often expensive as compared to thermoelectric generator devices (TEGs). Thermoelectric generator device (TEG) is a device that directly converts heat into electricity. Essentially, TEG is thermoelectric module (TEM), which

consists of thermopiles, that is, a set of thermocouples built by legs of p- and n-type semi-conductors, which are connected electrically in series and thermally in parallel [1, 2]. Thermo-couples built by legs of p- and n-type semiconductors are sandwiched between two ceramic plates, which are to be held at two different temperatures to realize generation regime. Temperature gradient induced between top and bottom ceramic plates originates voltage on TEG poles due to Seebeck effect in thermocouples built by legs of p- and n-type semiconductors.

Employing waste heat as heat source for TEGs is cost-effective due to waste heat being free of charge and already available. About 70% of the world energy production is known to be wasted into atmosphere through heat dissipation, which is one of significant contributions in global warming [3]. Therefore, the utilization of waste heat by converting into electricity using TEGs can contribute to energy savings and preservation of the environment as well. Thermoelectric device can also operate in reverse mode as thermoelectric cooler (TEC) and produce reverse temperature gradient between top and bottom ceramic plates due to Peltier effect, if electrical bias is applied. Depending on operation mode, applying bias voltage to thermoelectric module (TEM) and hence initiating flow of electrical current result in the production of temperature difference between top and bottom plates and TEM acts as thermoelectric cooler (TEC) and vice versa; the placement of TEM in temperature gradient results in the occurrence of voltage on TEM poles and TEM acts as heat pump with the function of thermoelectric generator (TEG) [4].

Thermoelectric devices possess various advantages compared to other power-generation systems [5]. TEGs are branded attractive power-generation systems, because they are silent solid-state devices with no moving parts, environmental friendly, scalable from small to giant heat sources, and highly reliable. They also have extended lifetime and ability to utilize low-grade thermal energy to generate electrical energy.

2. TEG-working principle

2.1. Seebeck effect

Seebeck effect describes the induction of voltage, when junctions of two different conducting materials are maintained at different temperatures as shown in **Figure 1**. Seebeck effect increases in magnitude, when Seebeck coefficient of conducting materials and/or temperature difference between their connections increases. Voltage induced through Seebeck effect is defined as below:

$$V = \alpha \Delta T, \tag{1}$$

where α is Seebeck coefficient and ΔT is the temperature difference between hot junction and cold junction.

2.2. Peltier effect

Peltier effect describes heat dissipation or absorption at the connection of two conducting materials, when current flows through the junction as shown in **Figure 2**. Depending on the direction of current flow, heat is either absorbed or dissipated at connection.

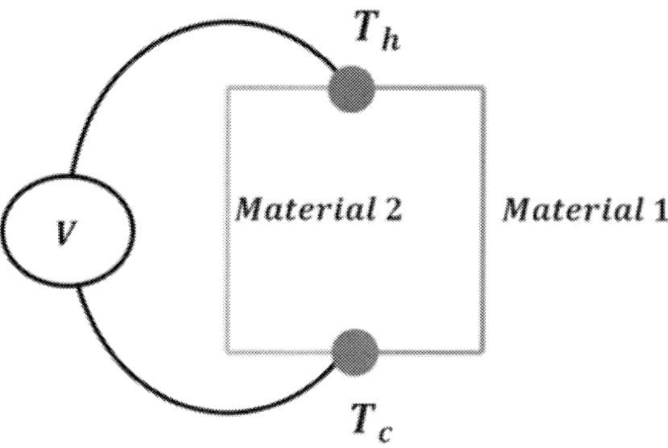

Figure 1. Seebeck effect.

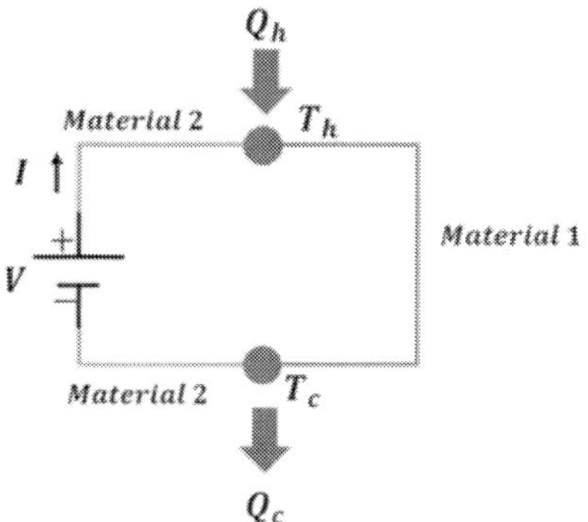

Figure 2. Peltier effect.

2.3. Thomson effect

Thomson effect describes the dissipation or absorption of heat, when electric current passes through a circuit composed of a single material, which has temperature variation along its length, as shown in **Figure 3**. ΔQ represents heat dissipation, when electrical current flows through a homogeneous conductor. Thomson coefficient is given by second Kelvin relationship [6–9]:

$$\mu = T \frac{d\alpha}{dT}, \tag{2}$$

where μ and T, respectively, symbolize Thomson coefficient and temperature. If Seebeck coefficient, α, is temperature independent, then Thomson coefficient is equal to zero.

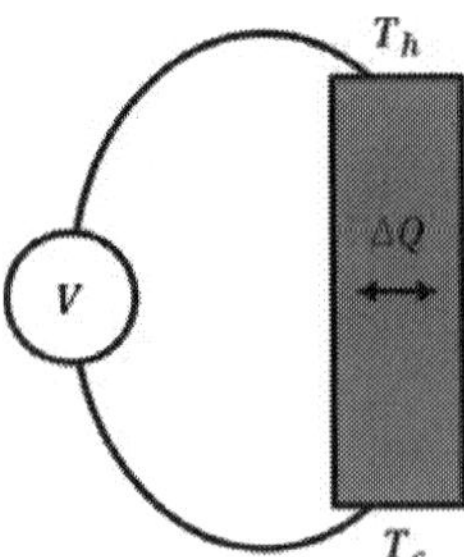

Figure 3. Thomson effect.

2.4. Joule heating

Joule-heating effect defines heat dissipated by material with nonzero electrical resistance in the presence of electrical current, as shown in **Figure 4**,

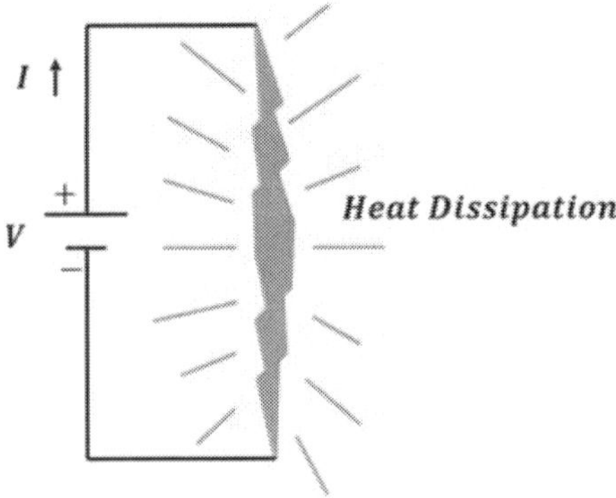

Figure 4. Joule heating.

3. Structure of TEG

3.1. Three-dimensional representation of comprehensive operation of TEG

TEGs are composed of numerous legs (slabs) made of p- and n-type semiconductors forming thermocouples, all connected electrically in series and thermally in parallel. Semiconductor legs are connected to each other through conductive copper tabs, and they are sandwiched between two ceramic plates, which conduct heat, but behave as insulators to electrical current. Schematic diagram of three-dimensional (3-D) multielement thermoelectric generator is shown in **Figure 5**.

Waste heat from various sources, such as automobile engines exhaust, industry and infrastructure-heating activities, geothermal, and others, can be supplied to top ceramic plate of TEGs.

As shown in **Figure 5**, heat flows through ceramic plate and copper-conductive tabs before reaching the top surface of p- and n-type legs made of proper semiconductors, which is defined as the hot side of TEG. Heat flows through both semiconductor's legs and then again through copper-conductive tabs and bottom ceramic plate. Through heat sink, the bottom ceramic plate is maintained at significantly lower temperature than top ceramic in order to produce high-temperature gradient, which will lead to high-power output. Allowed temperature applied on top and bottom ceramic plates depends on materials of p- and n-type legs. Also, p- and n-type materials are designed to possess low thermal conductivity in order to restrict, as much as possible, heat flow through semiconductors and maintain temperature difference between hot and cold sides of TEG.

Pictorial distribution of temperature along legs of TEG at conditional difference of temperature ΔT between hot and cold sides is shown in **Figure 6**.

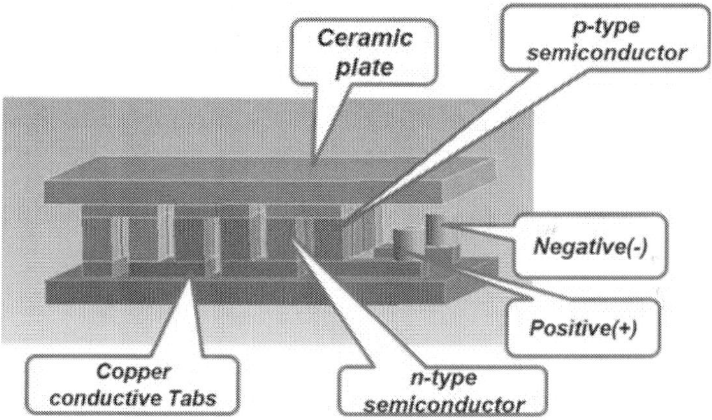

Figure 5. 3-D schematic of multielement TEG.

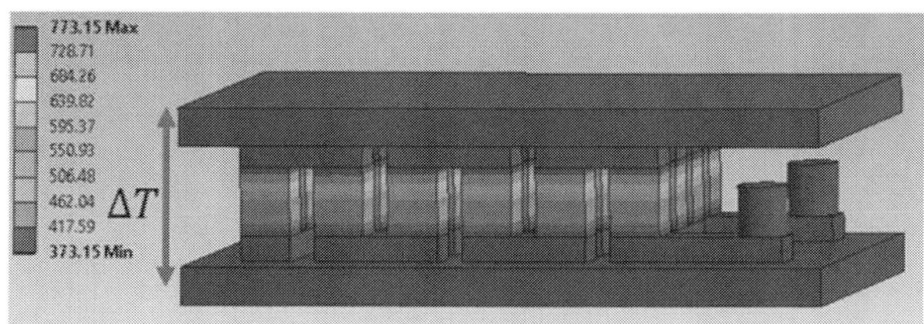

Figure 6. Temperature gradient within TEG.

After temperature gradient has been induced between hot and cold sides of TEG, voltage occurred on TEG-positive and -negative poles due to Seebeck effect, as depicted in **Figure 7**.

Voltage generated in TEG due to Seebeck effect induces the movement of charge carriers within p-and n-type semiconductor legs and, hence, electrical current in electrical circuit including load resistor R_L connected to TEG poles, current density formed, is displayed in **Figure 8**.

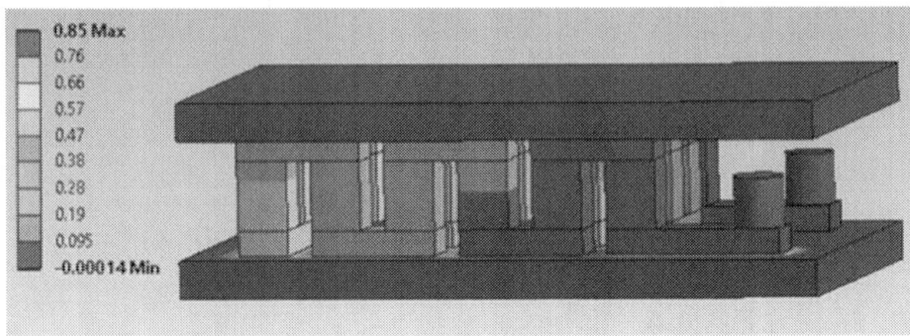

Figure 7. Voltage distribution within TEG.

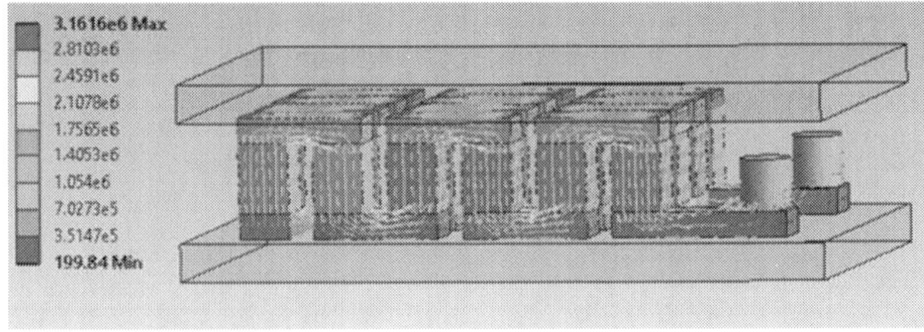

Figure 8. Current density within TEG.

3.2. 1-D representation of TEG

Establishing one-dimensional (1-D) representation of TEG is helpful in determining analytical expressions of heat absorbed and heat rejected, as the power output of TEG is defined as the difference between heat absorbed and heat rejected. **Figure 9** represents 1-D schematic of TEG with heat source and heat sink, respectively, applied on top and bottom sides of TEG.

T_H, Q_H, and K_H are, respectively, heat source temperature, heat supplied from heat source to TEG, and thermal conductance of hot side of TEG. T_L, Q_L, and K_L are, respectively, heat sink temperature, heat rejected from TEG to heat sink, and thermal conductance of TEG cold side. T_h and Q_h define the temperature of hot junction of thermocouples and heat flow through hot junctions of TEG. T_c and Q_c describe the temperature at cold junction of thermocouples and heat flow through cold junctions of TEG. Assuming thermoelectric properties to be temperature independent, α, k, ρ can, respectively, be defined as constant Seebeck coefficient, constant thermal conductivity, and constant electrical resistivity.

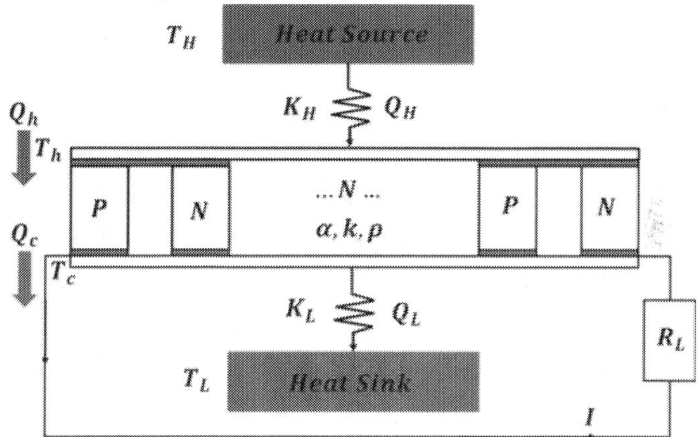

Figure 9. 1-D schematic of multielement TEG.

3.3. Electrical network resistance

Electrical resistance network of TEG is shown in **Figure 10**. P-type and n-type semiconductor legs are connected to each other electrically in series through copper-conductive tabs.

R_p and R_n are electrical resistance associated, respectively, with p- and n-type semiconductor legs. R_{cpeh}, R_{cpec}, and R_L are, respectively, electrical resistance of copper-conductive strips on the hot side, electrical resistance of copper-conductive strips on the cold side, and external load resistance.

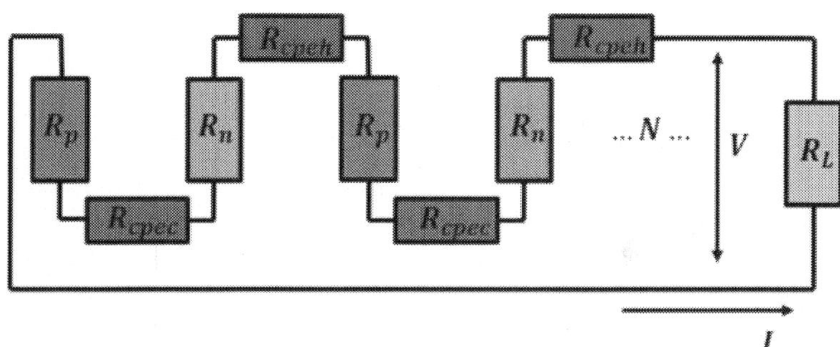

Figure 10. Electrical network resistance.

3.4. Thermal network resistance

Thermal resistance of TEG is shown in **Figure 11** and it assists in determining heat transfer rate through ceramic plates, copper strips, and p- and n-type semiconductor legs. The number of thermocouples is N.

T_{eceh}, T_{iceh}, and R_{ceh} are, respectively, external temperature of hot ceramic plate, internal temperature of hot ceramic plate, and thermal resistance associated with ceramic plate on the hot side. T_h, R_{cph}, and R_{teg} are, respectively, the temperature at the hot junction of p- and n-type semiconductor legs, thermal resistance of copper strip on the hot side, and thermal resistance of both p- and n-type semiconductor legs. T_c, R_{cpc}, and T_{icec}, are, respectively, the temperature at cold junction of p- and n-type semiconductor legs, thermal resistance of ceramic plate on the cold side, and internal temperature of cold ceramic plate. R_{cec} and T_{ecec} are, respectively, thermal resistance of cold ceramic plate and external temperature of cold ceramic plate.

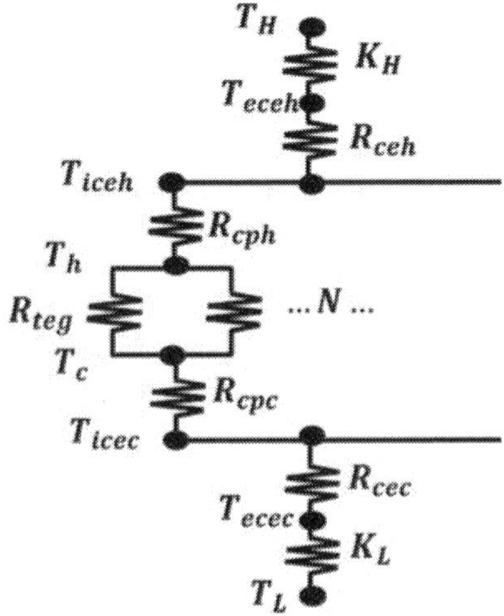

Figure 11. Thermal resistance network.

4. Theoretical model

4.1. Analysis of thermoelectric material properties and geometry of TEG

Thermoelectric materials of TEG legs, p- and n-type semiconductors, are characterized by parameter called the figure of merit Z, which measures the ability of thermoelectric materials to convert heat into electrical power. The figure of merit is expressed as follows:

$$Z = \frac{\alpha^2}{\rho k},\tag{3}$$

where α, ρ, and k are, respectively, Seebeck coefficient, electrical resistivity, and thermal conductivity of thermoelectric materials. Great thermoelectric materials possess high Seebeck coefficient, low electrical resistivity, and low thermal conductivity [10].

In order to obtain maximum figure of merit, when designing TEG, the geometry of semiconductor legs and properties of thermoelectric materials need to satisfy the following equation [1, 11]:

$$\frac{A_p^2 L_n^2}{A_n^2 L_p^2} = \frac{k_n \rho_p}{k_p \rho_n},\tag{4}$$

where A_p, A_n, L_p, L_n, k_p, k_n, ρ_p, and ρ_n are, respectively, the cross-sectional area, length, thermal conductivity, and electrical resistivity of p- and n-type semiconductor legs.

To reduce manufacturing costs, p- and n-type semiconductor legs are fabricated with the same geometry, that is, $A_p = A_n = A$, and $L_p = L_n = L$. Similarly, p- and n-type semiconductor legs are made of doped alloys to produce the same thermoelectric properties, that is, $\rho_p = \rho_n$, $k_p = k_n$, and $\alpha_p = -\alpha_n$ [12].

4.2. TEG performance analysis

In order to obtain expressions describing TEG performance, thermocouple built by legs of p- and n-type semiconductors is extracted from **Figure 9** and represented in **Figure 12**. **Figure 12** represents heat transfer within single thermocouple. The length and cross-sectional area of both p- and n-type semiconductor legs are equal and symbolize as L and A, respectively. The junction of thermocouple is fixed at thermal conducting and electrical-insulating ceramic plate.

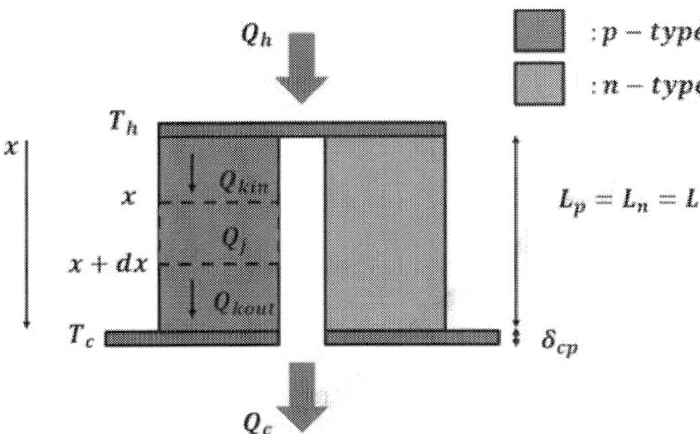

Figure 12. Heat transfer within TEG thermocouple.

Q_h, Q_c, Q_{kin}, Q_{kout}, Q_j, L_p, L_n, and δ_{cu} are, respectively, heat absorbed at hot junction, heat rejected at cold junction, Fourier heat conduction transferred inside of control volume, Fourier heat conduction transferred out of control volume, Joule heating generated within control volume, the length of p- and n-type legs, and the thickness of copper electrical-conducting strips.

Employing the conservation of energy and assuming one-dimensional steady-state condition, the energy equation of differential control volume inside of p-type semiconductor leg can be expressed as follows:

$$Q_{\text{kin}} - Q_{\text{kout}} + Q_j = 0, \tag{5}$$

$$Q(x) - Q(x + dx) + Q_j = 0. \tag{6}$$

Using Taylor expansion:

$$Q(x) - \left(Q(x) + \frac{\partial Q(x)}{\partial x} dx \right) + \frac{I^2 \rho_p}{A_p} dx = 0, \tag{7}$$

I represents electrical current induced within TEG device:

$$-\frac{\partial Q(x)}{\partial x} dx + \frac{I^2 \rho_p}{A_p} dx = 0. \tag{8}$$

Fourier's law of conduction for one-dimensional heat conduction states:

$$Q(x) = -k_p A_p \frac{\partial T_p}{\partial x}. \tag{9}$$

Substituting Eq. (9) into Eq. (8):

$$-\frac{\partial}{\partial x} \left(-k_p A_p \frac{\partial T_p}{\partial x} \right) dx + \frac{I^2 \rho_p}{A_p} dx = 0. \tag{10}$$

Provided that thermoelectric properties are temperature independent, k_p can be taken out of derivative and Eq. (10) can be expressed as follows:

$$k_p A_p \frac{d^2 T_p}{dx^2} dx + \frac{I^2 \rho_p}{A_p} dx = 0. \tag{11}$$

Integrating Eq. (11):

$$\int_0^x k_p A_p \frac{d^2 T_p}{dx^2} dx + \int_0^x \frac{I^2 \rho_p}{A_p} dx = 0, \tag{12}$$

$$k_p A_p \left(\frac{dT_p}{dx} \Big|_x - \frac{dT_p}{dx} \Big|_0 \right) + \frac{I^2 \rho_p}{A_p} x = 0, \tag{13}$$

$$x = 0 \rightarrow T_p(0) = T_h, \tag{14}$$

$$k_p A_p \frac{dT_p}{dx} \Big|_0 = Q_p(0), \tag{15}$$

where $Q_p(0)$ is Fourier heat conduction transferred inside of top p-type leg:

$$\int_0^{L_p} k_p A_p \frac{dT_p}{dx}\,dx + \int_0^{L_p} \frac{I^2 \rho_p}{A_p} x\,dx = -\int_0^{L_p} Q_p(0)\,dx, \tag{16}$$

$$k_p A_p \left(T_p(L_p) - T_p(0)\right) + \frac{I^2 \rho_p}{A_p}\frac{L_p^2}{2} = -Q_p(0)L_p, \tag{17}$$

$$x = 0 \;\rightarrow\; T_p(0) = T_h, \tag{18}$$

$$x = L_p \;\rightarrow\; T_p(L_p) = T_c, \tag{19}$$

$$Q_p(0) = \frac{k_p A_p}{L_p}(T_h - T_c) - 0.5\frac{I^2 \rho_p L_p}{A_p}. \tag{20}$$

Considering Peltier effect happening at the hot junction of p-type leg:

$$Q_{ph} = \alpha_p I T_h + \frac{k_p A_p}{L_p}(T_h - T_c) - 0.5\frac{I^2 \rho_p L_p}{A_p}, \tag{21}$$

where Q_{ph} is the total heat absorbed at the hot junction of p-type leg.

Employing the same procedure with the same boundary conditions to derive heat flow through n-type leg leads to the expression of Q_{nh} as follows:

$$Q_{nh} = -\alpha_n I T_h + \frac{k_n A_n}{L_n}(T_h - T_c) - 0.5\frac{I^2 \rho_n L_n}{A_n}, \tag{22}$$

where Q_{nh} is the total heat absorbed at the hot junction of n-type leg. The total heat absorbed at the hot junction of both p- and n-type semiconductor legs is, therefore:

$$Q_h = Q_{ph} + Q_{nh}, \tag{23}$$

$$Q_h = (\alpha_p - \alpha_n)I T_h + \left(\frac{k_p A_p}{L_p} + \frac{k_n A_n}{L_n}\right)(T_h - T_c) - 0.5\left(\frac{\rho_p L_p}{A_p} + \frac{\rho_n L_n}{A_n}\right)I^2. \tag{24}$$

We use the same method to derive expression for heat rejected at the cold junction of p-type and n-type legs. Consequently, the following expression is obtained:

$$Q_c = (\alpha_p - \alpha_n)I T_c + \left(\frac{k_p A_p}{L_p} + \frac{k_n A_n}{L_n}\right)(T_h - T_c) + 0.5\left(\frac{\rho_p L_p}{A_p} + \frac{\rho_n L_n}{A_n}\right)I^2. \tag{25}$$

4.3. TEG performance expressions

TEG is characterized by numerous performance expressions, including heat absorbed on the hot side, heat rejected on the cold side, power output, voltage induced, and current flowing in the electrical circuit with load resistor. Defining symbols below from Eqs. (24) and (25):

$$K = \frac{k_p A_p}{L_p} + \frac{k_n A_n}{L_n}, \tag{26}$$

$$r = \frac{\rho_p L_p}{A_p} + \frac{\rho_n L_n}{A_n}, \tag{27}$$

$$\alpha = \alpha_p - \alpha_n. \tag{28}$$

Expressions of heat flow through the hot and cold junctions for N semiconductor thermocouples can therefore be expressed as follows:

$$Q_h = N(\alpha I T_h - 0.5 r I^2 + K(T_h - T_c)), \tag{29}$$

$$Q_c = N(\alpha I T_c + 0.5 r I^2 + K(T_h - T_c)). \tag{30}$$

As stated previously, the power generated by TEG is defined as the difference between heat absorbed at the hot junction and heat rejected at the cold junction:

$$P = Q_h - Q_c = N(\alpha I(T_h - T_c) - r I^2). \tag{31}$$

Optimal current generated in TEG is obtained by first deriving Eq. (31) with respect to current as follows:

$$\frac{dP}{dI} = N(\alpha(T_h - T_c) - 2Ir). \tag{32}$$

Eq. (32) is equated to zero to determine the following expression of optimal current:

$$I_{opt} = \frac{\alpha(T_h - T_c)}{2r}. \tag{33}$$

Generally speaking, voltage, current, and output power induced in TEG consisting of set of thermocouples similar to the one represented in **Figure 9** are, respectively, defined as:

$$I = \frac{\alpha(T_h - T_c)}{r + R_L}, \tag{34}$$

$$P = I^2 R_L = \left(\frac{\alpha(T_h - T_c)}{r + R_L}\right)^2 R_L, \tag{35}$$

$$V = I R_L = \frac{\alpha(T_h - T_c)}{r + R_L} R_L, \tag{36}$$

where R_L, is the external resistance load. To get optimum electrical current induced, and output power generated in the electrical circuit with TEG consisting of set of thermocouples, external resistance needs to be equal to the total internal electrical resistance of p- and n-type semiconductor legs. The efficiency of TEG is given by:

$$\eta = \frac{P}{Q_h}.\tag{37}$$

In actual TEG, two thermoelectric materials are used, that is, p- and n-type semiconductors. The maximum efficiency provided by TEG is expressed as follows:

$$\eta_{\max} = \left(1 - \frac{T_h}{T_c}\right)\frac{\sqrt{1 + Z\overline{T}} - 1}{\sqrt{1 + Z\overline{T}} + \frac{T_h}{T_c}},\tag{38}$$

where Z and $\overline{T}$ are, respectively, the figure of merit of p- and n-type semiconductors and averaged temperature between temperatures at the hot and cold sides.

4.4. Performance simulation example of a TEG

Numerical example is adopted in order to optimize and analyze effects of heat transfer governing equations on output power, efficiency, and induced voltage of TEG.

In numerical analysis, the following geometry is adopted (**Table 1**).

The following thermoelectric properties are adopted (**Table 2**).

Number of pairs (N)	Cross-sectional area (A)	Length (L)
10	$2.5 \times 2.5 \times 10^{-6}\,\mathrm{m}^2$	$2 \times 10^{-3}\,\mathrm{m}$

Table 1. Geometry of TEG.

α_p	α_n	$\rho_p = \rho_n$	$k_p = k_n$
$185 \times 10^{-6}\,V/K$	$-185 \times 10^{-6}\,V/K$	$1.65 \times 10^{-5}\,\mathrm{Ohm} \times m$	$1.47\,W/(mK)$

Table 2. Thermoelectric properties.

All obtained performance curves are computed at the hot-side temperature up to $T_h = 673$ K and the cold-side temperature of $T_c = 373$ K.

4.4.1. Power and efficiency as function of electrical current

By fixing the cold side at temperature $T_c = 373$ K and varying the hot-side temperature from 473 to 673 K with increment of 100 K, the power generated behaves as follow:

One can observe that the power as a function of current behaves as a parabola with optimum power value at specific current. **Figure 13** shows the existence of maximal current value, which corresponds to optimum power. Any current higher or lower than the maximum current value generates power output less than optimum power. Also, as temperature at the hot side increases, then power produced increases as well.

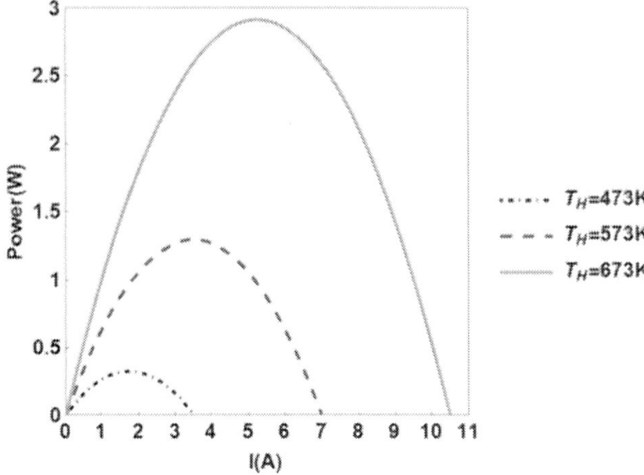

Figure 13. TEG output power as a function of electrical current.

Efficiency curves shown in **Figure 14** behave as parabola as well, with specific current value maximizing efficiency for each temperature difference. In real devices, TEGs are always operated at an optimal current. One thing to note is that the efficiency of TEG is still low compared to other energy-conversion techniques. A lot of effort has been made to enhance efficiency [13, 14]. Given that heat sources are plenty and free, TEGs could be promising solutions, when they are employed to harvest waste heat from industry activities and central-heating systems.

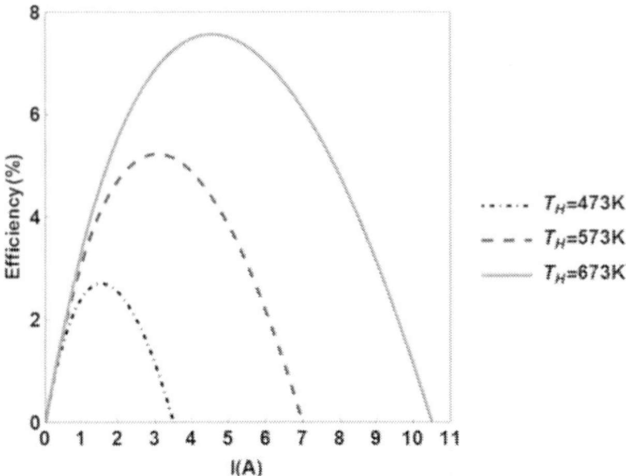

Figure 14. Efficiency as a function of current.

4.4.2. I-V dependences

Employing various temperature differences, while maintaining the cold-side temperature at 373 K, voltage induced as a function of current behaves as shown in **Figure 15**.

One can observe from **Figure 15** that voltage induced for each temperature difference is decreasing and the linear function of output electrical current. Slopes of I–V dependences are the same.

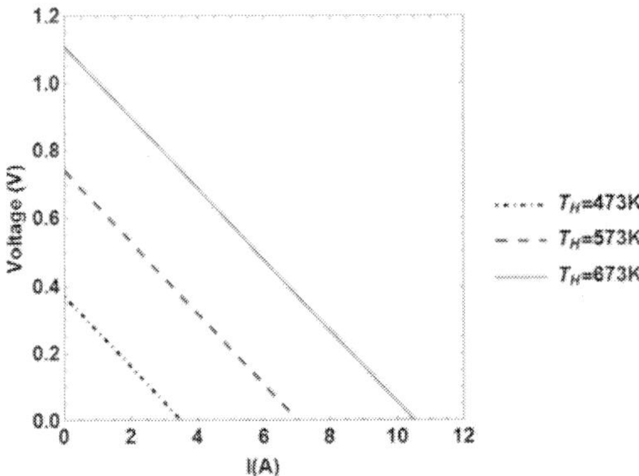

Figure 15. Voltage as a function of current (*I-V* dependences of TEG).

4.4.3. Power and efficiency as a function of hot-side temperature

While still maintaining the cold side at a temperature of 373 K and replacing current in output power equation (Eq. (31)) by optimal current expression (Eq. (33)), power expression becomes a function of temperature at the hot side, and **Figure 16** shows the behavior of output power as a function of the hot-side temperature.

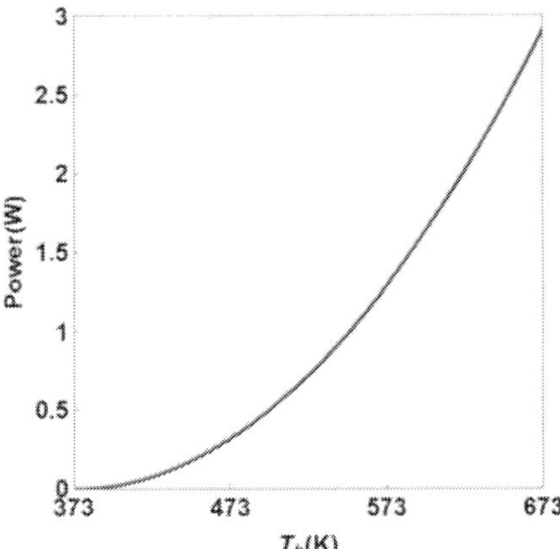

Figure 16. Power as a function of hot-side temperature.

Output power as a function of hot-side temperature behaves as nonlinear curve increasing as the hot-side temperature increases.

The efficiency of TEG as a function of hot-side temperature is shown in **Figure 17**.

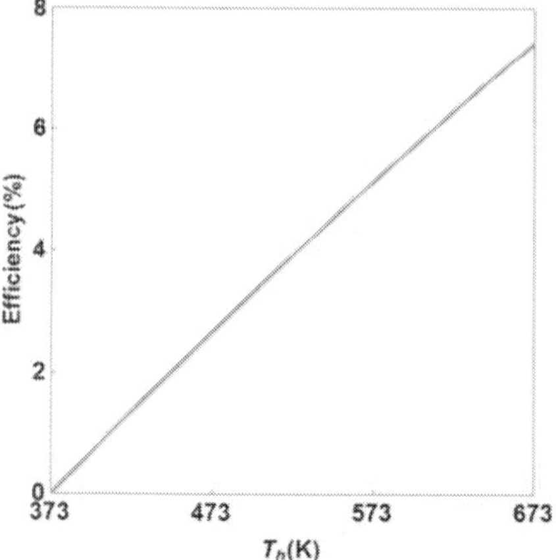

Figure 17. Efficiency of TEG as a function of hot-side temperature.

4.4.4. Power as a function of external load resistance

Figure 18 depicts variations of output power as a function of external load resistance. Eq. (35) is used to obtain dependences shown in **Figure 18**.

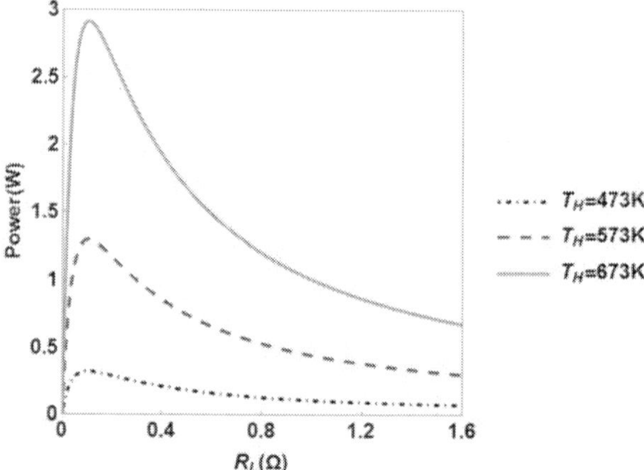

Figure 18. Output power as a function of external load resistance.

Optimal output power occurs when load resistance equates to internal electrical resistance of the total number of p- and n-type semiconductor legs.

4.4.5. Efficiency as a function of the figure of merit (ZT)

ZT value is modified figure of merit, where T represents averaged temperature between the hot-side and cold-side temperatures. For each temperature difference, efficiency increases as ZT value increases. Therefore, employing thermoelectric materials possessing high ZT values leads to great TEG efficiency (**Figure 19**).

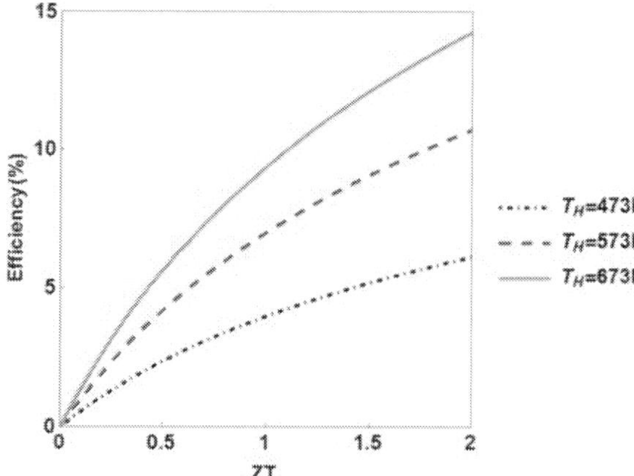

Figure 19. Efficiency as a function of ZT value.

5. Conclusion

In this chapter, the basics of thermoelectric generator devices are covered including phenomena that guide their operation. State-of-the-art modeling efforts are summarized. The presented modeling is crucial for comprehensive understanding of heat to electric energy conversion in TEGs. Simulation results are very useful in predicting the maximum ratings of TEGs during operation under different ambient conditions.

Acknowledgements

This work was funded by the startup fund from Virginia Polytechnic Institute and State University.

Author details

Eurydice Kanimba and Zhiting Tian*

*Address all correspondence to: zhiting@vt.edu

Virginia Polytechnic Institute and State University, USA

References

[1] Rowe DM. CRC handbook of thermoelectric, CRC press, Boca Raton, 1995.

[2] Rowe DM and Bhandari CM. Modern thermoelectrics, Prentice Hall, Upper Saddle River, 1983.

[3] Zevenhoven R and Beyene A. The relative contribution of waste heat from power plants to global warming. Energy, 2011. **36**(6): p. 3754–3762. DOI: 10.1016/j.energy.2010.10.010

[4] Moh'd AA-N, Tashtoush BM and Jaradat AA. Modeling and simulation of thermoelectric device working as a heat pump and an electric generator under Mediterranean climate. Energy, 2015. **90**: p. 1239–1250.

[5] Mamur H and Ahiska R. A review: Thermoelectric generators in renewable energy. International Journal of Renewable Energy Research (IJRER), 2014. **4**(1): p. 128–136.

[6] Ioffe A, Kaye J and Welsh JA. Direct conversion of heat to electricity. 1960: John Wiley and Sons, Inc.

[7] Sutton GW. Direct energy conversion, McGraw-Hill, New York, 1966.

[8] Decher R. Direct energy conversion: fundamentals of electric power production, Oxford University Press on Demand, Oxford, 1997.

[9] Riffat SB and Ma X. Thermoelectrics: A review of present and potential applications. Applied Thermal Engineering, 2003. **23**(8): p. 913–935. DOI: 10.1016/S1359-4311(03) 00012-7

[10] Dziurdzia P. Modeling and simulation of thermoelectric energy harvesting processes, InTech Open Access Publisher, Croatia, 2011.

[11] Thomas JP, Qidwai MA and Kellogg JC. Energy scavenging for small-scale unmanned systems. Journal of Power Sources, 2006. **159**(2): p. 1494–1509. DOI: 10.1016/j. jpowsour.2005.12.084

[12] Meng F, Chen L, and Sun F. A numerical model and comparative investigation of a thermoelectric generator with multi-irreversibilities. Energy, 2011. **36**(5): p. 3513–3522. DOI: http://dx.doi.org/10.1016/j.energy.2011.03.057

[13] Ebling D, et al. Module geometry and contact resistance of thermoelectric generators analyzed by multiphysics simulation. Journal of Electronic Materials, 2010. **39**(9): p. 1376–1380. DOI: 10.1007/s11664-010-1331-0

[14] Priya S and Inman DJ. Energy harvesting technologies. Vol. 21, Springer, New York 2009.

Chapter 19

Calculation Methods for Thermoelectric Generator Performance

Fuqiang Cheng

Additional information is available at the end of the chapter

http://dx.doi.org/10.5772/65596

Abstract

This chapter aims to build one-dimensional thermoelectric model for device-level thermoelectric generator (TEG) performance calculation and prediction under steady heat transfer. Model concept takes into account Seebeck, Peltier, Thomson effects, and Joule conduction heat. Thermal resistances between heat source, heat sink, and thermocouple are also considered. Then, model is simplified to analyze influences of basic thermal and electrical parameters on TEG performance, when Thomson effect is neglected. At last, an experimental setup is introduced to gauge the output power and validate the model. Meantime, TEG simulation by software ANSYS is introduced briefly.

Keywords: thermoelectric generator, thermoelectric model, output power, thermoelement

1. Introduction

Output power P_{out} and energy conversion efficiency η are the primary parameters to characterize TEG performance. They are intensively influenced by such factors as temperature of heat source and sink, thermoelectric materials physical properties, thermocouple geometries, thermal and electrical contact properties, and load factor. Therefore, it is necessary to build physical model formulating these factors concisely, to conduct realistic TEG design. At present, many significant works have been undertaken for modeling device-level TEG precisely [1–3]. In addition, comprehensive three-dimensional (3D) thermoelectric model has been successfully developed in software ANSYS [4]. In Refs. [5–7], quasi-one-dimensional thermoelectric model is established, where Thomson effect and thermal resistances between thermocouple and heat source, heat sink are neglected. In Ref. [8], improved one-dimensional model

including Thomson coefficient and thermal resistances is used to analyze the matched load, the limit of energy conversion efficiency, and the influence of Peltier effect. It shows, that expression of matched load contains not only the inner electrical resistance of TEG, but also the terms resulting from Peltier and Joule effects. In Ref. [9], one-dimensional model to analyze the influence of Thomson heat is built and experimentally validated.

In this chapter, Seebeck, Peltier, Thomson effect, and Joule conduction heat are formulated in thermoelectric generation module model. By model simplification, analytical expressions of output power and energy efficiency are introduced. Essential factors for enhancing the output power are extracted. Then, an experimental setup is built to measure the output power and validate the model. And TEG simulation by software ANSYS is presented.

2. Thermoelectric model for device-level TEG

2.1. TEG cell structure

TEG cell consisting of thermocouple is shown in **Figure 1**, where basic thermoelectric effects including Peltier and Joule heat and a circuit with load R_L are included. The p and n thermo-elements are cuboids of the same thickness and bridged by an electrode in series. Practical devices usually make use of thermoelectric modules containing a number of TEG cells connected electrically in series and thermally in parallel. Cross-sectional area and thickness of thermocouple are marked as A and l. Subscripts 'n' and 'p' are used to discriminate conductivity type of thermoelements. Temperature of heat source and heat sink is T_1 and T_0, and that of hot and cold side of thermocouple is T_h and T_c. $\Delta T_g = T_h - T_c$ is temperature difference on thermoelements, and $\Delta T = T_1 - T_0$ is the one of heat source and heat sink.

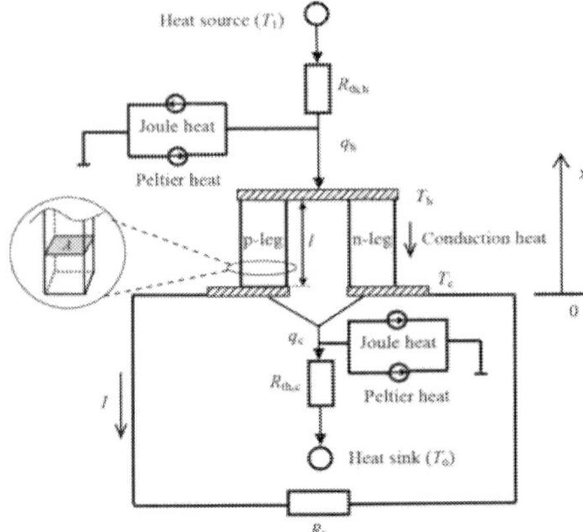

Figure 1. Structure and circuit sketch of TEG cell.

There are Joule heat flowing out and Peltier heat flowing in at hot end of thermoelements, and at cold end, Peltier heat flows out and Joule heat flows out. In addition, there is thermal resistance $R_{th,h}$ and $R_{th,c}$ between thermoelements, heat source and heat sink. Heat flow q_h passes from heat source to hot side of thermocouple and the counterpart q_c outflows from cold side of thermocouple to heat sink.

2.2. Basic model

It is assumed, that thermoelements are physically homogeneous and insulated from the surroundings both electrically and thermally, except at junction-reservoir contacts [8–9]. Variable x is defined as location in the thickness direction of thermoelements. According to nonequilibrium thermodynamics under steady heat transfer, energy conservative equations of temperature distributions $T_n(x)$ and $T_p(x)$ are:

$$\begin{cases} K_n l_n \dfrac{d^2 T_n(x)}{dx^2} - \tau_n I \dfrac{dT_n(x)}{dx} + \dfrac{R_n I^2}{l_n} = 0 \\[2mm] K_p l_p \dfrac{d^2 T_p(x)}{dx^2} + \tau_p I \dfrac{dT_p(x)}{dx} + \dfrac{R_p I^2}{l_p} = 0 \end{cases} \tag{1}$$

Three terms in the above equations represent thermal conduction, Thomson and Joule heat. K, R, and τ are thermal conductance, electrical resistance and Thomson coefficient (V·K^{-1}), respectively. Relationship of K, R and A, l is $K = \frac{\lambda A}{l}$ and $R = \frac{\rho l}{A}$, where λ and ρ are thermal conductivity and electrical resistivity of thermoelectric materials. To solve Eq. (1) analytically, material parameters K, R, and τ are considered to be constant. The boundary conditions of Eq. (1) are:

$$T_n(0) = T_p(0) = T_c, \tag{2}$$

$$T_n(l_n) = T_p(l_p) = T_h. \tag{3}$$

Electrical current I is determined by formula:

$$I = \frac{U_0}{R_L + R_g}, \tag{4}$$

$$U_0 = \int_{T_c}^{T_h} \alpha(T)dT, \tag{5}$$

where U_0 is the voltage of thermocouple, R_g is the electrical resistance of TEG cell, which contains resistance of thermocouple and contact resistance, and $\alpha(T) = \alpha_p(T) - \alpha_n(T)$ is Seebeck coefficient $(V \cdot K^{-1})$ of thermocouple.

In practice, temperature of heat source T_1 and heat sink T_0 can be measured and determined. To acquire T_h and T_c, relationship of T_1, T_0 and T_h, T_c is necessary. That is:

$$\begin{cases} T_1 - T_h = R_{th,h}q_h \\ T_c - T_0 = R_{th,c}q_c \end{cases}. \tag{6}$$

In Eq. (6), heat flows q_h and q_c are:

$$q_h = K_n l_n \left.\frac{dT_n(x)}{dx}\right|_{x=l_n} + K_p l_p \left.\frac{dT_p(x)}{dx}\right|_{x=l_p} + \alpha(T_h)T_h I - I^2 R_{ch}, \tag{7}$$

$$q_c = K_n l_n \left.\frac{dT_n(x)}{dx}\right|_{x=0} + K_p l_p \left.\frac{dT_p(x)}{dx}\right|_{x=0} + \alpha(T_c)T_c I + I^2 R_{cc}, \tag{8}$$

wherein R_{ch} and R_{cc} are contact electrical resistances at hot and cold side of the thermocouple. Thermal conduction heat, Peltier heat (the third term), and contact Joule heat are within Eqs. (7) and (8). By solving Eq. (1) with Eqs. (2)–(5), $T_n(x)$ and $T_p(x)$ only relating to T_h, T_c and R_L can be obtained. And flows q_h and q_c can be formulated with T_h, T_c and R_L in Eqs. (7) and (8). Then, T_h and T_c can be determined for a given R_L by solving Eq. (6) numerically, which is presented in detail [9].

Finally, the output power P_{out} and energy conversion efficiency η are calculated by the basic equations of thermoelectricity:

$$P_{out} = q_h - q_c = I^2 R_L, \tag{9}$$

$$\eta = \frac{P}{q_h}. \tag{10}$$

When neglecting Thomson heat, the problem will be much simplified. By solving Eqs. (1)–(8) with $\tau = 0$, an cubic equation about ΔT_g can be yielded as:

$$a_1 \Delta T_g^3 + b_1 \Delta T_g^2 + c_1 \Delta T_g + 1 = 0, \tag{11}$$

where:

$$a_1 = \frac{\bar{\alpha}^4 R_L R_{th,c} R_{th,h}}{\left(R_g + R_L\right)^3 \Delta T},$$

$$b_1 = -\frac{\bar{\alpha}^2 R_{th,c} R_{th,h}}{\left(R_g + R_L\right)^2 \Delta T}\left[R_g\left(\frac{\varepsilon}{R_{th,c}} + \frac{\varepsilon - 1}{R_{th,h}}\right) + R_L\left(\frac{1}{R_{th,c}} - \frac{1}{R_{th,h}}\right)\right],$$

$$c_1 = -\frac{R_{th,c} R_{th,h}}{\Delta T}\left[\left(\frac{T_1}{R_{th,h}} + \frac{T_0}{R_{th,c}}\right)\frac{\alpha^2}{R_g + R_L} + \frac{K_g}{R_{th,h}} + \frac{K_g}{R_{th,c}} + \frac{1}{R_{th,c} R_{th,h}}\right],$$

$$\varepsilon = \frac{0.5 R_0 + R_{cc}}{R_g},$$

and Seebeck coefficient α becomes a constant. Equation (11) suits thermoelectric module consisting of m thermocouples, as well, where α and K_g are m times of those of a single thermocouple, but $R_{th,h}$ and $R_{th,c}$ are exactly on the contrary.

Generally, c_1 is far larger than a_1 and b_1 in absolute value. Mainly, because of practical module, Seebeck coefficient α has a very small value of about 10^{-2} V·K^{-1}, which is much less than unity. For example, taking module TEG-127-150-9 in Ref. [8], $\alpha = 0.05$ V·K^{-1}, $R_g = 3.4$ Ohm, $R_L = 4$ Ohm, $R_{th,c} = 6$ K·W^{-1}, $R_{th,h} = 0.1$ K·W^{-1}, $K_g = 2.907$ W·K^{-1}, $\varepsilon \approx 0.5$, $T_0 = 297$ K, and $T_1 = 323$ K, calculation result is $a_1 \approx 1.423 \times 10^{-9}$ K^{-3}, $b_1 \approx 5.905 \times 10^{-5}$ K^{-2}, and $c_1 \approx -0.7461$ K^{-1}. So, the terms with ΔT_g order higher than unity can be neglected. At last, here is:

$$\Delta T_g = \frac{\Delta T}{1 + R_{th,c} K_g + R_{th,h} K_g + \alpha^2 \dfrac{\left(R_{th,c} T_1 + R_{th,h} T_0\right)}{R_g + R_L}}. \tag{12}$$

It can be seen, that ΔT_g is influenced not only by thermal resistances $R_{th,c}$ and $R_{th,h}$, but also by Peltier effect, which is presented in the last term of the denominator and functions to decrease ΔT_g. Because it is tantamount to accelerate heat conduction in thermocouple, Peltier heat flows in and out on two sides of thermocouple. By combing Eqs. (12) and (4), (5), (9), the output power P_{out} is:

$$P_{out} = \frac{\alpha^2 \Delta T_g^2 R_L}{\left(R_g + R_L\right)^2} = \frac{\alpha^2 \Delta T^2 R_L}{\left(1 + R_{th,c} K_g + R_{th,h} K_g\right)^2 \left[R_g + R_L + \dfrac{\alpha^2 \left(R_{th,c} T_1 + R_{th,h} T_0\right)}{1 + R_{th,c} K_g + R_{th,h} K_g}\right]^2}. \tag{13}$$

R_L, T_1, T_0, $R_{th,c}$, $R_{th,h}$, α, R_g and K_g directly affect P_{out}. In those parameters, α, R_g and K_g are TEG internal factors, and R_L, T_1, and T_0 are the external ones, and $R_{th,c}$ and $R_{th,h}$ originate from both the internal and external. From the form of Eq. (13), it is obvious, that reducing T_1, T_0, $R_{th,c}$, and $R_{th,h}$ can increase P_{out}, if ΔT is constant, owing to influence of Peltier effect on ΔT_g. On the other hand, P_{out} has a maximum along with R_g and K_g.

2.3. Matched load, output power and energy efficiency

First of all, influence of R_L on P_{out} is analyzed. In Eq. (13), P_{out} reaches maximum, when R_L is:

$$R_L = R_g + \frac{\alpha^2 \left(R_{th,c} T_1 + R_{th,h} T_0 \right)}{1 + R_{th,c} K_g + R_{th,h} K_g}, \tag{14}$$

which is the matched load and marked as $R_{L,m}$. Indeed, $R_{L,m}$ is slightly larger than R_g due to the very small value of α^2. It means, that existence of Peltier effect increases irreversible heat in thermoelectric module. And reducing T_1 and T_0 helps to cut down this irreversible heat. When $K_g \rightarrow +\infty$, $R_{L,m}$ is equal to R_g, since at this moment heat conduction in thermocouple runs under infinitesimal temperature difference and the irreversibility of heat transfer disappears. However, this irreversibility exists with finite K_g, leading to heat loss in thermocouple, that is equivalent to increase in internal resistance. Define R_L/R_g as the load factor s_L. So, s_L is:

$$s_L = 1 + \frac{ZK_g \left(R_{th,c} T_1 + R_{th,h} T_0 \right)}{1 + R_{th,c} K_g + R_{th,h} K_g}, \tag{15}$$

when R_L is equal to matched load and $Z = \dfrac{\alpha^2}{K_g R_g}$ is the figure of merit. For thermoelectric module, the output power is:

$$P_{out} = \frac{m^2 \alpha^2 \Delta T^2 R_L}{\left(1 + R_{th,c} K_g + R_{th,h} K_g\right)^2 \left\{ R_L + m \left[R_g + \dfrac{\alpha^2 (R_{th,c} T_1 + R_{th,h} T_0)}{1 + R_{th,c} K_g + R_{th,h} K_g} \right] \right\}^2}, \tag{16}$$

where m is the number of thermocouples. And the corresponding matched load is

$$R_{L,m} = m \left[R_g + \frac{\alpha^2 (R_{th,c} T_1 + R_{th,h} T_0)}{1 + R_{th,c} K_g + R_{th,h} K_g} \right].$$

As for energy efficiency η, by Eq. (7), which can be expressed as function of ΔT_g and Eqs. (12) and (13), it is:

$$\eta = \frac{P}{q_h} = \frac{\alpha^2 \Delta T^2 R_L}{c_2^2\left(R_L + R_g\right)^2 \left\{ \dfrac{\alpha^2 \Delta T \left[T_h\left(R_L + R_g\right) - \dfrac{\varepsilon \Delta T R_g}{c_2} \right]}{c_2\left(R_L + R_g\right)^2} + \dfrac{\Delta T K_g}{c_2} \right\}},$$

(17)

where

$$c_2 = 1 + R_{th,c} K_g + R_{th,h} K_g + \frac{\alpha^2 \left(R_{th,c} T_1 + R_{th,h} T_0 \right)}{R_L + R_g}.$$

By solving Eq. (17) about the partial derivative of R_L, it can be obtained, that when load factor s_L is:

$$s_L = \sqrt{1 + ZT_h + \frac{Z\varepsilon\Delta T + ZK_g\left(1 + ZT_h\right)\left(R_{th,c} T_1 + R_{th,h} T_0\right)}{1 + R_{th,c} K_g + R_{th,h} K_g}},$$

(18)

then η reaches maximum. Equation (18) is downright different from Eq. (15) in the expressions, so achieving maximum of output power and energy efficiency simultaneously is impossible. Actually, the corresponding load factor of the former is smaller than that of the latter. When the ideal state is considered ($R_{th,c} = R_{th,h} = 0$), $s_L = 1$ is for the former and $s_L = \sqrt{1 + ZT_h + Z\varepsilon\Delta T}$, which is larger than 1, is for the latter.

2.4. Influence of K_g on TEG performance

K_g is important internal factor that influences the output performance in TEG. When matched load is reached, the corresponding output power $P_{out,m}$ is:

$$P_{out,m} = \frac{ZK_g \Delta T^2}{4\left(1 + R_{th,c} K_g + R_{th,h} K_g\right)^2 \left[1 + \dfrac{ZK_g\left(R_{th,c} T_1 + R_{th,h} T_0\right)}{1 + R_{th,c} K_g + R_{th,h} K_g} \right]}.$$

(19)

For a common thermoelectric module, thermoelements have the same size, $l_n = l_p$ and $A_n = A_p$, so the figure of merit Z is not related to their size, but material physical parameters. From Eq. (19), we can see, that increase in Z will enhance the output power. By solving Eq. (19) regarding the partial derivative of K_g, when:

$$K_g = \frac{1}{\sqrt{\left(R_{th,c} + R_{th,h}\right)^2 + Z\left(R_{th,c}T_1 + R_{th,h}T_0\right)\left(R_{th,c} + R_{th,h}\right)}} = \left(\lambda_p + \lambda_n\right)\frac{A_e}{l_e}, \tag{20}$$

then $P_{out,m}$ reaches maximum, where l_e and A_e are thickness and cross-sectional area of thermoelements. Since $R_{th,c}$ and $R_{th,h}$ are related to A_e, but not to l_e, there is an optimal l_e to maximize $P_{out,m}$:

$$P_{out,m} = \frac{Z\Delta T^2}{4c_3\left(1 + \dfrac{R_{th,c}}{c_3} + \dfrac{R_{th,h}}{c_3}\right)^2\left[1 + \dfrac{Z\left(R_{th,c}T_1 + R_{th,h}T_0\right)}{c_3\left(1 + \dfrac{R_{th,c}}{c_3} + \dfrac{R_{th,h}}{c_3}\right)}\right]}, \tag{21}$$

and

$$c_3 = \sqrt{\left(R_{th,c} + R_{th,h}\right)^2 + Z\left(R_{th,c}^2 T_1 + R_{th,h}^2 T_0 + R_{th,c}R_{th,h}T_1 + R_{th,c}R_{th,h}T_0\right)}.$$

2.5. Influence of Peltier effect on TEG performance

When Peltier effect is neglected, the relation of ΔT_g and ΔT is:

$$\Delta T_g = \frac{R_{th,g}}{R_{th,g} + R_{th,h} + R_{th,c}}\Delta T, \tag{22}$$

and the corresponding output power with matched load $R_L = R_g$ and constant Seebeck coefficient is:

$$P_{out,m} = \frac{\left(\alpha\Delta T\right)^2}{4R_g\left(1 + R_{th,h}K_g + R_{th,c}K_g\right)}. \tag{23}$$

Meantime, the output power Eq. (13) is for the condition without Peltier effect, and the ratio of Eq. (23) and Eq. (19), η^{Pelt}, reflects influence degree of Peltier effect on the output power:

$$\eta_{Pelt} = \left(1 + \frac{ZT_0}{2}\frac{R_{th,c} + R_{th,h}}{R_{th,g} + R_{th,h} + R_{th,c}} + \frac{Z}{2}\frac{R_{th,c}\Delta T}{R_{th,g} + R_{th,h} + R_{th,c}}\right)^2. \tag{24}$$

It is known, that when $R_{th,g} \ll R_{th,c} + R_{th,h}$, η_{Pelt} is approximately $(1 + 0.5ZT_0 + 0.25Z\Delta T)^2$ with $R_{th,c} \approx R_{th,h}$, and even with $\Delta T \to 0$, the output power calculated without Peltier effect is more than the output power considering Peltier effect, by over 120% for a common Bi_2Te_3-based module with $ZT \approx 1$. That means, the influence of Peltier effect must be considered. Similar status is obtained, where the difference is more than 50%, when $R_{th,g} \approx R_{th,c} + R_{th,h}$. On the contrary, when $R_{th,g} \gg R_{th,c} + R_{th,h}$, η_{Pelt} is approximately equal to 1, which means the influence of Peltier effect is negligible. Hence, the smaller the thermal resistance of thermocouple $R_{th,g}$, the stronger is the influence of Peltier effect.

Eventually, basic factors for enhancing TEG output power are summarized as:

1. Enhancing ZT of thermoelectric materials and ΔT, decreasing $R_{th,c}$ and $R_{th,h}$.

2. When ΔT is fixed, lower T_0 and T_1 can reduce irreversible heat to elevate output power.

3. Matched load is a little larger than the inner electrical resistance of TEG.

4. There exists an optimal thermocouple thickness to maximize output power.

3. Test validation

3.1. Materials property

P-type and n-type Bi_2Te_3-based materials are, respectively, $Bi_{0.5}Sb_{1.5}Te_3$ and $Bi_2Te_{2.85}Se_{0.15}$, which are prepared by mechanical alloy + spark plasma sintering method. Seebeck coefficient and resistivity of the materials are tested by HGTE-II thermoelectric material performance test system (Chinese patent no. ZL200510018806.4) with test temperature up to 1073 K, relative error of not more than 6%. Thermal conductivity of the materials is measured by laser perturbation method (Type TC-7000 of ULVAC RIKO®). As shown in **Table 1**, parameters are obtained by polynomial fitting of the experimental data. In temperature range 273 K < T < 493 K, Seebeck coefficient value α is between 170×10^{-6} V·K^{-1} and 220×10^{-6} V·K^{-1}, decreasing with rising temperature. Electrical resistivity ρ is $(8.3–20.0) \times 10^{-6}$ Ohm·m and thermal conductivity λ is 1.4–2.1 W·m^{-1}·K^{-1}, which both show obvious increase with temperature rise.

Material	Parameters	Values
p-$Bi_{0.5}Sb_{1.5}Te_3$	α/V·K^{-1}	$-1.791 \times 10^{-11}T^3 + 1.763 \times 10^{-8}T^2 - 5.714 \times 10^{-6}T + 8.304 \times 10^{-4}$
	ρ/Ohm·m	$-7.929 \times 10^{-13}T^3 + 7.992 \times 10^{-10}T^2 - 1.947 \times 10^{-7}T + 1.728 \times 10^{-5}$
	λ/W·m^{-1}·K^{-1}	$3.342 \times 10^{-5}T^2 - 2.24 \times 10^{-2}T + 5.118$
n-$Bi_2Te_{2.85}Se_{0.15}$	α/V·K^{-1}	$1.321 \times 10^{-11}T^3 - 1.383 \times 10^{-8}T^2 + 4.81 \times 10^{-6}T - 7.774 \times 10^{-4}$
	ρ/Ohm·m	$-7.618 \times 10^{-13}T^3 + 8.098 \times 10^{-10}T^2 - 2.537 \times 10^{-7}T + 3.207 \times 10^{-5}$
	λ/W·m^{-1}·K^{-1}	$3.264 \times 10^{-5}T^2 - 2.228 \times 10^{-2}T + 5.302$

Table 1. Physical parameters of Bi_2Te_3-based materials (273 K < T < 493 K).

In practice, it is difficult to measure thermal resistances $R_{th,c}$ and $R_{th,h}$, and contact electrical resistances r_{cc} and r_{ch}. Their values are determined according to empirical formulas. Contact electrical resistivity ρ_c (Ohm·m^2) at leg-strap junctions and thermal conductivity λ_c (W m^{-1} K^{-1}) of thermal conductive layer ($\approx$1.2 mm thick) are according to the empirical formulas given by Rowe et al. [5]:

$$\frac{2\rho_c}{\rho} = 0.1 \text{ mm}, \frac{\lambda}{\lambda_c} = 0.2. \tag{25}$$

Here ρ and λ are electrical resistivity and thermal conductivity of thermoelements, respectively. In our calculation, the mean values of ρ and λ over the temperature range are taken as references. As ρ and λ vary with temperature, values of ρ_c and λ_c are also different as the temperature varies. Experiments under four temperature conditions are carried out and the corresponding parameter values are shown in **Table 2**.

Temperatures, °C Parameters	$T_0 = 23$ $T_1 = 81$	$T_0 = 23$ $T_1 = 111$	$T_0 = 27$ $T_1 = 147$	$T_0 = 27$ $T_1 = 177$
r_{cc}/mOhm	0.6	0.7	0.7	0.8
r_{ch}/mOhm	0.6	0.7	0.7	0.8
$R_{th,c}$/K·W^{-1}	31.0	31.5	30.9	31.0
$R_{th,h}$/K·W^{-1}	23.5	23.3	22.8	23.0

Table 2. Thermal resistances and contact electrical resistances under different temperatures.

3.2. Test setup

System for measuring output performance of thermoelectric modules was established, mainly including electric heating plate controlled by PID, adjustable load, circulatory cooling unit, thermal imaging device, temperature and voltage data acquisition units, etc., with its basic structure as shown in **Figure 2**. Electric heating plate is used as heat source, with temperature control precision of ±0.1 K and temperature ranging from room temperature to 773 K. Cooling unit, which consists of heat sink, water tank, flow meter and flow valve, etc., takes cold water as the coolant. Heat sink is made of red-copper and its temperature could be adjusted by controlling flowrate of cooling water. In addition, some thermal conductive filler is pasted on both sides of module to reduce thermal resistance between module, heat source and heat sink. Electrical current in the circuit is obtained via measuring voltage on both ends of sampling resistor (metal film precision resistor: 0.2 Ohm, precision of ±1%).

Voltage and temperature signal are acquired by 9207 and 9214 acquisition card of National Instruments (NI) Company, with precision of ±0.5%. Data to be acquired include as follows: (1) temperature of heat source and heat sink; (2) temperature of the coolant (water) inside heat sink and water tank; (3) voltage on adjustable load and sampling resistor. K-type thermocouples with diameter of 1 mm are inserted in heat source and heat sink to measure temperature values. Actually, even though electric heating plate is controlled by PID, heat source temperature still fluctuates during the change of load resistance. In order to eliminate impacts of such transient effect, data shall not be acquired until the heat source and heat sink temperatures are stable.

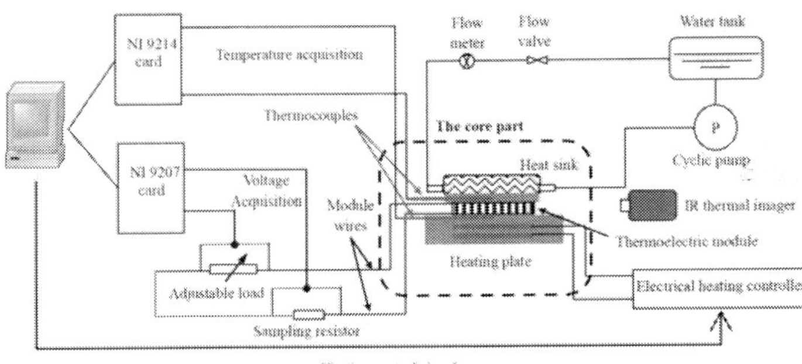

Figure 2. Configuration of the output performance test system for the thermoelectric modules.

Test of energy efficiency is not undertaken due to its complexity, where the heat flow into the hot side of the module must be measured or evaluated. An effective way is to adopt heat flux sensor and bury it just under the module. But that would impact heat conduction between heat source and the module, leading to higher thermal resistance. And heat flux sensors of high temperature enduring are really costly. Another useful method is by calculating electrically generated heat in heat source, and at the same time, radiation and convective heat loss must be subtracted, as is introduced in Ref. [10].

3.3. Comparison of test results with calculation

Figures 3 and **4** show variations of output power P_{out} with load R_L at four temperature conditions, acquired by physical model calculation, ANSYS simulation and experiment. ANSYS method will be introduced in the next part. **Figure 3** shows data at heat sink temperature $T_0 = 300$ K, while **Figure 4**—at $T_0 = 296$ K. **Figures 5** and **6** are the corresponding current-voltage (I-V) characteristics. R_L results are disposed in the same way. From the results, it is found, that the output power has maximum value with the increase of load. And current is linearly related to voltage. Calculation results are well coincident with ANSYS results, and they are both a little higher than experimental data. Under the four temperature conditions, values of maximum output power are 2.5, 2.6, 2.8 and 1.1% higher than experimental results with T_1 changing from high to low. They are especially coincident well, when temperature difference ΔT is small. From the analysis follows, deviation of calculated results is caused mainly by taking thermal conductivity and electrical resistivity as constant (i.e., using the mean values), when solving Eq. (1). When ΔT is small, then material physical parameters vary within a narrow range. So, values of parameters are close to the real values. Otherwise, when ΔT is large, material physical parameters change within a large scale, leading to a great deviation of calculations.

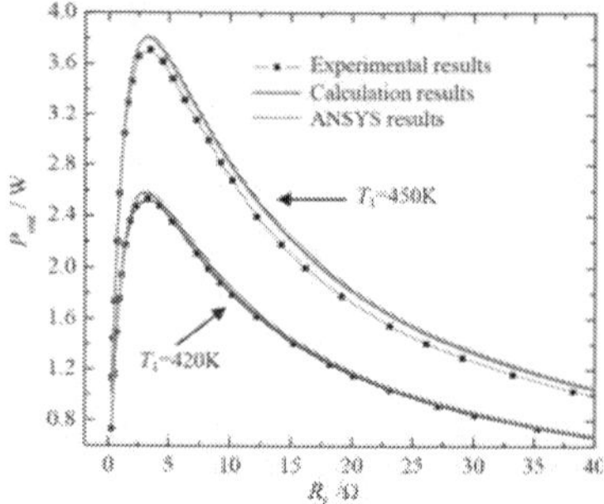

Figure 3. Dependences of output power on load resistance: calculations, experiments and ANSYS at $T_0 = 300$ K.

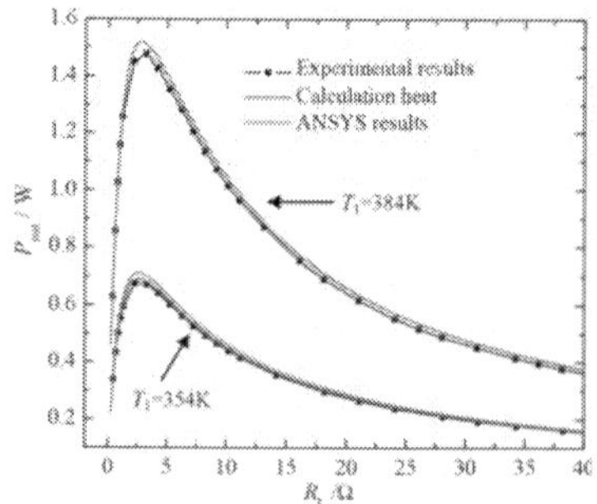

Figure 4. Dependences of output power on load resistance: calculations, experiments and ANSYS at $T_0 = 296$ K.

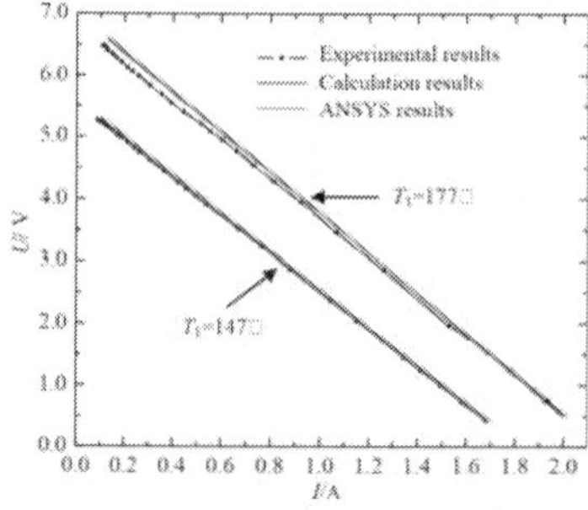

Figure 5. I-V characteristics of thermoelectric module: calculations, experiments and ANSYS at $T_0 = 300$ K.

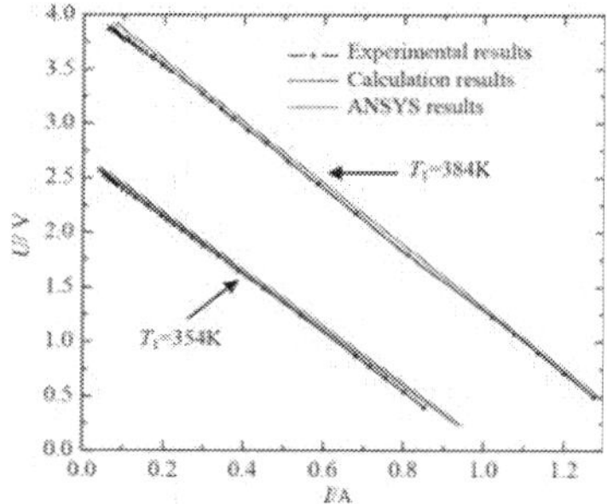

Figure 6. I-V characteristics of thermoelectric module: calculations, experiments and ANSYS at $T_0 = 296$ K.

4. Introduction to TEG simulation in ANSYS

4.1. TEG cell model

By software simulation, TEG performance can be achieved both in thermal and in electrical aspects. But it is not direct to cognize and understand the influence of thermoelectric effects, when compared with the above physical model. In this part, TEG cell model is set up by ANSYS, and geometry and meshing methods are illustrated in **Figure 7**. Thickness and cross-sectional area of thermoelements are 1.6 mm and 1.4 mm × 1.4 mm, respectively. Other geometry parameters are shown in **Figure 7**. Thermoelectric module consists mainly of p-n thermoelements, current-conducting copper straps and ceramic substrates for heat conducting and electric insulation. Thermoelements and copper strap are meshed by element SOLID226 in ANSYS. This type of element contains 20 nodes with voltage and temperature as the degrees of freedom. It can simulate 3D thermal-electrical coupling field. Element SOLID90 is used to mesh ceramic substrate. It has 20 nodes with temperature as the degree of freedom. Load resistance is simulated by element CIRCU124.

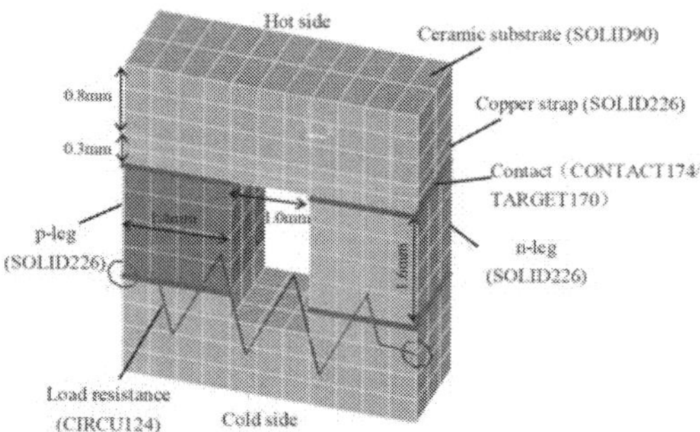

Figure 7. The geometry of TEG cell in ANSYS and its mesh.

Contact properties of the leg-strap junction are implemented with element pairs CONTACT174/TARGET170. Detailed finite element formulations in ANSYS are introduced in [4], and the range of contact thermal conductivity and electrical resistivity is explicated in [11].

4.2. APDL codes for TEG simulation

ANSYS Parametric Design Language (APDL) is widely used for programed simulation. The following APDL codes have taken temperature variation of materials properties, thermal contact and thermal radiation (although its influence is very weak) into consideration. According to the practical requirements, the readers could use the code more concisely by neglecting certain physical effects. The unit referring to length is meter and the temperature unit is Celsius.

```
! defining the TEG cell dimensions

ln=1.6e-3 ! n-type thermoelement thickness

lp=1.6e-3 ! p-type thermoelement thickness

wn=1.4e-3 ! p-type thermoelement width

wp=1.4e-3 ! p-type thermoelement width

d=1.0e-3 ! Distance between the thermoelements

hs=0.2e-3 ! copper strap thickness

hc=1e-3 !substrate thickness

! definition of several physical parameters

rsvx=1.8e-8 ! copper electrical resistivity

kx=200 ! copper thermal conductivity

kxs=24 !substrate thermal conductivity

T1=250 ! temperature of heat source

T0=30 ! temperature of heat sink

Toffst=273 ! temperature offset

! defining TEG output parameters and the load

*dim,P0,array,1 ! defining P0 as the output power

*dim,R0,array,1 ! defining R0 as the load

*dim,Qh,array,1 ! defining Qh as the heat flow into the TEG cell

*dim,I,array,1 ! defining I as the current

*dim,enta,array,1 ! defining enta as the energy efficiency
```

```
*vfill,R0(1),ramp,0.025 ! setting the load (Ohm)
! pre-processing before calculation, defining element type, building the structure and meshing
/PREP7
toffst,Toffst ! set temperature offset
et,1,226,110 ! 20-node thermoelectric brick element
et,2,shell57 ! shell57 element for radiation simulation
et,3,conta174 ! conta174 element for contact simulation
et,4,targe170 ! target170 element for contact simulation
keyopt,3,1,4 ! taking temperature and voltage as the degree of freedom
keyopt,3,9,0
keyopt,3,10,1
keyopt,4,2,0
keyopt,4,3,0
! Temperature data points
mptemp,1,25,50,75,100,125,150
mptemp,7,175,200,225,250,275,300
mptemp,13,325,350
! Seebeck coefficient of the n-type material (V·K⁻¹)
mpdata,sbkx,1,1,-160e-6,-168e-6,-174e-6,-180e-6,-184e-6,-187e-6
mpdata,sbkx,1,7,-189e-6,-190e-6,-189e-6,-186.5e-6,-183e-6,-177e-6
mpdata,sbkx,1,13,-169e-6,-160e-6
! electrical resistivity of the n-type material (Ohm*m)
mpdata,rsvx,1,1,1.03e-5,1.06e-5,1.1e-5,1.15e-5,1.2e-5,1.28e-5
mpdata,rsvx,1,7,1.37e-5,1.49e-5,1.59e-5,1.67e-5,1.74e-5,1.78e-5
mpdata,rsvx,1,13,1.8e-5,1.78e-5
! thermal conductivity of the n-type material (m* K⁻¹)
mpdata,kxx,1,1,1.183,1.22,1.245,1.265,1.265,1.25
mpdata,kxx,1,7,1.22,1.19,1.16,1.14,1.115,1.09
mpdata,kxx,1,13,1.06,1.03
! Seebeck coefficient of the p-type material (V·K⁻¹)
```

```
mpdata,sbkx,2,1,200e-6,202e-6,208e-6,214e-6,220e-6,223e-6
mpdata,sbkx,2,7,218e-6,200e-6,180e-6,156e-6,140e-6,120e-6
mpdata,sbkx,2,13,101e-6,90e-6
! electrical resistivity of the p-type material (Ohm*m)
mpdata,rsvx,2,1,1.0e-5,1.08e-5,1.18e-5,1.35e-5,1.51e-5,1.7e-5
mpdata,rsvx,2,7,1.85e-5,1.98e-5,2.07e-5,2.143e-5,2.15e-5,2.1e-5
mpdata,rsvx,2,13,2.05e-5,2.0e-5
! thermal conductivity of the p-type material (m* K⁻¹)
mpdata,kxx,2,1,1.08,1.135,1.2,1.25,1.257,1.22
mpdata,kxx,2,7,1.116,1.135,1.13,1.09,1.12,1.25
mpdata,kxx,2,13,1.5,2.025
! material property for cooper strap
mp,rsvx,3,rsvx
mp,kxx,3,kx
! material property for the substrate
mp,kxx,4,kxs
!radiation property for the p-n materials
mp,emis,5
! contact friction coefficient
mp,mu,6,0
! build the TEG cell structure
block,d/2,wn+d/2,-ln,0,,t
block,-(wp+d/2),-d/2,-lp,0,,t
block,d/2,wn+d/2,,hs,,t
block,-(wp+d/2),-d/2,,hs,,t
block,-d/2,d/2,,hs,,t
block,-(wp+d/2),-d/2,-lp,-(lp+hs),,t
block,d/2,wn+d/2,-ln,-(ln+hs),,t
block,-(wp+d/2),wn+d/2,hs,hs+hc,,t
block,-(wp+d/2),wn+d/2,-(lp+hs),-(lp+hs+hc),,t
```

```
! glue the copper strap and the substrate
vsel,s,loc,y,0,hs
vsel,a,loc,y,hs,hc+hs
vglue,all
allsel
vsel,s,loc,y,-lp-hs,-lp
vsel,a,loc,y,-lp-hs-hc,-lp-hs
vglue,all
allsel
! meshing the TEG cell structure
numcmp,all
mshape,0,3d
mshkey,1
type,1
mat,3
lsel,s,loc,x,-d/2,d/2
lsel,r,loc,y,0
lsel,r,loc,z,t
lesize,all,d/3
vsel,s,loc,x,-d/2,d/2
vsel,r,loc,y,0,hs
vsweep,all
allsel
esize,ww/3
type,1
mat,3
vsel,s,loc,y,0,hs
vsel,u,loc,x,-d/2,d/2
vsweep,all
vsel,s,loc,y,-lp-hs,-lp
```

```
vsweep,all
type,1
mat,1
vsel,s,loc,x,d/2,d/2+wn
vsel,r,loc,y,-ln,0
vmesh,all
mat,2
vsel,s,loc,x,-(wp+d/2),-d/2
vsel,r,loc,y,-lp,0
vmesh,all
type,1
mat,4
vsel,s,loc,y,hs,hs+hc
vsel,a,loc,y,-lp-hs-hc,-lp-hs
vsweep,all
allsel
! defining the contact parameters
r,5 ! selecting the thermal contact conductivity and resistivity
RMORE,
rmore,,7e5 ! setting the thermal contact conductivity
rmore,0.67e8,0.5 ! setting the thermal contact resistivity
! defining the contact layer between p-leg and upper copper strap
vsel,s,loc,y,0,hs
asel,s,ext
asel,r,loc,y,0
nsla,s,1
nsel,r,loc,x,-(wp+d/2),-d/2
type,3
mat,6
real,5
```

```
esurf

allsel

! defining the target layer between p-leg and upper copper strap

vsel,s,mat,,2

asel,s,ext

asel,r,loc,y,0

nsla,s,1

type,4

mat,6

esurf

allsel

! defining the contact layer between n-leg and upper copper strap

vsel,s,loc,y,0,hs

asel,s,ext

asel,r,loc,y,0

nsla,s,1

nsel,r,loc,x,d/2,d/2+wn

type,3

mat,6

real,5

esurf

allsel

! defining the target layer between n-leg and upper copper strap

vsel,s,mat,,1

asel,s,ext

asel,r,loc,y,0

nsla,s,1

type,4

mat,6

esurf
```

```
allsel
! defining the contact layer between p-leg and bottom copper strap
vsel,s,loc,y,-hs-lp,-lp
vsel,r,loc,x,-wp-d/2,-d/2
asel,s,ext
asel,r,loc,y,-lp
nsla,s,1
type,3
mat,6
real,5
esurf
allsel
! defining the target layer between p-leg and bottom copper strap
vsel,s,mat,,2
asel,s,ext
asel,r,loc,y,-lp
nsla,s,1
type,4
mat,6
esurf
allsel
! defining the contact layer between n-leg and bottom copper strap
vsel,s,loc,y,-hs-ln,-ln
vsel,r,loc,x,d/2,d/2+wn
asel,s,ext
asel,r,loc,y,-ln
nsla,s,1
type,3
mat,6
real,5
```

```
esurf

allsel

! defining the target layer between n-leg and bottom copper strap

vsel,s,mat,,1

asel,s,ext

asel,r,loc,y,-ln

nsla,s,1

type,4

mat,6

esurf

allsel

! defining the shell element for radiation simulation, outputting radiation matrix

! defining the shell element for copper strap

type,2

aatt,3,,2

asel,s,loc,x,-(wp+d/2),wn+d/2

asel,r,loc,y,0,hs

asel,u,loc,y,0

asel,u,loc,y,hs

amesh,all

allsel

asel,s,loc,x,-d/2,d/2

asel,r,loc,y,0

amesh,all

allsel

aatt,3,,2

asel,s,loc,x,-(wp+d/2),wn+d/2

asel,r,loc,y,-lp-hs,-lp

asel,u,loc,y,-lp

asel,u,loc,y,-lp-hs
```

```
amesh,all
allsel
aatt,4,,2
asel,s,loc,x,-d/2,d/2
asel,r,loc,y,-lp-hs
amesh,all
! defining the shell element for p-n thermoelements
allsel
aatt,5,,2
asel,s,loc,x,-(wp+d/2),wn+d/2
asel,r,loc,y,-lp,0
asel,u,loc,y,-lp
asel,u,loc,y,0
amesh,all
! defining the space node for radiation simulation
n,10000,0,0,3e-3
fini
! using radiation matrix method
/aux12
emis,3,1 ! setting the emissivity
emis,4,1
emis,5,1
allsel
geom,0
stef,5.68e-8 ! setting the Stefan-Boltzmann constant
vtype,hidden
space,10000
write,teg,sub ! outputting the radiation super element
fini
/prep7
```

```
! deleting the shell elements and the corresponding mesh
allsel
asel,s,type,,2
aclear,al
etdele,2
allsel
et,5,matrix50,1 ! defining radiation matrix element
! defining boundary conditions and the load
nsel,s,loc,y,hs+hc ! TEG cell hot side
cp,1,temp,all ! coupling of temperature degree of freedom
nh=ndnext(0) ! getting the master node
d,nh,temp,Th ! setting the temperature constraint to the hot side
nsel,all
nsel,s,loc,y,-(ln+hs+hc) ! selecting the TEG cell cold side
d,all,temp,Tc ! setting the temperature constraint to the cold side
nsel,s,loc,y,-(ln+hs),-ln
nsel,r,loc,x,d/2+wn
cp,3,volt,all ! electrical coupling
nn=ndnext(0) ! getting the master node
d,nn,volt,0 ! setting the ground connection node
nsel,all
nsel,s,loc,y,-(lp+hs),-lp
nsel,r,loc,x,-(wp+d/2)
cp,4,volt,all ! ! electrical coupling
np=ndnext(0) ! getting the master node
nsel,all
type,5
allsel
d,10000,temp,300 ! setting the temperature of the space node
se,teg,sub ! reading the radiation super element
```

```
et,6,CIRCU124,0 ! setting the load resistor element
fini
/prep7
! setting the load value and property
r,1,R0(1)
type,6
real,1
numcmp,all
e,np,nn
esel,s,type,,6
circu_num=elnext(0) !getting circuit element number
allsel
fini
! starting the calculation
/SOLU
antype,static ! solution type
cnvtol,heat,1,1.e-3 ! setting the converging value for heat condition
cnvtol,amps,1,1.e-3 ! setting the converging value for the current
neqit,50 ! calculation iteration step
solve ! starting solving
fini
*get,P0(1),elem,circu_num,nmisc,1 ! getting the output power of the TEG cell
*get,Qh(1),node,nh,rf,heat ! getting the heat flow into the TEG cell
*get,I(1),elem,circu_num,smisc,2 ! getting the current
*voper,enta,P0,div,Qh ! calculating the energy efficiency of the TEG cell
```

5. Conclusions

The built one-dimensional model, which is validated by test results, can calculate TEG output power and energy efficiency accurately. By simplifying this model, it is convenient to analyze influences of different thermal and electrical parameters on TEG performance. And basic factors to enhance TEG output power and energy efficiency are extracted. At last, ANSYS simulation considering thermal contact and radiation effects for TEGs is introduced briefly, and basic APDL codes are shared.

Author details

Fuqiang Cheng

Address all correspondence to: chengfq101@aliyun.com

1 Key Laboratory of Fault Diagnosis and Maintenance of In-orbit Spacecraft, Xi'an Satellite Control Center, Xi'an, China

2 Equipment Academy, Beijing, China

References

[1] Kim S.: Analysis and modelling if effective temperature differences and electrical parameters of thermoelectric generators. Applied Energy. 2103;102:1458–1463.

[2] Lee H.: Optimal design of thermoelectric devices with dimensional analysis. Energy. 2013;106:79–88.

[3] Gou X., Yang S., Ou Q.: A dynamic model for thermoelectric generator applied in waste heat recovery. Energy. 2013;52:201–209.

[4] Antonova E. E., Looman D. C.: Finite elements for thermoelectric device analysis in ANSYS. In: 24th International Conference on Thermoelectrics; IEEE; 2005. p. 200–203.

[5] Rowe D. M., Gao M.: Design theory of thermoelectric modules for electrical power generation. IEE Proceedings of Science, Measurement and Technology. 1996;143(6): 351–356.

[6] Glatz W., Muntwyler S., Hierold C.: Optimization and fabrication of thick flexible polymer based micro thermoelectric generator. Sensor and Actuator A. 2006;132:337–345.

[7] Strasser M., Aigner R., Lauterbach C., Sturm T. F., Franosch M., Wachutka G.: Micro-machined CMOS thermoelectric generators as on-chip power supply. Sensor and Actuators A. 2004;114:362–370.

[8] Freunek M., Müller M., Ungan T., Walker W., Reindl L. M.: New physical model for thermoelectric generators. Journal of Electronic Materials. 2009;38(9):1214–1220.

[9]   Cheng F., Hong Y., Zhu C.: A physical model for thermoelectric generators with and without Thomson heat. Journal of Energy Resources Technology. 2014;136(4):2280–2285.

[10] Tatarinov D., Wallig D., Bastian G.: Optimized characterization of thermoelectric generators for automotive application. Journal of Electronic Materials. 2012;41(96): 1706–1712.

[11] Ziolkowski P., Poinas P., Leszczynski J., Karpinski G., Muller E.: Estimation of thermoelectric generator performance by finite element modeling. Journal of Electronic Materials. 2010;39(9):1934–1943.

Performance Analysis of Composite Thermoelectric Generators

Alexander Vargas Almeida,

Miguel Angel Olivares-Robles and Henni Ouerdane

Additional information is available at the end of the chapter

http://dx.doi.org/10.5772/66143

Abstract

Composite thermoelectric generators (CTEGs) are thermoelectric systems composed of different modules arranged under various thermal and electrical configurations (series and/or parallel). The interest for CTEGs stems from the possibility to improve device performance by optimization of configuration and working conditions. Actual modeling of CTEGs rests on a detailed understanding of the nonequilibrium thermodynamic processes at the heart of coupled transport and thermoelectric conversion. In this chapter, we provide an overview of the linear out-of-equilibrium thermodynamics of the electron gas, which serves as the working fluid in CTEGs. The force-flux formalism yields phenomenological linear, coupled equations at the macroscopic level, which describe the behavior of CTEGs under different configurations. The relevant equivalent quantities—figure of merit, efficiency, and output power—are formulated and calculated for two different configurations. Our results show, that system performance in each of these configurations is influenced by combination of different materials and their ordering, that is, position in the arrangement structure. The primary objective of our study is to contribute new design guidelines for development of composite thermoelectric devices that combine different materials, taking advantage of the performance of each in proper temperature range and type of configuration.

Keywords: thermoelectric energy conversion, thermoelectric devices, thermodynamic constraints on energy production, thermoelectric figure of merit, thermoelectric optimization, efficiency

1. Introduction

Thermoelectric devices are heat engines, which may operate as generators under thermal bias or as heat pumps. For waste energy harvesting and conversion, thermoelectricity offers quite

appropriate solutions, when temperature difference between heat source and heat sink is not too large. The physics underlying this type of energy conversion is based on the fundamental coupling between electric charge and energy that each mobile electron carries. The coupling strength is given by the so-called Seebeck coefficient or thermoelectric power [1]. The performance of thermoelectric system is usually assessed against the so-called figure of merit [2]: a dimensionless quantity denoted ZT, which combines the system's thermal and electrical transport properties, as well as their coupling at temperature T.

To qualify as a good thermoelectric, a material (semiconductor or strongly correlated) must boast the following characteristics: small thermal conductivity and large electrical conductivity on the one hand, so that, it behaves as a phonon glass—electron crystal system [2], and large thermoelectric power on the other hand. All these properties, which can be optimized, are temperature-dependent, so they may take interesting values only in a particular temperature range. Improvement of thermoelectric devices in terms of performance and range of applications is highly desired, as their conversion efficiency is not size-dependent, and the typical device does not contain moving parts. Much progress in the field of thermoelectricity has been achieved since the early days, which saw the pioneering works of Seebeck [3] and Peltier [4], but decisive improvement of the energy conversion efficiency, typically 10% of the efficiency of ideal Carnot thermodynamic cycle, is still in order.

In a general manner, transport phenomena are irreversible processes: the generation of fluxes within the system, upon which external constraints are applied, are accompanied by energy dissipation and entropy production [5]. Therefore, thermoelectric effects may be viewed as the result of the mutual interaction of two irreversible processes, electrical transport, and heat transport, as they take place [6]. Not too far from equilibrium, these transport phenomena obey linear phenomenological laws; so, general macroscopic description of thermoelectric systems is, in essence, phenomenological. Linear nonequilibrium thermodynamics provides the most convenient framework to characterize the device properties and the working conditions to achieve various operation modes.

A thermoelectric generator (TEG) is under the influence of two potentials: electrochemical (μ_e) and thermal (T); for each of which there is a flux and a force (as shown in examples of **Table 1**). If force is capable of getting the system to state close to equilibrium after perturbation, then the linear regime may characterize the situation, and approximation in this case is the linear response theory (LRT). In this chapter, we will review and discuss these issues considering thermoelectric system composed of different modules: we are particularly interested in the performance analysis of composite thermoelectric generator (CTEG). For this purpose, we will use a framework based on LRT, which allows to derive a set of linear coupled equations, which contain the system's thermoelectric properties: Seebeck coefficient (α), thermal conductivity (κ), and electrical resistivity (ρ), which are combined to form the effective transport parameters of CTEG in different thermal and electrical arrangements.

The present chapter is organized as follows: as thermoelectric conversion results primarily from nonequilibrium thermodynamic processes, a brief overview of some of the basic concepts and tools developed by Onsager [7, 8] and Callen [6] is very instructive, and we will see, that the force-flux formalism is perfectly suited for a description of thermoelectric processes [9]. Then, we will turn our attention to the physical model of composite thermoelectric generators, deriving and analyzing the figure of merit, the conversion efficiency and maximum output power. The chapter ends with a discussion and concluding remarks.

Variables	Transport coefficient	Expression and name
Particle flux and density	Diffusion coefficient	$\mathbf{J}_N = -D\nabla n$ Fick's law
Energy flux and temperature	Thermal conductivity	$\mathbf{J}_E = -\kappa\nabla T$ Fourier's law
Electrical current density and electric field	Electrical conductivity	$\mathbf{J} = \sigma\mathbf{E} = -\sigma\nabla\varphi$ Ohm's law

Table 1. Linear thermodynamic phenomenological laws—illustrative examples of forces and fluxes.

2. Basic notions of linear nonequilibrium thermodynamics

2.1. Instantaneous entropy

The thermodynamic formulation presented here is that of Callen [10]. To each set of extensive variables associated to a thermodynamic system, there is a counterpart, that is, a set of intensive variables. The thermodynamic potentials are constructed from these variables. At the macroscopic scale, the equilibrium states of a system may be characterized by a number of extensive variables X_i macroscopic by nature. As one may assume that a macroscopic system is made of several subsystems, which may exchange matter and/or energy among themselves, the values taken by the variables X_i correspond to these exchanges, which occur as constraints are imposed and lifted. When constraints are lifted, relaxation processes take place until the system reaches a thermodynamic equilibrium state, for which a positive and continuous function S differentiable with respect to the variables X_i can be defined as follows:

$$S : X_i \mapsto S(X_i). \tag{1}$$

The function S, called entropy, is extensive; its maximum characterizes equilibrium as it coincides with the values that the variables X_i finally assume after the relaxation of constraints. Note, that extensive variables X_i differ from microscopic variables because of typical time scales, over which they evolve: the relaxation time of microscopic variables is extremely fast, while the variables X_i are slow in comparison. To put it simply, relaxation time toward local equilibrium τ_{relax} is much smaller than the time necessary for the evolution toward the macroscopic equilibrium τ_{eq}. Hence, one may define an instantaneous entropy, $S(X_i)$, at each step of the relaxation of the variables X_i. The differential of the function S is as follows:

$$dS = \sum_i \frac{\partial S}{\partial X_i} dX_i = \sum_i F_i dX_i, \tag{2}$$

where each quantity F_i is the intensive variable conjugate of the extensive variable X_i.

2.2. Thermodynamic forces and fluxes

Examples of well-known linear phenomenological laws are given in **Table 1**. These laws establish a proportionality relationship between forces, which derive from potentials, and fluxes. Proportionality factors are transport coefficients, as fluxes are the manifestation of transport phenomena. Indeed, the system's response to externally applied constraints is transport, and when these are lifted, the system relaxes toward an equilibrium state.

Following the introductory discussion of this section, we now see in more detail how these forces and fluxes appear. The notions, which follow, are easily introduced considering the case of a discrete system like, for instance, two separate homogeneous systems initially prepared at two different temperatures and then put in thermal contact through a thin diathermal wall. The thermalization process triggers a flow of energy from one system to the other. So, assume now an isolated system composed of two weakly coupled subsystems, to which an extensive variable taking the values X_i and $X_{i'}$, is associated. One has $X_i + X_{i'} = X_i^{(0)} = constant$ and $S(X_i) + S(X_{i'}) = S(X_i^{(0)})$. Then, the equilibrium condition reads:

$$\frac{\partial S^{(0)}}{\partial X_i}\Big|_{X_i^{(0)}} = \frac{\partial (S + S')}{\partial X_i}\,dX_i\Big|_{X_i^{(0)}} = \frac{\partial S}{\partial X_i} - \frac{\partial S'}{\partial X_{i'}} = F_i - F_i' = 0, \tag{3}$$

as it maximizes the total entropy. Therefore, if the difference $\mathcal{F}_i = F_i - F'_i$ is equal to zero, the system is in equilibrium; otherwise, irreversible process takes place and drives the system to equilibrium. The quantity $\mathcal{F}_i$ is the affinity or generalized force allowing the evolution of the system toward equilibrium. Further, we also introduce the variation rate of the extensive variable X_i, as it characterizes the response of the system to the applied force:

$$I_i = \frac{dX_i}{dt}. \tag{4}$$

The relationship between affinities and fluxes characterizes the changes due to irreversible processes: non-zero affinity yields non-zero conjugated flux, and a given flux cancels, if its conjugate affinity cancels.

In local equilibrium, fluxes depend on their conjugate affinity, but also on the other affinities; so, we see, that there are direct effects and indirect effects. Therefore, the mathematical expression for the flux I_i, at a given point in space and time $(\mathbf{r},t)$, shows a dependence on the force $\mathcal{F}_i$, but also on the other forces $\mathcal{F}_{j \neq i}$:

$$I_i(\mathbf{r},t) \equiv I_i(\mathcal{F}_1, \mathcal{F}_2, \ldots). \tag{5}$$

Close to equilibrium $I_i(\mathbf{r},t)$ can be written as Taylor expansion:

$$I_k(\mathbf{r},t) = \sum_j \frac{\partial I_k}{\partial \mathcal{F}_j} \mathbf{F}_j + \frac{1}{2!} \sum_{i,j} \frac{\partial^2 I_k}{\partial \mathcal{F}_i \mathcal{F}_j} \mathcal{F}_i \mathcal{F}_j + \ldots = \sum_k L_{jk} \mathcal{F}_k + \frac{1}{2} \sum_{i,j} L_{ijk} \mathcal{F}_i \mathcal{F}_j + \ldots . \tag{6}$$

The quantities L_{jk} are the first-order kinetic coefficients; they are given by the equilibrium values of intensive variables F_i. The matrix $[\mathcal{L}]$ of kinetic coefficients characterizes the linear response of the system. Onsager put forth the idea that there are symmetry and antisymmetry relations between kinetic coefficients [6, 7]: the so-called reciprocal relations must exist in all thermodynamic systems, for which transport and relaxation phenomena are well described by linear laws. The main results can be summarized as follows [5]: (1) Onsager's relation: $L_{ik} = L_{ki}$; (2) Onsager-Casimir relation: $L_{ik} = \varepsilon_i \varepsilon_k L_{ki}$; (3) generalized relations: $L_{ik}(\mathbf{H}, \mathbf{\Omega}) = \varepsilon_i \varepsilon_k L_{ki}(-\mathbf{H}, -\mathbf{\Omega})$, where $\mathbf{H}$ and $\mathbf{\Omega}$ denote, respectively, the magnetic field and angular velocity associated with Coriolis field; the parameters ε_i denote the parity with respect to time reversal: if the quantity studied is invariant under time reversal transformation, it has parity $+1$; otherwise, this quantity changes sign, and it has parity -1.

3. Thermoelectric forces and fluxes

3.1. Coupled fluxes of heat and electrical charges

The thermoelectric effect results from the mutual interference of two irreversible processes occurring simultaneously in the system, namely heat transport and charge carriers transport. The Onsager force-flux derivation is obtained from the laws of conservation of energy and matter:

$$\mathbf{I}_E = \mathbf{I}_Q + \mu_e \mathbf{I}_N, \tag{7}$$

where $\mathbf{I}_E$ is energy flux, $\mathbf{I}_Q$ is heat flux, and $\mathbf{I}_N$ is particle flux. Each flux is the conjugate variable of its potential gradient. Considering the electron gas, correct potentials for particles and energy are μ_e/T and $1/T$, and related forces are as follows: $\mathbf{F}_E = \nabla(1/T)$ and $\mathbf{F}_N = \nabla(-\mu_e/T)$, where μ_e is the electrochemical potential [1]. Then, the linear coupling between forces and fluxes may simply be described by a linear set of coupled equations involving the so-called kinetic coefficient matrix $[\mathcal{L}]$:

$$\begin{pmatrix} \mathbf{I}_N \\ \mathbf{I}_E \end{pmatrix} = \begin{pmatrix} L_{NN} & L_{NE} \\ L_{EN} & L_{EE} \end{pmatrix} \begin{pmatrix} \nabla(-\mu_e/T) \\ \nabla(1/T) \end{pmatrix}, \tag{8}$$

where $L_{NE} = L_{EN}$. Now, to treat properly heat flow and electrical current, it is more convenient to consider $\mathbf{I}_Q$ instead of $\mathbf{I}_E$. Using $\mathbf{I}_E = \mathbf{I}_Q + \mu_e \mathbf{I}_N$, we obtain:

$$\begin{pmatrix} \mathbf{I}_N \\ \mathbf{I}_Q \end{pmatrix} = \begin{pmatrix} L_{11} & L_{12} \\ L_{21} & L_{22} \end{pmatrix} \begin{pmatrix} -\nabla(\mu_e/T) \\ \nabla(1/T) \end{pmatrix} \tag{9}$$

with $L_{12} = L_{21}$. Since $\nabla(-\mu_e/T) = -\mu_e \nabla(1/T) - 1/T \nabla(\mu_e)$, then heat flow and electrical current read:

$$\begin{pmatrix} \mathbf{I}_N \\ \mathbf{I}_Q \end{pmatrix} = \begin{pmatrix} L_{NN} & L_{NE} - \mu_e L_{NN} \\ L_{NE} - \mu_e L_{NN} & -2L_{NE}\mu_e + L_{EE} + \mu_e^2 L_{NN} \end{pmatrix} \begin{pmatrix} \nabla(-\mu_e/T) \\ \nabla(1/T) \end{pmatrix} \tag{10}$$

with the following relationship between kinetic coefficients:

$$L_{11} = L_{NN}, \tag{11}$$

$$L_{12} = L_{NE} - \mu_e L_{NN}, \tag{12}$$

$$L_{22} = L_{EE} - 2\mu_e L_{EN} + \mu_e^2 L_{NN}. \tag{13}$$

Note, that since electric field derives from electrochemical potential, we also obtain:

$$\mathcal{E} = -\frac{1}{e}\nabla\mu_e. \tag{14}$$

3.2. Thermoelectric transport coefficients

The thermoelectric transport coefficients can be derived from the expressions of electron and heat flux densities depending on applied thermodynamic constraints: isothermal, adiabatic, electrically open or closed circuit conditions. Under isothermal conditions, electrical current may be written in the form:

$$\mathbf{I}_N = \frac{-L_{11}}{T}\nabla(\mu_e). \tag{15}$$

This is expression of Ohm's law, since with $\mathbf{I} = e\mathbf{I}_N$ we obtain the following relationship between electrical current density and electric field:

$$e\mathbf{I}_N = \mathbf{I} = e\frac{-L_{11}}{T}\nabla(\mu_e) = \sigma_T\left(-\frac{\nabla(\mu_e)}{e}\right) = \sigma_T\mathcal{E}, \tag{16}$$

which contains the definition for isothermal electrical conductivity expressed as follows:

$$\sigma_T = \frac{e^2}{T}L_{11}. \tag{17}$$

Now, if we consider the heat flux density in the absence of any particle transport or, in other words, under zero electrical current, we get:

$$\mathbf{I}_N = 0 = -L_{11}\left(\frac{1}{T}\nabla(\mu_e)\right) + L_{12}\nabla(\frac{1}{T}), \tag{18}$$

so that, the heat flux density under zero electrical current, $\mathbf{I}_{Q_{I=0}}$, reads:

$$\mathbf{I}_{Q_{I=0}} = \frac{1}{T^2}\left[\frac{L_{21}L_{12} - L_{11}L_{22}}{L_{11}}\right]\nabla(T). \tag{19}$$

This is Fourier's law, with thermal conductivity under zero electrical current given by:

$$\kappa_I = \frac{1}{T^2}\left[\frac{L_{11}L_{22} - L_{21}L_{12}}{L_{11}}\right]. \tag{20}$$

We can also define the thermal conductivity κ_E under zero electrochemical gradient, that is, under closed circuit conditions:

$$\mathbf{I}_{Q_{E=0}} = \frac{L_{22}}{T^2}\,\nabla(T) = \kappa_E \nabla(T). \tag{21}$$

It follows, that thermal conductivities κ_E and κ_I are simply related through:

$$\kappa_E = T\alpha^2\sigma_T + \kappa_I. \tag{22}$$

As thermal and electric processes are coupled, the actual strength of the coupling is given by Seebeck coefficient:

$$\alpha \equiv \frac{-\frac{1}{e}\nabla(\mu_e)}{\nabla(T)} = \frac{1}{eT}\frac{L_{12}}{L_{11}}, \tag{23}$$

defined as the ratio of two forces that derive from electrochemical potential for one and from temperature for the other.

The analysis and calculations developed above allow to establish complete correspondence between kinetic coefficients and transport parameters:

$$L_{11} = \frac{\sigma_T}{e^2}\,T, \tag{24}$$

$$L_{12} = \frac{\sigma_T S_I T^2}{e^2}, \tag{25}$$

$$L_{22} = \frac{T^3}{e^2}\sigma_T S_I^2 + T^2\kappa_I, \tag{26}$$

so that, expressions for electronic current and heat flow may take their final forms:

$$\mathbf{I}_N = \frac{\sigma_T}{e^2}T\left(-\frac{\nabla(\mu_e)}{T}\right) + \frac{\sigma_T S_I T^2}{e^2}\left(\nabla(\tfrac{1}{T})\right), \tag{27}$$

$$\mathbf{I}_Q = \frac{\sigma_T S_I}{e^2}T^2\left(-\frac{\nabla(\mu_e)}{T}\right) + \left[\frac{T^3}{e^2}\sigma_T S_I^2 + T^2\kappa_I\right]\left(\nabla(\tfrac{1}{T})\right). \tag{28}$$

Since $\mathbf{I} = e\mathbf{I}_N$, it follows that:

$$\mathbf{I} = \sigma_T\mathbf{E} - \frac{\sigma_T S_I}{e}\nabla(T), \tag{29}$$

from which we obtain:

$$\mathbf{E} = \rho_T\mathbf{I} + \alpha\nabla(T), \tag{30}$$

where ρ_T is the isothermal conductivity. This is a general expression of Ohm's law.

4. Formulation of physical model for thermoelectric generators

For TEG performance analysis, we have applied the model given by [11, 12], associating thermal circuit for heat transport and electrical circuit for charge carriers transport, see **Figure 1**.

Electrical current and heat flow, I_i and I_{Q_i}, are functions of generalized forces [11], related to differences in voltage, ΔV_i, and temperature, ΔT_i, of thermoelectric generator:

$$\begin{pmatrix} I_i \\ I_{Q_i} \end{pmatrix} = \begin{pmatrix} 1/R_i & \alpha_i(1/R_i) \\ \alpha_i(1/R_i)T & \alpha_i^2(1/R_i)T + K_i \end{pmatrix} \begin{pmatrix} \Delta V_i \\ \Delta T_i \end{pmatrix}, \tag{31}$$

where T is average temperature.

In this model, TEG is characterized by its internal electrical resistance, R, thermal conductance under open electrical circuit condition, K, and Seebeck coefficient, α. Physical conditions assumed for this model are as follows: (i) thermoelectric properties are independent on temperature, (ii) the only electrical resistance taken into account is that of the legs, (iii) there is no thermal contact resistance between the ends of the legs and heat source, and (iv) in this model, doping of the legs (p- or n-type) is not taken into account, so that, TEG can be seen as only one leg.

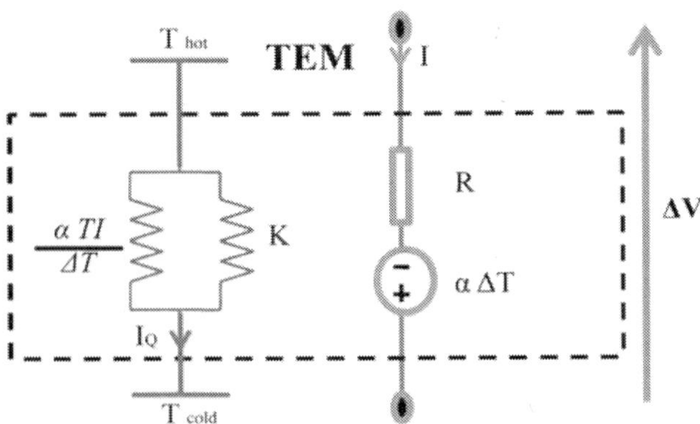

Figure 1. Circuit model for thermoelectric generator, red (thermal circuit), blue (electrical circuit), where ΔV, voltage; R, electrical resistance; K, thermal conductance; T_{cold}, temperature of the cold side; T_{hot}, temperature of the hot side; ΔT, temperature difference; α, Seebeck coefficient; and T, average temperature.

5. Heat balance equation

The heat balance in TEG is governed by the following equations; basically, there are two extreme points: one in contact with the heat source (incoming point):

$$Q_{in} = \alpha T_h I - \frac{1}{2} R_{in} I^2 + K(T_h - T_c),\qquad(32)$$

the other point is point, where heat is rejected:

$$Q_{re} = \alpha T_c I + \frac{1}{2} R_{in} I^2 + K(T_h - T_c),\qquad(33)$$

where $\alpha T_i I$ is Seebeck heat, $\frac{1}{2} R_{in} I^2$ is Joule heat, and $K(T_h - T_c)$ is thermal conduction heat; in terms of these quantities, electrical power is defined as:

$$P_{electrical} = Q_{in} - Q_{re} = \alpha I(T_h - T_c) - RI^2.\qquad(34)$$

6. Composite thermoelectric generator (CTEG)

We consider a composite thermoelectric generator, which is composed of three thermoelectric elements (TEGs) in different configurations, each TEG is made of a different thermoelectric material, see **Figure 2**. The configurations considered are as follows: (A) two-stage thermally and electrically connected in series (TES-CTEG); (B) segmented TEG, conventional TEG, thermally and electrically connected in parallel (PSC-CTEG). Also, we consider the effect of the arrangement of the materials on the performance of the composite system. Thus, for each of the systems (A, B), we have the following arrangements:

a. TEG 1 = material one, TEG 2 = material two, TEG 3 = material three;

b. TEG 1 = material three, TEG 2 = material one, TEG 3 = material two;

c. TEG 1 = material two, TEG 2 = material three, TEG 3 = material one.

In the following sections, we analyze and show results for CTEG by applying the conditions listed above in order to contribute to development of new design guidelines for thermoelectric systems with news architectures and even to provide some clues to the search for new physical conditions in the area of science and engineering of thermoelectric materials.

6.1. Formulation of equivalent figure of merit for CTEG

To analyze CTEG performance, equivalent quantities are defined, which contain the overall contribution of individual properties of each TEG building up composite system. These quantities are as follows: equivalent Seebeck coefficient (α_{eq}), equivalent electrical resistance (R_{eq}), and equivalent thermal conductance (K_{eq}), in terms of which it is possible to have equivalent figure of merit (Z_{eq}). We show the impact of the configuration of the system on Z_{eq} for each of configuration (A, B) listed in Section 6, and we suggest the optimum configuration. In order to justify the effectiveness of the equivalent figure of merit, the corresponding efficiency has been calculated for each configuration.

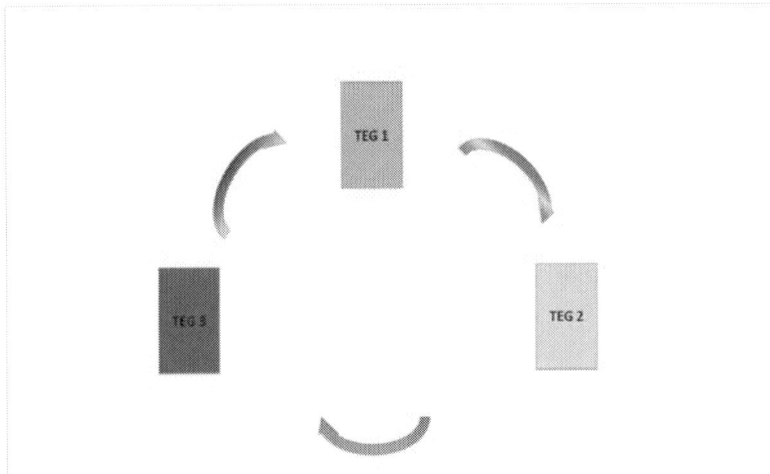

Figure 2. Composite thermoelectric generator (CTEG) (components are three TEGs, each made of different material).

6.1.1. Two-stage thermally and electrically connected in series

Schematic view of this system is shown in **Figure 3**. The first stage (bottom stage) consists of two different thermoelectric modules (TEG), while the top stage consists of only one TEG. Each of components is characterized by proper thermoelectric properties (α_i, R_i, K_i) [13].

Using Eq. (31), the heat flux within any segment in TEGs is:

$$I_{Q_i} = \alpha_i T I_i + K_i \Delta T_i. \tag{35}$$

By continuity of the heat flux through the interface between stages of TES-CTEG:

$$I_{Q1} = I_{Q2} + I_{Q3}$$

$$K_1(T_{hot}-T_i) + \alpha_1 T I = K_2(T_i-T_{cold}) + \alpha_2 T I + K_3(T_i-T_{cold}) + \alpha_3 T I, \tag{36}$$

from which we obtain the average temperature at the interface between stages [12]:

$$T_i = \frac{K_1 T_{hot} + (K_2 + K_3)T_{cold} + (\alpha_1-\alpha_2-\alpha_3)TI}{K_1 + K_2 + K_3}. \tag{37}$$

Since all components are electrically connected in series, the total voltage is given by:

$$\Delta V = -\alpha_1(T_{hot}-T_i)-\alpha_2(T_i-T_{cold})-\alpha_3(T_i-T_{cold}) + (R_1 + R_2 + R_3)I, \tag{38}$$

substituting the value of T_i in the last equation, we have:

$$\Delta V = \left[\frac{-(\alpha_2 + \alpha_3)K_1 - \alpha_1 K_2 - \alpha_1 K_3}{K_1 + K_2 + K_3}\right][T_{hot} - T_{cold}] +$$
$$+ \left[\frac{(\alpha_1 - \alpha_2 - \alpha_3)^2 T}{K_1 + K_2 + K_3} + (R_1 + R_2 + R_3)\right] I. \tag{39}$$

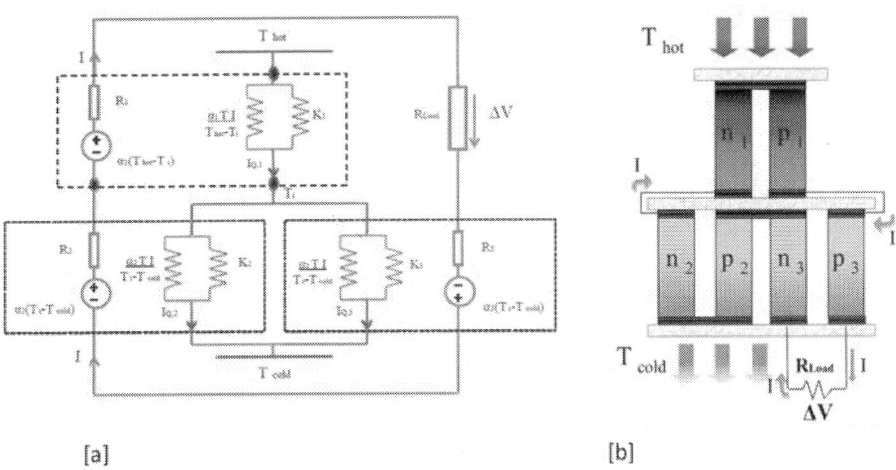

Figure 3. Schematic representation of thermoelectric system composed of two stages thermally and electrically connected in series (TES-CTEG). (a) Equivalent circuit for TES-CTEG, where ΔV is the voltage, R_i is the electrical resistance, K_i is the thermal conductance, T_{cold} is the temperature of the cold side, T_{hot} is the temperature of the hot side, ΔT is the temperature difference, α_i is the Seebeck coefficient, T is the average temperature, R_{load} is the load; (b) practical device related to TES-CTEG, where n_i is the ith n-type material, p_i is the ith p-type material.

From Eq. (39), we identified the equivalent series Seebeck coefficient, α_{eq-TES}, and equivalent series electrical resistance, R_{eq-TES}, as follows:

$$\alpha_{eq-TES} = \frac{-(\alpha_2 + \alpha_3)K_1 - \alpha_1 K_2 - \alpha_1 K_3}{K_1 + K_2 + K_3}, \tag{40}$$

$$R_{eq-TES} = R_1 + R_2 + R_3 + R_{relax}, \tag{41}$$

where

$$R_{relax} = \frac{(\alpha_1 - \alpha_2 - \alpha_3)^2 T}{K_1 + K_2 + K_3}. \tag{42}$$

Considering open circuit condition for the system, $I = 0$, we find, that equivalent thermal conductance for the whole system:

$$K_{eq-TES} = \frac{K_1(K_2 + K_3)}{K_1 + K_2 + K_3}. \tag{43}$$

We define the figure of merit in terms of equivalent quantities [12]:

$$Z_{eq} = \frac{\alpha_{eq}^2}{R_{eq}K_{eq}}. \tag{44}$$

By replacing the results obtained in Eqs. (40)–(43), we have:

$$Z_{eq\text{-}TES} = \frac{\left[\frac{-(\alpha_2+\alpha_3)K_1-\alpha_1K_2-\alpha_1K_3}{K_1+K_2+K_3}\right]^2}{\left[\frac{(\alpha_1-\alpha_2-\alpha_3)^2 T}{K_1+K_2+K_3} + (R_1+R_2+R_3)\right]\left[\frac{K_1(K_2+K_3)}{K_1+K_2+K_3}\right]}. \tag{45}$$

6.1.2. Segmented TEG-conventional TEG thermally and electrically connected in parallel

In this section, we consider CTEG system, which is composed by segmented TEG and conventional TEG. These TEGs are thermally and electrically connected in parallel (PSC-CTEG), as is shown in **Figure 4**.

In the composite system, there are two currents, I_s for *TEG 1* and *TEG 2*, I_c for *TEG 3*. If the electrical current is conserved, then [13]:

$$I_{eq} = I_s + I_c. \tag{46}$$

The heat flux through the whole system is the sum of the heat flux flowing through segmented generator and the heat flux in conventional generator. Thus:

$$I_{Q\text{-}eq} = I_{Q_s} + I_{Q_c}. \tag{47}$$

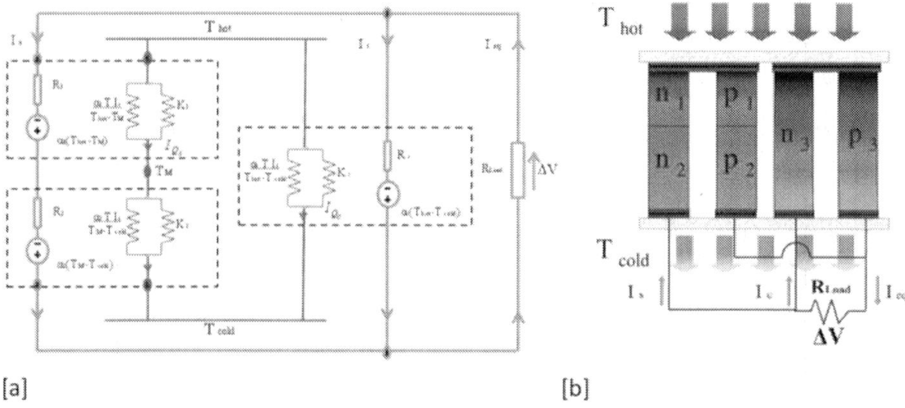

Figure 4. Schematic representation of (PSC-CTEG). (a) Thermal-electrical circuit, where ΔV is the voltage, R_i is the electrical resistance, K_i is the thermal conductance, T_{cold} is the temperature of the cold side, T_{hot} is the temperature of the hot side, ΔT is the temperature difference, α_i is Seebeck coefficient, T is the average temperature, R_{load} is the load resistance, T_M is the intermediate temperature; (b) structure design, where n_i is the ith n-type material, p_i is the ith p-type material.

To obtain the equivalent electrical resistance, $R_{eq\text{-}PSC}$, using Eq. (45), the isothermal condition, $\Delta T = 0$, is required. Under this condition, we recover the usual expression of equivalent electrical resistance for an ohmic circuit. Thus, we get:

$$R_{eq\text{-}PSC} = \frac{R_s R_c}{R_s + R_c},\tag{48}$$

where R_c is the internal electrical resistance of conventional TEG and R_s is the electrical resistance of the segmented TEG:

$$R_s = R_1 + R_2 + R_{relax}\tag{49}$$

and

$$R_{relax} = \frac{(\alpha_1 - \alpha_2)^2 T}{K_1 + K_2}.\tag{50}$$

Assuming the condition of closed circuit, $\Delta V = 0$, and applying Eq. (45), we have for equivalent Seebeck coefficient [13]:

$$\alpha_{eq\text{-}PSC} = \frac{R_c \alpha_s + R_s \alpha_c}{R_s + R_c},\tag{51}$$

where

$$\alpha_s = \frac{K_2 \alpha_1 + K_1 \alpha_2}{K_1 + K_2}.\tag{52}$$

To determine equivalent thermal conductance, K_{eq}, we use the open circuit condition, $I_{eq} = 0$, which is satisfied when $I_s = -I_c = I$, and, due to preservation of heat flow:

$$K_{eq\text{-}PSC} = K_s + K_c + \frac{(\alpha_s - \alpha_c) T I}{\Delta T},\tag{53}$$

where

$$K_s = \frac{K_2 K_1}{K_1 + K_2}.\tag{54}$$

Under open circuit condition, $I_{eq} = 0$, so that, $\Delta V = -\alpha_{eq} \Delta T$. Applying this result, we have for I:

$$I = \frac{1}{R_s + R_c}(\alpha_s - \alpha_c)\Delta T.\tag{55}$$

Using this last result in Eq. (53), we have:

$$K_{eq\text{-}PSC} = K_s + K_c + (\alpha_s - \alpha_c)^2 T \frac{1}{R_s + R_c}.\tag{56}$$

Now, we can write the figure of merit for this PSC-CTEG system:

$$Z_{eq\text{-}PSC} = \frac{\alpha^2_{eq\text{-}PSC}}{R_{eq\text{-}PSC}K_{eq\text{-}PSC}}. \tag{57}$$

Using the results obtained in Eqs. (48), (51), and (56), we have:

$$Z_{eq\text{-}PSC} = \frac{\left(\frac{R_c\alpha_s + R_s\alpha_c}{R_s+R_c}\right)^2}{\left[\frac{R_s R_c}{R_c+R_s}\right]\left[K_s + K_c + (\alpha_s-\alpha_c)^2\, T\, \frac{1}{R_s+R_c}\right]}. \tag{58}$$

6.1.3. Analysis of equivalent figure of merit for composite systems

Equivalent figure of merit (Z_{eq}) is calculated in this section for *TES* and *PSC* systems. For performing calculations, the best known thermoelectric materials for commercial applications have been selected: BiTe, PbTe, and SiGe (experimental data taken from Refs. [14–16] have been used as numerical values of thermoelectric parameters). It has also been calculated equivalent maximum efficiency ($\eta_{eq\text{-}max}$).

It is important to emphasize, that in this study we analyzed also the behavior of Z_{eq} and η_{eq}, when ordering of materials in the composite system changes (i.e., change its position).

Table 2 shows, that performance of composite system is affected by the type of thermal and electrical connection, as well as ordering of materials. For example, PSC case reaches the highest value of Z_{eq} and η_{eq} with the ordering *TEG 1 = PbTe, TEG 2 = SiGe, TEG 3 = BiTe*.

To analyze the performance of the composite system, with each of the different orderings, we have built plots (**Figure 5a, b**), that show variation of equivalent figure of merit with Seebeck coefficients ratio α_j/α_i.

6.2. Maximum efficiency

The figure of merit measures the performance of materials in thermoelectric device, but, if we measure the performance when the TEG is operating under a temperature difference, then the value called thermal efficiency quantifies the ability of TEG to utilize the supplied heat effectively.

TEG 1	TEG 2	TEG 3	$Z_{eq\text{-}TES}$	$Z_{eq\text{-}PSC}$	$\eta_{eq\text{-}TES}$	$\eta_{eq\text{-}PSC}$
BiTe	PbTe	SiGe	0.000433	0.000463	0.079936	0.084392
PbTe	SiGe	BiTe	0.000508	0.001905	0.091045	0.224724
SiGe	BiTe	PbTe	0.000574	0.000622	0.100217	0.106658

Table 2. Numerical values of Z_{eq} and η_{eq} in each equivalent thermoelectric system for different arrangements of the TE materials.

From thermodynamics, Carnot cycle thermal efficiency is known as:

$$\eta_{Carnot} = \frac{T_{hot} - T_{cold}}{T_{hot}}. \tag{59}$$

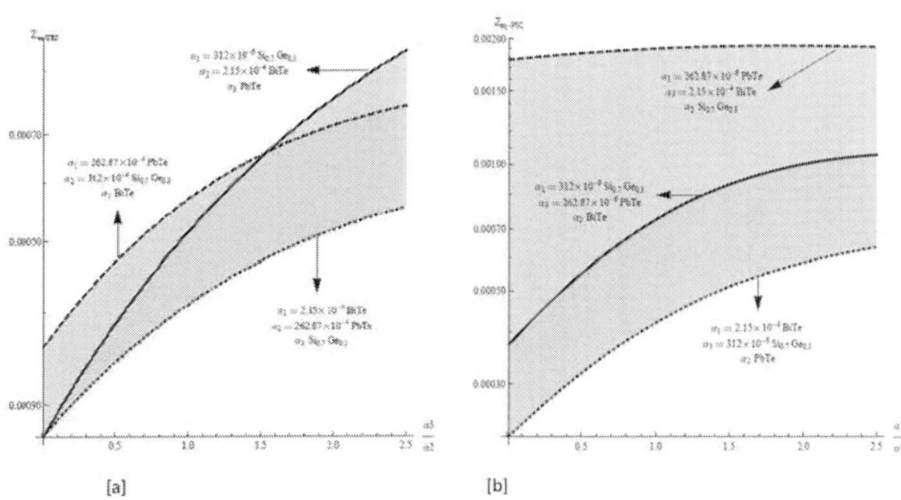

[a]   [b]

Figure 5. (a) Z_{eq-TES} vs. ratio α_3/α_2, maintaining α_1 and α_2 constant; (b) Z_{eq-PSC} vs. ratio, α_2/α_1, maintaining α_1 and α_3 constant.

In terms of η_{Carnot} and Z_{eq}, the maximum efficiency of thermoelectric device is defined by the next equation (with thermoelectric properties (α, R, κ) constant with respect to temperature) [2]:

$$\eta_{max-j} = \frac{\Delta T}{T_{hot}} \cdot \frac{\sqrt{1 + Z_{eq-j}T} - 1}{\sqrt{1 + Z_{eq-j}T} + \frac{T_{cold}}{T_{hot}}}, \tag{60}$$

where Z_{eq-j} with $j = TES, PSC$ is given by Eqs. (45) and (58), respectively. Thus, we have for the maximum efficiency of TES-CTEG system:

$$\eta_{eq-TES} = \frac{\Delta T}{T_{hot}} \cdot \frac{\sqrt{1 + Z_{eq-TES}T} - 1}{\sqrt{1 + Z_{eq-TES}T} + \frac{T_{cold}}{T_{hot}}}. \tag{61}$$

For the maximum efficiency of *PSC-CTEG* system:

$$\eta_{eq-PSC} = \frac{\Delta T}{T_{hot}} \cdot \frac{\sqrt{1 + Z_{eq-PSC}T} - 1}{\sqrt{1 + Z_{eq-PSC}T} + \frac{T_{cold}}{T_{hot}}}. \tag{62}$$

Our results are shown in **Figure 6**.

Plots in **Figure 6** show typical dependences of CTEGs efficiency on the properties of component materials. The presented results of maximum efficiency reached by the thermoelectric device approach the limit established by Bergman's theorem for composite materials [17]: the efficiency of composite thermoelectric system cannot be greater than the module's component with highest efficiency.

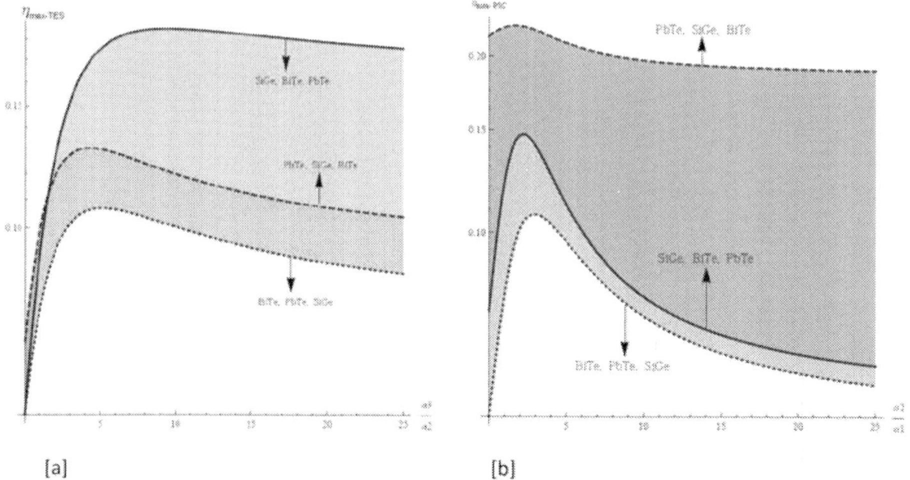

Figure 6. (a) $\eta_{max-TES}$ vs. ratio α_3/α_2. (b) $\eta_{max-PSC}$ vs. ratio α_2/α_1.

The maximum efficiencies achieved by studied CTEGs, see plots in **Figure 6**, are of similar order of magnitude as CTEG systems investigated in some works, e.g. [18], where reported efficiencies from 17 to 20%.

6.3. CTEG: maximum output power

We analyze also the maximum output power of the studied CTEG system, again, assuming configurations and physical conditions shown in Section 6. The obtained results have been compared with some analytical work and numerical simulations.

For the case of thermoelectric generator connected to load resistor R_{load} (**Figure 7**), the power delivered to R_{load} is given by the following equation [19]:

$$P_{out-m} = \frac{[\alpha(T_H - T_C)]^2 m}{(m+1)^2 R},$$
(63)

The strategy consists of defining the optimal ratio $m = R_{load}/R$ and then by applying the method of maximizing variable to obtain the value of the load resistance, which maximizes power. It yields $R_{load} = R$, and in this case, the maximum output power is:

$$P^{max} = \frac{\alpha^2 (T_H - T_C)^2}{4R}.$$
(64)

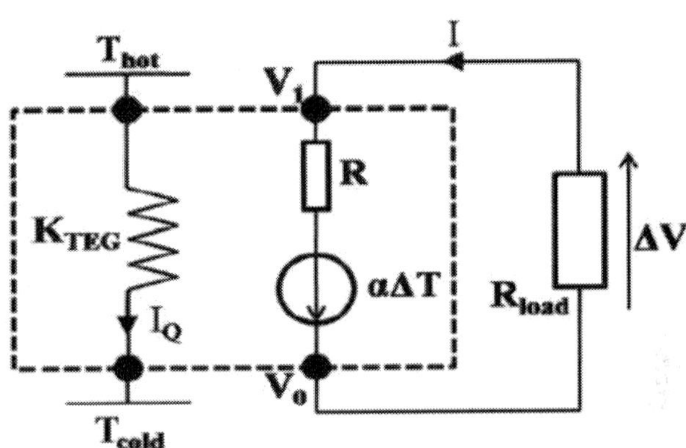

Figure 7. Thermal-electrical circuit for *TEG* delivering power to the load, where ΔV is the voltage, R_i is the electrical resistance, K_i is the thermal conductance, T_{cold} is the temperature of the cold side, T_{hot} is the temperature of the hot side, ΔT is the temperature difference, α_i is Seebeck coefficient, R_{load} is the load resistance.

6.3.1. Formulation of output power for CTEG

Here, in similar way as in previous sections, formulating of output power will be considered using thermoelectric equivalent quantities, see Sections 6.1.1, 6.1.2 [20]. Thus, using Eqs. (62, 63) in terms of α_{eq} and R_{eq}, we can write:

$$P_{out-eq-m} = \frac{[\alpha_{eq}(T_H-T_C)]^2}{R_{eq}} \frac{m}{(m+1)^2}, \tag{65}$$

$$P_{eq}^{max} = \frac{\alpha_{eq}^2(T_H-T_C)^2}{4R_{eq}}. \tag{66}$$

Application of the formalism described above Eqs. (64, 65) give the output power for each configuration as follows.

Two-stage thermoelectric system connected in series:

$$P_{out-eq-(TES-CTEG)-m} = \frac{\left(\left[\frac{-(\alpha_2+\alpha_3)K_1-\alpha_1 K_2-\alpha_1 K_3}{K_1+K_2+K_3}\right](T_H-T_C)\right)^2}{\left[R_1 + R_2 + R_3 + \frac{(\alpha_1-\alpha_2-\alpha_3)^2 \overline{T}}{K_1+K_2+K_3}\right]} \frac{m}{(m+1)^2} \tag{67}$$

and the maximum power is given by:

$$P_{eq-(TES-CTEG)}^{max} = \frac{\left(\left[\frac{-(\alpha_2+\alpha_3)K_1-\alpha_1 K_2-\alpha_1 K_3}{K_1+K_2+K_3}\right](T_H-T_C)\right)^2}{4\left[R_1 + R_2 + R_3 + \frac{(\alpha_1-\alpha_2-\alpha_3)^2 \overline{T}}{K_1+K_2+K_3}\right]}. \tag{68}$$

Segmented-conventional thermoelectric system in parallel (PSC-CTEG):

$$P_{out-eq-(PSC)-m} = \frac{\left(R_c\left[\frac{K_2\alpha_1+K_1\alpha_2}{K_1+K_2}\right] + \left[R_1 + R_2 + \left[\frac{(\alpha_1-\alpha_2)^2\overline{T}}{K_1+K_2}\right]\right]\alpha_c\right)^2(T_H-T_C)^2}{\left[\left[R_1 + R_2 + \frac{(\alpha_1-\alpha_2)^2\overline{T}}{K_1+K_2}\right]R_c(R_s + R_c)\right]}\frac{m}{(m+1)^2}, \tag{69}$$

and using Eqs. (51, 48) and Eq. (66), the maximum power of this system obtained is:

$$P_{eq-(PSC)}^{max} = \frac{1}{4}\frac{\left(R_c\left[\frac{K_2\alpha_1+K_1\alpha_2}{K_1+K_2}\right] + \left[R_1 + R_2 + \left[\frac{(\alpha_1-\alpha_2)^2\overline{T}}{K_1+K_2}\right]\right]\alpha_c\right)^2(T_H-T_C)^2}{\left[\left[R_1 + R_2 + \frac{(\alpha_1-\alpha_2)^2\overline{T}}{K_1+K_2}\right]R_c(R_s + R_c)\right]}. \tag{70}$$

6.3.2. Analysis of output power

We show the behavior of the electrical output power delivered in each CTEG configuration using the data of Section 6.1.3. **Figure 8**, panels (a) and (b), shows the output power as a function of the ratio between the electrical resistance of the load and the electrical resistance of the thermoelectric system $m = \frac{R_{load}}{R}$.

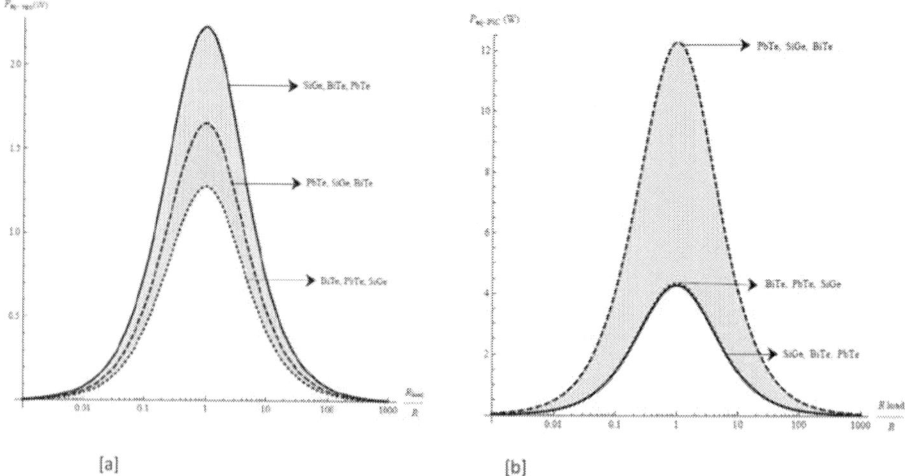

Figure 8. (a) Plot for output power delivered by TES-CTEG system as function of ratio R_{load}/R; combination, producing the highest output power, is (TEM 1=SiGe, TEM 2=BiTe, TEM 3=PbTe); (b) plot for output power delivered by the PSC-CTEG system as function of ratio R_{load}/R; combination, producing the highest output power, is (TEM 1=PbTe, TEM 2= SiGe, TEM 3=BiTe).

Plots in **Figure 8** show, that similarly to the equivalent figure of merit and equivalent efficiency (Sections 6.1.3 and 6.2), the output power of a composite system is also influenced by the type of thermal-electrical connection and ordering of materials, and again, PSC-CTEG case shows the highest performance quantified by generated output power. This result is consistent with the results obtained by Vargas-Almeida et al. [20], and the behavior of the output power for each array of equivalent TES-CTEG is consistent with the results obtained by Apertet et al. [11]. **Table 3** shows the comparison of maximum output power values for different types of connections and possible arrangements.

TEG 1	TEG 2	TEG 3	$P_{max-eq-TES}$	$P_{max-eq-PSC}$
BiTe	PbTe	SiGe	1.27618	4.34854
PbTe	SiGe	BiTe	1.65563	12.2877
SiGe	BiTe	PbTe	2.22968	4.28067

Table 3. Numerical values of maximum output power, in terms of equivalent amounts of each compound of CTEG, evaluated for each order of building TEGs.

To confirm the validity of our results, we have built plots for CTEG output power using ΔT values of some work: [21] (experimental) and [22, 23] (analytical). Plots in **Figure 9** were produced using the temperature difference of Ref. [21].

The results for comparisons with [22, 23] are shown in [24].

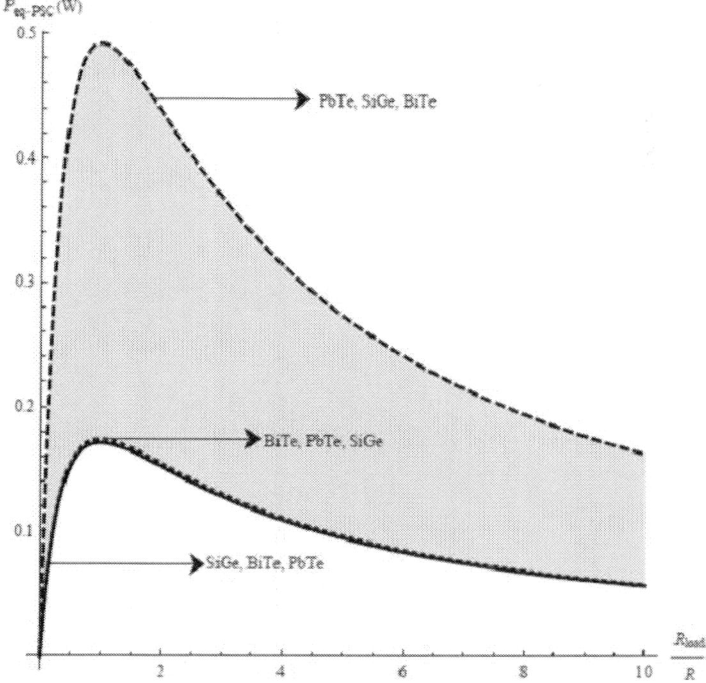

Figure 9. Output power $P_{Out-eq-PSC}$ delivered by composed *PSC* system vs ratio R_{load}/R. At temperature difference ΔT = 20 K, curves behave similarly to the plots shown in Ref. [21]. This figure is consistent with the result obtained by Abdelkefi [21]. Our results have also been compared to other published works [22, 23].

7. Opportunity analysis to improve CTEG design by varying configuration

In this section, we generalize results shown in previous sections by formulating corollary and including some results with realistic approaches, for example, consideration of contact thermal conductance. To achieve this goal, we combine physical conditions imposed in Section 6 with the next options: (1) the whole system is formed of the same thermoelectric material (α_1, K_1, $R_1 = \alpha_2$, K_2, $R_2 = \alpha_3$, K_3, R_3); (2) the whole system is constituted by only two different thermoelectric materials (α_i, K_i, $R_i = \alpha_j$, K_j, $R_j \neq \alpha_l$, K_l, R_l), where i, j, l can be 1, 2 or 3, [25].

7.1. Case A: homogeneous thermoelectric properties, configuration effect

We consider configurations of CTEG with the same thermoelectric material, $(\alpha_1, K_1, R_1) = (\alpha_2, K_2, R_2) = (\alpha_3, K_3, R_3)$. In this case, equivalent figure of merit Z_{eq}^h is as follows,

for homogeneous TES-CTEG:

$$Z_{eq\text{-}TES}^h = \frac{\left(\frac{-4\alpha_i}{3}\right)^2}{\left(\frac{(-\alpha_i)^2 T}{3K_i} + 3R_i\right)\left(\frac{2K_i}{3}\right)}, \tag{71}$$

for homogeneous PSC-CTEG:

$$Z_{eq\text{-}PSC}^h = \frac{(\alpha_i)^2}{\left(\frac{2R_i}{3}\right)\left(\frac{3K_i}{2}\right)}, \tag{72}$$

where $i = (BiTe, PbTe, SiGe)$.

Table 4 shows numerical values of equivalent figure of merit Z_{eq}^h obtained by us for CTEG with considered configurations.

Material	$Z_{eq\text{-}TES}^h$	$Z_{eq\text{-}PSC}^h$
BiTe	0.00212133	0.00305269
PbTe	0.00055109	0.000657238
SiGe	0.000287562	0.00033337

Table 4. Numerical values of Z_{eq}^h, for each of three configurations with different materials.

It is important to note, that fulfillment condition *TEG 1=TEG 2=TEG 3* evidences the fact, that although composite system is made of single material, the figure of merit reaches different values depending on type of connection.

7.2. Case B: two different materials in CTEG

CTEG is made of two same materials and the other one different. Thus, two TEGs include same semiconductor material and the other one different semiconductor material. In this case, equivalent figure of merit Z_{eq}^h is as follows, for heterogeneous TES-CTEG:

$$Z_{eq\text{-}TES}^{Inh} = \frac{\left(\frac{-(\alpha_j+\alpha_l)K_i-\alpha_i(K_j+K_l)}{K_i+K_j+K_l}\right)^2}{\left(\frac{(\alpha_i-\alpha_j-\alpha_l)^2 T}{K_i+K_j+K_l} + R_i + R_j + R_l\right)\left(\frac{K_i(K_j+K_l)}{K_i+K_j+K_l}\right)}, \tag{73}$$

for heterogeneous PSC-CTEG:

$$Z_{eq\text{-}PSC}^{Inh} = \frac{\left(\frac{R_l\left(\frac{K_j\alpha_i+K_i\alpha_j}{K_i-K_j}\right) + \left(R_i-R_j-\frac{(\alpha_i-\alpha_j)^2 T}{K_i-K_j}\right)\alpha_l}{R_i+R_j+R_l-\frac{(\alpha_i-\alpha_j)^2 T}{K_i-K_j}}\right)^2}{\left(\frac{R_l\left(R_i-R_j-\frac{(\alpha_i-\alpha_j)^2 T}{K_i-K_j}\right)}{R_l-R_i+R_j+\frac{(\alpha_i-\alpha_j)^2 T}{K_i-K_j}}\right)\left(\frac{K_jK_i}{K_i+K_j} + K_l + \left(\frac{(K_j\alpha_i+K_i\alpha_j)}{K_i-K_j}-\alpha_l\right)^2\frac{T}{R_i-R_j-R_l+\frac{(\alpha_i-\alpha_j)^2 T}{K_i-K_j}}\right)}. \tag{74}$$

Eqs. (72) and (73) are applied with condition $TEG_i = TEG_j$, that is, two TEGs are made of the same thermoelectric material, and third TEG_l is made of different thermoelectric material. Thus, we have three possibilities (*TEG 1=TEG 2≠TEG3, TEG 1=TEG 3≠TEG 2, TEG 2=TEG 3≠ TEG 1*) for each configuration [25]. Note that, each arrangement has six different combinations, if the cyclical order of the material is taken into account.

The behavior of the equivalent figure of merit as a function of the ratio of the thermal conductivities of the two component materials is shown in **Figure 10**. This step is important, because it shows numerical values, that CTEG maker must meet for both component materials to reach the highest value of Z_{eq}.

Table 5 shows maximum values of equivalent figure of merit of CTEG with material arrangements in every configuration, when $TEG_i = TEG_j{\neq}TEG_l$.

Table 6 shows each configuration with the most efficient material arrangements for every *TEG*.

Results show again, that the most efficient system of three configurations is PSC with corresponding material arrangement, namely *TEG 1=TEG 2=PbTe≠TEG 3=BiTe*; see **Figure 11**.

Again, it is important to note, that this result proves, that although the performance of composite systems is affected by combination of different materials, it is affected by the position of such materials in the system structure as well.

7.3. Performance analysis with realistic approximations

The results of the previous sections have argued, that application of output power and efficiency as quantities to measure performance of the system is reasonable; however, in this new section, we extend the analysis of these quantities using realistic considerations. Numerical treatment is performed with $Z_{eq\text{-}PSC}^{Inh}$.

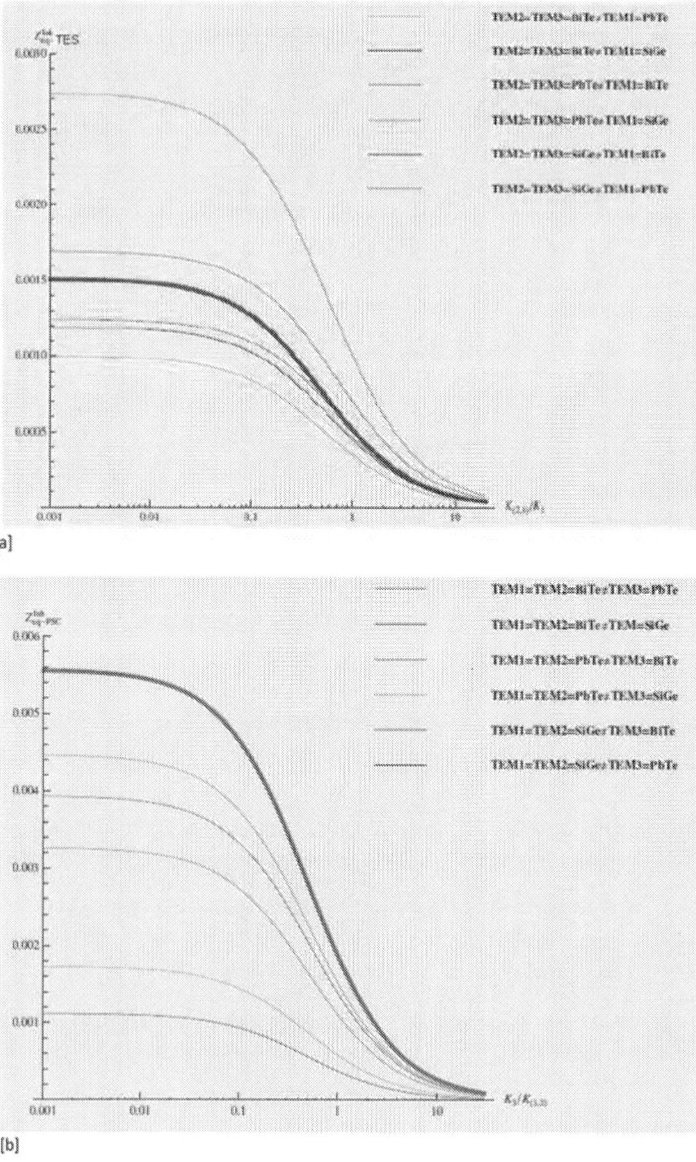

Figure 10. (a) Equivalent figure of merit for heterogeneous *TES-CTEG*, under condition *TEG 2=TEG 3≠TEG 1*, the highest numerical value is corresponding to *TEG 2=TEG 3=BiTe≠TEG 1=PbTe*; (b) Equivalent figure of merit for heterogeneous *PSC-CTEG* under condition *TEG 1=TEG 2≠TEG 3*, the highest numerical value is corresponding to *TEG 1=TEG 2=PbTe≠ TEG 3=BiTe*.

TEG 1	TEG 2=TEG 3	$Z^{Inh}_{eq-TES-max}$
BiTe	PbTe	0.00168734
BiTe	SiGe	0.0012388
PbTe	BiTe	0.00273649
PbTe	SiGe	0.00118802
SiGe	BiTe	0.00150947
SiGe	PbTe	0.000994534
TEG 3	**TEG 1=TEG 2**	$Z^{Inh}_{eq-PSC-max}$
BiTe	PbTe	0.0055567
BiTe	SiGe	0.00325841
PbTe	BiTe	0.00445846
PbTe	SiGe	0.0011157
SiGe	BiTe	0.00392902
SiGe	PbTe	0.00172358

Table 5. Maximum values of equivalent figure of merit of *CTEG* with material arrangements in every configuration, when $TEM_i = TEM_j{\neq}TEM_l$.

System	Arrangement
TES	TEG 2=TEG 3=BiTe≠TEG 1=PbTe
PSC	TEG 1=TEG 2=PbTe≠TEG 3=BiTe

Table 6. Most efficient material arrangements $TEG_i = TEG_j{\neq}TEG_l$ for TES and PSC-CTEG systems.

7.3.1. Maximum output power

In the following analysis, we consider thermoelectric modules as isolated units only. Although this is usually considered as an ideal situation, such an approach is useful to study the performance of materials in the composite system. However, for real applications, modules must be coupled to heat exchangers, which produces thermal conductance of contact (K_c) at the coupling points. This affects system performance and reflects in the output power. Here, the maximum output power is calculated using the maximum value of the equivalent figure of merit (Z^{Inh}_{eq-PSC}) [23]:

$$P_{max-PSC} = \frac{(K_c\Delta T)^2}{4(K_{I=0} + K_c)\overline{T}}\frac{Z^{Inh}_{eq-PSC}\overline{T}}{1 + Z^{Inh}_{eq-PSC}\overline{T} + K_c/K_{I=0}}. \tag{75}$$

Figure 12a shows maximum output power values for PSC system as function of ratio $K_{I=0}/K_c$, that is, in terms of internal thermal conductance $K_{I=0}$ and contact thermal conductance K_c, under condition *TEG 1=TEG 2≠TEG 3*.

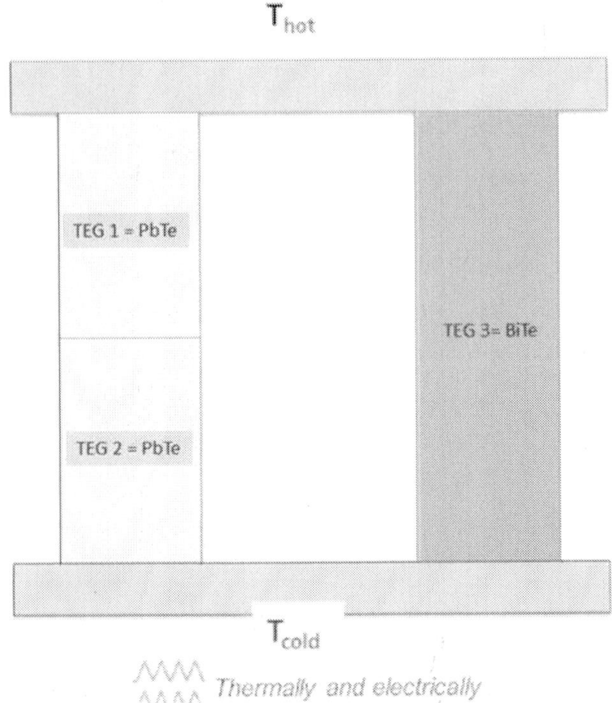

Figure 11. Optimal configuration corresponds to *PSC-CTEG* with arrangement *TEG 1=TEG 2=PbTe≠TEG 3=BiTe.*

7.3.2. Efficiency

To calculate the efficiency of PSC systems with TEG 1=TEG 2=PbTe≠TEG 3=BiTe arrangement, we applied the equation:

$$\eta_{eq\text{-}PSC}^{Inh} = \frac{\Delta T}{T_H}\frac{\sqrt{1 + Z_{eq\text{-}PSC}^{Inh}\overline{T}}-1}{\sqrt{1 + Z_{eq\text{-}PSC}^{Inh}\overline{T}} + \frac{T_C}{T_H}}. \tag{76}$$

Finally, for an ideal TEG, that is, without taking into account heat exchangers, we can analyze
TEG efficiency considering intrinsic thermal conductances ratio $(K_3/K_{1,2})$ and electrical resistances ratio $(R_3/R_{1,2})$.

Figure 12b shows contour plot for different values of $\eta_{eq\text{-}PSC}^{Inh}$ as function of ratios, $K_3/K_{1,2}$ and
$R_3/R_{1,2}$. We can see, that the range of optimal values for the best efficiency of $PSC-CTEG$ lies
in intervals 0.1–1.0 and 0.1–0.5 for $K_3/K_{1,2}$ and $R_3/R_{1,2}$, respectively. It is remarkable, that
thermal conductances ratio shows a wider range of good values in comparison with electrical
resistances ratio, which shows narrower range.

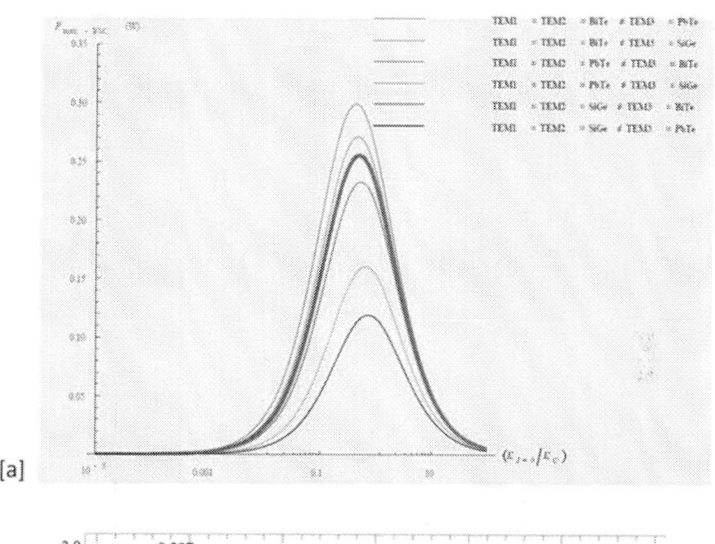

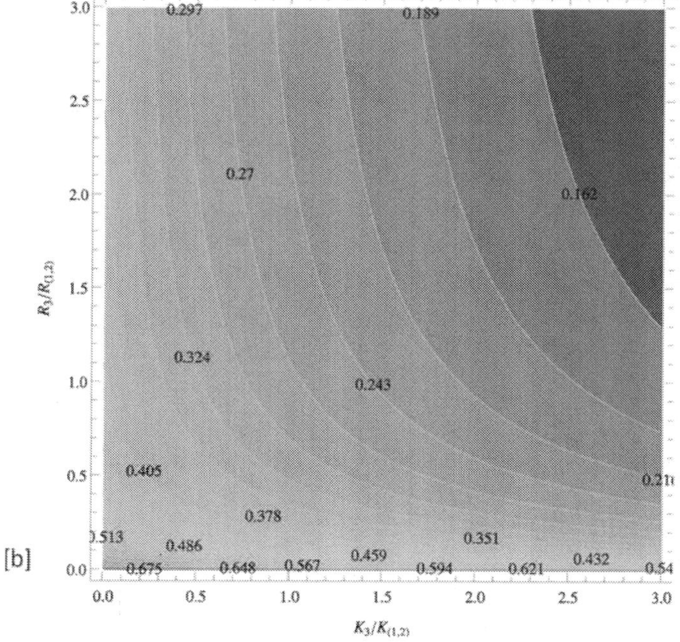

Figure 12. (a) Maximum power of *PSC* system under condition *TEG 1=TEG 2≠TEG 3*, the highest numerical value corresponding to arrangement *TEG 1=TEG 2=PbTe≠TEG 3=BiTe*. (b) Contour plot: efficiency of PSC system under condition *TEG 1=TEG 2≠TEG 3*, assuming the maximum value of efficiency $Z^{Inh}_{eq\text{-}PSC}$ for arrangement *TEG 1=TEG 2= PbTe≠TEG 3=BiTe*.

7.3.3. Corollary: maximum efficiency Z_{eq} for composite thermoelectric generator

Based on the progress presented in this paper, we have been formulated the following corollary: two features of design must be met to ensure the maximum value of Z_{eq} of CTEG:

- If the material is the same in all components, CTEG reaches the maximum value of Z_{eq} with a specific type of thermal—electrical connection.

- When components of TEGs composing CTEG are made of different materials, $TEG_i \neq TEG_j \neq TEG_l$ where $i; j; l$ can be 1, 2, or 3; then, for a given thermal-electrical connection, there exists an optimal arrangement of thermoelectric materials for which Z_{eq} is maximum.

8. Conclusions

The main objective of this chapter was to present new ideas for designing more complex thermoelectric systems taking into account the effects of electrical and thermal connection, combination of different materials and ordering of materials in *CTEG*. For this purpose, we considered the framework of linear response theory for nonequilibrium thermodynamic processes, and we used the constant parameter model. Through the definition of equivalent parameters α_{eq}, R_{eq}, and K_{eq}, we have shown the significant impact of these parameters on the system's properties, which characterize the performance of *CTEG*, namely Z_{eq}, η_{eq}, and P_{eq}. The numerical results show, that the optimal configuration for *CTEG* considered here is the thermal and electrical connection in parallel with arrangement (*PbTe, SiGe and BiTe*). For completeness, we have shown the effect of contact thermal conductance on the parameter Z^{Inh}_{eq-PSC} for the most efficient case—*PSC-CTEG* system, in terms of both ratio $K_3/K_{1,2}$ (intrinsic thermal conductances) and $R_3/R_{1,2}$ (intrinsic electrical resistance). Although in this study, the composite system is restricted to only three components, the results can be generalized to systems consisting of N modules, either analytically by extension of the mathematical model or through numerical simulations; guidelines for this purpose are provided by the corollary 7.3.3.

Author details

Alexander Vargas Almeida[1], Miguel Angel Olivares-Robles[2]* and Henni Ouerdane[3,4]

*Address all correspondence to: molivares67@gmail.com

1 Departamento de Termofluidos, Facultad de Ingenieria, Universidad Nacional Autonoma de Mexico, Mexico

2 Instituto Politecnico Nacional, SEPI-Esime Culhuacan, Coyoacan, Ciudad de Mexico, Mexico

3 Russian Quantum Center, Skolkovo, Moscow Region, Russian Federation

4 UFR Langues Vivantes Etrangeres, Universite de Caen Normandie, Esplanade de la Paix, Caen, France

References

[1] Apertet Y, Ouerdane H, Goupil C, Lecoeur Ph: A note on the electrochemical nature of thermoelectric power. EPJ Plus. 2016;131:76. doi:10.1140/epjp/i2016-16076-8

[2] Ioffe AF: Semiconductor Thermoelements and Thermoelectric Cooling. London: Infosearch; 1957.

[3] Seebeck TJ: Treatises Royal Academy of Sciences, Berlin. 1821:289.

[4] Peltier JCA: Annals of Chemical Physics. 1834;56:371–386.

[5] Pottier N: Out-of-equilibrium statistical physics, linear irreversible processes irréversibles linéaires. Paris: EDP Sciences/CNRS Editions: 2007.

[6] Callen HB: The application of Onsager's reciprocal relations to thermoelectric, thermomagnetic, and galvanomagnetic effects. Phys. Rev. 1948;73:1349–1358. doi:10.1103/PhysRev.73.1349

[7] Onsager L. Reciprocal relations in irreversible processes. I. Phys. Rev. 1931;37:405–426. doi:10.1103/PhysRev.37.405

[8] Onsager L. Reciprocal relations in irreversible processes. II. Phys. Rev. 1931;38:2265–2279. doi:10.1103/PhysRev.38.2265

[9] Domenicali CA. Irreversible thermodynamics of thermoelectricity. Rev. Mod. Phys. 1954;26:237–275. doi:10.1103/RevModPhys.26.237

[10] Callen HB. Thermodynamics and an introduction to thermostatistics, 2nd revised ed. New York: John Wiley & Sons; 1985.

[11] Apertet Y, Ouerdane H, Goupil C, Lecoeur Ph. Internal convection in thermoelectric generator models. J. Phys. Conf. Series. 2012;395:012103. doi:10.1088/1742-6596/395/1/012103

[12] Apertet Y, Ouerdane H, Goupil C, Lecoeur Ph. Equivalent parameters for series thermoelectrics. Energy Convers. Manag. 2015;93:160–165. doi:10.1016/j.enconman.2014.12.07

[13] Apertet Y, Ouerdane H, Goupil C, Lecoeur PH. Thermoelectric internal current loops inside inhomogeneous systems. Phys. Rev. B. 2012;85:033201. doi:10.1103/PhysRevB.85.033201

[14] Hsu CT, Huang GY, Chu HS, Yu B, Yao DJ. An effective Seebeck coefficient obtained by experimental results of a thermoelectric generator module. Appl. Energy. 2011;88:5173–5179. doi:10.1016/j.apenergy.2011.07.33

[15] Barron KC. Experimental studies of the thermoelectric properties of microstructured and nanostructured lead salts. [Bachelor's thesis]. Cambridge: Massachusetts Institute of Technology; 2005, p. 28.

[16] Hurwitz EN, Asghar M, Melton A, Kucukgok B, Su L, Orocz M, Jamil M, Lu N, Ferguson IT. Thermopower study of GaN-based materials for next-generation thermoelectric devices and applications. J. Electron. Mater. 2011;40:513–517. doi:10.1007/s11664-010-1416-9

[17] Shakouri A. Recent developments in semiconductor thermoelectric physics and materials. Annu. Rev. Mater. Res.. 2011;41:399–431. doi:10.1146/annurev-matsci-062910-100445

[18] Ouyang Z, Li D. Modelling of segmented high-performance thermoelectric generators with effects of thermal radiation, electrical and thermal contact resistances. Sci. Rep. 2016;6:24123. doi:10.1038/srep24123. .

[19] Goupil C, Seifert W, Zabrocki K, Muller E, Snyder GJ. Thermodynamics of thermoelectric phenomena and applications. Entropy. 2011;3:1481–1517. doi:10.3390/e13081481

[20] Vargas A, Olivares MA, Camacho P. Thermoelectric system in different thermal and electrical configurations: its impact on the figure of merit. Entropy. 2013;15:2162–2180. doi:10.3390/e15062162

[21] Abdelkefi A, Alothman A, Hajj MR. Performance analysis and validation of thermoelectric energy harvesters. Smart Mater. Struct. 2013;22:095014. doi:10.1088/0964-1726/22/9/095014

[22] Nemir D, Beck J. On the significance of the thermoelectric figure of merit Z. J. Electron. Mater. 2010;39:1897–1901. doi:10.1007/s11664-009-1060-4

[23] Apertet Y, Ouerdane H, Glavatskaya O, Goupil C, Lecoeur P. Optimal working conditions for thermoelectric generators with realistic thermal coupling. EPL. 2012;97:28001. doi:10.1209/0295-5075/97/28001

[24] Vargas A, Olivares MA, Camacho P. Maximum power of thermally and electrically coupled thermoelectric generators. Entropy. 2014;16:2890–2903. doi:10.3390/e16052890

[25] Vargas A, Olivares MA, Mendez Lavielle F. Performance of composite thermoelectric generator with different arrangements of SiGe, BiTe and PbTe under different configurations. Entropy. 2014;16:2890. doi:10.3390/e17117387

Discussion Panel

Chapter 21

Non-Stationary Thermoelectric Generators

John G. Stockholm

Additional information is available at the end of the chapter

http://dx.doi.org/10.5772/66421

Abstract

A review of theoretical publications on non-steady thermoelectrics is given. Review concerns different aspects of non-stationary and pulsed processes in thermoelectric materials and devices. Theoretical analysis of dynamic behaviour of thermoelectric devices, including analysis of small and large signals of thermoelectric generator, is given and details of concepts of quasi-equilibrium thermoelectricity are discussed as well. Special attention is paid to theoretical study of the non-routine regime of non-steady thermoelectricity—fast-time dependence of thermoelectric properties when material or device is well out of equilibrium. Theoretical findings of fast-time dependence give reason to believe that it can increase the output electrical power of thermoelectric generator compared to stationary regime of operation. We also present experimental results obtained with first non-stationary thermoelectric generator prototype, which was designed for operation in fast-time dependence mode. Several research teams are presently making and testing devices to confirm that more electrical power can be obtained in AC mode (AC frequency about hundreds of kHz) than in DC mode. Descriptions with an analysis are given.

Keywords: thermoelectricity, electricity generation, quasi-stationary systems, out-of-equilibrium transient modes, ultra-fast conduction, microscopic processes

1. Introduction

Operation of thermoelectric (TE) heat pumps used for cooling (TECs) and thermoelectric generators (TEGs) are closely related. Results of first theoretical studies of thermoelectric

devices operating in non-steady regimes were published for thermoelectric coolers by Stilbans and Fedorovich in the year 1958 [1]. Then, Gray presented in the year 1960 [2] results of analytical consideration concerning small and large signals of thermoelectric generator. In 1961, Landecker [3] published results concerning fast transient behaviour of Peltier junctions during temperature measurements with thermocouples. Then, in the year 1971 Babin and Iordanishvili studied quasi-stationary thermoelectric generation and in the year 1983, Iordanishvili and Babin published a monograph [4] of non-stationary processes in thermo-electric and thermomagnetic energy conversion systems. It presents the fundamentals of pulsed current generation, pulsed thermoelectric generator, quasi-steady-state generator and single-pulse thermoelectric generator.

There is also non-steady, well-out-of-equilibrium, fast-time dependence. This subject has not been extensively developed and few papers were published on this subject. Such non-steady processes have the potential to transfer higher electrical powers, than in steady state. Currently ongoing experimental work is presented.

2. Theoretical aspects of pulsing

2.1. Two different types of non-stationary phenomena

2.1.1. Quasi-stationary thermoelectricity

Quasi-stationary thermoelectricity, that is, quasi-equilibrium thermoelectricity, assumes a slight time dependence and is allowed for all the laws of the equilibrium, which means that this time dependence is only slow in comparison with the microscopic processes of heat transfer and electric conduction. This is considered in detail by Gray [2] and Iordanishvili and Babin [4]. Goupil [5, 6] studied non-stationary operation with Onsager approach and calculates for thermoelectric device the important parameters for transient regimes: time constant τ and capacitance C.

2.1.2. Far from equilibrium fast-time dependence

Thermoelectric device developed by Strachan, published by Aspen in the year 1992 [7, 8], was operated in MHz range; unfortunately, functional physics was never established and the unit operated in heat-pump mode and in electricity-generating mode, as well.

The first person who studied the fast-time dependence or well-out-of-equilibrium regime was Apostol [9]. His work can be considered as the beginning of new era in thermoelectric power generation. The physical mechanism is ultra-fast conduction, but this mechanism can be obtained only under certain circumstances, viz.:

- sharp electrical current impulse, associated with shorting of electrical circuit, which produced with appropriate frequency, 'amperage' pulse does not have the time to dissipate, that is, widen, during the time it takes to go around the circuit.

2.2. Quasi-stationary thermoelectricity

2.2.1. Dynamic thermoelectric processes

The study of dynamic behaviour of thermoelectric devices was presented in 1960 by Gray [2]. The object is behaviour of TE energy conversion devices under dynamic conditions. TE generator and TE heat pump also are non-linear distributed-parameter systems. Therefore, it is difficult to exactly analyse their dynamic behaviour. In this study, the dynamic behaviour is investigated by means of linear small-signal-distributed-parameter models, for which useful analytical results are obtained. Analysis allows predicting the dynamic behaviour of thermo-electric device both in frequency-domain and in time-domain equalization algorithms. This can be used to assess quantitative effects of device parameters with regard to dynamic response of the device. It is possible to linearize system's equations in small-signal variations in order to obtain linear mathematical models, more amenable to analysis. This approach is of limited usefulness in computing large-signal dynamic behaviour, such as transient processes, when TE generator is turned on. The study of complete small-signal behaviour of TE generators is developed, the results are expressed in terms of frequency-domain transfer functions and procedures are developed, which allow direct calculation of time-domain transients for the case of input and disturbance time functions, which are Fourier transformable. Experimental observation of small-signal dynamic behaviour of simple heat pump is presented and it confirmed theoretical model. No measurements were made on TE generator, but the accuracy of description should be as good, because mathematical model and analytical techniques are identical.

Comparison of experimentally obtained large-signal static and small-signal dynamic behaviour of TE heat pump with corresponding theoretical results indicates a disagreement at large currents due to neglecting of temperature dependence of parameters.

Complete set of equations for the case of large-signal dynamic analysis for step changes in current is presented.

To conclude, all this work with many equations is the basis for all studies related to small and large signals and it is written in view for appropriate control of TE generators.

2.2.2. Non-stationary processes in thermoelectricity

The subject of non-stationary thermoelectricity is also presented in detail in 1983 by Iorda-nishvili and Babin [4]. Several very interesting examples of TE generators are presented, including the case of solid body fuel (SF), capable of producing uniform heat production throughout the volume.

Thus, SF must maintain solid-state properties after heat production process (combustion).

Using the model of semi-infinite rods, assume that at time $t = 0$ there is instantaneous heat production, after which SF temperature becomes equal to T_ω (initial temperature of two bodies is T_0). The system temperature as a function of position and time was calculated and the graph is given (**Figure 1**). Such a generator has high speed and stability of the output characteristics.

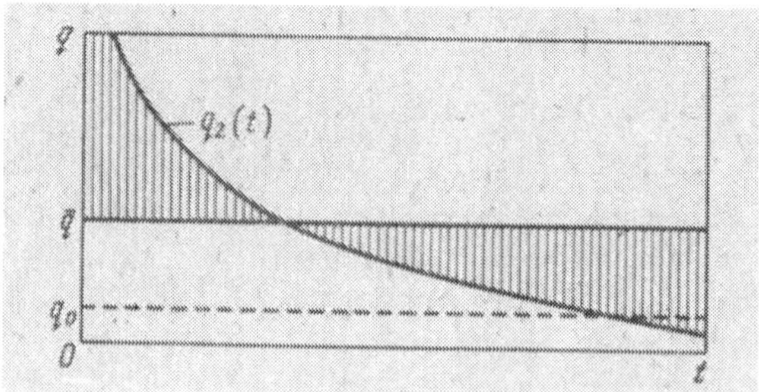

Figure 1. Heat flux q versus time t, $\bar{q}$ is the average effective heat flow during time t, q_0 is the average effective heat flow in stationary mode, $\bar{q}/q_0 \approx 3$.

This graph $q(t)$ shows that in this model, due to thermal conductivity, average, during time t, effective value of heat flow $\bar{q}$ is three times higher than the average effective heat flow q_0 in classic TEG, operating in stationary mode.

Consideration of non-stationary mode includes the following parameters: energy capacity, speed of response, response time improvement, temperature dependence of physical parameters and output characteristics, accounting for Peltier heat production, effectiveness of side-surfaces thermal insulation and thermal stabilization based on the model of finite length. All this information can be very useful for researchers working in the field of transients.

2.3. Thermoelectricity in systems far from equilibrium

M. Apostol in the year 2000 has presented theory of ultrafast thermoelectric conduction [9]. The theory was developed and described [10–12]. It is about moving heat pulse along a rod of thermoelectric material of length 1 mm with on and off connection along a wire 1 cm long. The electrical power is extracted along this wire. A sketch of the pulses is below (**Figure 2**).

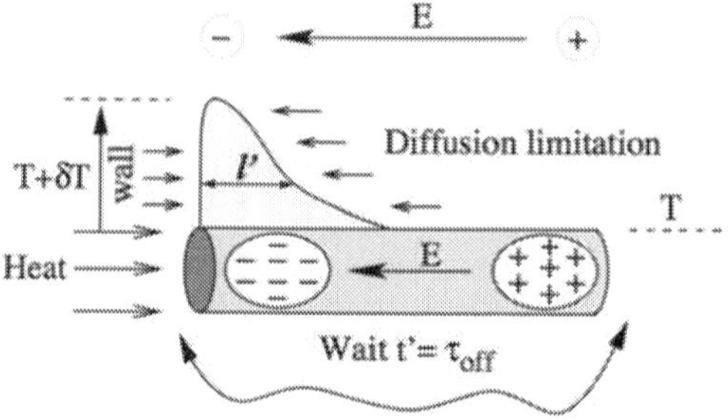

Figure 2. The pulse is deflated of heat at the cold end and another pulse is constructed at the hot end. T is the temperature, δT is the variation in temperature, t' is time off, which is the time between 2 pulses.

The shape of the wave as a function of time t and distance x is $n(x,t)$ [10]:

$$n(x,t) = \frac{V \times \delta_n}{\sqrt{2\pi \times \Lambda \times v \times \tau}} \times \exp\left(-\frac{x^2}{2\Lambda vt}\right),$$ (1)

here V original volume of δ peak; δ_n is quasi-particle density in δ peak, Λ is mean free path of quasi-particles; t is time; v is transport velocity.

The width of the pulse is expressed as:

$$l = \sqrt{\Lambda \times v \times \tau_{\text{off}}},$$ (2)

where l = width of pulse, τ_{off} = time off.

He gives the relation of maximum power output $P_{\text{ext}}^{\text{max}}$ with pulses of ultrafast conduction compared to maximum power in DC mode $P_{\text{dc}}^{\text{max}}$ as:

$$P_{\text{ext}}^{\text{max}} = 2\frac{l}{\sqrt{\Lambda l}} \times P_{\text{dc}}^{\text{max}},$$ (3)

$$P_{\text{ext}}^{\text{max}} = 60 \times P_{\text{dc}}^{\text{max}}.$$ (4)

The ratio of 60 is theoretical value calculated with some assumptions. Even when the experimental ratio is much smaller than 60, then the output power $P_{\text{ext}}^{\text{max}}$ can be many times greater than the DC mode value. This can be the major breakthrough that we are searching for to 'beat' material limitation of $ZT = 2$.

3. Pulsed mode thermoelectric generators

3.1. Thermoelectric generator working on MHz frequency

3.1.1. Primary patents and papers

Priority on thermoelectric energy conversion in high-frequency mode was filed by Aspen and Strachan in the years 1990–1991 [13]. In the year 1992, Aspen and Strachan published results of development and demonstration tests of MHz thermoelectric generator [7, 8]. This is the first example of thermoelectric device operating far from equilibrium.

3.1.2. Description of thermoelectric device operating far from equilibrium

Strachan develops a vibrator to break kidney stones. He discovered that his device could operate as a heat pump (to produce ice) or, with temperature difference between two sides, it could generate electrical power sufficient to operate a small fan. Experimental study was done in collaboration with Oxborrow. It was headed by John Stockholm of Marvel Thermoelectrics and was financed by automobile company PSA.

John Stockholm and M. Almeida from Supélec visited Scott Strachan in his office at the Technology Transfer Centre, University of Edinburgh, Scotland, in February 1996. The device built by Scott Strachan over a year ago was no longer operational, as it had degraded with time, but he showed a video and gave us a copy of it.

He was extremely open and gave a great deal of advice on how to make a unit.

A 'still' from the video is shown in **Figure 3**.

Figure 3. Scott Strachan operating the device.

Thermoelectric generator is located below his fingers, below the square white box, into which Scott Strachan pours hot water to generate ΔT across both sides of TE generator.

It is a small multi-layered device 30 × 2.5 mm (ΔT is across dimension 2.5 mm) consisting of stack: Al film—Ni film—piezoelectric film (PVDF)—ethyl cyanoacrylate adhesive—Fe film on BASF recording tape—Mylar, BASF recording tape—ethyl cyanoacrylate adhesive—Al film; repeat periodically (**Figure 4**). A device of several cubic centimetres in volume is vibrated by PZT at 500 kHz. It produces 0.14 W of electrical power from heat throughput of 3.7 W.

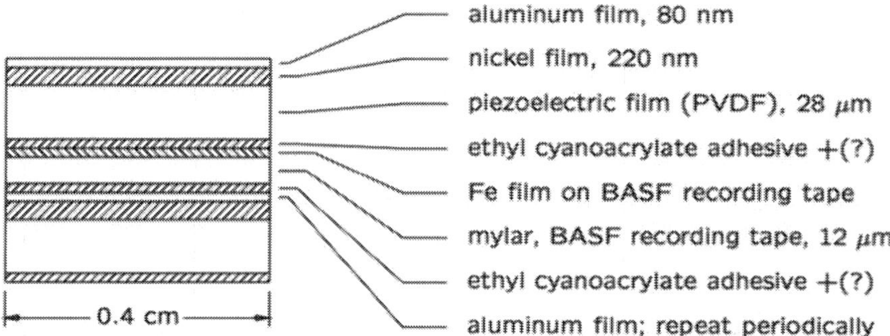

Figure 4. Strachan device-stack of different materials.

Energy conversion efficiency calculated from the above values is equal to 4%. This is very high, considering that ΔT is below 75 K (boiling water and ambient air), compared to bismuth telluride, which for similar ΔT would give much less than 2%.

Two prototypes were made by Mark Oxborrow.

Prototypes are very difficult to make, in particular with 'super glue', if the sheet is badly placed, one has to start the assembly process all over again.

The one, he sent to Supélec, was destroyed during testing and John Stockholm terminated the study, as it was apparent that the device was not adaptable to powers required for automobiles.

Our objective was to test the device, then to make similar more simple devices to determine the key components, but we were not successful in even reproducing the device.

Many people showed an interest in Strachan's device. Details of TE generator were given, in presentation by Maynard and Mahan of Penn State at ICT2005 in Clemson SC [14]. Also, people from MIT showed interest in Strachan's device. But nobody was able to make the device following all advices from Scott Strachan. All those involved have doubts about the theoretical aspects described by Harold Aspen, who by profession was a patent attorney.

It is incredible that Strachan's work could not be reproduced and devices made to determine the key technical parameters. He was pioneer in the realization of far-from-equilibrium regime of TE device by pulsing in MHz range.

3.2. Schroeder's pulsed TE generator

In the year 1994, Jon Schroeder announced a ring-shaped TE generator pulsed at 60 Hz and he published a paper about it in the year 1999 [15].

A prototype was shown to visitors, but only measurement on one thermoelectric couple was made. This measurement was used to calculate the output electrical power of the ring, but thermoelectric generator did not work.

Only in 2004, TE generator prototype with bismuth telluride elements, pulsed at 200 kHz, was demonstrated. But only the output electrical power of prototype was measured. So, we did not know if pulsed-operated TE generator produced greater efficiency than standard DC-operated TE generator with bismuth telluride couples operated within temperature limits of bismuth telluride.

In 2008, Marin Nedelcu with a friend visited Jon Schroeder in Leander, Texas. A video was made showing Jon Schroeder operating the device.

The electrical conversion efficiency was calculated as exceeding two times the DC mode efficiency. A photograph taken from the video made by Marin Nedelcu is shown in **Figure 5**.

Figure 5. Schroeder's pulsed TE generator.

The unit weighs about 12 kg. Heat source is from propane in a bottle and cold source is from Blades facing downwards immersed in water.

Schroeder ring. It consists of a ring designed by Jon Schroeder, described in a patent [16]. The heat is produced in the centre by hot gases; gases can be combustion gases from natural gas. The heat is transferred by convection and radiation to the radial blades 'hot' (**Figure 6**).

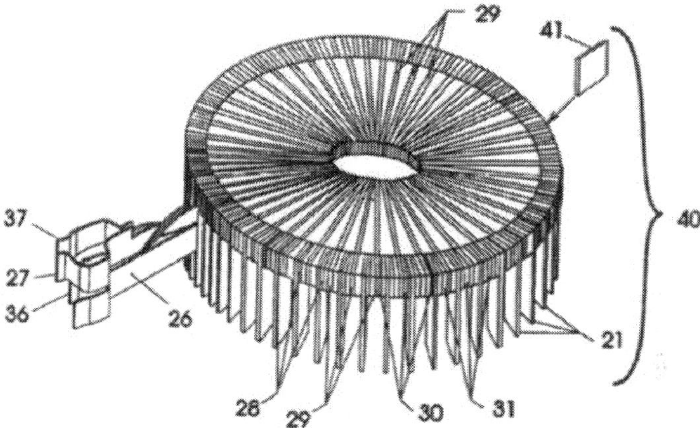

Figure 6. Schematic of Schroeder's ring. The ring is 250 mm diameter. The cold blades 21 are oriented downwards and are cooled by forced convection. Voltage outlets 36, 37.

Figure 7 shows cross section, with a central burner; hot blades marked 4 are connected to the hot side of the pieces of TE and cold blades are marked 3. The propane is burned in volume 14 and the hot combustion gases exit through the central hole 20.

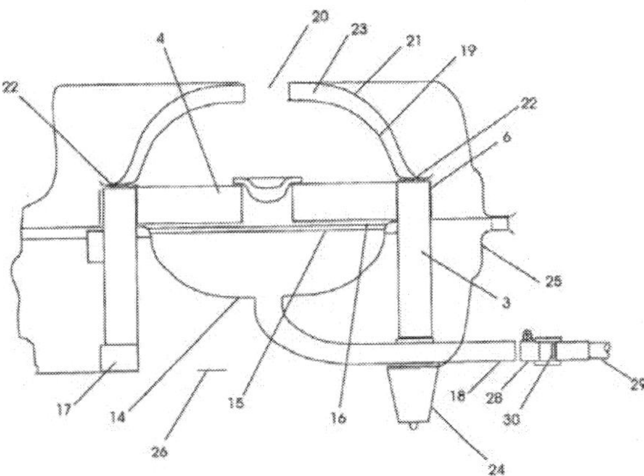

Figure 7. Cross section of the unit.

This schematic shows that radial blades 4 are heated by hot gas; blades 3 facing downwards are air-cooled. Thermoelectric material: bismuth telluride (20 × 20 × 1 mm), alternatively n- and p-type, is placed between heated and cooled blades. With temperature difference between hot and cold blades, voltage is created at outlets 36 and 37 (**Figure 6**) that are connected to the primary winding of a transformer through MOSFETs, which operate in 100-kHz range.

Output connections of TE generator are connected to two circuits in parallel (not shown).

The secondary winding of a transformer increases the voltage. The switching between two circuits (primaries wound in different directions of a transformer) is done before opening the circuit, so one primary winding is always connected to TE generator. In this way, electrical current is pulsed at around 200 kHz, then rectified and can be converted into AC current, for example, 220 V.

4. Impedance of thermoelectric devices

We examine values obtained from three different approaches of the impedance:

- First approach is theoretical. Finite time thermodynamic (FTT) approach is used by Goupil [5]. The development of Novikov-Curzon-Ahlborn description of TEG has demonstrated that there exists a close feedback effect between the output electrical load value and entering heat flow.

- Second approach is experimental. Low-frequency impedance spectroscopy technique of thermoelectric modules was proposed by García-Cañadas [17].

- Third approach is theoretical; it is proposed by Apostol [9].

4.1. Onsager equations

Onsager equations are for stationary systems; equations can be extended to non-stationary conditions as presented by Goupil [5].

Assuming that the entropy per charge carrier expression is:

$$A = -e \times \alpha, \tag{5}$$

α is Seebeck coefficient, e is charge of the electron.

The others terms derive directly from the thermo-elastic coefficients of the electronic gas,

$\tau_0 = \dfrac{e^2 \chi_T}{G}, \quad l = \dfrac{C_N}{T \chi_T}, \quad L = \dfrac{e^2 K}{GT}$, where K and G are, respectively, thermal and electrical conductance of the system. We also have the specific heat:

$$C_N = \frac{T}{N} \times \left(\frac{\partial S}{\partial T} \right)_N, \tag{6}$$

χ_T is the isothermal compressibility, S is the entropy:

$$\chi_T = \frac{1}{N} \times \left(\frac{\partial N}{\partial \mu} \right)_S ,$$

(7)

with N the number of carriers and μ the electrochemical potential. After calculation of the eigenvalues and eigenvectors, we get the complete expressions of the transient response of the system. In most of the cases, the thermal and electrical time constants, respectively, τ_{thermal} and τ_{elec}, are mixed together in a complicated way, but two extreme cases can be considered:

– Under very weak coupling, $A < < L/l$, we find classical results, $\tau_{\text{elec}} = \tau_0$ and:

$$\tau_{\text{thermal}} = \tau_0 \frac{l}{L};$$

(8)

– Under strong coupling condition, $A > > L/l$, thermal time constant diverges and electrical time constant becomes:

$$\tau_{\text{elec}} = \frac{C_N}{\chi_T e^2 \alpha^2 T} \tau_0 = \frac{C_N}{\alpha^2 GT}.$$

(9)

Since G is the resistive part of the impedance, the reactive part gives the expression of the so-called 'thermoelectric capacitance':

$$C_{TE} = \frac{C_N}{\alpha^2 T}.$$

(10)

Thermoelectric time constant of the system is then finally:

$$\tau_{TE} = R_{TE} C_{TE}.$$

(11)

Reference [5] gives all the intermediate steps and the result is thermoelectric capacitance:

$$C_{TE} = \frac{3Nk}{\alpha^2 T}.$$

(12)

Assuming standard sample with parameters close to room temperature, ($V' = 1$ cm^3, $n = 5 \times 10^{19}$ cm^{-3}, $T = 400$ K, $\alpha = 160$ μV K^{-1}), we get:

$$C_{TE} = \frac{3Nk}{\alpha^2 T} \approx 202 \text{ F}.$$

(13)

Complete internal resistance of the system is now replaced by internal impedance:

$$Z_{in} = R_{TE} // C_{TE} + R. \tag{14}$$

This result shows that the internal resistance is larger at zero frequency (stationary conditions), // means that R_{TE} and C_{TE} are electrically in parallel, (f=0) = R_{TE} +R , than it is for high frequencies, $Z_{in}(f = 0) = R_{TE} + R$, then it is for high frequencies, since R_{TE} term becomes short-circuited by C_{TEG} giving $Z_{in}(f > > 0) = R$.

$R_{TE} \approx R \approx 10^{-4}$ Ohm; $\tau_{TE} \approx R_{TE}C_{TE} \approx 2 \times 10^{-1}$ s, which gives low-pass filtering at really low frequencies. Depending on precise values of internal isothermal resistance, low-pass filtering effect may be found in the range from 1 kHz to 1 MHz.

4.2. Spectroscopic impedance

Spectroscopy analysis of thermoelectric devices has been performed by García-Cañadas [17] and Gao Min [18]. They have studied thermodynamics of thermoelectric phenomena by frequency-resolved methods, which show that TE devices have a capacitive component that can be exploited to improve performance.

4.3. Basics of pulsed-operating thermoelements

Apostol and Cune [11] show how electrical circuit of thermoelectric device can be represented with electrical resistance in series and capacitor in parallel.

5. Experimental devices

5.1. Extraction of electrical power

Electrical power from DC-operated TE systems is most commonly extracted on electrical load resistance. Recently, pulse-width modulation (PWM) with matched load has been used, especially, when powers and voltages are small. In the case of pulsed systems, we have already AC system, so PWM is the logical way to extract electrical power.

When we have DC-operated thermoelectric generator, then power is extracted on resistive load. The maximum power is extracted, when external load resistance R_L is equal to the resistance of power source R_S. It was shown in 2001 by M. Nedelcu and J. Stockholm that when electrical current was pulsed at 50 kHz, then the electrical output power was constant at $R_L < R_S$. This observation showed that small electrical resistance, such as impedance, might be of interest to extract electrical power efficiently.

Equivalent electrical circuit contains electrical resistance of thermoelectric material, capacitance in parallel, to which one can add an inductance composed of the primary winding of a transformer to extract electrical power at voltages of usable values. Today, there are no mathematical models to dimension such circuits.

Spectroscopic impedance measurements made on commercial thermoelectric modules (e.g., 40 × 40 mm with 127 couples) have shown very high values of capacitance in Farad range. We are dealing with super-capacitors, which can be operated at more than 100 kHz. Several teams are designing and making different units to check ultra-fast conduction theory.

5.2. Devices

Jon Schroeder built the first high-frequency-operated generator, but, unfortunately, he has not been able to make a second unit. There are explanations. The ring design is the ideal design, but it has major flaws: when a generator is operated and stopped, the ring expands and contracts creating shear stresses on a thermoelectric material; these stresses degrade bismuth telluride material and interfaces with copper. The second flaw is that all thermoelectric discs must be soldered in one operation; chances of success are very small. So, prototypes being built now have column structure, where TE elements are in line and compressed between heat sources, which are placed laterally or in line, when liquid is used.

5.2.1. Column structure

Several teams are working on devices based on the column structure.

5.2.1.1. Column design by Nedelcu

A team headed by Marin Nedelcu in Bucharest started by building ring units with individual water cooling for each bismuth telluride disc, but following problems of water leaks abandoned the ring and resorted to column structure. M. Nedelcu used in column design bismuth telluride discs: $D = 24$ mm and 1 mm thick; he built several columns, the latest is shown in **Figure 8**.

Figure 8. Thermoelectric column built by Nedelcu.

Nedelcu has built eight columns the same as this one. Each column has seven TE couples plus one at the end, so that both ends are cold. Approximate dimensions of cold blocks are 30 × 35 × 8 mm with two holes with a diameter of 4 mm. Hot plate is attached to the top flat surfaces; water flows through the holes through the cold side, which are connected by rubber u-bends, not shown. The water is distilled and has high electrical resistance. The assembly is on-going and testing is planned.

5.2.1.1.1. Commercial module design

Recently, Nedelcu built TE generator using four commercial electricity generating modules connected electrically in parallel, with total electrical resistance around 30 mOhm. Modules are assembled between water-cooled plate at T ~38°C and heated plate at T ~188°C, heating power with 170 V and 5.9 A = 1003 W.

Electrical current output was pulsed using MOSFETs at around 200 kHz. The output current from the transformer is rectified. The load is 100 W filament electrical light bulb. Measured voltage is 210 V and current is 0.4 A, so electrical power output is 84 W. So, overall efficiency (including heat losses) is 8.4%; this is about two times higher conversion efficiency, when operated in DC mode.

These measurements seem to confirm that pulsing can improve efficiency due to ultra-fast conduction as predicted by Apostol. When dealing with such breakthrough results, one must be extremely cautious and obtain results from another laboratory.

This TE device made by Nedelcu is very different from Schroeder ring (120 slices of bismuth telluride with an area of 4 cm^2, thickness of 1 mm, all in series electrically); here we have about 71 couples of bismuth telluride (unknown dimensions, can be estimated: area 10 mm^2).

5.2.1.2. Pulsed TE generators by Krzysztof Wojciechowski team

A team headed by Krzysztof Wojciechowski of AGH Cracow and Institute of Electron Technology, Cracow, designed and built several small prototypes to detect high-conversion efficiencies by pulsing thermoelectric circuit. First prototype had water-cooled plate at the top and the bottom, see **Figure 9**.

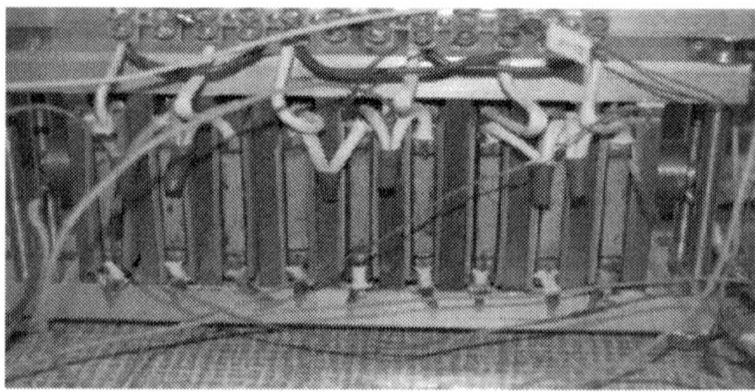

Figure 9. Unit with 18 bismuth telluride discs (n- and p-type).

Figure 10 shows a smaller unit with six TE discs with a diameter of 24 mm and a thickness of 1 mm, aluminium water-cooled plates above and below, electrical heaters at the front and back. Detail of column electrical heater 'red' at top and bottom without the water cooled plate, that would be located on the right side and on the left side (**Figure 11**).

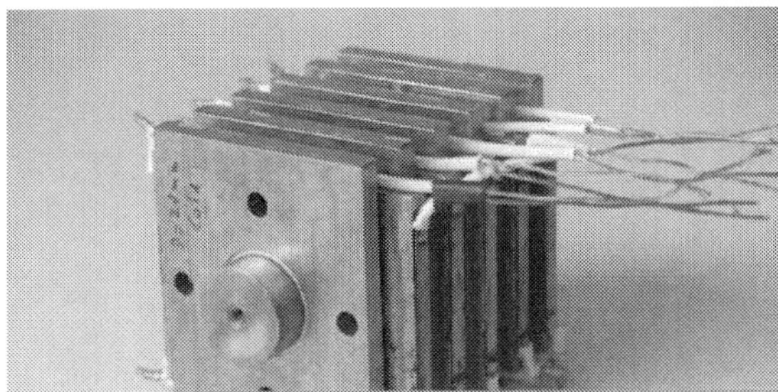

Figure 10. Unit with six bismuth telluride discs—electrical heaters front and back, compact water coolers, see **Figure 12**.

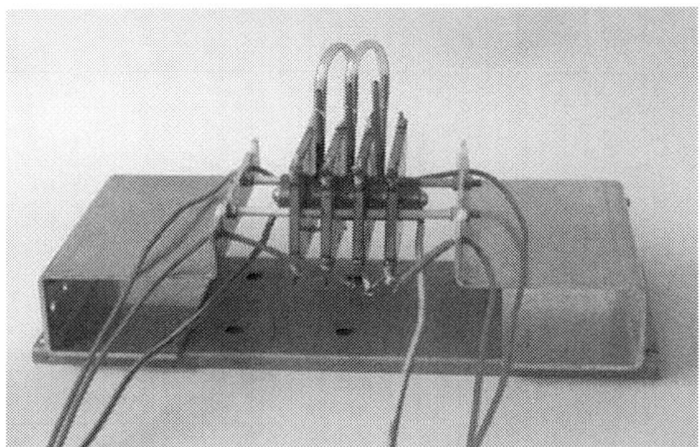

Figure 11. Unit with three TE discs of an outside diameter 24 mm and thickness 2 mm. The heaters are electrical strips on the left and right sides. The cooling is with compact cooler as shown in **Figure 12**.

Thermal and electrical measurements showed that produced voltage was insufficient to operate MOSFET switching. Once again, this confirms the difficulty to have sufficient heat flux through bismuth telluride material, as bismuth telluride has thermal conductivity of about 1.5 W/(m K), si in bismuth telluride disc with a cross section of 4.5 cm^2 and a thickness of 1 mm, heat flux power at $\Delta T = 100$ K is equal to 67.5 W (**Figure 12**).

5.2.1.3. Unit with column structure by Glick et al

A team with John Stockholm, James Glick and Brad Vier are also building a TE generator unit, which has a column structure. First prototype P1 was built with five couples of bismuth telluride discs with a diameter of 24 mm and a thickness of 1 mm. It was operated with propane combustion gases and cooled by ambient air. The objective was to measure temperatures at the interface of bismuth telluride.

A schematic of pulsed-width modulation electrical circuit is shown in **Figure 13**.

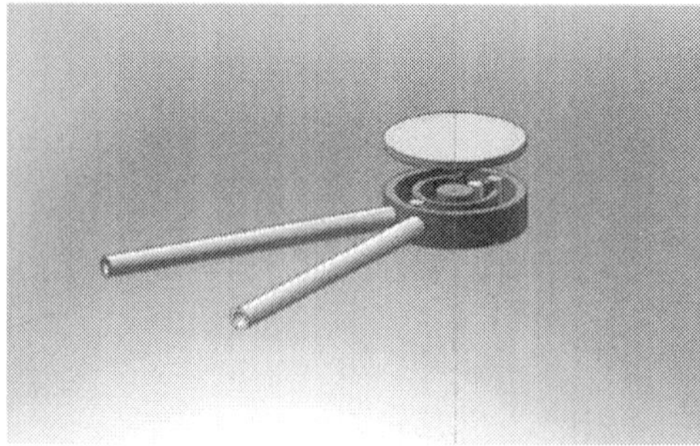

Figure 12. Detail of compact water cooler with an outside diameter of 25 mm placed on the axis of the column.

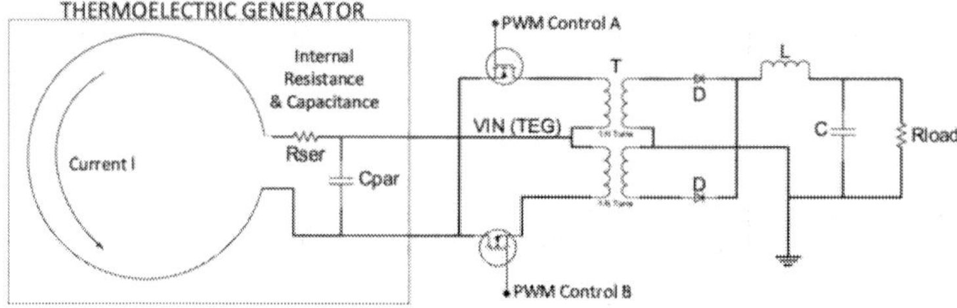

Figure 13. Schematic of PWM controller electrical circuit.

A photograph in **Figure 14** shows a lateral view of the inside of TE generator.

Figure 14. Lateral view of inside of P1 prototype TE generator, left side at the bottom is the infrared burner, combustion gases go upwards; in the centre at top, column structure-heated blades; to the left, cooled blades by ambient air blown upwards (fan is not shown).

Figure 15 shows the top view of TE generator: on the left are heated blades and on the right are blades cooled by air, the white material is to insulate thermally thermoelectric column from hot combustion air. Bismuth telluride disc with copper caps is shown in **Figure 16**.

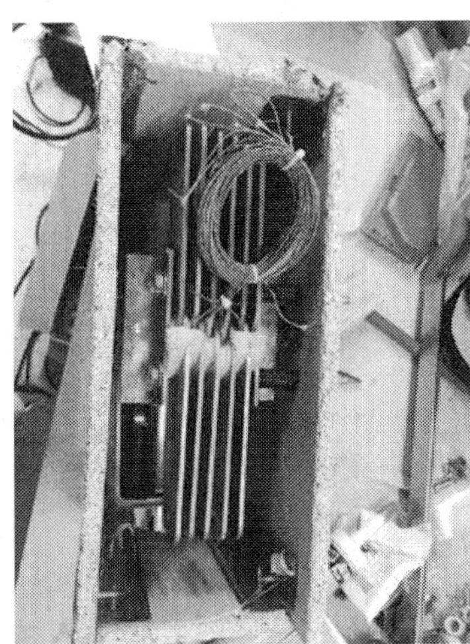

Figure 15. Top view of inside of P1 prototype.

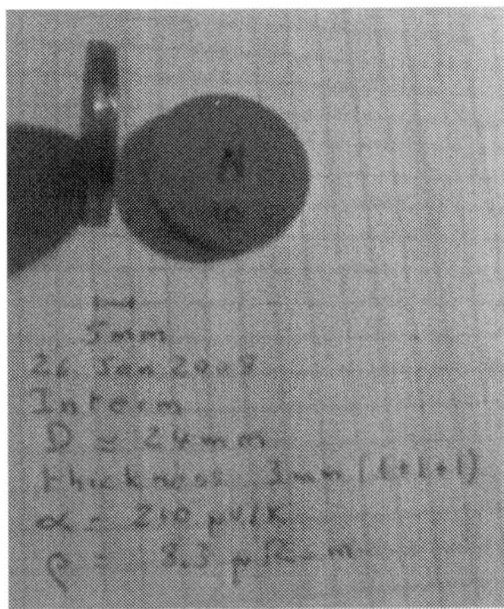

Figure 16. Bismuth telluride disc with copper caps.

When the unit was operated, the propane combustion gases heated and forced ambient air cooled. However, measured temperature difference at the interface with bismuth telluride was insufficient to operate TE generation, but valuable information was obtained to improve the design.

Stockholm presented in 2014 results concerning different aspects of pulsed TE generator operation, such as high-efficiency pulse-width modulator thermoelectric generator with an active load [19] and transient thermoelectric generator: an active load story [20].

Development is on-going. A prototype designed for measurements is being made.

Figure 17 shows the 10 kW transformer to be installed. **Figure 18** shows four Hi-Z-14 modules.

Results will be published as soon as they have been repeated to ensure validity.

Figure 17. PWM 10 kW transformer dimensions 250 × 190 × 140 mm high.

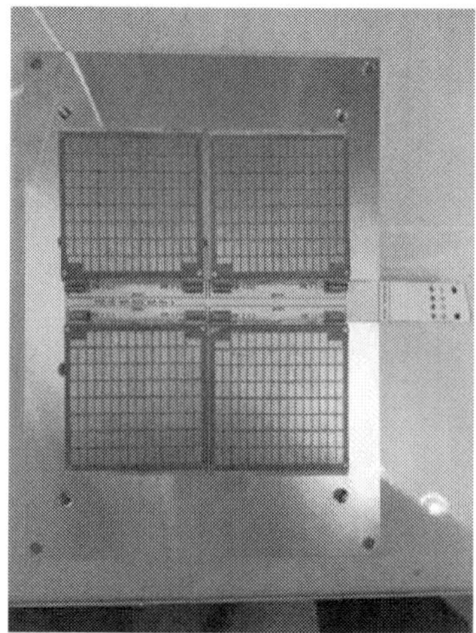

Figure 18. Four Hi-Z-14 modules on a water-cooled plate.

6. Potential of non-stationary far-from-equilibrium TE generators

Very few people have studied the microscopic phenomena of non-stationary, far-from-equilibrium thermoelectrics. Marian Apostol is the main contributor to the subject with many papers. He has shown great potential of this phenomenon. Why is it that 15 years after his first paper nobody has an industrial unit? There are many reasons. People don't believe new phenomena!

Technology of these systems is difficult, because of constraints: high heat fluxes and very low electrical resistances. Electrical currents can be enormous, more than 1000 A and voltages can reach kV. All this is confirmed by the numerous problems encountered by all experimentalists. Strachan had great voltage problems. High-frequency pulsing is not stable. Marian Apostol compares these systems to a sphere on top of a mountain! Prototypes are being built and tested to confirm these high-energy conversion efficiencies.

In 2014, measurements on the unit built by Schroeder were made in his workshop, on a unit not designed for any measurements inside the unit. The only possible measurements were the amount of propane burnt and electrical power output. Measured values of conversion efficiencies exceeded by a factor of more than 2, the values, of equivalent DC-operated unit with similar temperature differences.

Recently, Nedelcu has built a TE device with four commercial electricity-generating modules Hi-Z14 (56 × 56 mm) with 31 bismuth telluride couples. They are electrically in parallel so that the electrical resistance of the load is minimum about 40 mOhm placed between heated plate at 188°C and cold plate at 38°C. He claims that the conversion efficiency exceeds two times the DC conversion efficiency from thermal to electrical power. All these claims need to be repeated in a laboratory and fully documented in particular measurements of electrical pulses and power on primary circuit must be made.

There are also potential sources of degradation at interfaces. Anatol Casian at the University of Moldova is concerned about possible charge carriers' depletion.

Today, several research groups, on very small budgets, are working on devices to confirm experimentally that very high-performing generators are a reality; as soon as there is confirmation, research money will be available. Many theoretical researches are necessary before any transfer to industry.

7. Conclusions and outlook

The operation in non-stationary modes near equilibrium has been presented. Two books, one by Gray (1960) [2] and one by Iordanishvili [4], cover extensively the subject. Recently, Nedelcu et al. [21], Min [18] and García-Cañadas [17] have done experimental spectroscopic analysis of thermoelectric systems, which confirm that TE device is equivalent electrically with a capacitance in parallel with TE generator, which has internal electrical resistance. This is confirmed by theoretical papers by Apostol [9–11] and Goupil [5].

One important advantage of operating in a non-stationary mode is to obtain thermoelectric generators with better performances, than with DC operation. The present-day DC-operated TE generators are limited by ZT of TE material, which is less than 2.

Measurements need to be confirmed by other laboratories. The main interest is energy conversion efficiency. Prototypes have mainly used bismuth telluride, because it is the most suitable for generators operating with the cold side around ambient, but any thermoelectric material could be used.

Prototypes are being built and tested to confirm these high-energy conversion efficiencies, which are predicted by ultra-fast conduction theory of M. Apostol [9].

Acknowledgements

The author wishes to thank all the people who have contributed to his interest in non-stationary TE systems, especially Nikolai Kukhar, Valeriy Zhmurko, Lyudmyla Vikhor, Oleg Luste, Scott Strachan, Jon Schroeder, Marian Apsotol, Marin Nedelcu and Christophe Goupil.

Author details

John G. Stockholm

Address all correspondence to: johngstockholm@gmail.com

Marvel Thermoelectrics, Joachim du Bellay, Vernouillet, France

References

[1] Stilbans LS, Fedorov NA. On the work of cooling thermoelements in the transient regime. Journal Tekhn. Physics. 1958;XXVIII:489–492 (in Russian).

[2] Gray P. The dynamic behavior of thermoelectric devices. The Technology Press, MIT & John Wiley; 1960 (Library of Congress No. 60-12981).

[3] Landecker K, Findlay AW. Study of the fast transient behaviour of Peltier junctions. Solid-State Electronics. 1961;3:239–260.

[4] Iordanishvili EK, Babin VP. Non-stationary processes in thermoelectric and thermo-magnetic energy conversion systems. Moscow: Nauka; 1983. 213 p. (in Russian) English Edition available from: http://marvelte.com/biblio [Dec 2016].

[5] Goupil C, Ouerdane H, Stockholm JG, Apertet Y, Lecoeur Ph. Non-stationary response of thermoelectric generators, In: Proceedings of 12th European Conference on Thermoelectrics (ECT2014); 24–26 September 2014. Published in Materials Today Proceedings; 2014.

[6] Goupil C, Seifert W, Zabrocki K, Eckhard Muller E, Snyder J. Thermodynamics of thermoelectric phenomena and applications. Entropy. 2011;13:1481–1517; DOI: 10.3390/e13081481

[7] Aspen H. The electronic heat engine. In: Proceedings of 27th Intersociety energy conversion conference. 3–7 August 1992; San Diego: pp. 4357–4363 doi:10.4271/929474

[8] Aspen H, Strachan JS. Solid state thermoelectric refrigeration. In: Proceedings of 28th Intersociety energy conversion conference. July 1993: pp. 1891–1896: 2772-5/93/0028-10 American Chemical Society.

[9] Apostol M. Ultrafast thermoelectric conduction. Journal of Theoretical Physics. 2000;59, ISSN 1453-4428

[10] Apostol M. Transferring power to a resistive external circuit in a pulsed thermoelectric conduction. Journal of Theoretical Physics. 2002;81, ISSN 1453-4428.

[11] Apostol M, Cune LC. "Equivalent circuits" for pulse-operating thermoelements. Journal of Theoretical Physics. 2004;101, ISSN 1453-4428.

[12] Apostol M, Nedelcu M. Pulsed thermoelectricity. Journal of Applied Physics. 2010;108:023702.

[13] Aspen H. US patent 5,065,085.

[14] Maynard JD, Mayan GD. The 24th International Conference on Thermoelectrics. June 19–23 2005; Clemson University, Clemson SC. Not in the Proceedings, a copy of presentation was obtained directly from J. D. Mahan. http://www.its.org/ict2005 Contact: ict2005@its.org

[15] Schroeder JM. ENTELEC News; Winter 1999.

[16] Schroeder JM. US Patent US 8,101,846 Jan 24 2012.

[17] García-Cañadas J. Thermal dynamics of thermoelectric phenomena from frequency resolve method. AIP Advances. March 2016. DOI:10.1063/1.4943958

[18] Gao M. Improving the conversion efficiency of thermoelectric generators through "pulse mode" operation. AIP Conference Proceedings. 2012;1449:447. DOI: 10.1063/1.4731592

[19] Stockholm J, Wojciechowski K, Leszcznski J, Witek K, Guzdec P, Kulawik J. Thermoelectric generator with an active load. In: Materials Today Proceedings 12th European Conference on Thermoelectrics, Madrid; 2014.

[20] Stockholm J, Goupil C, Maussion P, Ouerdan H. Transient thermoelectric generator: an active load story. Journal of Electronic Materials. Proceedings of International Conference on Thermoelectrics; Nashville, TE 2014.

[21] Nedelcu M, Apostol J, Stockholm JG. Pulsed thermoelectric machine. Materialy Ceramiczne/Ceramic Materials. 2014;66:170–174.